Studies in Mechanobiology, Tissue Engineering and Biomaterials

Volume 21

Series editor

Amit Gefen, Ramat Aviv, Israel

More information about this series at http://www.springer.com/series/8415

Joaquim Miguel Oliveira
Rui Luís Reis

Editors

Regenerative Strategies for the Treatment of Knee Joint Disabilities

 Springer

Editors
Joaquim Miguel Oliveira
3B's Research Group—Biomaterials,
  Biodegradables and Biomimetics
University of Minho
Guimarães
Portugal

Rui Luís Reis
ICVS/3B's—PT Government Associate
  Laboratory
University of Minho
Guimarães
Portugal

ISSN 1868-2006   ISSN 1868-2014   (electronic)
Studies in Mechanobiology, Tissue Engineering and Biomaterials
ISBN 978-3-319-83135-0   ISBN 978-3-319-44785-8   (eBook)
DOI 10.1007/978-3-319-44785-8

Printed on acid-free paper

This Springer imprint is published by Springer Nature
The registered company is Springer International Publishing AG
The registered company address is: Gewerbestrasse 11, 6330 Cham, Switzerland

Preface

This book is composed of four main sections totalizing 20 chapters, where the progress on biomaterials and current knowledge on the clinical management and preclinical regenerative strategies for the treatment of knee joint lesions are overviewed. The role of different biomaterials from natural to synthetic, and processing methodologies to obtain scaffolds and patient-specific implants are deeply discussed. The current tissue engineering approaches and bioreactors technology are also revised. The advances on regenerative strategies and future trends based on expert opinion will be highlighted in dedicated chapters. Section one provides an updated and comprehensive discussion on articular cartilage tissue regeneration and role of synovial knee joint. Section two focuses on the important contributions for bone and osteochondral tissue engineering and highlights the importance of developing novel bilayered scaffolds and culturing conditions for enhancing osteogenesis and chondrogenesis, simultaneously. Section three overviews the recent advances on the treatment of meniscus lesions. A special focus is given both to preclinical strategies aiming to regenerate menisci and to the methodologies for the production of patient-specific meniscus implants. Finally, section four addresses the current strategies for treatment of ligament lesions. A next generation of biomaterials that hold great promise for improving the biofunctionality of engineered ligaments is also presented.

Guimarães, Portugal
Joaquim Miguel Oliveira
Rui Luís Reis

Contents

Part I
Articular Cartilage Lesions: Clinical Management and Regenerative Strategies

Chapter 1
Knee Articular Cartilage

Marta Ondrésik, Joaquim Miguel Oliveira and Rui Luís Reis

Abstract Articular cartilage (AC) is vital for the proper functioning of the knee. This smooth white connective tissue covers the joint surfaces and allows pain free movement for decades. Its high durability originates from its unique structure mainly composed of cells, macromolecules and water. The same structure allows the cartilage to transmit the load and act like a cushion in the harsh mechanical environment of the body's largest joint. Herein, it is discussed the origin, function and structure of knee AC. After briefly discussing its embryological development, the biochemistry and the related biomechanical properties of AC are also overviewed. The tissue components will be individually described and its role in the cartilage will be explained. It is also reviewed the different mechanical behaviours of the tissue. Finally, AC tissue homeostasis and maintenance is discussed, which is still somewhat requires a deeper knowledge. The anabolic and catabolic processes, namely tissue synthesis and degradation and the involved molecules and signalling pathways are also subjects that have been addressed in the current chapter.

1.1 Introduction and Types of Cartilage in the Human Body

The human body is composed of different cells, which essentially define tissues, and form organs. There are four different basic tissues namely, nervous, muscular, epithelial, and connective tissue. The type connective tissue is the most abundant, and most various. Fat tissue, blood, fibrous tissue, bone, bone marrow and cartilage are all belong to the group of connective tissues. Connective tissues are highly

M. Ondrésik (✉) · J.M. Oliveira · R.L. Reis
3B's Research Group—Biomaterials, Biodegradables and Biomimetics,
European Institute of Excellence on Tissue Engineering and Regenerative Medicine,
University of Minho, Avepark, Guimarães, Portugal
e-mail: marta.ondresik@dep.uminho.pt

M. Ondrésik · J.M. Oliveira · R.L. Reis
ICVS/3B's—PT Government Associate Laboratory, Braga, Guimarães, Portugal

© Springer International Publishing AG 2017
J.M. Oliveira and R.L. Reis (eds.), *Regenerative Strategies for the Treatment of Knee Joint Disabilities*, Studies in Mechanobiology, Tissue Engineering and Biomaterials 21, DOI 10.1007/978-3-319-44785-8_1

specialized, rich in matrix, and typically having cells located apart from each other. These tissues are to provide cohesion, support and protect the organs and parts of the body.

Cartilage is a flexible but extremely strong type of connective tissue found at many areas in the body. It has a dense matrix with gelatinous base, packed with collagens and elastic fibres. Depending on its structure, it is subdivided into several groups:

Hyaline cartilage—it has a bluish-white colour, the matrix is rich in collagens fibrils, and also abundant in glycosaminoglycans (GAGs). Hyaline cartilage is mainly associated with the skeletal system.

Elastic cartilage—found in the external ear and epiglottis owing a matrix with great networks of elastic fibres, and a rather yellowish appearance.

Fibrous cartilage—the strongest type of cartilage in the human body, its matrix has a sponge like structure, densely packed with collagen. Menisci, calli and inter-vertebral discs are all composed of this type of cartilage.

From all types, hyaline cartilage is the most widespread in the human body. A specialized type of hyaline cartilage, called 'articular cartilage' found in synovial joints. Articular cartilage (AC) is a vital part of the musculoskeletal system. Its function, origin, unique structure and biochemical properties are the subject of this chapter with special focus on the knee.

1.2   Anatomy of the Knee Joint

The knee joint is the most complex joint of the body. It is a modified hinge joint, which allows flexion and limited rotation, still provides a complete stability, which is very important considering the great load it is subjected to. It consists of two joints, the patellofemoral and the femorotibial joints. The osseous part is composed of the tibia, fibula, femur and patella, stabilized by the combination of ligaments, tendons, the capsule and musculature. The patella rests in the trochlear grove of the femur forming a sellar joint, helping to stabilize the knee during extension. The femorotibial joint is plainly comprised by the tibia and femur composing the largest joint of the body. Both the tibia and femur are divided into lateral and medial compartments. The femur has two condyles, a medial and lateral condyle. Both condyles has cam—shapes from lateral view, which perfectly fit into the slightly concaved tibial plateaus (also lateral and medial). The two bones are attached by the synovial membrane, a loose connective tissue. Between the femoral condyles and tibial plateaus located the menisci, which improve the alignment of the bones, and has crucial shock absorbing function in the joint. The gliding surface of the bones in the contact areas and further cushioning is provided by AC.

1.3 Gross Anatomy and Function of Articular Cartilage

AC is a thin layer of elastic tissue which plays fundamental role in the synovial joint functioning. It has an average thickness of 3–4 mm, but at some parts can be as thick as 7 mm. By providing lubrication and smoothness to the articular surfaces it allows the bones to slide over each other without any friction or pain. Given its rubbery like nature, it also contributes to the load transmitting, thus minimizes the contact stress occurring in the joint during movement, and thus protects the subchondral bone from a permanent damage. Its unique structure allows plastic deformation when loaded. Under normal circumstances AC is a tissue with very slow turnover, high resistance and with the potential to function for decades.

1.4 Embriology of Articular Cartilage

AC is originated from mesenchyme. It forms during foetal and early post-natal life by a complex multi-step process. Briefly, a mass of undifferentiated mesenchymal cells form the so called skeletal blastema. The blastema differentiates into a three layered structure comprising the chondrogenic, myogenic and osteogenic lineages. The cartilage anlagen are formed centrally by a four-step process, namely: (1) cell migration (2) aggregation and (3) condensation and (4) chondrification. The steps of cell recruitment, aggregation, and condensation are governed by mesenchymal–epithelial interactions [1]. The condensation of chondroprogenitor mesenchymal cells will result in the formation of the so called interzone, the first morphological sign of joint development [2, 3]. Eventually, cells of the interzone will give rise to the articular cartilage and also form all the other articular structures i.e. ligaments, synovial tissue. The interzone is a three layered structure: two outer layers composed of chondrogenic cells, and an inner layer of densely packed compact cells, the presumptive precursors of articular chondrocytes. The cell interconnectivity is assured by gap junctions [4] allowing cell communication and precise regulation of limb formation. Finally the interzone separates by a process called cavitation, giving rise to the synovial space and defining the articular surfaces.

The embryonic cartilaginous tissue has two fates: (1) it can either create the articular cartilage by retaining its phenotype, and (2) it can serve as a template for bone morphogenesis. This latter one occurs as a process called endochondral ossification when the cells progress through terminal differentiation, namely: intensive proliferation, maturation, hypertrophy, calcification, and apoptosis until the chondrocytes are ultimately replaced by bone. The cells of the intermedial interzone on the contrary will block terminal differentiation, and will acquire the phenotype of mature articular chondrocytes, without entering the hypertrophic stage. These cells will establish the articular cartilage, produce and deposit the extracellular matrix and maintain it throughout life.

The processes of joint formation and cartilage development are governed by the interplay of several signalling pathways, morphogens and regulators. Most importantly, Sox transcriptional factors (Sox9, Sox6 and Sox5) the master regulators of the early chondrogenesis and Runx (Runx2 and Runx3) transcriptional factors, which initiate chondrocyte hypertrophy [5]. The major signalling pathway of the early joint formation is the Wnt/β-catenin signalling [6]. Members of fibroblast growth factor (FGF), bone morphogenic proteins (BMP) and vascular endothelial growth factor (VEGF) play essential role in chondrocyte proliferation, differentiation, endochondral bone development and vascular invasion, respectively [7–10]. The chondrocyte proliferation is regulated by the parathyroid hormone-related peptide (PTHrP)/Indian hedgehog (Ihh) axis, while the articular cartilage development is aided by the *ets* transcription factor ERG (ets-related gene).

1.5   Structure and Biochemistry of Articular Cartilage

AC appears to be a simple tissue, as it has only one cell type; and lacks vascularisation, innervation and lymphatic drainage (Fig. 1.1). Nevertheless, cartilage is not fairly homogeneous, neither simple; in contrary AC represents a highly specialized connective tissue with a very sophisticated structure. The cells constitute 1–2 % of the whole tissue; the rest is composed of a complex network of fibrils, creating the extracellular matrix (ECM). The resident cells of the cartilage are the chondrocytes. These cells are responsible for the maintenance of the entire tissue [11]. They secrete the large macromolecules which compose the ECM as well as produce the catabolic and anabolic factors which regulate the tissue homeostasis. The ECM principally consists of the combination of proteoglycans and collagen fibres, forming an elastic tissue, which is abundant in water and possessing extraordinary mechanical properties. The cell morphology as well as the components of the ECM varies throughout the multiple layers of the cartilage from the surface down to the underlying bone (Fig. 1.2).

1.5.1   Chondrocytes

The constituent cells of the cartilage are the highly specialized chondrocytes. Although, chondrocytes make up only 1–5 % volume of the whole cartilage, they play important role in the maintenance of the tissue [11]. Chondrocytes are responsible for the matrix-generation and repair [12]. As AC has no blood supply the chondrocytes completely relay on nutrition provided by diffusion of synovial fluid from the synovial cavity. In fact, they entire metabolism is specialized to operate in low oxygen tension and to cover they energy requirement by glycolysis [13]. The cells are located in the so called lacunae, cytoplasmically separated from

Fig. 1.1 Safranin O—fast green staining of human knee AC, early OA

each other. The mature chondrocytes do not proliferate and are metabolically quiescent, unless during pathology, such as osteoarthritis (OA) when they tend to form cluster typically in the lower zones of the tissue. The chondrocytes are surrounded by the narrow pericellular matrix (PCM), forming a complex with it and termed as chondron [14, 15]. However, it is not entirely revealed what is the role of the PCM, supposedly besides protecting the chondrocytes it also regulates their microenvironment and serves as a transducer of both mechanical and biomechanical stimuli [14–16]. Indeed, chondrocytes are able to sense the mechanical environment, transform the mechanical signals into their metabolic events, and modify the matrix properties accordingly. Unfortunately, chondrocytes change caused by ageing. They lose both their mitotic and metabolic activity due to senescence, which decreases their ability to maintain the cartilage [17].

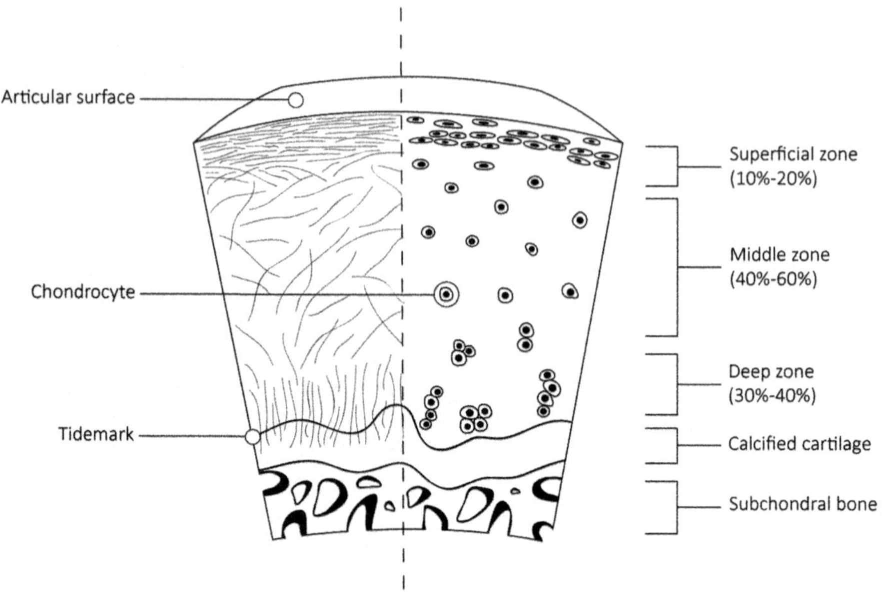

Fig. 1.2 Schematic representation of the AC structure

1.5.2 Constituents of Extracellular Matrix (ECM)

As it was mentioned already AC has a complex and highly organized structure, designed to withstand tremendous amount of loads during a whole lifetime. The ground substance of hyaline cartilage is mainly composed of a framework of structural macromolecules, e.g. collagens, proteoglycans, glycoproteins and non-collagenous proteins [17] which surrounds the cellular component, and amounting up to 25 % of the total volume of the cartilage. The rest of the tissue is water. In fact, water is the most abundant component of the cartilage, forming approximately 65–80 % of the wet weight, creating a movable interstitial fluid phase in the tissue. It provides resilient strength, flexibility and reversible deformability to the tissue. The water also ensures the proper nutrition supply of cartilage, and aids the lubrication system [18]. Only a small portion of the total volume of water resides intracellularly in the cartilage, the rest is associated with the macromolecular meshwork and occupies the interfibrillar space of the matrix. Thus, when the tissue is subjected to mechanical loading, the water is free to move. During pathology the level of water can exceed, for instance in case of osteoarthritis, when the tissue loses its integrity. The extracellular matrix is crucial, as the slightest disruption can cause total functional impairment of the tissue.

1.5.2.1 Proteoglycans and Gylcosaminoglycans

Proteoglycans and glycosaminoglycans (GAGs) are major components of the articular cartilage matrix [19]. There are three types of GAGs present in the AC, the hyaluronic acid (HA), keratin sulphate and chondroitin sulphate. The keratin sulphate and chondroitin sulphate form the subunits of the proteoglycan molecules, while the hyaluronan acts like a central filament, to which the proteoglycan monomers bind via a link protein. Fundamentally, the proteoglycan molecules are composed of a core protein and the attached GAGs chains forming, large protein-polysaccharide molecules [17]. Approximately 10–20 % of the wet weight of hyaline cartilage is composed of proteoglycans. The proteoglycans endow the cartilage with resilience and elasticity properties and they have immensely important role in the maintenance of AC fluid flow and nutrition transport [20]. There are two main groups of proteoglycans in the articular cartilage, the large aggregating proteoglycans and the small leucine-rich protegylcans (SLRP). The chief proteoglycan is the aggrecan, which belongs to the large aggregating monomers and exists in association with the hyaluronic acid, creating large multi-molecular aggregates. The aggregates are entangled in the collagen network, as the GAGs are cross-linked by collagen fibers, but no chemical bonding is in between the two, thus enabling smooth nutrient diffusion within the tissue. Among the SLRPs decorin, biglycan fibromodulin and lumican are the major ones. These molecules help to maintain tissue integrity, and to regulate its metabolism. Unlike aggrecan, they do not associate with the HA.

The ability of articular cartilage to retain large amount of water is mostly attributed to the hydrophilic properties of the proteoglycan molecules, especially to aggrecan. The water binds to the negatively charged sulphate and carboxylate groups of the GAG chains. This way the proteoglycans can bind water up to 50 times their own weight [21]. By attracting the positive molecules and repel the negatively charged molecules, the proteoglycans generate osmotic imbalance of free anions and cations in between the tissue and at the surroundings, resulting in the Gibbs–Donnan effect [22]. By creating swelling pressure, the proteoglycans provide compressive strength to the tissue, allowing mechanical load transitions without any permanent tissue deformation [23].

The loss of aggrecan from the cartilage is attributed to the action of aggrecanases or ADAMTS (disintegrins and metalloproteinases with thrombospondin motifs), the zinc-dependent metalloproteinases of the metzincin family [24]. Aggrecanase-1 (ADAMTS-4) and aggrecanase-2 (ADAMTS-5) are the two major aggrecanolytic enzymes of the cartilage that cleave the aggrecan and are markedly responsible for the pathological processes observed during early OA. Their action is down-regulated by tissue inhibitor metalloproteinases (TIMPs), particularly by TIMP3 [25] and the expression of ADAMTS-4 is enhanced by cytokines, such as interleukin (IL)1-β and tumor necrosis factor (TNF)-α.

1.5.2.2 Collagens

Collagens make up two-third of the dry weight of the adult AC. Fundamentally, collagens form the structural backbone of the tissue matrix, reinforcing the proteoglycan-water gel-like structure. The fibril orientation differs zone by zone creating an arcade-like architecture, which is believed to be critical for the tissue mechanical properties [26]. There are several different types of collagen present in the AC, but the three most relevant are collagen type II, XI and IX. Collagen type II is the most abundant from all, comprising the majority (90–95 %) of the overall collagen content of the AC. Together with collagen type IX and type XI it creates a heteropolymeric template, which aids the fibril formation and helps to create a randomized, loose network of macromolecules [27]. This provides tensile strength to the cartilage, and enables to entrap other structural molecules, such as the previously mentioned aggrecan and hyaluronan. The second most abundant collagen in the AC is the collagen type XI, which exist in association with the collagen type II by covalently binding its surface. Collagen type XI is also a fibril-forming collagen and was found to be an important regulator of the collagen type II fibril formation by inhibiting its lateral expansion. On the contrary type IX is seemingly dispensable for fibrillogenesis. Collagen type IX is the third major collagen component of the AC, accounts for around 1 % of the overall collagen content. Collagen type IX is a member of fibril-associated collagens (FACIT). Although it does not directly responsible for fibril formation, it is believed, that collagen type IX is responsible for matrix integrity by facilitating the anchorage of other macromolecules and regulating their distribution [28]. Along with collagen type VI collagen type IX is preferentially localized in the pericellular regions of the cartilage [29] typically present on the surface of collagen type II and collagen type XI molecules [30]. Further important collagen components of the AC are the collagen type VI and X. Collagen type VI is a less abundant collagen of the articular cartilage although it is found in all cartilage layers and comprises the major component of the pericellular area. Collagen type VI molecules establish a crucial connection between the chondrocytes and the extracellular matrix by interacting with both the chondrocytes and the fibrosus elements of the chondron [15]. Also important to mention the collagen type X, which is expressed by chondrocytes of the hypertrophic zone, thus typical for the growth plate and calcified cartilage located in the pericellular microenvironment. Collagen type X helps the mineralization processes. During pathological conditions as osteoarthritis its level is increasing, and it is used as a marker of hypertrophy.

The collagen molecules possess extreme stability, endowing the AC with great durability. The half-life of collagens was estimated to be more than 100 years [31]. This long-term stableness of collagens lies in their particular structure. Essentially, collagen molecules consists of a triplet of procollagen polypeptide chains (αI (II) chains), which are coiled in a triple helix, creating large collagen fibrils. The collagen molecules of the triple helix are cross-linked with other collagen molecules of other triple helixes further enhancing the fibrillar meshwork. This bonding prevents their dissociation when they are exposed to high tensile forces. The collagen triple helix is also remarkably resistant to proteolytic degradation [30]. Only few enzymes are able to initiate the cleavage of the interstitial collagens, namely cathepsin K and the collagenases. There are three mammalian collagenases, the collagenase 1 (MMP1), collagenase 2 (MMP8) and collagenase 3 (MMP13). However, they all display collagenase activity, they differ in efficacy, substrate specificity and biochemical properties. MMP13 is the most effective on collagen type II, as compared to the other collagenases, has cleaves on aggrecan as well, and its level and activity is up-regulated by many pathways, therefore it is considered to be the main collagenase of the AC. In healthy conditions low concentration of MMPs is expressed which modulates the tissue remodelling in balance with the anabolic activities; however during pathology, such as OA, the level of MMPs is exceeded upon cytokine induction, and causes abnormal digestion of the cartilage. The MMPs' effect is also suppressed by TIMPs.

1.5.2.3   Non-Collagenous Proteins of AC

Besides the large macromolecules, there are also other structural proteins, polypeptides present in the ECM. These molecules may comprise only small portion of the cartilage, still possess remarkable role in defining the AC matrix organisation and functioning. Cartilage oligometric matrix protein (COMP), cartilage matrix protein (CMP) or matrilin-1, and cartilage intermediate layer protein (CILP) are all non-collagenous, non-proteoglycan structural proteins uniquely present in the cartilage. Other more ubiquitous structural proteins of the cartilage are the fibronectin, fibrillin and elastin. Important to mention the collagen binding protein anchorin CII (annexin V) and the lubricin or surface zone protein (SZP) a secreted glycoprotein. These molecules help to stabilize and protect the matrix, anchor the chondrocytes to the collagen fibrils and aid their interactions with the ECM.

1.5.3   *Synovial Fluid*

A normal synovial joint contains approximately 2.5 mL synovial fluid. The synovial fluid is a viscous liquid secreted by the cells of the inner synovial membrane filtered from the blood plasma. Its viscosity originates from its high HA content. This fluid supplies the nutrition for the cartilage, transports the waste material as well as provides lubrication and reduces friction by covering the articulating surfaces. Besides hyaluronic acid it also contains lubricin, another important lubricating factor as well as proteases, collagenases and prostaglandins. Its cellular component includes type A and type B synovial cells, lymphocytes B and T, monocytes, and neutrophils. Over and above, occasionally macrophages and mesenchymal stem cells are also present in the synovial fluid, typically during pathological conditions as osteoarthritis and rheumatoid arthritis. With regard to the rheological properties, the synovial fluid is a non-Newtonian fluid, with non-linear viscosity.

1.6 Ultrastructure of Articular Cartilage

The healthy articular cartilage has a layered structure, which morphologically can be divided into four zones e.g. superficial zone (1), intermediate zone (2), deep zone (3) and calcified zone (4). The zones vary in composition, structure and function, including fibril alignment as well as chondrocytes morphology and behaviour.

1. The superficial layer, or tangential layer is the thinnest of all, makes up around 10–20 % of the whole volume of the AC. Therein the chondrocytes are flattened, and localize sporadically. They are expressing proteins providing ultimate surface lubrication and protective functions. Most importantly the previously mentioned lubricin, a mucous glycoprotein, which forms a thin film on the AC surface equipping it with exquisite gliding properties [32, 33]. Another prominent protein present in this layer is the superficial zone protein (SZP) [12] a large proteoglycan secreted in the synovial fluid. SZP is homologous with lubricin and together they form the lamina splendens [22, 34]. The extracellular matrix is rich in fibronectin and water. Proteoglycans, such as decorin and biglycan are also present, but in a lower concentration as compared to other layers. Both the collagen fibrils and the cells lie parallel to the surface. Any disruption of this layer will result in changed biochemical properties of AC, and eventually lead to the development of osteoarthritis.
2. The intermediate layer covers approximately the 40–60 % of the total volume of the AC. In this zone the chondrocytes are less dense, but slightly larger and posses' spheroid shapes. They are embedded in the extracellular matrix, which is abundant in aggrecan, and has thick collagen fibres [35]. The cells have a higher concentration of synthetic organelles herein, e.g. Golgi-membranes and endoplasmic reticulum. The fibril organization is random to the cartilage surface.
3. The deep or radial zone embraces the 30 % of the cartilage. This layer has the highest level of proteoglycan and the largest collagen fibrils, but has the least of cells. The chondrocytes are spherical, stack in columns and along with the collagen fibres lie perpendicular to the articular surface [17].
4. The basal layer of cartilage called the calcified cartilage forms a boundary between the AC, and the subchondral bone. This fourth layer of cartilage is separated from the previous three by a border called the tidemark. The calcified cartilage is crucial to attach the AC to the bone and to transmit the force in between the two. In the basal layer, the chondrocytes are smaller, showing a hypertrophic phenotype. They only contain a low amount of endoplasmic reticulum and Golgi membranes, which suggests their low metabolic activity [36]. They are also completely surrounded by the extracellular matrix. Typical collagen is the collagen type X in this layer, which has a crucial role in load transmission and calcification.

The tidemark is a narrow layer, which separates the non-calcified and the viscoelastic hyaline cartilage from the deep, calcified cartilage. It is a metabolically active part of the calcification [37]. This important interface assures the structural integrity between the soft articular cartilage and the more stiff calcified cartilage [38]. Together with the calcified cartilage, the tidemark plays a crucial role in damping the loads, and to transmit the mechanical forces to the subchondral bone.

1.6.1   Subchondral Bone

Underneath the calcified cartilage, there is the subchondral bone, which is an important shock absorber and together with the articular cartilage composes the osteochondral unit. They act together to transmit the loads through the joint, and provide friction free movement [39]. The properties of subchondral bone, which allow to both absorb and distribute the mechanical forces, are highly influenced by the mineral content and the composition of the bone matrix [40]. Regarding the architecture of subchondral bone, it consist of two distinct anatomic entities the subchondral bone plate, and the subchondral spongiosa or subchondral trabecular bone [41]. The subchondral bone is richly perforated by hollow spaces allowing the invasion of arterial and venous vessels, as well as nerves up to the calcified cartilage from the spongiosa. Thus, the subchondral bone also serves as an exchange area for the cartilage, providing around 50 % of its oxygen and nutrient requirements [42].

1.7   Cartilage Homeostasis

The AC homeostasis is dependent on the metabolic activity of chondrocytes. In healthy conditions these cells ensure tissue integrity by the balanced production of anabolic and catabolic agents. The normal turnover of AC relies on the coordinated effect of MMPs and TIMPs and the simultaneous production of the ECM macromolecules. Besides the prevalent role of chondrocytes, signals derived from the environment have a great impact on the cartilage maintenance too. Soluble mediators and biophysical factors influence the matrix turnover in a significant manner by affecting the chondrocytes through a complex network of signalling pathways.

1.7.1   Biomechanical Loading

In vivo, AC is subjected to a wide range of mechanical stimuli, and the chondrocytes are able to utilize these biomechanical signals in their metabolic activity. Mechanotransduction in the matrix causes plasma membrane deformation of chondrocytes, followed by changes in intracellular signalling and results in cellular response through altered gene expression [43]. The pericellular matrix serves as a transducer, while the mechanical sensors of chondrocytes include the primary cilium, which is a small organelle present on chondrocytes; as well as cell surface receptors such as integrin, syndecan and ion channels [44].

It is well established that mechanical stimuli are essential for normal tissue homeostasis. The mechanical load induced hydrostatic pressure facilitates the interstitial fluid flow, thus guarantees nutrition supply and waste removal in the cartilage. Furthermore, moderate mechanical forces facilitate anabolic responses, regulate the biosynthesis of various ECM macromolecules and positively affect cellular viability. It is also well known that individuals who regularly practice physical exercises are less likely to develop OA [45], and during in vitro studies cyclic loading showed beneficial effect on chondrocyte monolayers by inhibiting the catabolic and inflammatory effects of pro-inflammatory cytokines [46].

Clearly, the coordinated effect of mechanical signals along with other metabolic factors of the environment is fundamental for the development and maintenance of the AC. Immobilization, or on the contrary overloading has disadvantageous effects on cartilage growth. Inappropriate mechanical stimuli can disrupt tissue composition and induce degradative activity of chondrocytes [47]. Numerous studies have shown that higher load impact on cartilage causes collagen degradation and cell apoptosis [48–50]. The exact pathways by which mechanical stimuli control AC homeostasis are still being revealed, shedding light on them may help the development of new therapies.

1.7.2 Mechanical Behaviour of Articular Cartilage

It is common that the mechanical behaviour of AC strongly depends on the biochemical properties and physical interactions of the ECM components. In this regard, we distinguish different behaviours such as swelling, compression, tensile, shear, and viscoelastic behaviour which characterize the AC.

As a result of the non-homogeneous distribution of collagen and proteoglycan molecules, AC displays mechanical inhomogeneity and anisotropy. Fundamentally, it is a viscoelastic tissue, thus its response to constant load alters by time. The drag force of the intrinsic fluid flow and the time dependent deformation of the solid components are the two important mechanisms establishing the viscoelastic behaviour [51–53].

Proteoglycans and collagen molecules provide different intrinsic material properties to the cartilage by cause of their different biochemical composition. Proteoglycans rather contribute to the compressive stiffness of the tissue by providing swelling propensity. The swelling appears due to the presence of fixed charges in the GAG chains [54], which generate a higher cation concentration within the tissue than in the surroundings. This consequently creates a high Donnan osmotic pressure inside the tissue and induces swelling of cartilage [22]. The hydrophilic nature of GAGs is very important from the mechanical point of view as the movable fluid content contributes ability of cartilage to bear high compression and resist loading. On the contrary collagen molecules prevent the swelling of AC.

While compression and swelling behaviour depend on the proteoglycan content of the tissue, collagen molecules are primarily responsible for the tensile properties of cartilage [55]. Collagens are stiff in tension, and upon loading stretch along the loading axis in a strain-dependent manner [56]. Namely, small tensile stress will cause realignment of the collagen molecules while higher strain induces the linear stretching of the fibres [53, 56]. The collagen molecules' structure, organization, density and the extent of cross-linking are all important parameters which determine the tissue tensile modulus [53]. Human AC exhibits larger tensile strength in the collagen rich superficial layer as compared to lower layers.

Compressive and tension stress testing methods on AC enable us to obtain information on the equilibrium properties of the tissue solid matrix, while when testing shear stress under infinitesimal conditions where no interstitials fluid flow and thus volumetric change of the tissue occurs, we can study the intrinsic viscoelastic properties of the collagen-proteoglycan core. Although, the shear stiffness of AC is derived from the collagen content, proteoglycans also contribute via their interaction with collagens, mainly by pre-stressing the collagen network as a consequence of swelling [57].

Cartilage mechanical properties can dramatically alter as a cause of changes in the molecular structure and organization of macromolecules. Cartilage degeneration can typically be characterized by increased water content, increased thickness and permeability and decreased stiffness from the mechanical point of view [53, 58–63]. Cartilage degeneration primarily appears due to osteoarthritis (OA), which is the most common joint degenerative disease. Although we do not understand the precise process of OA, studies suggest that the initial steps are associated with the disruption of collagen network, which consequently leads to increased swelling. Intense synthesis of proteoglycans was detected in early OA, which potentially be another cause of the swelling. Superficial fibrillation is and later lesions are hallmark of OA. Cracks first appear on the superficial layers of the cartilage, but may extend even to the subchondral bone as the disease progresses [62, 63]. The most affected parts of the cartilage are the load bearing sites, where in advanced phases of the disease the cartilage may completely disappear from the bony surfaces. Besides cartilage degeneration, the knee suffers other changes as well during the onset of OA. The thickening of synovial membrane, osteophyte formation and joint space narrowing are also common phenomena observed in an OA knee articulation (Fig. 1.3).

1.7.3   Anabolic and Catabolic Factors

Cartilage synthesis or destruction strongly depends on the chondrocytes response to anabolic and catabolic factors. The main anabolic factors in AC are growth factors such as transforming growth factor β (TGF β), insulin growth factor-1 (IGFs), bone morphogenic factor (BMPs) and fibroblast growth factors (FGFs). Growth factors are polypeptide signalling molecules which are capable of stimulating cellular

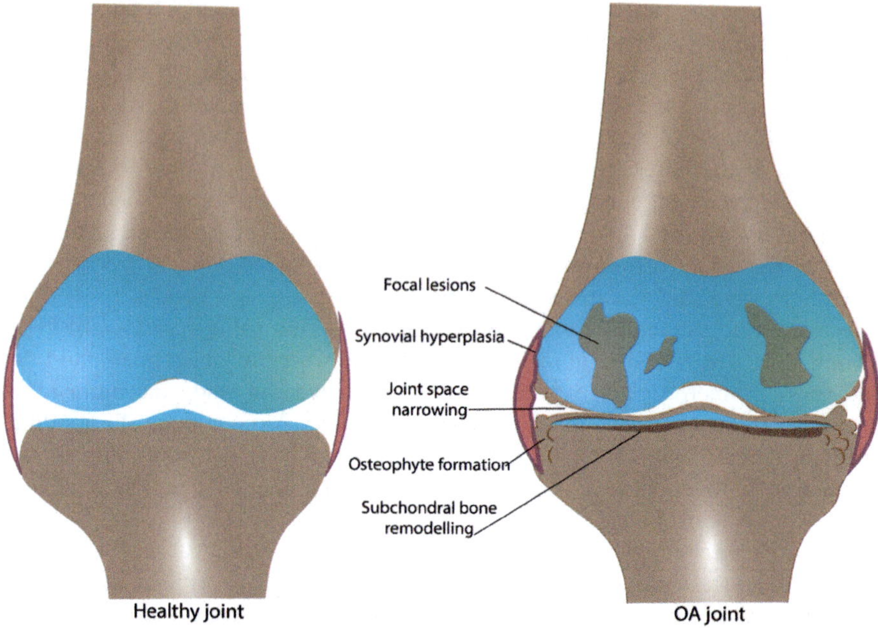

Fig. 1.3 Anatomy of healthy and OA knee joint

proliferation and enhance the anabolic activity of the cells. In healthy conditions their level is low, but enough to maintain the required macromolecule synthesis. Pro-inflammatory cytokines on the other hand enhance the production of matrix degrading enzymes, thus belong to the catabolic group of factors. The best studied cytokines of the AC are interleukin-1β (IL-1 β) and tumor necrosis factor α (TNF-α). These cytokines act synergistically to induce the production of tissue proteases and to inhibit the synthesis of ECM molecules [64–67]. They can also stimulate the production of other cytokines and chemokines thus inducing a catabolic cascade mechanism, which is a hallmark of osteoarthritis.

Both cytokines and growth factors are produced by the chondrocytes as well as other joint tissues, such as the synovium and subchondral bone. They act in an autocrin-paracrin manner and through a complex network of signalling transductions. The main pathways that regulate the AC homeostasis are TGF- β, BMP, FGF related signalling pathways, furthermore Wnt/β-catenin, nuclear factor kappa B (NF-κB), mitogen activated protein kinase (MAPK) and hedgehog (Hh) [68, 69]. Any disruption of the anabolic-catabolic balance leads to the loss of tissue integrity and eventually to tissue degeneration. Thus, the precise regulation of signalling molecules and the action of signalling pathways is critical for the adequate tissue homeostasis.

1.8 Final Remarks

There is no doubt that articular cartilage is an essential component of the musculoskeletal system. Despite its simple appearance it is a complex tissue with a critical role in the joints' proper functioning. As the life expectancy increases, to maintain its integrity is crucial. Unfortunately, at the age of 65 mostly everybody shows radiographic signs of OA, which later may lead to limited freedom of movement and in certain cases to total joint failure. Articular cartilage bears only a limited intrinsic healing capacity; hence unless OA is detected in early phase, cartilage damage becomes irreversible. Medical doctors and researchers who work on cartilage regeneration are facing many challenges. A better understanding on its structure, functioning and overall tissue homeostasis is therefore highly important. Cellular behaviour and molecular interactions are intensively studied to aid the work of pharmaceutical researchers in order to develop efficient treatments. Promising outcomes are shown on the field of tissue engineering and regenerative medicine (TERM), which are expected to become novel therapies. TERM uses biomaterials and stem cells apart or in combination to recreate and heal the damaged tissue. Despite the great efforts though, no regenerative methods could restore a durable AC yet. Therefore, to understand the mysterious nature of cartilage tissue and develop an efficient treatment modality remains an enormous challenge to be solved in the future.

References

1. Wu L, Bluguermann C, Kyupelyan L et al (2013) Human developmental chondrogenesis as a basis for engineering chondrocytes from pluripotent stem cells. Stem Cell Rep 1:575–589. doi:10.1016/j.stemcr.2013.10.012
2. Holder N (1977) An experimental investigation into the early development of the chick elbow joint. J Embryol Exp Morphol 39:115–127
3. Mitrovic D (1978) Development of the diarthrodial joints in the rat embryo. Am J Anat 151:475–485. doi:10.1002/aja.1001510403
4. Zimmermann B (1984) Assembly and disassembly of gap junctions during mesenchymal cell condensation and early chondrogenesis in limb buds of mouse embryos. J Anat 138(Pt 2):351–363
5. Lefebvre V, Smits P (2005) Transcriptional control of chondrocyte fate and differentiation. Birth Defects Res C Embryol Today 75:200–212. doi:10.1002/bdrc.20048
6. Macsai CE, Foster BK, Xian CJ (2008) Roles of Wnt signalling in bone growth, remodelling, skeletal disorders and fracture repair. J Cell Physiol 215:578–587. doi:10.1002/jcp.21342
7. Minina E, Kreschel C, Naski MC et al (2002) Interaction of FGF, Ihh/Pthlh, and BMP signaling integrates chondrocyte proliferation and hypertrophic differentiation. Dev Cell 3:439–449
8. Kobayashi T, Lyons KM, McMahon AP, Kronenberg HM (2005) BMP signaling stimulates cellular differentiation at multiple steps during cartilage development. Proc Natl Acad Sci USA 102:18023–18027. doi:10.1073/pnas.0503617102
9. Wu X, Shi W, Cao X (2007) Multiplicity of BMP signaling in skeletal development. Ann NY Acad Sci 1116:29–49. doi:10.1196/annals.1402.053

10. Haque T, Nakada S, Hamdy RC (2007) A review of FGF18: its expression, signaling pathways and possible functions during embryogenesis and post-natal development. Histol Histopathol 22:97–105

11. Stockwell RA (1979) Biology of cartilage cells. Cambridge University Press, Cambridge, 329 p; illustrated

12. Pearle AD, Warren RF, Rodeo SA (2005) Basic science of articular cartilage and osteoarthritis. Clin Sports Med 24:1–12. doi:10.1016/j.csm.2004.08.007

13. Archer CW, Francis-West P (2003) The chondrocyte. Int J Biochem Cell Biol 35:401–404. doi:10.1016/S1357-2725(02)00301-1

14. Poole CA, Flint MH, Beaumont BW (1988) Chondrons extracted from canine tibial cartilage: preliminary report on their isolation and structure. J Orthop Res 6:408–419. doi:10.1002/jor.1100060312

15. Poole CA, Ayad S, Gilbert RT (1992) Chondrons from articular cartilage V.* Immunohistochemical evaluation of type VI collagen organisation in isolated chondrons by light, confocal and electron microscopy. J Cell Sci 103:1101–1110

16. Guilak F, Alexopoulos LG, Upton ML et al (2006) The pericellular matrix as a transducer of biomechanical and biochemical signals in articular cartilage. Ann NY Acad Sci 1068:498–512. doi:10.1196/annals.1346.011

17. Buckwalter J (1983) Articular cartilage. AAOS. Instr Course Lect 120:349–370

18. Mow VC, Holmes MH, Michael Lai W (1984) Fluid transport and mechanical properties of articular cartilage: a review. J Biomech 17:377–394. doi:10.1016/0021-9290(84)90031-9

19. Timothy E, Hardingham AJF (1992) Proteoglycans: many forms and many functions. FASEB J 6:861–870

20. Kempson GE (1972) Mechanical properties of articular cartilage. J Physiol 223:23P

21. Geihan R, Agnes R, Earl B, Poole AR (1992) Studies of the articular cartilage proteoglycan aggrecan in health and osteoarthritis evidence for molecular heterogeneity and extensive molecular changes in disease. J Clin Invest 90:2268–2277. doi:10.1172/JCI116113.These

22. Buckwalter J, Mankin H (1997) Articular cartilage I: tissue design and chondrocyte-matrix interactions. J Bone Jt Surg 79-A:600–611

23. Lai WM, Hou JS, Mow VC (1991) A triphasic theory for the swelling and deformation behaviors of articular cartilage. J Biomech Eng 113:245–258

24. Fosang AJ, Rogerson FM, East CJ, Stanton H (2008) ADAMTS-5: the story so far. Eur Cell Mater 15:11–26

25. Kashiwagi M, Tortorella M, Nagase H, Brew K (2001) TIMP-3 is a potent inhibitor of aggrecanase 1 (ADAM-TS4) and aggrecanase 2 (ADAM-TS5). J Biol Chem 276:12501–12504. doi:10.1074/jbc.C000848200

26. Benninghoff A (1925) Form und Bau der Gelenkknorpel in ihren Beziehungen zur Funktion. Z Anat Entwicklungsgesch 76:43–63. doi:10.1007/BF02134417

27. Cells E, Eyre DR, Weis MA, Wu J (2006) Articular cartilage collagen: an irreplaceable framework. Eur Cell Mater 12:57–63

28. Blumbach K, Bastiaansen-Jenniskens YM, DeGroot J et al (2009) Combined role of type IX collagen and cartilage oligomeric matrix protein in cartilage matrix assembly: cartilage oligomeric matrix protein counteracts type IX collagen-induced limitation of cartilage collagen fibril growth in mouse chondrocyte cultures. Arthritis Rheum 60:3676–3685. doi:10.1002/art.24979

29. Hagg R (1998) Cartilage fibrils of mammals are biochemically heterogeneous: differential distribution of decorin and collagen IX. J Cell Biol 142:285–294. doi:10.1083/jcb.142.1.285

30. Martel-Pelletier J, Boileau C, Pelletier J-P, Roughley PJ (2008) Cartilage in normal and osteoarthritis conditions. Best Pract Res Clin Rheumatol 22:351–384. doi:10.1016/j.berh.2008.02.001

31. Verzijl N, DeGroot J, Thorpe SR et al (2000) Effect of collagen turnover on the accumulation of advanced glycation end products. J Biol Chem 275:39027–39031. doi:10.1074/jbc.M006700200

32. Swann BDA, Sotman S, Dixon M, Brooks C (1977) The isolation and partial characterization of the major glycoprotein (LGP-I) from the articular lubricating fraction from bovine synovial fluid. Biochem J 161:473–485
33. Swann DA, Hendren RB, Radin EL et al (1981) The lubricating activity of synovial fluid glycoproteins. Arthritis Rheum 24:22–30
34. Jay GD, Tantravahi U, Britt DE et al (2001) Homology of lubricin and superficial zone protein (SZP): products of megakaryocyte stimulating factor (MSF) gene expression by human synovial fibroblasts and articular chondrocytes localized to chromosome 1q25. J Orthop Res 19:677–687. doi:10.1016/S0736-0266(00)00040-1
35. Ayral X, Pickering EH, Woodworth TG et al (2005) Synovitis: a potential predictive factor of structural progression of medial tibiofemoral knee osteoarthritis—results of a 1 year longitudinal arthroscopic study in 422 patients. Osteoarthr Cartil 13:361–367. doi:10.1016/j.joca.2005.01.005
36. Temenoff JS, Mikos AG (2000) Review: tissue engineering for regeneration of articular cartilage. Biomaterials 21:431–440
37. Lemperg R (1971) The subchondral bone plate of the femoral head in adult rabbits. I. Spontaneus remodelling studied by microradiography and tetracycline labelling. Virchows Arch A Pathol Pathol Anat 352:1–13
38. Redler I, Mow VC, Zimny ML, Mansell J (1975) The ultrastructure and biomechanical significance of the tidemark of articular cartilage. Clin Orthop Relat Res 112:357–362
39. Eckstein F, Milz S, Anetzberger H, Putz R (1998) Thickness of the subchondral mineralised tissue zone (SMZ) in normal male and female and pathological human patellae. J Anat 192:81–90. doi:10.1046/j.1469-7580.1998.19210081.x
40. Goldring SR (2012) Alterations in periarticular bone and cross talk between subchondral bone and articular cartilage in osteoarthritis. Ther Adv Musculoskelet Dis 4:249–258. doi:10.1177/1759720X12437353
41. Duncan H, Jundt J, Riddle JM et al (1987) The tibial subchondral plate. A scanning electron microscopic study. J Bone Joint Surg Am 69:1212–1220
42. Imhof H, Sulzbacher I, Grampp S et al (2000) Subchondral bone and cartilage disease: a rediscovered functional unit. Invest Radiol 35:581–588
43. Mobasheri A, Barrett-Jolley R, Carter S et al (2005) Functional roles of mechanosensitive ion channels, ß1 integrins and kinase cascades in chondrocyte mechanotransduction—mechanosensitivity in cells and tissues—NCBI bookshelf. In: Mechanosensitivity cells tissues. Moscow Acad. http://www.ncbi.nlm.nih.gov/books/NBK7517/. Accessed 18 Nov 2014
44. Houard X, Goldring MB, Berenbaum F (2013) Homeostatic mechanisms in articular cartilage and role of inflammation in osteoarthritis. Curr Rheumatol Rep 15:375. doi:10.1007/s11926-013-0375-6
45. Bader DL, Salter DM, Chowdhury TT (2011) Biomechanical influence of cartilage homeostasis in health and disease. Arthritis 2011:979032. doi:10.1155/2011/979032
46. Leong DJ, Hardin JA, Cobelli NJ, Sun HB (2011) Mechanotransduction and cartilage integrity. Ann NY Acad Sci 1240:32–37. doi:10.1111/j.1749-6632.2011.06301.x
47. Sun HB (2010) Mechanical loading, cartilage degradation, and arthritis. Ann NY Acad Sci 1211:37–50. doi:10.1111/j.1749-6632.2010.05808.x
48. Gosset M, Berenbaum F, Levy A et al (2008) Mechanical stress and prostaglandin E2 synthesis in cartilage. Biorheology 45:301–320
49. Griffin TM, Guilak F (2005) The role of mechanical loading in the onset and progression of osteoarthritis. Exerc Sport Sci Rev 33:195–200
50. Buckwalter, Joseph A., Henry J. Mankin AJG, Buckwalter JA, Mankin HJ, Grodzinsky AJ (2005) Articular cartilage and osteoarthritis. Instr Course Lect Am Acad Orthop Surg 54:465
51. Mak AF (1986) The apparent viscoelastic behavior of articular cartilage—the contributions from the intrinsic matrix viscoelasticity and interstitial fluid flows. J Biomech Eng 108:123. doi:10.1115/1.3138591

52. Mak AF (1986) Unconfined compression of hydrated viscoelastic tissues: a biphasic poroviscoelastic analysis. Biorheology 23:371–383
53. Wilson W, van Donkelaar CC, van Rietbergen R, Huiskes R (2005) The role of computational models in the search for the mechanical behavior and damage mechanisms of articular cartilage. Med Eng Phys 27:810–826. doi:10.1016/j.medengphy.2005.03.004
54. Rosenberg L, Hellmann W, Kleinschmidt AK (1970) Macromolecular models of proteinpolysaccharides from bovine nasal cartilage based on electron microscopic studies. J Biol Chem 245:4123–4130
55. Akizuki S, Mow VC, Muller F et al (1987) Tensile properties of human knee joint cartilage. II. Correlations between weight bearing and tissue pathology and the kinetics of swelling. J Orthop Res 5:173–186. doi:10.1002/jor.1100050204
56. Nordin M, Frankel VH (2001) Basic biomechanics of the musculoskeletal system, 3rd edn. Lippincott Williams & Wilkins, Philadelphia, 450 p
57. Zhu W, Mow VC, Koob TJ, Eyre DR (1993) Viscoelastic shear properties of articular cartilage and the effects of glycosidase treatments. J Orthop Res 11:771–781. doi:10.1002/jor.1100110602
58. Herzog W, Diet S, Suter E et al (1998) Material and functional properties of articular cartilage and patellofemoral contact mechanics in an experimental model of osteoarthritis. J Biomech 31:1137–1145. doi:10.1016/S0021-9290(98)00136-5
59. Arokoski J, Kiviranta I, Jurvelin J et al (1993) Long-distance running causes site-dependent decrease of cartilage glycosaminoglycan content in the knee joints of beagle dogs. Arthritis Rheum 36:1451–1459. doi:10.1002/art.1780361018
60. Haut RC, Ide TM, De Camp CE (1995) Mechanical responses of the rabbit patello-femoral joint to blunt impact. J Biomech Eng 117:402. doi:10.1115/1.2794199
61. Atkinson PJ, Haut RC (1995) Subfracture insult to the human cadaver patellofemoral joint produces occult injury. J Orthop Res 13:936–944. doi:10.1002/jor.1100130619
62. Hollander AP, Pidoux I, Reiner A et al (1995) Damage to type II collagen in aging and osteoarthritis starts at the articular surface, originates around chondrocytes, and extends into the cartilage with progressive degeneration. J Clin Invest 96:2859–2869. doi:10.1172/JCI118357
63. Hollander AP, Heathfield TF, Webber C et al (1994) Increased damage to type II collagen in osteoarthritic articular cartilage detected by a new immunoassay. J Clin Invest 93:1722–1732. doi:10.1172/JCI117156
64. Lefebvre V, Peeters-Joris C, Vaes G (1990) Modulation by interleukin 1 and tumor necrosis factor alpha of production of collagenase, tissue inhibitor of metalloproteinases and collagen types in differentiated and dedifferentiated articular chondrocytes. Biochim Biophys Acta 1052:366–378
65. Goldring MB, Birkhead J, Sandell LJ et al (1988) Interleukin 1 suppresses expression of cartilage-specific types II and IX collagens and increases types I and III collagens in human chondrocytes. J Clin Invest 82:2026–2037. doi:10.1172/JCI113823
66. Goldring MB, Birkhead J, Sandell LJ, Krane SM (1990) Synergistic regulation of collagen gene expression in human chondrocytes by tumor necrosis factor-? and interleukin-1? Ann NY Acad Sci 580:536–539. doi:10.1111/j.1749-6632.1990.tb17983.x
67. Mueller MB, Tuan RS (2011) Anabolic/catabolic balance in pathogenesis of osteoarthritis: identifying molecular targets. PM&R 3:S3–S11. doi:10.1016/j.pmrj.2011.05.009
68. Mariani E, Pulsatelli L, Facchini A (2014) Signaling pathways in cartilage repair. Int J Mol Sci 15:8667–8698. doi:10.3390/ijms15058667
69. Vincenti MP, Brinckerhoff CE (2002) Transcriptional regulation of collagenase (MMP-1, MMP-13) genes in arthritis: integration of complex signaling pathways for the recruitment of gene-specific transcription factors. Arthritis Res 4:157–164

Chapter 2
Synovial Knee Joint

Isabel Oliveira, Cristiana Gonçalves, Rui Luís Reis
and Joaquim Miguel Oliveira

Abstract The knee is a synovial knee joint that allows flexion, extension and also a slight ability to rotate medially and laterally, being able to move on two planes, being known as a modified hinge joint. The main features of a synovial joint are the articulating cartilage, the joint capsule, the joint cavity, the bursae and the ligaments. The joint capsule is composed by the synovium and a fibrous capsule, keeping the synovial fluid that fills the joint cavity inside the joint. The most common injuries or diseases that affects this complex biomechanical system are arthritis, bursitis and dislocations. In fact, the synovium the central area of pathology in a number of inflammatory joint diseases, such as rheumatoid arthritis (RA) and spondyloarthritis (SpA).

2.1 Introduction

The knee is the largest synovial joint in the body and one the most complex biomechanical system known. The purpose of synovial joints, also known as diarthroses, is to allow movement. The knee supports flexion and rotation, promoting complete stability and control under large variety of conditions [1]. The knee is composed by four main bones—the femur or thigh bone, the tibia or shin bone, the

---

Isabel Oliveira and Cristiana Gonçalves have contributed equally to this chapter.

I. Oliveira (✉) · C. Gonçalves · R.L. Reis · J.M. Oliveira
3B's Research Group—Biomaterials, Biodegradables and Biomimetics, European Institute of Excellence on Tissue Engineering and Regenerative Medicine, University of Minho, Avepark, Parque Da Ciência E Tecnologia, Zona Industrial da Gandra, 4805-017 Barco, Gmr, Portugal
e-mail: isabel2oliveira@hotmail.com

C. Gonçalves
e-mail: cristiana.goncalves@dep.uminho.pt

I. Oliveira · C. Gonçalves · R.L. Reis · J.M. Oliveira
ICVS/3B's—PT Government Associate Laboratory, Braga, Guimarães, Portugal

© Springer International Publishing AG 2017

J.M. Oliveira and R.L. Reis (eds.), *Regenerative Strategies for the Treatment of Knee Joint Disabilities*, Studies in Mechanobiology, Tissue Engineering and Biomaterials 21, DOI 10.1007/978-3-319-44785-8_2

fibula or outer skin bone and patella or knee cap (Fig. 2.1). The foremost movements of the knee joint occur between the three bones: femur, patella and tibia.

The patella, located at top centre, is small and is flat triangular—shaped bone that moves and rotates with knee [2]. The distal end of the femur has a medial and a lateral condyle, the structure of these condyles is important in the movement of the tibia on the femur. The proximal end of the tibia form a plateau with medial and lateral sections. The menisci intensify the shape of plateaus and serve to increase the conformity of the joint and help the rotation of the knee [3] (Fig. 2.1). Therefore, the structure of the femur, tibia, and patella highly contributes to the stability, strength and flexibility of the knee joint, with static and dynamic restrictions of the ligaments and crossing the joint [1].

The knee is composed by tendons and ligaments that are essentially constituted by connective tissues. The former connect muscle to bones and the latter connect bone to bone. Whether the tendons as ligaments are made of strands of elastic proteins. Ligaments prevent bones from moving too far and tendons help in the movement of muscles. The patella is inside a tendon, the patellar tendon which annexes the quadriceps muscles on the front of thigh and cover the patella [4].

The knee has important ligaments, such as the lateral ligaments of the femorotibial joints. Accordingly, the external condyle stays in position on the superior articular surface of the tibia by the external lateral ligaments (fibular collateral) and the anterior cruciate ligament that represents an internal ligament. The internal condyle stays in position by internal lateral ligaments (tibial collateral) and the posterior cruciate ligament that represents an external lateral ligament [5].

Moreover, knee joint include the tibiofemoral joint, a condyloid joint between the condyles of the femur and tibia, and patellofemoral joint, that it is between the

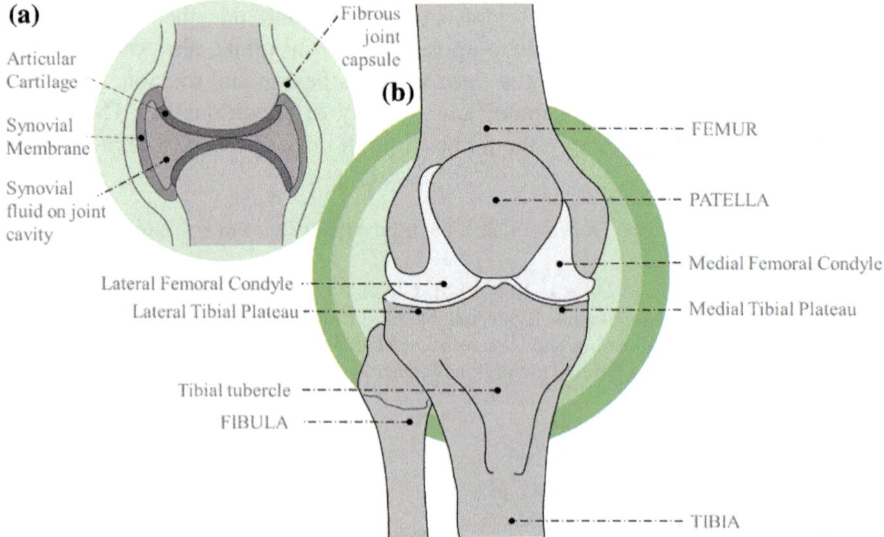

Fig. 2.1 Structure of synovial joint (**a**) and the anatomy of the knee (**b**)

posterior surface of the patella and the patellar surface of the femur. The upper tibiofibular joint often communicates with the femorotibial joint [6]. The tibiofemoral joint provides transmission of body weight from femur to the tibia, allowing a rotation along with a little degree of tibial axial rotation. The patellofemoral joint allows along with the tibialis anterior and ankle joint support the body to start the gait cycle [7].

The knee joint is separated in three articulations, one articulation between each femoral condoyle, corresponding to meniscus, other relative tibial condyle and other between the femur and patella.

The structure of synovial joint is different from cartilaginous (synchondroses and symphyses) and fibrous joints (such as gomphoses and syndesmoses). The key structural differences are related the existence of capsules surrounding the articulating surfaces of a synovial joint and the presence of lubricating synovial fluid within those capsules (synovial cavities).

The articular capsule that lining a synovial joint is constituted by fibrous capsule (thick outer layer) and synovial membrane (inner layer) [8]. The joint capsule is vital to the function of synovial joints. It protects the joint space, allowing stability by restricted movements, promoting active stability through nerve endings and may form articular surface for the joint, lined with synovium and forms a sleeve around the articulation bones to which it is attached [9]. The synovial membrane is highly vascular, it is constituted by connective tissue, it is responsible for the secretion of viscid synovial fluid that serve to lubricates and nourishes the joint [8].

The synovium (synovial membrane or stratum synovial) is a thin highly organized structure that is present between the joint cavity and the fibrous joint capsule [10], lining the joint capsule and producing the synovial fluid. This specialized mesenchymal soft tissue, essential for the appropriate function of the locomotor apparatus [11], covers the spaces of diarthrodial joints, tendon sheaths and bursae. The synovial fluid, secreted by the synovial membrane, is responsible by the (1) joint lubrication (which reduces friction and avoids shock between the surfaces of cartilage) and (2) nourishment (supplying oxygen and nutrients). This non-Newtonian fluid is composed by hyaluronic acid (synthesized by the synovial membrane) and lubricin (also named as Proteoglycan 4). Synovial cells, also known as synoviocyte, synovial lining cell or (synovial) intimal cell, comprise the main types A (showing ultrastructural resemblances to the macrophage series of cells) and B (with abundant endoplasmic reticulum and appear like fibroblasts synoviocytes) [12].

The synovial membrane includes the continuous surface layer of cells (intima) and the underlying tissue (subintima) [13]. Likewise, the lining layer is formed by condensed cells (macrophages and fibroblasts [10, 13]), one- to four-cells thick, and a loosely organized subliming layer, in contact with synovial fluid, that includes blood and lymphatic vessels, a cellular content of both resident fibroblasts and infiltrating cells in a collagenous extracellular matrix [10, 13]. In contrast to the compact layer of the synovial lining layer, the sublining is more amorphous, consisting of loose connective tissue that forms a microanatomic base for the synovial lining. The sublining allows for the transfer of both molecular and cellular elements from circulating blood to the synovial lining and the synovial fluid space

[10], since between the intimal surfaces is a small amount of fluid (usually rich in hyaluronan). Together, this structure provides a non-adherent surface between tissue elements. Synovium is derived from ectoderm and does not contain a basal lamina [13]. The synovium is the central area of pathology in a number of inflammatory joint diseases, including rheumatoid arthritis (RA) and spondyloarthritis (SpA) [13]. Synovitis is defined as inflammation of the synovium, the joint lining and responsible for the knee lubrication.

2.2   Synovial Membrane

The synovial membrane of the knee-joint is the largest in the body. This membrane start at the upper edge of the patella, on the lower part of the front of the femur that communicates with a bursa interposed between the tendon and the front of the femur [14].

The synovial membrane is a specialized mesenchymal tissue covering the spaces of diarthrodial joints, bursae, and tendon. This membrane has two layers, the inner layer, constituted by macrophages or synoviocytes, and outer layer, composed of two to three layers of synoviocytes over connective tissue with fibroblasts, secreting collagen, and other extracellular matrix proteins. The outer layer has few macrophages and lymphocytes, blood vessels with contain nutrients to the synovial membrane and the adjacent avascular cartilage and fat cells [15]. The synovial membrane is surrounding the cavity of joints, taking the space with synovial fluid. It has an important role in nutrition of the articular cartilage. The synovial fluid lubricates the ends of the bones making possible them to move often. The inefficiency to keep the level of metabolism starts destructive process [16, 17].

Most investigators believe that rheumatoid arthritis is primarily an inflammatory disease of synovial membrane of the joints [18]. The main characteristics of rheumatoid arthritis (RA) are chronic inflammation and progressive joint destruction. The synovial lining layer is thickened and hyperplastic, and synovial villi form. The sublining layer contains proliferating blood vessels and is invaded by inflammatory cells such as lymphocytes, plasma cells, and macrophages [19]. Aggressive resident synovial cells invading and destroying cartilage and bone in the joints of patients with RA is observed being the induction of apoptosis in synovial cells suggested as a successful strategy for the treatment of RA [19].

2.3   Synovitis

Synovitis is known as inflammation of the synovial membrane (synovium). Synovitis is characterized by thickening of the synovial membrane/capsule, increased synovial fluid and joint effusion, with the associated symptoms: pain, stiffness in the joint, swelling, warmth and redness over the area. Early synovitis is initially classified in clinical practice on the basis of the extent, location, and symmetry of the joint involvement [20]. This condition could play an important role in the pathophysiology of osteoarthritis [21, 22] and is regarded as a potential target for novel treatment strategies [23]. It can be either a result of rheumatoid arthritis, gout, cancer, and injury or occur independently on its own.

Synovitis is noticeably associated with a wide spectrum of infectious agents, and the mechanisms underlying this association are varied and complex, being the inflammatory process usually completely solved with the prompt and successful eradication of the organism [20].

The incidence of joint degeneration is higher for the knee joint [24]. The treatment of the inflammation depends on the cause, being the milder cases solely solved by conservative measures (such as rest, ice, compression and elevation) and the severe cases require an arthroscopic surgery (open synovectomy and arthroscopic synovectomy [25]). The treatment approaches to be applied are then related to the extent and level of injury, damage, from viscosupplementation methodologies to the joint replacement implants.

Chronic synovitis where conservative measures fail, can be treated effectively by the operation (open and arthroscopic synovectomy) through resecting the inflamed synovium. The arthroscopic synovectomy is the ideal operation owing to the fast recovery, less postoperative pain and cosmetic effect [25]. The use of the intra articular injection of a viscosupplement aims to support the synovial fluid function, mimicking its rheological behavior and also the biological functions. The most common material applied has been hyaluronic acid (HA), due to its leading characteristics, for instance it can reduce the production of pro-inflammatory mediators [26] and it is existent on normal synovial fluid, being responsible for the viscoelastic properties, important for the lubrication of the tissue surfaces in diarthrodial joints [27]. Nonetheless, due to its high costs [28] and some limitations, new studies with other promising materials are emerging to replace HA on this very challenging application [29].

Clinical trials are used to evaluate the safety and effectiveness of new treatments. ClinicalTrials.gov. is a website, maintained by the U.S. National Library of Medicine (NLM) at the National Institutes of Health (NIH), which provides access to a database with information on clinical studies, on a wide range of diseases and conditions. Information is provided and updated by the sponsor or principal investigator of the clinical study. A search on this database, using relevant keywords returned the registered clinical trials related with the application of biomaterials in knee joint diseases (Table 2.1). Table 2.1 summarizes some of those studies.

Table 2.1 A summary of clinical trials regarding synovial knee joint interventions with biomaterials

Trial ID	Denomination	Objectives	Interventions	Patients age (years)	Time-frame	Period time
NCT01211119	Novel one-step repair of knee meniscal tear using platelet-rich fibrin	Analyse if PRF implantation can help regeneration process of meniscectomized knee and T2 map MRI can evaluate the process in those patients with meniscal injuries	No platelet-rich fibrin (PRF)	20–45	1 month	2011–2013
NCT01425021	Retrieval and analysis of orthopedic implants at revision arthroplasty surgery	Evaluate safety, efficacy, performance and durability of biomaterial and implant designs used in joint replacements	NA	18–85	10 years	2010–2015
NCT02609074	Pilot clinical trial of CPC/rhBMP-2 (calcium phosphate cement scaffold/recombinant human bone morphogenetic protein-2) microffolds as bone substitute for bone regeneration	Analyse safety and preliminary efficacy of bone tissue repairing capacity and to test standard clinical and rehabilitation protocols	Procedure: minimally invasive internal fixation surgeries; Device: CPC/rhBMP-2 micro-scaffolds and CPC paste (group control)	16–70	1, 2, 3, 4, 6, 8 and 12 months post-operation	2013–2015
NCT00945399	Phase III protocol comparing a microfracture treatment to a CARTIPATCH® chondrocyte graft treatment in femoral condyle lesions	Compare the clinical improvement between the microfracture-treated group and CARTIPATCH® chondrocyte graft-treated group	Procedure: CARTIPATCH procedure and Microfracture	18–45	18 moths	2008–2014

2.4 Final Remarks

The synovial knee joint plays an important role as it provides unique movements (flexion, extension and a small degree of medial and lateral rotation) while supporting the body's weight. To do so, it is composed by different tissues such as bones (femur, tibia, and patella), strong ligaments, joint capsule, bursae, and articular fat pads. This complexity, which confers the foremost knee joint characteristics, also consents the incidence of some complications, mainly related with lesions and arthritis (such as rheumatoid arthritis, osteoarthritis, gout, bursitis, and tendonitis). Besides anti-inflammatory drugs, with an important role on relieving pain, inflammation, fever and even swelling, products capable of treating the problem from its source is of major interest to fully solve it. In fact, the current strategies for treatment of the inflammation can range from the simple conservative measures up to arthroscopic surgeries. Thus, biomaterials capable of mimic the affected structures, and even loaded with the necessary drugs for their recovery, are emergent and promising technologies currently being developed.

References

1. Goldblatt JP, Richmond JC (2003) Anatomy and biomechanics of the knee. Oper Tech Sports Med 11(3):172–186
2. Scott WN, Colman C, Gotlin RS (1996) Dr. Scott's knee book: symptoms, diagnosis, and treatment of knee problems including torn cartilage, ligament damage, arthritis, tendinitis, arthroscopic surgery, and total knee replacement. Simon and Schuster, New York
3. Blackburn TA, Craig E (1980) Knee anatomy a brief review. Phys Ther 60(12):1556–1560
4. Pujari A, Alton NS (2010) The healthy back book: a guide to whole healing for outdoor enthusiasts and other active people. Skipstone
5. Fullerton A (1916) The surgical anatomy of the synovial membrane of the knee–joint. Br J Surg 4(13):191–200
6. Saavedra MÁ, Navarro-Zarza JE, Villasenor-Ovies P, Canoso JJ, Vargas A, Chiapas-Gasca K, Hernández-Díaz C, Kalish RA (2012) Clinical anatomy of the knee. Reumatol Clin 8:39–45
7. Flandry F, Hommel G (2011) Normal anatomy and biomechanics of the knee. Sports Med Arthrosc Rev 19(2):82–92
8. Fenn S, Datir A, Saifuddin A (2009) Synovial recesses of the knee: MR imaging review of anatomical and pathological features. Skelet Radiol 38(4):317–328
9. Ralphs J, Benjamin M (1994) The joint capsule: structure, composition, ageing and disease. J Anat 184(Pt 3):503
10. Kiener HP, Karonitsch T (2011) The synovium as a privileged site in rheumatoid arthritis: cadherin-11 as a dominant player in synovial pathology. Best Pract Res Clin Rheumatol 25 (6):767–777
11. O'Connell JX (2000) Pathology of the synovium. Am J Clin Pathol 114(5):773–784
12. Revell P (1989) Synovial lining cells. Rheumatol Int 9(2):49–51
13. Smith MD (2011) The normal synovium. Open Rheumatol J 5(1):100–106
14. Gray H (1918) Anatomy of the human body. Lea & Febiger, Philadelphia
15. de Sousa EB, Casado PL, Neto VM, Duarte M, Aguiar DP (2014) Synovial fluid and synovial membrane mesenchymal stem cells: latest discoveries and therapeutic perspectives. Stem Cell Res Ther 5(5):112

16. Bobick J, Balaban N (2008) The handy anatomy answer book. Visible Ink Press, Michigan
17. Korochina K, Polyakova V, Korochina I (2016) Morphology of synovial membrane and articular cartilage in the knee joint in experimental chronic heart failure. Bull Exp Biol Med 160(3):376–380
18. Fujii K, Tsuji M, Tajima M (1999) Rheumatoid arthritis: a synovial disease? Ann Rheum Dis 58(12):727–730
19. Ospelt C, Gay S (2008) The role of resident synovial cells in destructive arthritis. Best Pract Res Clin Rheumatol 22(2):239–252
20. El-Gabalawy H (1999) The challenge of early synovitis: multiple pathways to a common clinical syndrome. Arthritis Res 1(1):31–36
21. de Lange-Brokaar B, Ioan-Facsinay A, Yusuf E, Visser A, Kroon H, Andersen S, Herb-van Toorn L, van Osch G, Zuurmond A-M, Stojanovic-Susulic V (2014) Degree of synovitis on MRI by comprehensive whole knee semi-quantitative scoring method correlates with histologic and macroscopic features of synovial tissue inflammation in knee osteoarthritis. Osteoarthr Cartil 22(10):1606–1613
22. Haugen I, Mathiessen A, Slatkowsky-Christensen B, Magnusson K, Bøyesen P, Sesseng S, van der Heijde D, Kvien T, Hammer H (2015) Synovitis and radiographic progression in non-erosive and erosive hand osteoarthritis: Is erosive hand osteoarthritis a separate inflammatory phenotype? Osteoarthr Cartil 24(4):647–654
23. Oei E, McWalter E, Sveinsson B, Alley M, Hargreaves B, Gold G (2014) Non-contrast diffusion weighted imaging for the assessment of knee synovitis: a comparative study against contrast-enhanced MRI. Osteoarthr Cartil 22:S252
24. Park J, Lakes RS (2007) Biomaterials: an introduction. Springer, Berlin
25. Pan X, Zhang X, Liu Z, Wen H, Mao X (2012) Treatment for chronic synovitis of knee: arthroscopic or open synovectomy. Rheumatol Int 32(6):1733–1736
26. Migliore A, Procopio S (2015) Effectiveness and utility of hyaluronic acid in osteoarthritis. Clin Cases Miner Bone Metab 12(1):31
27. Swann D, Radin E, Nazimiec M, Weisser P, Curran N, Lewinnek G (1974) Role of hyaluronic acid in joint lubrication. Ann Rheum Dis 33(4):318
28. Wen DY (2000) Intra-articular hyaluronic acid injections for knee osteoarthritis. Am Fam Phys 62(3):565–572
29. De Sousa RPRA, Gertrudes ACF, Correia C, da Silva Morais AJ, da Mota Martins Gonçalves C, Radhouani H, Gomes CAV, dos Santos TSXC, de Oliveira JMAC, do Sameiro Espregueira-Mendes JDC (2015) Viscosupplement composition comprising ulvan for treating arthritis. Google Patents

Chapter 3
Clinical Management of Articular Cartilage Lesions

Carlos A. Vilela, Cristina Correia, Joaquim Miguel Oliveira,
Rui Amandi Sousa, Rui Luís Reis and João Espregueira-Mendes

Abstract Articular cartilage is extremely sensitive to traumatic lesions and natural repair is very limited. When regeneration occur the tissue found in the lesion site is mostly fibrocartilage with poor mechanical properties, rendering a poor long-term clinical outcome. Cartilage lesion is a common problem with an impressive clinical and economic impact. With a difficult diagnosis in an initial disease stage, the cartilage lesion can progress to osteoarthritis and, therefore, a prompt diagnosis and treatment is required. Clinical management of cartilage lesions is a very demanding issue and the treatment is dependent of the extension, depth, location, chronicity of the lesions, patient's conditions and patients' expectations as well as associated lesions. In the present chapter, we present the clinical findings and diagnosis methodology to identify a cartilage lesion in an early stage. Finally, we discuss the indications, contra-indications, advantages, disadvantages and treatment decision-making as well as the outcomes of the available therapeutic approaches.

C.A. Vilela (✉) · J.M. Oliveira · R.L. Reis · J. Espregueira-Mendes
3B's Research Group, European Institute of Excellence on Tissue Engineering
and Regenerative Medicine, University of Minho, Guimarães, Portugal
e-mail: cvilelagomes@gmail.com

C.A. Vilela · J.M. Oliveira · R.L. Reis · J. Espregueira-Mendes
ICVS/3B's–PT Government Associate Laboratory, Braga, Guimarães, Portugal

C.A. Vilela · J. Espregueira-Mendes
Life and Health Sciences Research Institute (ICVS), School of Health Sciences,
University of Minho, Braga, Portugal

C.A. Vilela
Orthopedic Department, Centro Hospitalar do Alto Ave, Guimarães, Portugal

J. Espregueira-Mendes
Clínica do Dragão, Espregueira-Mendes Sports Centre, FIFA Medical Centre
of Excellence and Dom Henrique Research Center, FC Porto Stadium, Porto, Portugal

J. Espregueira-Mendes
Dom Henrique Research Centre, Porto, Portugal

C. Correia · R.A. Sousa · R.L. Reis
Stemmatters, Biotecnologia e Medicina Regenerativa SA, Guimarães, Portugal

© Springer International Publishing AG 2017

J.M. Oliveira and R.L. Reis (eds.), *Regenerative Strategies for the Treatment
of Knee Joint Disabilities*, Studies in Mechanobiology, Tissue Engineering
and Biomaterials 21, DOI 10.1007/978-3-319-44785-8_3

3.1 Introduction

Articular cartilage is a smooth, contact interface that lines the surface of two articulating bones of a diarthrodial joint. At the femur condyle the cartilage thickness ranges from 1.4 to 3.5 mm while at tibial plateau it ranges from 1 to 6 mm [16]. Although so thin, the cartilage presents excellent mechanical properties: providing a low-friction interface for the gliding articular surface and is able to support and distribute to underlying subchondral bone very high compressive and repetitive loads during a lifetime that can reach for the knee, 1.2 MPa in each step [10].

Without a vascular, neural or a lymphatic network and due to the lack of progenitor cells, the cartilage has a limited capacity for self-recover from a lesion and represent a very difficult challenge to the orthopedic surgeon. In fact, a cartilage lesion is frequent being found in 61–66 % of patients submitted to an arthroscopy [1, 29, 46, 96]. About 900,000 Americans are affected by a cartilage lesion each year and more than 200,000 surgical procedures are done to solve this problem annually [29]. According to McCormick et al. [68], the mean annual incidence is 90 surgeries per 10,000 patients with an annual incidence growth of 5 %. Commonly, progression of the cartilage lesion is the rule resulting in osteoarthritis at later stages [32, 50, 51, 54]. Radiographic knee osteoarthritis was found in 53 % of symptomatic and in 17 % of asymptomatic patients most commonly involving the medial and femoro-patellar compartment of the knee [67]. Total knee replacement is a poor and sad solution, especially for patients under 50 years old. Therefore, clinical and economic impact of cartilage lesions are significant: it is estimated that about 10–15 % of adults over 60 years old suffer from osteoarthritis with direct and indirect costs over $65 billion annually [53]. Early diagnosis and treatment of cartilage lesions may play an important role by avoiding osteoarthritis development, patient suffering and saving important economic resources [1, 54].

3.2 Clinical Findings

The mechanism of injury evolves an acute high-energy force or a shear and torsional force acting repetitively on the superficial articular surface [4]. The patient's history can offer some clues to the cartilage lesion diagnosis and evaluation. How the complaints started, how long are they affecting the patient, which activities are pain provoker, what was the lesion mechanism, what is and was the normal activity of the patient, what are the real expectations of the patient for his future and return to his normal daily-life or sport activity. Those are questions to answer in order to reach a good evaluation of the pathology. Pain is the main patient symptom and its intensity is variable and described in many ways. Usually, pain worsens with activity [1], have a mechanical rhythm and are related to a previous trauma [1]. In other occasions, like in inflammatory or degenerative disease, pain is not dependent of the physical activity. There is a very wide range of causes for complaints exacerbations in daily or sport activities. Swelling of the joint can also be present. For the knee, climbing or descending stairs, arising from a chair can cause pain. When a loose-body is in the joint due to a cartilage loose fragment, patients can refer symptoms like giving-way, locking and pseudo-locking. Sometimes patients have symptoms related to other pathology like a meniscal tear or a ligamentous injury [1, 64]. A careful and complete physical examination of the affected joint is required and the search for a swelling, hemarthroses, limitations of the range of joint motion, painful crepitation are mandatory, as are the specific tests for the examined joint. The examination is complete when compared with the contra-lateral joint.

Once symptomatic, the patient's pain and the functional impairment are likely to progress. During the course of this disease, the patient has some asymptomatic periods, but symptoms will return and worsen. Cartilage lesion evolution is not well known, but is believed that is dependent of the nature and type of lesions, associated lesions, patient gender, patient genetics and patient co-morbidities [54, 60]. Normally, when a cartilage lesion occurs, the repairing process produces a fibro-cartilage mostly with type I collagen and abnormal proteoglycans without the mechanical properties of the normal articular cartilage and consequently more susceptible to breakdown and to an early osteoarthritis [77]. Initially, when an articular cartilage injury occurs, a macromolecules loss is observed followed by a cartilaginous matrix rupture and finally by rupture of the bone matrix. Those three cartilage lesion evolutions stages must be taken in consideration when the therapeutic approach is chosen [96]. Therefore, the extension, deepness, and location of the lesion are important characteristics to define how serious the lesion is. Associated lesions as meniscal tears or ligamentous injuries [17, 45, 49, 54], misalignment of the limbs [30, 54], obesity [5, 27, 54, 69], and previous failed treatments are others factors to take in account to a correct evaluation of the cartilage lesion severity.

Outerbridge in 1961 created a four grade classification score to evaluate the macroscopic changes in cartilage lesions of the patella. In Grade 1, the less severe lesion, only softening and swelling of the cartilage is observed (Fig. 3.1). Grade 2 included cartilage lesions with fragmentation and fissuring in an area 1.5 cm or less in diameter (Fig. 3.2). For Grade 3 the lesions presented also fragmentation and fissuring but involving an area more than 1.5 cm in diameter. Grade 4, the most severe, included the lesions with erosion of cartilage down to bone (Fig. 3.3). Grades 1 and 2 are considered low grade lesions, but they do not heal and usually progress to a more severe stage [75]. This classification was adopted and became popular as a classification system for other cartilage lesions in the knee and for other joints. The Outerbridge score allow the distinction between a partial (Grades 1 and 2) versus nearly full or full-thickness cartilage defect (Grades 3 and 4); between a small (Grade 2) and larger (Grade 3) lesion; and describes a complete loss of cartilage (Grade 4). However, the Outerbridge classification has specific limitations: for example, an extensive partial thickness defect with a potentially bad prognosis,

Fig. 3.1 Cartilage softening (Grade 1)

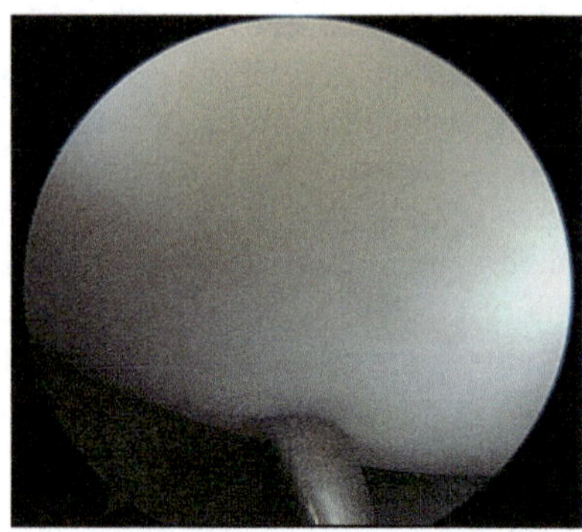

Fig. 3.2 Cartilage fragmentation and cartilage fissuring (Grade 2)

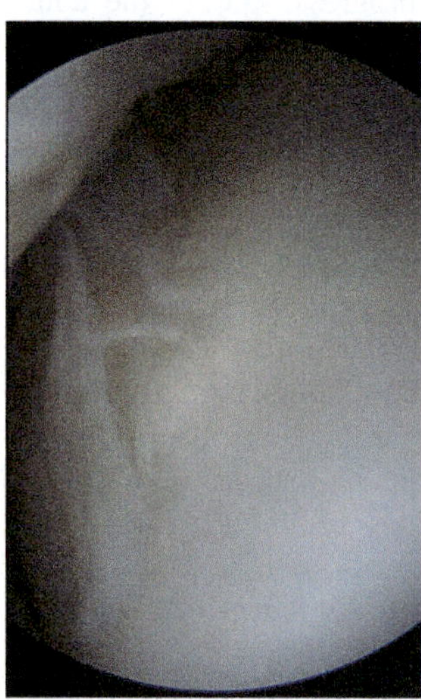

Fig. 3.3 Cartilage erosion (Grade 3)

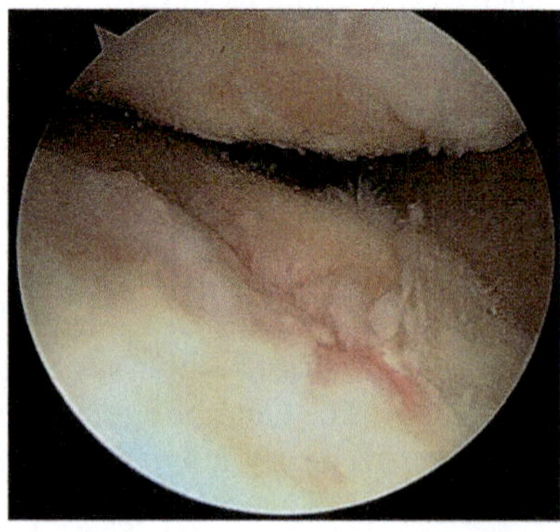

due to its size, is a Grade 1 defect. Whereas a direct cut or narrow fissure is a Grade 4 defect [11].

The International Cartilage Repair Society (ICRS) created a modified classification system that focuses on the depth of the cartilage injury combined with visual measurement (ICRS Cartilage Injury Evaluation Package). The ICRS grading score is a five grade score and intends a better macroscopic description of the defect and a better correlation with clinical outcome. Grade 0 relates to normal cartilage. In Grade1 are included superficial cartilage lesions, fissures, cracks and cartilage lesions with indentation (Grade 1A for the lesions with softening and/or superficial fissuring; Grade 1B when fissures and cracks were present. In Grade 2, fraying is found, lesions extending down to <50 % of cartilage depth. In Grade 3 the cartilage lesions present a partial loss of cartilage thickness and cartilage defects extending down >50 % of cartilage depth as well as down to the calcified layer. Grade 4 lesions relate to lesions with a complete loss of cartilage thickness and bone is visible [9].

Several other historical grading systems based in arthroscopy and/or MRI findings have been utilized: Insall (1976), Ficat et al. (1977), Casscells (1982), Beguin and Locker (1983), Bentley and Dowd (1984), Noyes and Stabler (1989), Frenche Society of Arthroscopy grading system (1994), Lewandrowski et al. (1996) [49].

According to ICRS classification, Grade 3 was the most common lesion found with 55 % of all patients submitted to an arthroscopy, followed by ICRS Grade 2 and ICRS grade 1 I lesions and only in 5 % of all patients presented an osteochondral grade 4 lesion [20, 46]. The majority of the cartilage lesions are single, affecting patients over 50 years old, with a mean area of 2.1 cm^2 (range between 0.5 and 12 cm^2), related to a previous trauma and affecting more frequently the medial condyle (in 58 % of the cases). Patella was the second place more frequent to have a cartilage lesion with 11 % of all the patients [46]. According the same authors, a concomitant meniscal, or ligamentous lesion was visible in 42 and 26 %, respectively. Others authors found similar results [1, 20, 27, 64]. According to Aroen et al. [1] arthroscopy review work, about 6 % of the patients submitted to an arthroscopy have an ICRS grade 3 or 4 cartilage defect with a size over 2 cm^2 and 11 % of all patients reviewed show cartilage defects suitable for cartilage repair procedures.

Most of the radiographic studies are normal and fail to reveal the majority of chondral lesions [14, 88, 101]. Although this evidence, radiographs can be very useful in patients with bigger osteochondral lesion, in patients affected by a severe osteoarthritis, osteochondritis-dissecans or a limb malalignment. For the study of the affected joint, at least an anteroposterior (AP) view and a lateral view are required. In some cases, a more specific view can be helpful. For the evaluation of the knee, the radiographic protocol includes a anteroposterior view, a standing anteroposterior view and a lateral view with the knee flexed 35°. A patellar view to study the patella is mandatory [67]. If the AP standing view or the clinical evaluation reveal a deformity in varus or in valgus a full-length standing radiograph can be useful. Rosemberg et al. concluded that a major chondral lesion was present

Fig. 3.4 Cartilage lesion—
RMI image

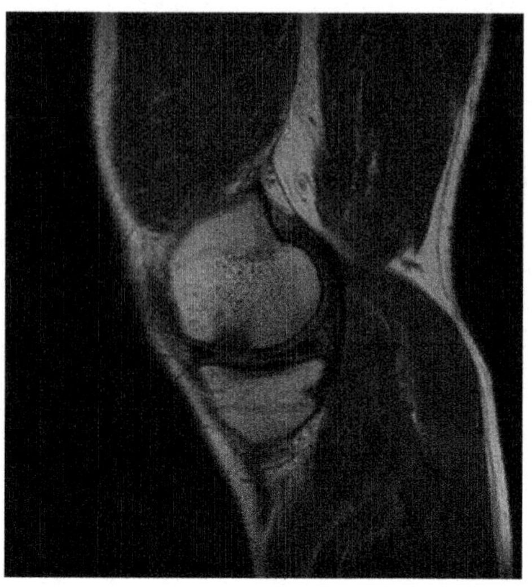

when a narrowing of the joint of more than 2 mm compared to contralateral knee space in a 45° posteroanterior weight-bearing view was visible [80, 86].

Magnetic resonance imaging (MRI) is the most appealing, powerful and important diagnostic procedure for the evaluation of cartilage lesions (Fig. 3.4). MRI is a noninvasive procedure and provides a more accurate information than radiographic studies and can document chondral lesions prior to radiographic changes and even prior to arthroscopy [12, 13]. In fact, MRI can detect metabolic and structural defects including the water content, before noticed in an arthroscopy [12]. Thus MRI is very useful for cartilage lesions diagnosis [13, 79, 98, 99], for monitoring the effects of chondral pharmacologic and surgical therapies, for study the cartilage disease evolution and in cartilage scientific research, namely in the semiquantitative and quantitative assessments of cartilage [12, 13]. However, for proposal therapies, although operator dependent, arthroscopy is yet the gold standard and the elected diagnose and validation procedure [12, 98].

For chondral diagnosis a High-magnetic-field-strength 1.5–3.0 tesla (T) scanner is needed (Fig. 3.5), which provides a higher signal-to-noise and contrast-to-noise ratios and, therefore, a thinner slice and a higher space resolution imaging. A minimum magnetic-field-strength of 1.0-T is needed for morphologic assessment of knee cartilage, but 1.5-T is currently used in cartilage evaluation. With a magnetic-field-strength of 3.0-T the time imaging acquisition is reduced, the image quality and image resolution is improved and, therefore the diagnose accuracy [57, 102]. However, the use of a higher strength magnetic field improves the magnetic susceptibility and the deposited energy in the tissues, images are more vulnerable to flow artifacts and the severity of chemical shift effects increases [18]. The 7.0-T MRI protocols was used in few studies and have not yet clearly shown advantages when compared with 3.0-T protocols [18, 90].

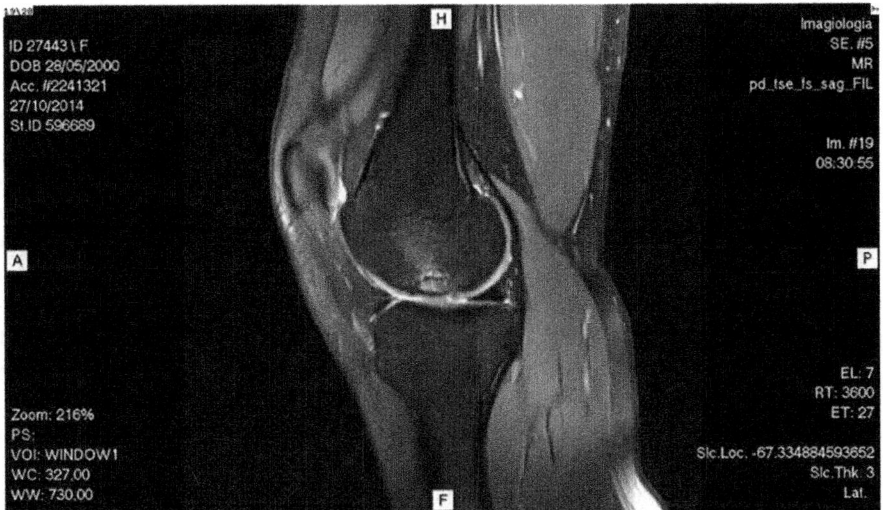

Fig. 3.5 Cartilage lesion—RMI image

A voxel is a rectangular volume element of the MRI images. The signal intensity of the voxel is a proportional sum of the signal of the composing tissues and the manipulation of the intensity of the contrast allows the highlight different tissue types. When an image is composed for two different tissues an artifact can occur in the interface of those tissues and is the reason for an incorrect evaluation of the cartilage lesions like the lesion dimension and cartilage thickness or even the diagnosis of a cartilage defect.

MRI imaging provides a morphologic characterization of the cartilage lesions and defines the deepness and extension of the lesions. Several acquisition techniques have been proposed: 2D and 3D fast spin-echo (FSE), 3D spoiled gradient–recalled echo (SPGR), 3D driven equilibrium fourier transform (3D-DEFT), 3D dual-echo steady state (3D-DESS), 3D balanced steady state free precession (3D-bSSFP), 3D fast SE sampling perfection with application-optimized contrast using different flip-angle evolutions (SPACE) [17].

2D-FSE is the most commonly used in the assessment of the joint cartilage and allows a good diagnosis of bone, menisci, or ligamentous injuries. For cartilage lesions, a good correlation and high sensitivity and specificity with arthroscopic technique was found [58]. 2D-FSE is recommended by the International Cartilage repair society for the evaluation of cartilage repair. 3D-FSE can reduce the time acquisition and has a diagnostic performance similar to 2D-FSE techniques but has not yet replaced the 2D-FSE in clinical practice [17, 85].

3D-SPGR is the gold standard technique for morphological knee cartilage evaluation. 3D-SPGR is very sensitive with a high accuracy to detect cartilage lesions and is very useful for cartilage thickness and volume measurements. Although those advantages, SPGR fails in the diagnosis of bone, menisci or ligamentous associated lesions. Besides, long time imaging is required and more metal artifacts are related [8, 17, 85]. Fast low-angle shot (FLASH) imaging is an SPGR technique useful for assessment of knee cartilage repair [18, 36, 85].

3D-DEFT increases contrast between fluid and cartilage and preserve the cartilage signal, resulting in a high signal intensity in both cartilage and synovial liquid [8, 18] but has a long acquisition times with a consequent vulnerability to motion artifacts. 3D-DEFT has a comparable performance to detect cartilage lesions when compared with FSE and SPGR techniques and is not reliable for assessing bone marrow [18, 85]. 3D-DESS has a shorter acquisition time than SPGR with similar accuracy for the detection of cartilaginous lesions. 3D-DESS allows a quantitative assessment of cartilage and decreased volume artifacts and has been validated for clinical use [8, 18]. 3D-bSSFP provides a good synovial fluid-cartilage contrast, decreased volume artifacts and is eventually useful for the study of other structures of the knee. VIPR Imaging is a SSFP derivative with shorter acquisition times and probably interesting for clinical and research practice [8, 18]. SPACE although the long acquisition times, has a good signal-to-noise ratio (SNR) and a high SNR efficiency but the capability to distinguishing cartilage and surrounding tissues is not as good as others techniques [18].

MRI provides also a compositional imaging of cartilage. The properties of the collagen and proteoglycan–associated glycosaminoglycan's macromolecular network of the hyaline cartilage, its content, electric charge, and status are assessed by MR imaging techniques. The current techniques that are available for the assessment of cartilage are focused on collagen and glycosaminoglycan content and include: T2 Mapping, dGEMERIC, T1p imaging, sodium imaging, and diffusion-weighted imaging [18, 85]. T2 Mapping reflects the interaction between the water content and the collagen network in a grey-scale map or in a color map and identifies the early stages of cartilage degeneration and the treatment effectiveness over time [2, 18]. T2 Mapping is clinically useful, well validated, but is a 2D-technique with a long acquisition time. dGEMERIC is related to the concentration of the negatively charged glycosaminoglycan molecules. The dGEMERIC acquisition RMI technique is well validated and clinically useful, but requires the administration of an intravenous contrast product and has a long acquisition time [18, 85]. The intravenous administration of Gd-DTPA 2- and is consequent and progressive concentration in cartilage is inversely proportional to the glycosaminoglycan content and is an evaluation method for monitoring cartilage repair procedures [18]. T1p Imaging is dependent of collagen network and glycosaminoglycan content and higher T1p Imaging values are indicative of a damaged cartilage. The use of T1p Imaging is limited to a few research centers and time consuming, thus limited for clinical use [18, 85]. Sodium imaging MRI acquisition technique is related to the glycosaminoglycan composition of the cartilage and can be useful in the differentiation of early stage of cartilage pathology. Sodium imaging MRI is available in few centers and need a special hardware. Diffusion-weighted imaging is based on the motion of water content which is related to the cartilage architecture and cartilage biochemical structure, thus dependent of the collagen and glycosaminoglycan content. This technique can be useful for the implants follow up [18].

With a FSE acquisition technique the lesion appears to be brighter than adjacent cartilage but with an SPGR acquisition technique the lesion is darker than adjacent normal cartilage [85]. The estimated area, depth, the presence and the volume of the bone attached to the cartilage defect, the exact location of the defect and edema bone marrow signal should be reported and crucial for the treatment plan and can provide some prognosis hints.

MRI is progressing and becoming a more sensitive and specific diagnosis procedure. MRI is also a more common monitoring cartilage progress repair procedure. [77, 85].

3.3 Treatment

Surgical cartilage lesions treatment has a long history and very early attempts to treat cartilage were described in the last century. One of the first osteoarticular transplants was described in 1925 and most of the current marrow stimulation procedures derived from initial studies of Pridie (1959) and Ficat (1979) [29].

The goals of cartilage repair are to diminish pain and swelling in afflicted patients; improve function and sports activities; prevent progression towards osteoarthritis; achieve these goals with lowest cost to society and lowest co-morbidity possible to the patient. For the clinicians' decision in cartilage repair, is important the characteristics of the lesion and etiology, the defect thickness, location and size. It is also important the containment, a ligamentous or meniscus injury, previous treatment, physiologic age and systemic disease. Therefore, the treatment must be focused on the specific patient we want to treat and there is no consensus regarding the best method to repair a cartilage defect [65] and about the long-term results [70]. The choice of the best treatment for our patient demands more rigorous prospective, adequately powered and randomized clinical trials. Besides, there are no long-term studies comparing the treated lesion with the untreated lesion which means that the cost-benefit ratio is yet unknown [96].

Some factors are related to a better clinical treatment result: age under 30 years, body mass index <30, a correct limb alignment and integrity of menisci or ligamentous. Lesions in the medial femoral condyle have better results than lesions in the lateral condyle, tibia plateau or in the patella [96].

3.3.1 Paliative Treatments

The articular lavage/debridement using a saline solution inserted into a joint with a needle or during an arthroscopy associated with the removal of chondral fragments and osteophytes, lose bodies, degenerated menisci and redundant synovia [73] is an empirical therapeutic approach without a solid scientific or biological basis for the symptomatic beneficial effects reported in few reported studies. Besides, the relative pain and symptoms relief is not consensual and some authors did not found this supposed clinical improvement with the articular lavage [56, 62]. This procedure is in decline [56, 62] and the indications are, eventually, limited to a patient with locking symptoms due to a loose body, in cases of unstable cartilage or with a concurrent meniscal tear [96].

Electrocautery, laser or radiofrequency energy (RFE) devices have been used for the treatment of cartilage lesions. In the electrocautery procedure, the tissue electrical tissue resistance to a high-frequency current, produce tissue destruction. Various devices are available and the results could not be better than simple chondroplasty. Laser was introduced to arthroscopy surgery in the 1990's and the effects produced when the laser energy touches a tissue are reflection, scatter, absorption and transmission. Absorption is the predominant effect that causes tissue heating. RFE systems for clinical application can be monopolar or bipolar and the experimental and clinical reported results are controversial and contradictory [24, 26, 94]. RFE is relatively inexpensive, safe and simple to use in an arthroscopic surgery and almost replaced the laser and electrocautery procedures for thermal chondroplasty. Various systems are available and under development: Vulcan EAS TM, Linvatec, VAPR TM, ArthroCareTM, UltrAblator Electrode. Although the clinical outcome regarding the use of RFE in cartilage lesions treatment are few, encouraging results have been reported with significant pain-relief but concerns about costs and security related to osteonecrosis, cartilage loss, proteoglycan loss and avascular necrosis limits its use [48, 94, 103].

3.3.2 Reparative Treatments

Stimulating bone marrow techniques include arthroscopic abrasion arthroplasty or simple abrasion chondroplasty, Pridie drilling and microfracture (MF) techniques popularized by Steadman et al. Spongialization is also a stimulating bone marrow procedure adopted for patellar lesions described by Ficat and colleagues. The rationale behind this concept is to stimulate a spontaneous and natural repair reaction by penetrating the subchondral bone and consequent spongeous bleeding with the resulting blood clot, promoting the recruitment of bone marrow cells to enhance a natural healing [66]. The usual repaired tissue is not a normal hyaline cartilage [66, 95].

Arthroscopic abrasion arthroplasty or simple abrasion chondroplasty is a salvage and palliative arthroscopic debridement procedure. Using a motorized burr or a curette the surgeon removes the superficial dead bone of the cartilage lesion, exposing viable bleeding bone, and leaving untouched the normal surrounding cartilage. The major clinical indication is the advanced and extended grade III/IV lesion or in an severe degenerative arthritis in an older patient, usually around 60 years old who is seeking for an alternative to a total knee arthroplasty. The surgery has been used for more than 25 years and the results are controversial. Some authors reported a clinical improvement [52] and a deferred knee joint arthroplasty for more than 5 years with a long durability of the repaired tissue. Other studies did not find this clinical improvement and concluded that arthroscopic results are no better than medical treatments or even placebo treatments [47, 71]. Although this lack of scientific evidence, chondroplasty and debridement procedures are the most performed procedure in the United States [68].

MF technique, as described by Steadman et al., included the complete cartilage lesion identification, debridement of all remaining cartilage fragments till the healthy cartilage limit creating a vertical stabilized shoulder (Fig. 3.6). The calcified cartilage layer of this well delimited area is removed. Using a small Awl and a mallet, perpendicular holes to the subchondral bone of 3–4 mm depth separated by 3–4 mm are done [91]. A specific rehabilitation program is required and usually a variable non-weight bearing period is demanded. This procedure become popular and widely used as a cartilage restoration procedure [29] and comparative technique with all other techniques [96]. Despite a good short-time clinical outcome [66, 91, 95], the repaired tissue is not hyaline cartilage but a fibrocartilage with poor mechanical properties [23, 25] and the results of few long-term studies following microfracture treatment for cartilage lesions are not conclusive [33, 66, 77, 95]. According to McCormick et al., MF and drilling are the second restorative cartilage procedure more often performed in the United Sates [68] and good clinical and imaging results were reported [61] even when compared with other procedures more expensive and demanding like mosaicoplasty or autologous chondrocyte implantation (ACI) [65, 95]. The indications are still under debate. The benefits of this technique according the location of the lesion, patient's age and the upper lesion size limit is not clear and controversial results were reported. According to Steadman [91] an improved outcome is expected in all knee compartments in patients with cartilage lesions greater than 4 cm^2 in patients under 45 years old and even better, under the age of 35 years old [91]. Others authors did not find the same good results in lesions greater than 4 cm^2 or in patella chondral lesions [29]. Goyal et al. [39, 41] in a systematic review observed that MF in patients with a small lesion and low activity had a good short-term outcome and beyond 5 years post-operatively a failure of the treatment could be expected. Chondral lesions with subchondral bone intact with an area lesion lesser than 2–2.5 cm^2 [3, 19, 32, 66] in a patient younger than 35 years old with a

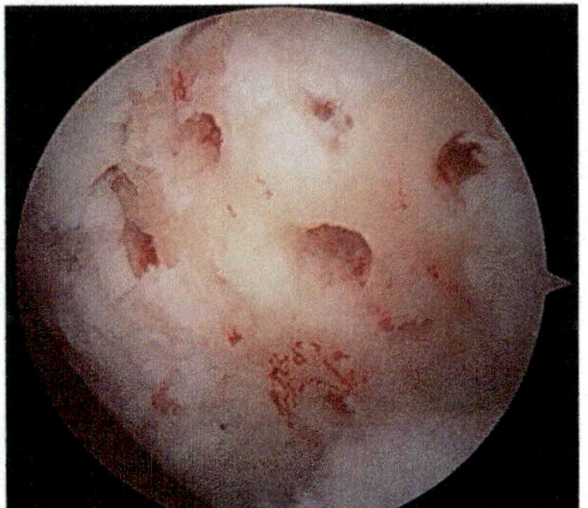

Fig. 3.6 Cartilage treatment according to the MF technique

body mass index under 25 kg/m^2 and a knee cartilage lesion with no more than 12 months evolution appeared to be the best indication for the microfrature technique [3, 96].

The association of those techniques with growth factors, platelet-rich plasma (PRP) or genetic engineering have been studied and may provide in the future an alternative and an improvement in cartilage treatments [34, 63, 93]. Another approach is the AMIC (autologous matrix-induced chondrogenesis) technique consisting of covering a cartilage lesion initially treated with MF with a collagen (I/III) membrane. This technique can be associated with the application of PRP or concentrated bone marrow (BMAC—bone marrow aspirate concentrate). The published results showed a clinical improvement, but the filling of the defect in the MRI analysis is not conclusive [97, 15, 37, 38].

Dr. Craig Morgan, Dr. Vladimir Bobic and Dr. Lazlo Hangody popularized the osteochondral autograft transfer technique [29]. Osteochondral autograft transfer (OAT) is an alternative cartilage procedure: harvesting a cartilage autologous plug from a non-weight bearing area of the knee or from other joint [28] to repair the cartilage defect (Fig. 3.7). In the knee, the most frequent donor site is the medial and lateral border of the condyles, the intercondylar notch or the sulcus terminalis of the femoral chondyle [37, 38, 66]. Useful for symptomatic small chondral or osteochondral defect, between 2.5 cm and 4 cm^2 in the weight bearing of a young patient [82, 6, 37, 38] and has a better outcome in lesions located in the condyles than in patella or tibial plateau [96]. In an ACI or MF failed procedure, OAT can be an alternative treatment [37, 38]. The OAT is a surgical demanding procedure and has other limitations: morbidity of the donor site and limitations of the available graft, congruency of the repaired surface specially when is necessary more than one plug to fit a more extensive lesion in a technique so called mosaicplasty (Fig. 3.8) [6, 82]. The osteochondral relocated plug has a good bone-to-bone potential healing, but rarely heals completely with the surrounding healthy cartilage. Good results have been reported [74, 81, 82, 96] even when compared to debridement and MF

Fig. 3.7 Cartilage harvesting from tibio-fibular proximal joint

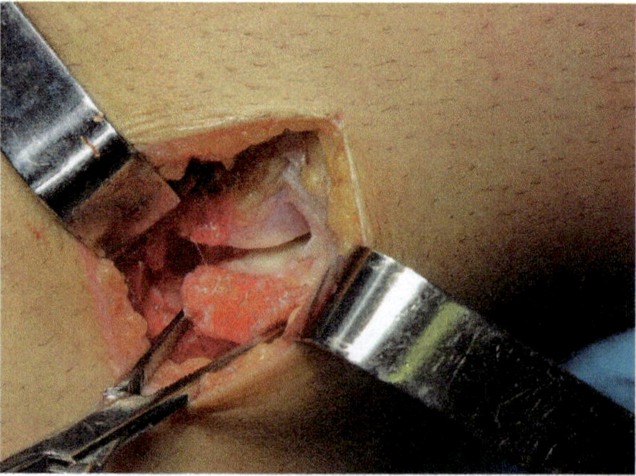

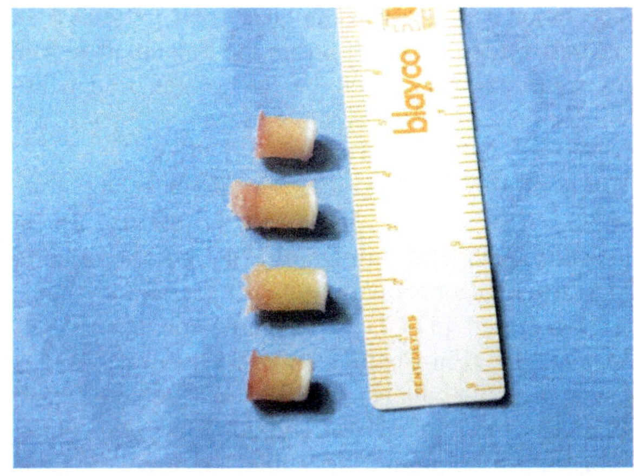

Fig. 3.8 Prepared osteochondral grafts before the application in a mosaicplasty

techniques [40, 42]. Solheim et al. [89] reported a poor outcome or even failure of the mosaicplasty in 40 % of the patients in a long-term follow up. The best results with mosaicplasty technique were reported in a small deep lesion under 2.5 cm^2 located in the medial condyle [97] an CAIS—Depuy-Mitek, Raynham MA.

A frozen or fresh osteochondral allograft can be the treatment of choice for a patient aged up to 50 years and/or active patient with an extensive chondral or osteochondral lesion, usually greater than 2.5 cm and when arthroplasty is the alternative [37, 38, 96]. Other indications for this procedure include salvage of previous cartilage procedure, osteochondritis dissecans, osteonecrosis of the femoral condyle or a reconstructive post-traumatic surgery [37, 38]. The surgical technique is demanding to achieve a good fixation and congruency with the healthy adjacent bone and cartilage. Medico-legal reasons, the risk of potential disease transmission, the low availability and the difficulties in preserving and managing the allografts, are serious drawbacks limiting the choice of this treatment. Immunogenicity of the allograft is also limiting and a percentage of patients become antibody positive with a less favorable outcome when compared with antibody negative patients [29]. Good clinical outcomes in medium/long term have been reported [29, 35] but others studies did not confirm these results and reported a high reoperation and failure rate [44].

Allogenic cartilage grafts are a recent therapy for cartilage lesions with the advantage of a lower immunological response. A morselized cartilage allograft or an allogenic chondrocyte implant is available for cartilage repair (DeNovo^R NT—Zimmer, Warshaw, Indiana). The surgical technique is similar to the MF/Autologous chondrocyte implantation (ACI) and clinical improvement were reported in a few studies already reported [29, 37, 38]. The cartilage autograft implantation system (CAIS—Depuy-Mitek, Raynham MA) is a single-stage procedure utilizing a glued autologous morselized cartilage onto a synthetic bio absorbable scaffold instead an allogenic graft.

Acellular three-dimensional scaffolds made up of more than one layer to mimic normal cartilage structure, have been proposed as cartilage regenerative procedure.

The rational of these techniques is to provide a structural support for immature reparative tissue resulting from the bone marrow stimulation. The simplicity of the procedure and the possibility of combination with cells or growth factors, make this technique an interesting approach. However, pain and persistent swelling have been reported [96]. TruFit[R] (Smith & Nephew, Andover, MA) is a bilayer porous PLGA-calcium-sulfate biopolymer that was proposed for cartilage regeneration. The early reports were good but the repaired tissue seemed to be heterogeneous, with cyst formation in the subchondral bone and any evidence of bone ingrowth, osteoconductivity, or integration. Unfavorable mid-term MRI results were reported [37, 38]. The commercialization of this scaffold was suspended. Maioregen[R] (Fin-Ceramica S.p.A., Faenza, Italy) is a nanostructured biomimetic hydroxyapatite-collagen scaffold with a porous 3-D tri-layer composite structure available for clinical use. Initial good clinical results were re-ported, but the follow-up is short and the studies are few [97, 21, 37, 38].

3.3.3 Regenerative Treatments

Autologous chondrocyte implantation (ACI)/Autologous chondrocyte transplantation (ACT) is a regenerative two-step cartilage therapy introduced in Sweden in the late 1980's by Peterson and Brittberg to resurface a symptomatic patient with a cartilage lesion. The first step is the assessment of the joint and performance of a cartilage biopsy. The procedure begins with the cartilage harvesting of approximately 200–300 mg of articular cartilage from a healthy and non-bearing area of the donor. The harvested cartilage fragment is processed to achieve chondrocyte isolation and expansion to a high chondrocyte density, usually between 5 and 10 million cells over a period of 4–6 weeks [77]. In the second step, a periosteal flap, harvested from the proximal tibia and 2 mm larger than the lesion, is sutured to the healthy borders of the prepared, non-bleeding and clean cartilage lesion. The covered lesion is than sealed with glue usually collagen or hyaluronan secured with fibrin glue or is self-adhering. Finally, expanded chondrocytes are implanted into the closed lesion. [4, 37, 38, 96].

The cartilage defect coverage in ACI first generation (ACI-1st generation) is made with a periosteal flap. For the ACI second generation (ACI-2nd generation), a membrane made often of collagen type I/III is the choice to cover the defect. The pointed advantages are decreased surgical exposition, reduced operating time and reduction of the complications related to the periosteal use [4, 23] despite the reported asymptomatic graft hypertrophy [23]. In ACI third generation (ACI-3rd generation) a matrix is seeded with cells and implanted in the cartilage lesion, The ACI-3rd generation is so called MACI (matrix-assisted chondrocyte implantation), but this was adopted as a trademark of Genzyme Biosurgery [32]. These treatments use a chondroinductuive or chondroconductive matrix usually seeded with autologous cells in controlled conditions to improve mechanical properties before the surgery. It is believed that ACI-3rd generation has an even chondrocyte distribution and there is no need of sutures or either a coverage which reduces the time of the surgery and the surgical exposure [4].

The indications for a ACI/ACT treatment in a knee cartilage lesion are well motivated patients under 55 years old, with pain, swelling, locking or catching with a grade II or IV cartilage lesion. ACI has been used to restore focal defects between 2 and 12 cm^2. However, it has been used in lesions up to 26.6 cm^2. In defects under 2 cm^2, ACI is indicated as a salvage procedure with poor reported outcomes. The best location is the femoral or patellar articular surface without a kissing lesion in the opposite articular surface. ACI is contra-indicated in patients with an inflammatory arthritis or with an articular infection associated lesions described above must be considered and included in the treatment plan [4, 32, 77, 96]. For the talus an ACI treatment is recommended in patients with a lesion greater than 2.5 cm in diameter and as alternative autograft or allograft transplantation could be chosen as an option [44].

Complications for the ACI 1st generation are related to the periosteal flap: the graft and/or periosteal delamination and periosteal hypertrophy were related. Technical difficulties, large exposition and stiffness of the joint are also drawbacks of this technique due to the large tissue exposition [4, 32]. Other complications have ben also reported: device rejection and migration, immune reaction, delamination, swelling, fever and joint stiffness [37, 38]. Better results were reported with ACI-2nd generation in a systematic review [39, 41].

ACI become a popular technique treatment to repair cartilage lesions and has been performed on an estimated 35,000 patients worldwide [77]. The bibliography review of ACI-1st generation studies show an improved clinical, histological and mechanical results where, even in long term follow up [4, 7, 77, 96]. good to excellent clinical outcomes were reported in studies with patients older than 45 years or in patients refractory to prior treatments [4]. A clinical improvement and lesser complications related to the periosteal graft was reported using type I/III collagen membranes [32, 92] but in other studies this finding was not confirmed [32]. Comparing ACI-1st generation with a ACI-2nd generation and ACI-3rd generation the results are not conclusive but some reported results are better with ACI-3rd generation treatment [25, 33, 96]. Although the reported good results [37, 38, 83, 87] it has not been possible the regeneration of hyaline cartilage in a consistently way [29, 32]. Comparing ACI procedures with others techniques (arthroscopic debridement, mosaicplasty) as Batty summarized, better significant clinical outcomes were found with ACI treatments [4, 22, 77, 83]. For the comparison of the ACI procedures with microfracture procedure, although a better clinical outcomes with ACI treatment, studies do not find always a statistically significant improvement [4, 22, 23, 72, 77, 78, 83, 96]. However, ACI was associated to a better structural repair, in more recent studies, a better long-term outcome was found with ACI 3rd generation [4, 96]. Although ACI 3rd generation has been used since 2000, few studies aimed to correlate the arthroscopic findings of a second-look arthroscopy with the histological appearance of the ACI biopsy [25]. As Enea et al. [25] demonstrated a nearly normal cartilage appearance of the repaired tissue with an ACI 3rd generation in a second-look arthroscopy in 80 % of cases and is not related with the histological findings of hyaline cartilage in 20 % of cases or with the functional patient's status. An important drawback of ACI techniques is the costing analysis as the Medical Service Advisory Committee of Australia reported. ACI procedures are more expensive than either MF or mosaicplasty due to the chondrocyte cell culture [83] but the cost-benefit over MF technique in terms of quality-adjusted life gained and osteoarthritis–related costs was also documented [23].

While chondrocytes are of great interest but due to its small initial number of cells and due to the de-differentiation on expansion to a fibroblast-like phenotype and consequent decreased proteoglycan synthesis and type II collagen expression and increased type I collagen expression, new alternatives cell sources are gaining increasing interest in the last years. Bone marrow stem cells (bMSCs), adipose stem cells (ASCs), Muscle derived stem cells (MDSCs), synovial stem cells (SSCs) and embryonic stem cells (ESCs) have been investigated in in vivo and pre-clinical studies [55, 59] but, to our knowledge, chondrocytes are the only cell source currently approved for clinical use. A great amount of cells that easily can be obtained without any adverse effect in the donor site, and the easy differentiation and expansion are pointed advantages for these alternative sources of cells as are the immunosuppressive and anti-inflammatory properties [22, 43, 59]. MSCs could be a suitable treatment option for cartilage repair and the few available data shows pre-clinical and clinical outcomes, at least, similar to the use of chondrocytes [84]. More and better quality studies and long-term follow up are needed to find the final conclusion [31]. Some questions need more affirmative answers: the optimal MSC source, the quality and durability of the repair tissue, the resistance to bone replacement and the integration with the surrounding normal non-treated cartilage and the heterogeneity and lack of standardized bioprocessing [31, 84]. The potential tumorogenesis in long follow-up was not confirmed by recent studies and the use of MSCs is probably a safe procedure [31, 76, 100].

The culture of these stem cells in a matrix and subsequent application onto a cartilage defect to treat a cartilage lesion is a new promising therapeutic approach so called a MASI procedure (matrix autologous stem-cells implantation) and can be classified as the 4th ACI generation. DeNovoR ET (Zimmer, Warshaw, Indiana) is a chondroconductive off-the-shelf matrix seeded with allogenic fetal chondrocytes is already available and clinical trials are underway. CARTSISTEM is a hyaluronate based gel seeded with mesenchymal stem cells from umbilical cord blood for a one step cartilage repair procedure and clinical trials are also underway [22].

Commonly, the defect location, the size, type of the lesion and activity of the patient are the most important factors for the treatment choice. The diagram in (Fig. 3.9). is an algorithm for the decision making in chondral and osteochondral knee injuries. In Table 3.1 we can find some of the commercial available repair sytems.

Signaling molecules including transforming growth factor (TGF-β), insulin-like growth factor (IGF), bone morphogenetic proteins (BMPs), and to a lesser extent fibroblast growth factors (FGFs) and epidermal growth factor (EGF), can improve chondrogenesis in in vitro studies when used isolated or associated. Important developments are expected in this field [59].

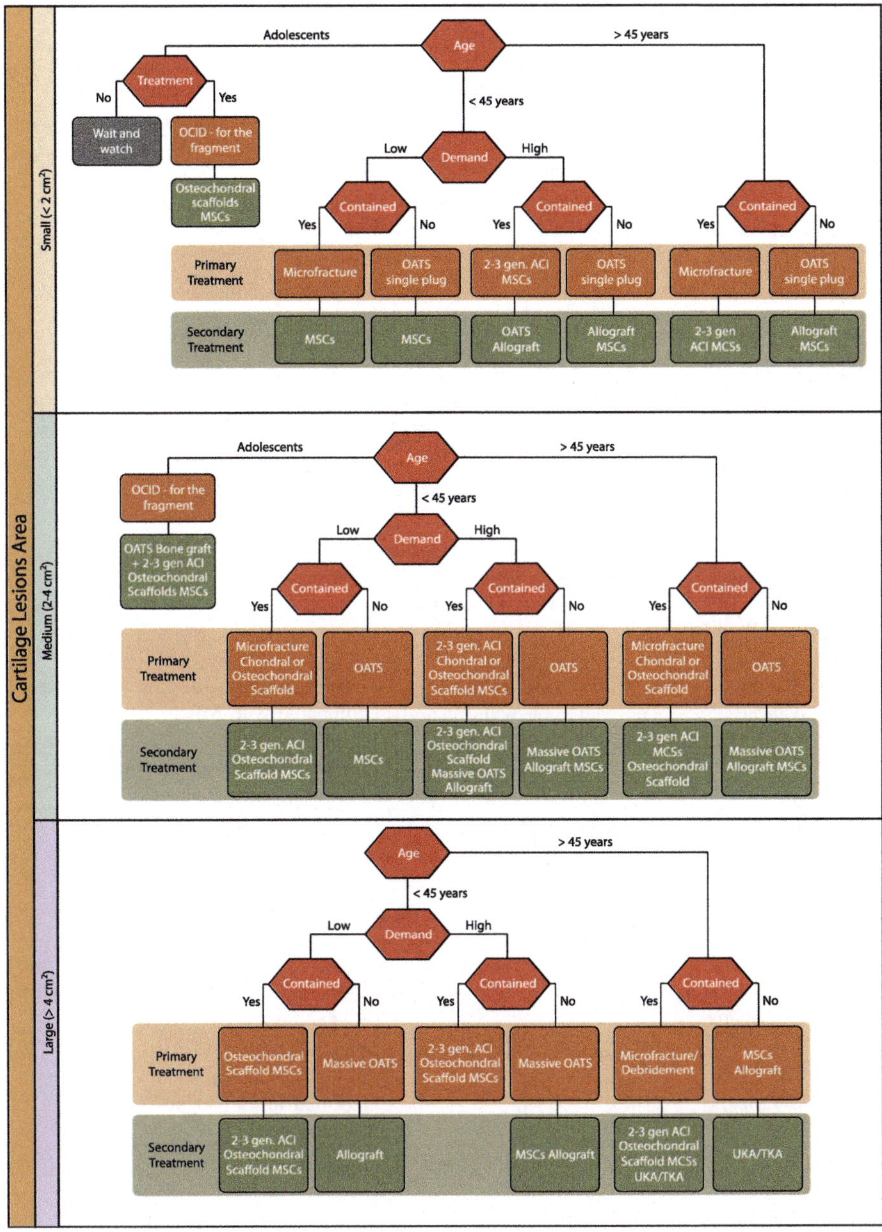

Fig. 3.9 Algorithm according to the International Cartilage Repair Society recomendations

Table 3.1 Comercial available cartilage repair sytems

Product name	Main material	Trials
ACI procedures		
ChondroCelect® TiGenix, Leuven, Belgium	10,000 cells/μl suspension (Dulbecco's modified eagles medium)	First approved cell-based product in Europe
Carticel® Genzyme Biosurgery, Cambridge, MA	12 million cells suspension	First FDA-approved cell therapy product
Chondro-Gide® GeistlichBiomaterials, Wolhusen, Switzerland	Colagen	Improved clinical outcome associated to MF or as an ACI procedure
MACI® Genzyme Biosurgery, Cambridge, MA	Porcine type I/III collagen	Phase III trials Improved outcome in case series in comparison with OAT and MF
CaReS® Ars Arthro, Esslingen, Germany	Rat-tail type I collagen	Improved clinical outcomes in a multicenter study with 116 patients/follow up: 30 months
NeoCart® Histiogenics Corporation, Waltham, MA	Bovine-type I collagen Chondrocyte culture in a bioreactor	Phase III trials
Hyalograft C® Fidia Advanced Biopolymers, Abano Terme, Italy	HYAFF 11-esterified derivative of hyaluronate	Improved clinical results even when compared with MF Improved clinical outcome in case series reported in 62 patients/follow up: 7 years
Cartipatch® Tissue Bank of France	Agarose-alginate	Phase III trials Improved clinical outcome in case series reported in 17 patients/follow up: 24 months
Bioseed C® BioTissue Technologies, GmbH, Freiburg, Germany	Copolymer of PGA, PLA and PDS—fibrin glue	Phase III trial Improved clinical outcomes in in case series reported in 52 patient/follow up: 4 years
BioCart II ProChon BioTech Ltd., Ness Ziona, Israel	Fibrinogen + Hyaluronan	Phase II trial Improved clinical results incase series reported in 31 patients/follow up: 17 months
DeNovo ET® Zimmer, Warshaw, Indiana	Matrix + allogenic fetal chondrocytes	Phase III trial
Cartsystem	Sodium hyaluronate + allogeneic umbilical cord MSCs	Phase II trial

(continued)

Table 3.1 (continued)

Product name	Main material	Trials
Graft		
DeNovo NT® Zimmer, Warshaw, Indiana	Matrix + allogenic chondrocytes	Good clinical outcomes in few studies reported
CAIS® Depuy-Mitek, Raynham MA	Glue + autologous morcelleied cartilage	Phase III trial
Cell free scaffold		
TruFit® Smith & Nephew, Andover, MA	PLGA-calcium-sulfate biopolymer bilayer porous	Suspended commercialization
BST-CarGel® Biosyntech, Quebec, Canada	chitosan + glycerol phosphate	Phase III trial Better outcomes than MF treatment in a 5 years follow-up
CaReS-1S® Arthro-Kinetics, Esslingen, German	Rat-tail type I collagen	Animal trials Short case series in adults
MaioRegen® Fin-Ceramica S.p.A., Faenza, Italy	Hydroxyapatite-collagen 3D tri-layers	Few studies

In conclusion, as we wait for new improved treatment techniques to achieve hyaline cartilage in the repaired tissue, more rigorous prospective, adequately powered and randomized clinical trials with the available treatments are needed to find the best cost-effective treatment for our patients. Despite the extensive efforts to develop an effective solution over the last century, there is still a paucity of clinical options of treatment for the cartilage lesions. We hope and believe that recent developments in the field of tissue engineering with all those materials, cellular approaches and repair cartilage enhancer's will find new solutions useful for cartilage lesion treatments.

References

1. Aroen A, Loken S, Heir S, Alvik E, Ekeland A, Granlund OG et al (2004) Articular cartilage lesions in 993 consecutive knee arthroscopies. Am J Sports Med 32(1):211–215
2. Ashton E, Riek J (2013) Advanced MR techniques in multicenter clinical trials. J Magn Reson Imaging 37(4):761–769
3. Asik M, Ciftci F, Sen C, Erdil M, Atalar A (2008) The microfracture technique for the treatment of full-thickness articular cartilage lesions of the knee: midterm results. Arthroscopy 24(11):1214–1220
4. Batty L, Dance S, Bajaj S, Cole BJ (2011) Autologous chondrocyte implantation: an overview of technique and outcomes. ANZ J Surg 81(1–2):18–25

5. Baum T, Joseph GB, Nardo L, Virayavanich W, Arulanandan A, Alizai H et al (2013) Correlation of magnetic resonance imaging-based knee cartilage T2 measurements and focal knee lesions with body mass index: thirty-six-month followup data from a longitudinal, observational multicenter study. Arthritis Care Res 65(1):23–33

6. Bentley G, Bhamra JS, Gikas PD, Skinner JA, Carrington R, Briggs TW (2013) Repair of osteochondral defects in joints—how to achieve success. Injury 44(Suppl 1):S3–S10

7. Biant LC, Bentley G, Vijayan S, Skinner JA, Carrington RW (2014) Long-term results of autologous chondrocyte implantation in the knee for chronic chondral and osteochondral defects. Am J Sports Med 42(9):2178–2183

8. Braun HJ, Dragoo JL, Hargreaves BA, Levenston ME, Gold GE (2013) Application of advanced magnetic resonance imaging techniques in evaluation of the lower extremity. Radiol Clin N Am 51(3):529–545

9. Brittberg M, Winalski CS (2003) Evaluation of cartilage injuries and repair. J Bone Joint Surg Am 85-A(Suppl 2):58–69

10. Brown TD, Shaw DT (1984) In vitro contact stress distribution on the femoral condyles. J Orthop Res 2(2):190–199

11. Casscells SW (1990) Outerbridge's ridges. Arthroscopy 6(4):253

12. Casula V, Hirvasniemi J, Lehenkari P, Ojala R, Haapea M, Saarakkala S et al (2014) Association between quantitative MRI and ICRS arthroscopic grading of articular cartilage. Knee Surg Sports Traumatol Arthrosc 24:2046–2054

13. Chan DD, Neu CP (2013) Probing articular cartilage damage and disease by quantitative magnetic resonance imaging. J R Soc Interface 10(78):20120608

14. Chan WP, Lang P, Stevens MP, Sack K, Majumdar S, Stoller DW et al (1991) Osteoarthritis of the knee: comparison of radiography, CT, and MR imaging to assess extent and severity. AJR Am J Roentgenol 157(4):799–806

15. Chung JY, Lee DH, Kim TH, Kwack KS, Yoon KH, Min BH (2014) Cartilage extra-cellular matrix biomembrane for the enhancement of microfractured defects. Knee Surg Sports Traumatol Arthrosc 22(6):1249–1259

16. Cohen ZA, McCarthy DM, Kwak SD, Legrand P, Fogarasi F, Ciaccio EJ et al (1999) Knee cartilage topography, thickness, and contact areas from MRI: in-vitro calibration and in-vivo measurements. Osteoarthr Cartil 7(1):95–109

17. Crema MD, Felson DT, Roemer FW, Wang K, Marra MD, Nevitt MC et al (2013) Prevalent cartilage damage and cartilage loss over time are associated with incident bone marrow lesions in the tibiofemoral compartments: the MOST study. Osteoarthr Cartil 21(2):306–313

18. Crema MD, Roemer FW, Marra MD, Burstein D, Gold GE, Eckstein F et al (2011) Articular cartilage in the knee: current MR imaging techniques and applications in clinical practice and research. Radiographics 31(1):37–61

19. Cucchiarini M, Madry H, Guilak F, Saris DB, Stoddart MJ, Koon Wong M et al (2014) A vision on the future of articular cartilage repair. Eur Cell Mater 27:12–16

20. Curl WW, Krome J, Gordon ES, Rushing J, Smith BP, Poehling GG (1997) Cartilage injuries: a review of 31,516 knee arthroscopies. Arthroscopy 13(4):456–460

21. Delcogliano M, de Caro F, Scaravella E, Ziveri G, De Biase CF, Marotta D et al (2014) Use of innovative biomimetic scaffold in the treatment for large osteochondral lesions of the knee. Knee Surg Sports Traumatol Arthrosc 22(6):1260–1269

22. Dewan AK, Gibson MA, Elisseeff JH, Trice ME (2014) Evolution of autologous chondrocyte repair and comparison to other cartilage repair techniques. Biomed Res Int 2014:272481

23. Dhollander AA, Verdonk PC, Lambrecht S, Verdonk R, Elewaut D, Verbruggen G et al (2012) Short-term outcome of the second generation characterized chondrocyte implantation for the treatment of cartilage lesions in the knee. Knee Surg Sports Traumatol Arthrosc 20(6):1118–1127

24. Dutcheshen N, Maerz T, Rabban P, Haut RC, Button KD, Baker KC et al (2012) The acute effect of bipolar radiofrequency energy thermal chondroplasty on intrinsic biomechanical properties and thickness of chondromalacic human articular cartilage. J Biomech Eng 134(8):081007

25. Enea D, Cecconi S, Busilacchi A, Manzotti S, Gesuita R, Gigante A (2012) Matrix-induced autologous chondrocyte implantation (MACI) in the knee. Knee Surg Sports Traumatol Arthrosc 20(5):862–869
26. Enochson L, Sonnergren HH, Mandalia VI, Lindahl A (2012) Bipolar radiofrequency plasma ablation induces proliferation and alters cytokine expression in human articular cartilage chondrocytes. Arthroscopy 28(9):1275–1282
27. Eskelinen AP, Visuri T, Larni HM, Ritsila V (2004) Primary cartilage lesions of the knee joint in young male adults. Overweight as a predisposing factor. An arthroscopic study. Scand J Surg 93(3):229–233
28. Espregueira-Mendes J, Pereira H, Sevivas N, Varanda P, da Silva MV, Monteiro A et al (2012) Osteochondral transplantation using autografts from the upper tibio-fibular joint for the treatment of knee cartilage lesions. Knee Surg Sports Traumatol Arthrosc 20(6): 1136–1142
29. Farr J, Cole B, Dhawan A, Kercher J, Sherman S (2011) Clinical cartilage restoration: evolution and overview. Clin Orthop Relat Res 469(10):2696–2705
30. Felson DT, Niu J, Gross KD, Englund M, Sharma L, Cooke TD et al (2013) Valgus malalignment is a risk factor for lateral knee osteoarthritis incidence and progression: findings from the Multicenter Osteoarthritis Study and the Osteoarthritis Initiative. Arthritis Rheum 65(2):355–362
31. Filardo G, Madry H, Jelic M, Roffi A, Cucchiarini M, Kon E (2013) Mesenchymal stem cells for the treatment of cartilage lesions: from preclinical findings to clinical application in orthopaedics. Knee Surg Sports Traumatol Arthrosc 21(8):1717–1729
32. Foldager CB (2013) Advances in autologous chondrocyte implantation and related techniques for cartilage repair. Dan Med J 60(4):B4600
33. Foldager CB, Bunger C, Nielsen AB, Ulrich-Vinther M, Munir S, Everland H et al (2012) Dermatan sulphate in methoxy polyethylene glycol–polylactide–co-glycolic acid scaffolds upregulates fibronectin gene expression but has no effect on in vivo osteochondral repair. Int Orthop 36(7):1507–1513
34. Furuzawa-Carballeda J, Lima G, Llorente L, Nunez-Alvarez C, Ruiz-Ordaz BH, Echevarria-Zuno S et al (2012) Polymerized-type I collagen downregulates inflammation and improves clinical outcomes in patients with symptomatic knee osteoarthritis following arthroscopic lavage: a randomized, double-blind, and placebo-controlled clinical trial. Sci World J 2012:342854
35. Giorgini A, Donati D, Cevolani L, Frisoni T, Zambianchi F, Catani F (2013) Fresh osteochondral allograft is a suitable alternative for wide cartilage defect in the knee. Injury 44(Suppl 1):S16–S20
36. Glaser C, Tins BJ, Trumm CG, Richardson JB, Reiser MF, McCall IW (2007) Quantitative 3D MR evaluation of autologous chondrocyte implantation in the knee: feasibility and initial results. Osteoarthr Cartil 15(7):798–807
37. Gomoll AH, Filardo G, Almqvist FK, Bugbee WD, Jelic M, Monllau JC et al (2012) Surgical treatment for early osteoarthritis. Part II: allografts and concurrent procedures. Knee Surg Sports Traumatol Arthrosc 20(3):468–486
38. Gomoll AH, Filardo G, de Girolamo L, Espregueira-Mendes J, Marcacci M, Rodkey WG et al (2012) Surgical treatment for early osteoarthritis. Part I: cartilage repair procedures. Knee Surg Sports Traumatol Arthrosc 20(3):450–466
39. Goyal D, Goyal A, Keyhani S, Lee EH, Hui JH (2013) Evidence-based status of second- and third-generation autologous chondrocyte implantation over first generation: a systematic review of level I and II studies. Arthroscopy 29(11):1872–1878
40. Goyal D, Keyhani S, Goyal A, Lee EH, Hui JH, Vaziri AS (2014) Evidence-based status of osteochondral cylinder transfer techniques: a systematic review of level I and II studies. Arthroscopy 30(4):497–505
41. Goyal D, Keyhani S, Lee EH, Hui JH (2013) Evidence-based status of microfracture technique: a systematic review of level I and II studies. Arthroscopy 29(9):1579–1588

42. Gudas R, Gudaite A, Mickevicius T, Masiulis N, Simonaityte R, Cekanauskas E et al (2013) Comparison of osteochondral autologous transplantation, microfracture, or debridement techniques in articular cartilage lesions associated with anterior cruciate ligament injury: a prospective study with a 3-year follow-up. Arthroscopy 29(1):89–97
43. Gupta PK, Das AK, Chullikana A, Majumdar AS (2012) Mesenchymal stem cells for cartilage repair in osteoarthritis. Stem Cell Res Ther 3(4):25
44. Haene R, Qamirani E, Story RA, Pinsker E, Daniels TR (2012) Intermediate outcomes of fresh talar osteochondral allografts for treatment of large osteochondral lesions of the talus. J Bone Joint Surg Am 94(12):1105–1110
45. Hirose J, Nishioka H, Okamoto N, Oniki Y, Nakamura E, Yamashita Y et al (2013) Articular cartilage lesions increase early cartilage degeneration in knees treated by anterior cruciate ligament reconstruction: T1rho mapping evaluation and 1-year follow-up. Am J Sports Med 41(10):2353–2361
46. Hjelle K, Solheim E, Strand T, Muri R, Brittberg M (2002) Articular cartilage defects in 1,000 knee arthroscopies. Arthroscopy 18(7):730–734
47. Howell SM (2010) The role of arthroscopy in treating osteoarthritis of the knee in the older patient. Orthopedics 33(9):652
48. Huang Y, Zhang Y, Ding X, Liu S, Sun T (2014) Working conditions of bipolar radiofrequency on human articular cartilage repair following thermal injury during arthroscopy. Chin Med J 127(22):3881–3886
49. Hunt N, Sanchez-Ballester J, Pandit R, Thomas R, Strachan R (2001) Chondral lesions of the knee: a new localization method and correlation with associated pathology. Arthroscopy 17(5):481–490
50. Hunziker EB (1999) Articular cartilage repair: Are the intrinsic biological constraints undermining this process insuperable? Osteoarthr Cartil 7(1):15–28
51. Hunziker EB (2002) Articular cartilage repair: basic science and clinical progress. A review of the current status and prospects. Osteoarthr Cartil 10(6):432–463
52. Ibarra-Ponce de Leon JC, Cabrales-Pontigo M, Crisostomo-Martinez JF, Almazan-Diaz A, Cruz-Lopez F, Encalada-Diaz MI et al (2009) Results of arthroscopic debridement and lavage in patients with knee osteoarthritis. Acta Ortop Mex 23(2):85–89
53. Jackson DW, Simon TM, Aberman HM (2001) Symptomatic articular cartilage degeneration: the impact in the new millennium. Clin Orthop Relat Res 391(Suppl):S14–S25
54. Johnson VL, Hunter DJ (2014) The epidemiology of osteoarthritis. Best Pract Res Clin Rheumatol 28(1):5–15
55. Johnstone B, Alini M, Cucchiarini M, Dodge GR, Eglin D, Guilak F et al (2013) Tissue engineering for articular cartilage repair–the state of the art. Eur Cell Mater 25:248–267
56. Katz JN, Brownlee SA, Jones MH (2014) The role of arthroscopy in the management of knee osteoarthritis. Best Pract Res Clin Rheumatol 28(1):143–156
57. Kijowski R, Blankenbaker DG, Davis KW, Shinki K, Kaplan LD, De SmetAA (2009) Comparison of 1.5- and 3.0-T MR imaging for evaluating the articular cartilage of the knee joint. Radiology 250(3):839–848
58. Kijowski R, Blankenbaker DG, Munoz Del Rio A, Baer GS, Graf BK (2013) Evaluation of the articular cartilage of the knee joint: value of adding a T2 mapping sequence to a routine MR imaging protocol. Radiology 267(2):503–513
59. Kock L, van Donkelaar CC, Ito K (2012) Tissue engineering of functional articular cartilage: the current status. Cell Tissue Res 347(3):613–627
60. Kreuz PC, Muller S, von Keudell A, Tischer T, Kaps C, Niemeyer P et al (2013) Influence of sex on the outcome of autologous chondrocyte implantation in chondral defects of the knee. Am J Sports Med 41(7):1541–1548
61. Kuni B, Schmitt H, Chloridis D, Ludwig K (2012) Clinical and MRI results after microfracture of osteochondral lesions of the talus. Arch Orthop Trauma Surg 132(12): 1765–1771
62. Lazic S, Boughton O, Hing C, Bernard J (2014) Arthroscopic washout of the knee: a procedure in decline. Knee 21(2):631–634

63. Lee GW, Son JH, Kim JD, Jung GH (2013) Is platelet-rich plasma able to enhance the results of arthroscopic microfracture in early osteoarthritis and cartilage lesion over 40 years of age? Eur J Orthop Surg Traumatol 23(5):581–587

64. Lewandrowski KU, Muller J, Schollmeier G (1997) Concomitant meniscal and articular cartilage lesions in the femorotibial joint. Am J Sports Med 25(4):486–494

65. Lim HC, Bae JH, Song SH, Park YE, Kim SJ (2012) Current treatments of isolated articular cartilage lesions of the knee achieve similar outcomes. Clin Orthop Relat Res 470(8): 2261–2267

66. Marcacci M, Filardo G, Kon E (2013) Treatment of cartilage lesions: What works and why? Injury 44(Suppl 1):S11–S15

67. McAlindon TE, Snow S, Cooper C, Dieppe PA (1992) Radiographic patterns of osteoarthritis of the knee joint in the community: the importance of the patellofemoral joint. Ann Rheum Dis 51(7):844–849

68. McCormick F, Harris JD, Abrams GD, Frank R, Gupta A, Hussey K et al (2014) Trends in the surgical treatment of articular cartilage lesions in the United States: an analysis of a large private-payer database over a period of 8 years. Arthroscopy 30(2):222–226

69. Mezhov V, Ciccutini FM, Hanna FS, Brennan SL, Wang YY, Urquhart DM et al (2014) Does obesity affect knee cartilage? A systematic review of magnetic resonance imaging data. Obes Rev 15(2):143–157

70. Mollon B, Kandel R, Chahal J, Theodoropoulos J (2013) The clinical status of cartilage tissue regeneration in humans. Osteoarthr Cartil 21(12):1824–1833

71. Moseley JB, O'Malley K, Petersen NJ, Menke TJ, Brody BA, Kuykendall DH et al (2002) A controlled trial of arthroscopic surgery for osteoarthritis of the knee. N Engl J Med 347(2):81–88

72. Negrin LL, Vecsei V (2013) Do meta-analyses reveal time-dependent differences between the clinical outcomes achieved by microfracture and autologous chondrocyte implantation in the treatment of cartilage defects of the knee? J Orthop Sci 18(6):940–948

73. Nukavarapu SP, Dorcemus DL (2013) Osteochondral tissue engineering: current strategies and challenges. Biotechnol Adv 31(5):706–721

74. Ollat D, Lebel B, Thaunat M, Jones D, Mainard L, Dubrana F et al (2011) Mosaic osteochondral transplantations in the knee joint, midterm results of the SFA multicenter study. Orthop Traumatol Surg Res 97(8 Suppl):S160–S166

75. Outerbridge RE (1961) The etiology of chondromalacia patellae. J Bone Joint Surg Br 43-B:752–757

76. Peeters CM, Leijs MJ, Reijman M, van Osch GJ, Bos PK (2013) Safety of intra-articular cell-therapy with culture-expanded stem cells in humans: a systematic literature review. Osteoarthr Cartil 21(10):1465–1473

77. Perera JR, Gikas PD, Bentley G (2012) The present state of treatments for articular cartilage defects in the knee. Ann R Coll Surg Engl 94(6):381–387

78. Petri M, Broese M, Simon A, Liodakis E, Ettinger M, Guenther D et al (2013) CaReS (MACT) versus microfracture in treating symptomatic patellofemoral cartilage defects: a retrospective matched-pair analysis. J Orthop Sci 18(1):38–44

79. Reed ME, Villacis DC, Hatch GF 3rd, Burke WS, Colletti PM, Narvy SJ et al (2013) 3.0-Tesla MRI and arthroscopy for assessment of knee articular cartilage lesions. Orthopedics 36(8):e1060–e1064

80. Resnick D, Vint V (1980) The "Tunnel" view in assessment of cartilage loss in osteoarthritis of the knee. Radiology 137(2):547–548

81. Reverte-Vinaixa MM, Joshi N, Diaz-Ferreiro EW, Teixidor-Serra J, Dominguez-Oronoz R (2013) Medium-term outcome of mosaicplasty for grade III–IV cartilage defects of the knee. J Orthop Surg 21(1):4–9

82. Robert H (2011) Chondral repair of the knee joint using mosaicplasty. Orthop Traumatol Surg Res 97(4):418–429

83. Rodriguez-Merchan EC (2013) Regeneration of articular cartilage of the knee. Rheumatol Int 33(4):837–845
84. Roelofs AJ, Rocke JP, De Bari C (2013) Cell-based approaches to joint surface repair: a research perspective. Osteoarthr Cartil 21(7):892–900
85. Rogers AD, Payne JE, Yu JS (2013) Cartilage imaging: a review of current concepts and emerging technologies. Semin Roentgenol 48(2):148–157
86. Rosenberg TD, Paulos LE, Parker RD, Coward DB, Scott SM (1988) The forty-five-degree posteroanterior flexion weight-bearing radiograph of the knee. J Bone Joint Surg Am 70(10):1479–1483
87. Schuttler KF, Schenker H, Theisen C, Schofer MD, Getgood A, Roessler PP et al (2014) Use of cell-free collagen type I matrix implants for the treatment of small cartilage defects in the knee: clinical and magnetic resonance imaging evaluation. Knee Surg Sports Traumatol Arthrosc 22(6):1270–1276
88. Sharma L, Chmiel JS, Almagor O, Dunlop D, Guermazi A, Bathon JM et al (2014) Significance of preradiographic magnetic resonance imaging lesions in persons at increased risk of knee osteoarthritis. Arthritis Rheumatol 66(7):1811–1819
89. Solheim E, Hegna J, Oyen J, Harlem T, Strand T (2013) Results at 10 to 14 years after osteochondral autografting (mosaicplasty) in articular cartilage defects in the knee. Knee 20(4):287–290
90. Stahl R, Krug R, Kelley DA, Zuo J, Ma CB, Majumdar S et al (2009) Assessment of cartilage-dedicated sequences at ultra-high-field MRI: comparison of imaging performance and diagnostic confidence between 3.0 and 7.0 T with respect to osteoarthritis-induced changes at the knee joint. Skelet Radiol 38(8):771–783
91. Steadman JR, Briggs KK, Rodrigo JJ, Kocher MS, Gill TJ, Rodkey WG (2003) Outcomes of microfracture for traumatic chondral defects of the knee: average 11-year follow-up. Arthroscopy 19(5):477–484
92. Takazawa K, Adachi N, Deie M, Kamei G, Uchio Y, Iwasa J et al (2012) Evaluation of magnetic resonance imaging and clinical outcome after tissue-engineered cartilage implantation: prospective 6-year follow-up study. J Orthop Sci 17(4):413–424
93. Tuan RS, Chen AF, Klatt BA (2013) Cartilage regeneration. J Am Acad Orthop Surg 21(5):303–311
94. Turker M, Cetik O, Cirpar M, Durusoy S, Comert B (2015) Postarthroscopy osteonecrosis of the knee. Knee Surg Sports Traumatol Arthrosc 23(1):246–250
95. Ulstein S, Aroen A, Rotterud JH, Loken S, Engebretsen L, Heir S (2014) Microfracture technique versus osteochondral autologous transplantation mosaicplasty in patients with articular chondral lesions of the knee: a prospective randomized trial with long-term follow-up. Knee Surg Sports Traumatol Arthrosc 22(6):1207–1215
96. Vaquero J, Forriol F (2012) Knee chondral injuries: clinical treatment strategies and experimental models. Injury 43(6):694–705
97. Versier G, Dubrana F (2011) Treatment of knee cartilage defect in 2010. Orthop Traumatol Surg Res 97(8 Suppl):S140–S153
98. von Engelhardt LV, Kraft CN, Pennekamp PH, Schild HH, Schmitz A, von Falkenhausen M (2007) The evaluation of articular cartilage lesions of the knee with a 3-T magnet. Arthroscopy 23(5):496–502
99. von Engelhardt LV, Schmitz A, Burian B, Pennekamp PH, Schild HH, Kraft CN et al (2008) 3-T MRI vs. arthroscopy for diagnostics of degenerative knee cartilage diseases: preliminary clinical results. Orthopade 37(9):914
100. Wakitani S, Okabe T, Horibe S, Mitsuoka T, Saito M, Koyama T et al (2011) Safety of autologous bone marrow-derived mesenchymal stem cell transplantation for cartilage repair in 41 patients with 45 joints followed for up to 11 years and 5 months. J Tissue Eng Regen Med 5(2):146–150

101. Wirth W, Duryea J, Hellio Le Graverand MP, John MR, Nevitt M, Buck RJ et al (2013) Direct comparison of fixed flexion, radiography and MRI in knee osteoarthritis: responsiveness data from the Osteoarthritis Initiative. Osteoarthr Cartil 21(1):117–125
102. Wong S, Steinbach L, Zhao J, Stehling C, Ma CB, Link TM (2009) Comparative study of imaging at 3.0 T versus 1.5 T of the knee. Skelet Radiol 38(8):761–769
103. Yan Z, Yang L, Guo L, Wang F (2014) Effectiveness of arthroscopic bipolar radiofrequency energy for lateral meniscus tear and cartilage injury. Zhongguo Xiu Fu Chong Jian Wai Ke Za Zhi 28(1):13–16

Chapter 4
Osteoarthritis

Marta Ondrésik, Joaquim Miguel Oliveira and Rui Luís Reis

Abstract Osteoarthritis (OA) is the most common form of arthritis and the most relevant musculoskeletal disorder worldwide. It is considered to be a joint degenerative disease with imbalanced homeostasis. It can potentially affect any joint in the body, but most often the knees, hip, hand and the lower back suffers from OA. Due to the strong functional association of the joint components OA affects all tissues, present in the joint to a certain degree. Because of its wide occurrence, and great impact on the society OA receives a lot of attention in clinical research, although the molecular background of OA remains incompletely understood. The complexity of the joint structure, and the limited knowledge we have on OA sets many obstacles to the development of effective therapies. As of current state there is no cure for OA.

4.1 Introduction

Despite, it was for long considered to be the disease of the articular cartilage (AC) now it is agreed that OA involves the entire joint. The intense functional association of the joint components consequently results in their common involvement in the OA caused degenerative processes [1]. Several risk factors have been associated with the development of OA (Fig. 4.1). Its incidence is increasing with the age, as caused by local factors, such as mechanical stimuli, and natural wear and tear. Furthermore, obesity, occupational influences and lifestyle, all

M. Ondrésik (✉) · J.M. Oliveira · R.L. Reis
3B's Research Group—Biomaterials, Biodegradables and Biomimetics,
Headquarters of the European Institute of Excellence on Tissue Engineering
and Regenerative Medicine, University of Minho, AvePark, Guimarães, Portugal
e-mail: marta.ondresik@dep.uminho.pt

M. Ondrésik · J.M. Oliveira · R.L. Reis
ICVS/3B's—PT Government Associate Laboratory, Braga, Guimarães, Portugal

© Springer International Publishing AG 2017
J.M. Oliveira and R.L. Reis (eds.), *Regenerative Strategies for the Treatment of Knee Joint Disabilities*, Studies in Mechanobiology, Tissue Engineering and Biomaterials 21, DOI 10.1007/978-3-319-44785-8_4

contribute to the development of OA. As stated by the Johnson County Osteoarthritis Project data, a lifetime risk to develop symptomatic knee OA is approximately 44.7 % for average population, while in cases of obese people, who are more prone to develop arthritis, this number increases to 60.5 % [2, 3].

Apart from the classic concept, that OA is a result of aging and overuse, recent progress in OA research suggests several other important aspects of the disease origin. Accordingly, now it is commonly accepted that OA might be initiated by different factors, and be associated with metabolic syndromes. Either case, ultimately leads to a common phenotype, causing a complete joint failure. In this regard, OA has subtypes, where it can be resulted from (i) ageing, or (ii) genetic predisposition [4]; caused by (iii) biochemical/inflammatory changes [5]; or (iv) inflicted by injuries, and developed as a secondary disease [5]. Nevertheless, the exact pathological sequence of its development is still poorly understood. Hallmarks of OA are, focal lesions in the cartilage, calcification of the cartilage, joint space narrowing, osteophyte formation and synovial inflammation, involving the role of various cytokines (Fig. 4.1) [6]. Ergo, OA represents a complex disease entity, which makes it difficult to diagnose, or to heal [7].

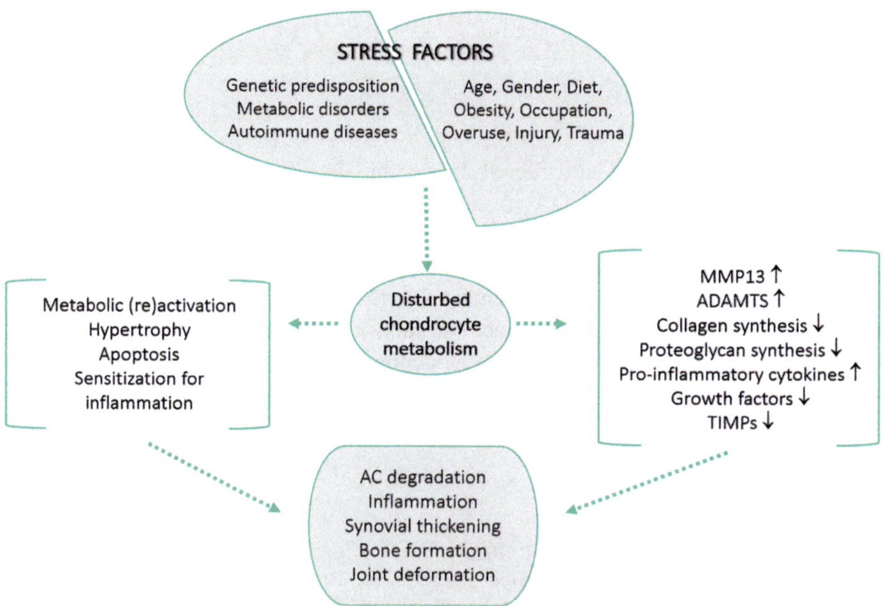

Fig. 4.1 Schematic representation of OA development

4.2 Epidemiology and Socioeconomic Impact

The estimation of international OA prevalence is largely varying as it depends on the studied age, sex and geographical terrain, as well as the definition of OA used in the assessments as epidemiologic studies can be based on either clinical OA, radiological or self-reported OA.

According to current reports, clinical OA is present among 241 825 million worldwide [8]. So far, it was estimated that the number of people suffered disability due to OA, globally increased by 71.9 % between 1990 and 2013, and this number is expected to further expand [9, 10]. In 2010 only the hip and knee joint affected by OA were responsible for 2.2 % of all years lived with disabilities (YLDs) globally, and comprised 10 % of all the disability adjusted life-years (DALYs) [11–13]. In the European population, the prevalence of symptomatic knee OA is ranging between 5.4 and 29.8 %., this number was 7 and 17 % in the US population as assessed by the Framingham study and Johnston County Osteoarthritis Project, respectively [3]. The same studies showed a higher percentage for the presence of radiographic OA, ranging between 19 and 28 % when the studied population is older than 45 years [3].

The prevalence of OA is also geographically influenced. Insomuch, that the clinically diagnosed OA is present 1.4 % among the urban Filipinos, but raises up to 19.3 % among the rural Iranian communities, according to the Community Oriented Program for Control of Rheumatic Disorders (COPCORD) [14]. Furthermore, COPCORD investigators showed sex distribution of knee OA with high female preponderance [14], suggesting the involvement of sex hormones in the disease processes.

With regard to the economic burden and associated costs, since OA is an irreversible disease, once is detected it requires indefinite medical care. Hospital visits, surgeries, therapies and required medication are associated with high health care expenses globally; the same time, non-healthcare related costs are also relevant [15, 16]. Considering that OA is responsible for a great percentage of work disability worldwide, substantial impact originates from the loss of economic productivity, along with the incapacity benefit, claimed by those who lost their working ability. Additional indirect costs emerge from the disability adjusted life costs, such as home-care costs, medical devices, specific household instruments, caregivers, and others [15, 17, 18]. OA associated social costs reach up to 0.25 and 0.50 % of a country's GDP with the annual healthcare costs of €705–€19,715; and non-healthcare costs ranging between €432 to €11,956 [13]. The exceeding trend of OA, and growing costs associated with it, urges the need of more efficient preventative care, and more efficient approaches for the evaluation of therapeutics, and OA management. Comparability, and standardization of techniques appraising OA treatment could help to reduce the costs, as it was suggested by the European Society for Clinical and Economic Aspects of Osteoporosis and Osteoarthritis (ESCEO) [19].

4.3 Pathobiology of OA

4.3.1 Early Changes

4.3.1.1 Chondrocyte Hypertrophy

During OA, the knee joint, most of all the AC undergoes several changes. Chondrocyte hypertrophy and cluster formation is a primary event in OA. The typically quiescent chondrocytes undergo certain activation process [20]. This alteration of the chondrocytes is a well-known phenomenon, and a natural process of endochondral ossification. But unlike in healthy AC, where only the growth plate and the lowest calcified cartilage contain hypertrophic chondrocytes, in OA the number of these cells is increased. Hypertrophic chondrocytes trigger disturbed AC homeostasis by promoting cartilage degeneration, and facilitating bone remodeling, calcium deposition in the extracellular matrix (ECM), and osteophyte formation [21, 22]. Typical markers of hypertrophic chondrocytes are type X collagen, matrix metalloproteinases (MMP)-13, and Runx2, but the elevated levels of the molecules osteopontin, osteocalcin, Indian Hedgehog [23, 24], vascular endothelial growth factor (VEGF) [25], transglutaminase-2 [26], and alkaline phosphatase [27] were also associated with chondrocyte hypertrophy but are typical for the later stages.

4.3.1.2 Changes in Cartilage ECM Synthesis

Chondrocyte dysregulation consequently results in altered ECM synthesis as well (Fig. 4.2). The loss of balance between their anabolic and catabolic activity is ascribed to the effect of inflammatory milieu, and abnormal mechanical stress, typical in OA [28]. The shift in homeostatic processes favors catabolism, and results in elevated levels of MMPs, and a disintegrin and metalloproteinase with thrombospondin motifs (ADAMTs); while proteoglycan (PG) synthesis is decreased, along with collagen type II production. Interestingly, the level of collagen type I is increased in early OA, resulting in shifted overall collagen/PG ratio [29, 30]. However, collagen type I possesses different structural properties than collagen type II, thus does not meet the mechanical demands of the articulation. Due to the high amount of hydroxylisine, glucosyl and galactosyl residues, collagen type II interacts, and coated with PG, which prevents its cleavage by MMPs [29]. This sophisticated synergy is a key for matrix integrity, and unique mechanical properties of the AC.

Undermining tissue integrity, aggrecan shedding can progress into OA, in fact, deemed as its early event [31]. Aggrecan is the most studied substrate of the ADAMTs; but the SLRPs, namely byglican, decorin, fibromodulin and keratocan, were also detected to suffer cleavage by these proteases in vitro [31–34]. ADAMT-4, and -5 are both cleaving aggrecan, but according to recent studies, ADAMT-5 is the most active [35], therefore also a potential target for OA therapy

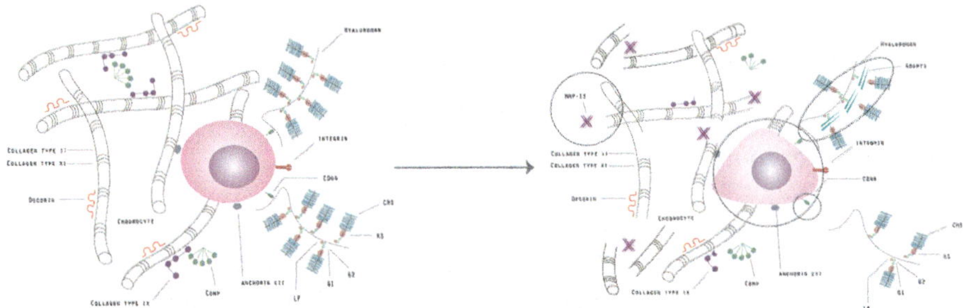

Fig. 4.2 Illustration of the articular cartilage extracellular matrix degradation during the onset of osteoarthritis. **a** In healthy state the extracellular matrix is intact, having a large collagen (mainly collagen type II) network richly interwoven by aggregates of proteoglycans, such as the aggrecan. The cells are embedded in the matrix by matrix-binding cell surface receptors, such as CD44 and integrin molecules. There are present also non-collagenous proteins, which ensure the integrity of the tissue, such as the cartilage oligomeric matrix protein (COMP). **b** During osteoarthritis, there are several key processes leading to the degeneration of cartilage (marked with the *dashed circles*). The overproduction of proteases, such as MMP-13 and aggrecanases (ADAMTs) leads to cleavage of collagen molecules and proteoglycan aggregates respectively, which eventually causes the loss of matrix integrity. This is partly the result of shifted cellular behaviour and altered phenotype. The shedding of receptors results in the loss of contact between the chondrocytes and the extracellular matrix, which may lead to increased rate of apoptosis. The formed debris accumulates in the synovial fluid and contributes to the development of inflammation

[36, 37]. By cleaving the aggrecan at the interglobular domain (IGD), namely between the N-terminal G1 and G2 globular domains [38, 39], the chondroitin sulphate chain bearing region of the molecule gets liberated from the tissue. This consequently leads to the alteration of fixed charge density and osmotic pressure in the ECM, leading to the swelling of the cartilage [40].

After the release of PG, the collagen meshwork becomes more vulnerable to the effect of MMPs [22]. MMPs are endopeptidases, which act extracellularly; display a large variety, and substrate specificity. Most well described is the MMP-13, or also called collagenase-3, as it is the major protease of collagen type II, the primary macromolecule of AC. MMP-13 is overexpressed during the onset of OA [41, 42], and believed to facilitate the activation of other MMPs as well [43–45]. Besides MMP-13, MMP-1, -8, and -14 are considered the key collagenases of AC [46–48]. Furthermore, the gelatinases, MMP-2, -9 the stromelysins, MMP-3, -10, the matrylisin MMP-7, and the cathepsin K, a cysteine protease, are regarded as main orchestrators of cartilage damage [49, 50]. Their effects in healthy stage are regulated by the tissue inhibitors of metalloproteinases (TIMPs). TIMP-1, and -2 are the two most relevant MMP inhibitors [51]. Their level was recently shown to be increased upon osteopontin induction as potential chondroprotective approach [52]. Moreover, owing anti-angiogenic properties, TIMP-2 could potentially help inhibiting angiogenesis, thus preventing OA [53]. As it was mentioned before, enhancement of ECM catabolism, and the MMP expression is triggered by mechanical overloading, inflammation, as well as other stress factors. Meanwhile, the fragments of cleaved macromolecules feedback amplify the MMP production by interacting with cell surface receptors e.g., integrin [20].

4.3.1.3 Early Morphological Alterations

The earliest morphological changes in OA are the degenerative changes on the surface and in the structure of AC; these include a mild fibrillation in the superficial zone [54], disruption of a macromolecular framework and a lower level of the aggrecan content. The glycosaminoglycans (GAGs) remain homogeneously distributed [54], but there is a decrease of their chain length. Elevated water level, and increased matrix permeability is also typical in the early OA, as well as changes in the cellular structure and morphology as it was mentioned earlier. Synovial tissue alterations are also present already at this stage of the disease. These involve the thickening of the synovial membrane, synovial hypertrophy and hyperplasia of the lining cells, as well as the infiltration of the underlying tissue and fibrosis [6, 55]. Pathological changes appear within the subchondral bone too already in early OA, however it is still controversial whether these changes precede the biomechanical alterations in the AC, or develop as secondary adaptation [56].

4.4 Advanced Stage of OA

4.4.1 Chondrocyte Apoptosis

The terminal differentiation of chondrocytes consequently results in apoptosis. Cell death is a hallmark of OA, including both apoptosis and necrosis. Besides terminal differentiation, cell death is attributed to the action of several factors, such as (i) abnormal mechanical stress, (ii) mitochondrial dysfunction, (iii) perturbation of pro-apoptotic/anti-apoptotic factors, and the (iv) significant presence of reactive oxygen species (ROS)/pro-inflammatory cytokines [57–60]. Apoptosis is a programmed cell death. It either proceeds through death receptor mediated pathway or trough mitochondrial pathway, upon binding the death ligands Fas (FasL) and Tumor necrosis factor-α (TNF-α) (extrinsic stimuli), or due to intracellular damage, and stress (intrinsic stimuli), respectively [61–63]. The biomolecular background of cell apoptosis in OA is still poorly understood though, and requires further investigation. It was also suggested that chondrocyte apoptosis may appear as secondary in OA; as because during OA development the cartilage ECM molecules are greatly degraded, and so, these anchorage dependent cells lose their contact with the tissue, and may suffer apoptosis due to the loss of adhesion; therefore, it is speculated, that apoptosis might be both a cause, or the consequence of OA [61]. The chondrocyte apoptosis in OA is a well-investigated phenomenon, as it contributes largely to tissue damage. Factors with potential anti-apoptotic effect are being studied for OA management. Recently reported neuropeptide urocortin (Ucn) was shown to have chondroprotective effects against nitric oxide (NO), but not against TNF-α in chondrocyte cell line studies [64]. Curcumin is another potent agent with cell survival effects. Curcumin was long studied for its anti-inflammatory effects, and was recently shown to prevent Interleukin-1β (IL-1β) induced swelling of the mitochondria, thus apoptosis; moreover, inhibited the expression of pro-apoptotic factors, such as caspase-3, while stimulated the synthesis of anti-apoptotic factors, such as B-cell lymphoma 2 (Bcl-2), B cell lymphoma—extra-large (Bcl-xL), and TNF receptor associated factor 1 (TRAF1) [65, 66]. Similarly to curcumin, resveratrol is also a polyphenolic natural compound, evidenced to possess anti-inflammatory effects [66]. Resveratrol is being tested now in OA animal models for its anti-inflammatory, and cartilage–protective effects [67–70].

4.4.2 Morphological Changes

As the disease progresses the loss of AC becomes more explicit. The disruption of the superficial zone is prominent; it becomes rough, and has cleavages and fissures, which can reach down even to the calcified cartilage [71, 72]. The loss of proteoglycan induced by enzymatic degradation is also more significant. The attrition of the cartilage is accompanied by alterations of the subchondral bone as well, e.g. increase of the subchondral bone density, thickening, sclerosis, appearance of bone cysts, bone marrow lesions.

The subchondral bone is knowingly important in OA development due to its previously mentioned intense cross-talk with the AC [89, 90]. The correlation between changes in the subchondral bone and AC degeneration has long been established [91, 92], and yet its exact role is still unknown. Chronic overloading of the joint is one important factor, which cause increased bone density at the load bearing sites, and eventually result in micro-cracks in both the calcified cartilage and trabecular bone, which initiate bone remodelling, and are associated with the presence of bone marrow lesions [93, 94]. Because of the micro-cracks, soluble products generated in the bone marrow can exert deleterious effects on both the bone and cartilage [95]. Furthermore, as the bone becomes stiffer, will not be able to transmit the loads as efficiently, and thus the AC will suffer more mechanical pressure [96]. Osteophytes are part of the first definitive marks of OA detectable by imaging techniques. Their formation appears as secondary phenomenon, supposedly as an attempt for joint stabilization, and the development process is very similar to the endochondral ossification, observed in embryogenesis. Osteophytes are produced by chondrocytes who undergo a secondary terminal ossification, and cause the deformation of the joint [97]. There is a large group of factors that contribute to osteophyte formation, including growth factors, e.g. Transforming growth factor-β (TGF-β), Bone morphogenetic protein-2 (BMP-2), VEGF and Insulin-like growth factor 1 (IGF-1), as well as the mechanical stimuli utilized by the cells [98, 99]. Osteophyte development is also linked to the act of synovial macrophages [100]. Furthermore, invasion of blood vessels from the subchondral bone to the tidemark and its penetration to the calcified cartilage is also common at the late stages of the disease [73]. This vascular invasion is known to be responsible for the mentioned cellular hypertrophy in the lower levels of the cartilage, which eventually will increase the tissue stiffness, thus also the mechanical stress in the cartilage [74]. The advancement, and duplication of the tidemark was also attributed to the vascularization process, however its mechanism has not yet been completely identified. At the late progression of the OA, typically the cartilage disappears completely from the weight bearing regions of the articulation, leaving the subchondral bone exposed. The loss of AC also leads to secondary changes in all the other structures of the joint, e.g. synovium, ligaments, capsules and the muscle, and finally eventuate a whole joint deformation.

4.4.3 Molecular Alterations During the Later Stages of OA

While the morphologic changes of osteoarthritic AC have been broadly described, the underlying molecular base remains incompletely known. The maintenance of AC is a very complex and poorly understood mechanism, which relies on the chondrocytes anabolic and catabolic activity driven by complex molecular networks. Interesting aspect is the contribution of the subchondral bone, and the cross-talk between the AC and subchondral bone during the evolution of OA too.

Recent studies showed the significant presence of the receptor activator of nuclear factor κ-B ligand (RANKL), which is known to contribute to osteoclastogenesis [101]. Furthermore, vessel formation, induced by VEGFs [102, 103], and subsequent formation of sensory nerves into the subchondral bone and focal layers of AC is also common in late OA. The vascularisation contributes to the progression of the disease, by facilitating the biochemical communication between the bone, and cartilage; and the presence of nerves is associated with pain experienced in OA. The coordinated signalling between the articulating bone and cartilage is widely investigated, but the involvement of subchondral bone in OA progression is not entirely clear yet [89, 91]. Several studies have demonstrated similar patterns between the subchondral bone and hypertrophic cartilage, as well as the noxious action of soluble mediators derived from bone in the AC during OA [104, 105]. For instance, Wnt signalling is normally activated in subchondral bone, as it is a major regulator of bone formation in early development, and is responsible for tissue maintenance, but was also shown to recapitulate the differentiation program in chondrocytes, leading to their hypertrophy [22]. Inhibition of cell response to Wnt signalling by syndecan-4 did result in cell protection from OA induced changes [106]. Similarly inhibiting VEGF expression in osteoblasts using Dickkopf (DKK) 1 slows down AC degradation and ossification [107]. Both DKK1, sclerostin (SOST) are produced by OA chondrocytes along with gremlin 1, which is a Wnt and BMP antagonist [108, 109], also suggesting a cross talk between the bone–cartilage unite. But Wnt signalling is not the only example for the existing cross talk between cartilage and bone. Other hypertrophy related pathways were studied, such as the TGF-β/BMP signalling, which is a crucial pathway for the healthy bone–cartilage homeostasis, but does also interfere with OA processes [110, 111]. Insomuch as, elevated levels of TGF-β were detected in the synovial space and subchondral bone of human OA patients, and were associated with early and late signs of OA, such as bone marrow lesions and osteophyte formation [112, 113]. On the other hand, TGF-β ablation lead to reduced osteophyte development in advanced OA [114], but the lack of TGF-β in OA cartilage was also found to lead PG loss and cause AC degeneration [114]. BMPs belong to the TGF-β superfamily as well [115]. BMP signalling is responsible for both early chondrogenesis, and chondrocyte terminal differentiation [116]; as well as important in the regulation of osteoclasts, and osteoblasts, thus a key regulator of both cartilage formation, and bone remodelling [117]. BMP-7 was shown to induce the production of matrix ECM molecules and to inhibit the effect of pro-catabolic factors on chondrocytes [118, 119]. Thus, BMP-7 is a potent candidate of cartilage regeneration [120, 121]. Similar investigations were performed on BMP-2, another member of the BMP family. BMP-2 is barely present in the healthy cartilage, but was detected to be over expressed during OA, especially around the AC lesions. It was speculated, that might be an attempt of chondrocytes for repair, and was tested as potential reparative agent [122]. However, the latest studies of Blaney Davidson could not detect any alteration in the course of OA as an effect of the elevated BMP-2, but found explicit osteophyte formation and aggravation of cartilage lesions [123].

4.4.4 Inflammation in OA

The recognition of the importance of inflammation, present in OA joints, was essential. Synovitis is another radiographic hallmark of OA. The thickened and inflamed synovial membrane allows the infiltration of several immune cells, which then, along with the synoviocytes and chondrocytes, produce a wide variety of pro-inflammatory factors [124, 125]. The pro-inflammatory factors, such as cytokines and chemokines now considered major players of OA both in early and late stages. IL-1β and TNF-α represents the two most important cytokines, which impact AC homeostasis by triggering the catabolic cascade [126]. The degree of contribution of these pro-inflammatory factors in the cartilage degeneration is widely investigated, but due to complexness of OA, still incompletely understood. Evidence strengthened the premises regarding the key roles of cytokines, as observed in the studies of Orita et al., who demonstrated a strong correlation of Kellgren-Lawrence grade and WOMAC score with the presence of pro-inflammatory factors in the synovial fluid [127]. Furthermore, Furman et al. recently showed that local inhibition of IL-1 can reduce post-traumatic OA development [128]. Moreover, injurious mechanical stress have been shown to trigger the same pathways, as the cytokines, i.e. Nuclear factor-κB (NF-κB), p38,

Mitogen-Activated Protein Kinase (MAPK), Extracellular signal-Regulated Kinase ½ (ERK1/2), and c-Jun N-terminal (JNK) signalling [50], making them potential targets for biologic therapies. In fact, using receptor antagonists to block the effect of TNF-α and IL-1β it is possible to reduce the activity of NF-κB signalling, and inhibit the catabolic cascade on this pathway. Similarly, thalidomide and hydrochloride can suppress the activity of NF-κB pathway induced by TNF-α, and IL-1β respectively [129–132]. But, exercising combined with diet also has beneficent effect on OA, as it was shown to suppress local inflammation in OA joint [133, 134]. Other relevant mediators of inflammation are the interleukins IL-6, IL-7, IL-17, prostaglandin E2 (PGE2), oncostantin M (OSM), and leukemia inhibitor factor (LIF) [135]. The interplay of these mediators, does not only leads to enhanced ECM degradation, but also promotes pain by inducing pain-pathways [136]. The relation of pain and OA is nicely overviewed in [136]. Among the chemokines, IL-8/CXCL-8, GROα/CXCL, MCP-1/CCL-2, RANTES/CCL5, MIP-1α/CCL3, and MIP-1β/CCL-4 were found as relevant in OA [137, 138].

4.5 Final Remarks

OA is a relevant and complex disease without a doubt. The impact of OA on the society and its economic burden in the medical care is a major issue of today's health system. Its incidence is high and is growing by the ageing population. It is associated with reduced mobility, pain and eventual work inability. Although, OA has been for long conceptualized as the disease of the cartilage, new researches has proved it to affect the entire joint, and now is considered as an organ disease. It was also established that OA is a multifactorial disease, and accordingly can develop due to a large number of extrinsic and intrinsic factors. These factors may have biochemical, genetic or biomechanical nature. As of now OA is detected once displays macroscopic changes, although it is developed due to a complex interplay of molecular factors, and potentially could be diagnosed—and treated—much before it reaches to the tissue scale. The interrelationship between the catabolic and anabolic processes, and their imbalance in OA is of large interest in medical and biological research fields. It involves rich molecular signalling with many key players of which many had been identified already, and stand as potential biomarkers for the early diagnosis of OA or possible targets. Moreover, the involvement of subchondral bone as potentatiator of the degenerative processes or on the contrary, source of regenerative capacity is well studied too. Inflammation is also the hallmark of OA, but there is still a debate whether is the origin or the consequence of OA development. More efforts should be devoted to entangle the underlying molecular networks, and to understand more on the subtypes of OA, only then we can develop targeted treatment approaches as well. The field is also open to new approaches, among which the most promising might be the techniques which aim to deliver molecular inhibitors, growth factors, stem cells or engineered cells to restore the molecular balance in the osteoarthritic environment, and tackle regeneration of the damaged tissues. Nevertheless, it requires the orchestrated work of researchers from various fields, such as tissue engineering, regenerative medicine, genetic engineering, biological and biomedical science, as well as the contribution of medical experts, and many more.

References

1. Goldring MB (2012) Chondrogenesis, chondrocyte differentiation, and articular cartilage metabolism in health and osteoarthritis. Ther Adv Musculoskelet Dis 4:269–285. doi:10.1177/1759720X12448454
2. Murphy L, Schwartz TA, Helmick CG et al (2008) Lifetime risk of symptomatic knee osteoarthritis. Arthritis Rheum 59:1207–1213. doi:10.1002/art.24021
3. Neogi T (2013) The epidemiology and impact of pain in osteoarthritis. Osteoarthr Cartil 21:1145–1153. doi:10.1016/j.joca.2013.03.018
4. Valdes AM, Spector TD (2010) The clinical relevance of genetic susceptibility to osteoarthritis. Best Pract Res Clin Rheumatol 24:3–14. doi:10.1016/j.berh.2009.08.005
5. Kapoor M, Martel-Pelletier J, Lajeunesse D et al (2011) Role of proinflammatory cytokines in the pathophysiology of osteoarthritis. Nat Rev Rheumatol 7:33–42. doi:10.1038/nrrheum.2010.196
6. Pelletier JP, Martel-Pelletier J, Ghandur-Mnaymneh L et al (1985) Role of synovial membrane inflammation in cartilage matrix breakdown in the Pond-Nuki dog model of osteoarthritis. Arthritis Rheum 28:554–561
7. Felson DT (2014) Osteoarthritis: priorities for osteoarthritis research: much to be done. Nat Rev Rheumatol 10:447–448. doi:10.1038/nrrheum.2014.76
8. Lawrence RC (2008) NIH public access. Arthritis Rheum 58:26–35. doi:10.1002/art.23176. Estimates
9. Kurtz S, Ong K, Lau E, Manley M (2011) Current and projected utilization of total joint replacements. Compr Biomater 6:1–9
10. Lohmander LS (2013) Knee replacement for osteoarthritis: facts, hopes, and fears. Medicographia 35:181–188
11. Vos T, Flaxman AD, Naghavi M et al (2012) Years lived with disability (YLDs) for 1160 sequelae of 289 diseases and injuries 1990–2010: a systematic analysis for the Global Burden of Disease Study 2010. Lancet 380:2163–2196. doi:10.1016/S0140-6736(12)61729-2
12. Murray CJL, Lopez AD (2013) Measuring the global burden of disease. N Engl J Med 369:448–457. doi:10.1056/NEJMra1201534
13. Puig-Junoy J, Ruiz Zamora A (2015) Socio-economic costs of osteoarthritis: a systematic review of cost-of-illness studies. Semin Arthritis Rheum 44:531–541. doi:10.1016/j.semarthrit.2014.10.012
14. Haq SA, Davatchi F (2011) Osteoarthritis of the knees in the COPCORD world. Int J Rheum Dis 14:122–129. doi:10.1111/j.1756-185X.2011.01615.x
15. Bitton R (2009) The economic burden of osteoarthritis. Am J Manag Care 15:S230–S235. doi:10.1002/art.1780290311
16. Gupta S, Hawker GA, Laporte A et al (2005) The economic burden of disabling hip and knee osteoarthritis (OA) from the perspective of individuals living with this condition. Rheumatology 44:1531–1537. doi:10.1093/rheumatology/kei049
17. Vos T, Barber RM, Bell B et al (2015) Global, regional, and national incidence, prevalence, and years lived with disability for 301 acute and chronic diseases and injuries in 188 countries, 1990–2013: a systematic analysis for the Global Burden of Disease Study 2013. Lancet. doi:10.1016/S0140-6736(15)60692-4

18. Hunter DJ, Schofield D, Callander E (2014) The individual and socioeconomic impact of osteoarthritis. Nat Rev Rheumatol 10:437–441. doi:10.1038/nrrheum.2014.44

19. Hiligsmann M, Cooper C, Guillemin F et al (2014) A reference case for economic evaluations in osteoarthritis: an expert consensus article from the European Society for Clinical and Economic Aspects of Osteoporosis and Osteoarthritis (ESCEO). Semin Arthritis Rheum 44:271–282. doi:10.1016/j.semarthrit.2014.06.005

20. Houard X, Goldring MB, Berenbaum F (2013) Homeostatic mechanisms in articular cartilage and role of inflammation in osteoarthritis. Curr Rheumatol Rep 15:375. doi:10.1007/s11926-013-0375-6

21. Van Donkelaar CC, Wilson W (2012) Mechanics of chondrocyte hypertrophy. Biomech Model Mechanobiol 11:655–664. doi:10.1007/s10237-011-0340-0

22. Van der Kraan PM, van den Berg WB (2012) Chondrocyte hypertrophy and osteoarthritis: role in initiation and progression of cartilage degeneration? Osteoarthr Cartil 20:223–232. doi:10.1016/j.joca.2011.12.003

23. Wei F, Zhou J, Wei X et al (2012) Activation of Indian hedgehog promotes chondrocyte hypertrophy and upregulation of MMP-13 in human osteoarthritic cartilage. Osteoarthr Cartil 20:755–763. doi:10.1016/j.joca.2012.03.010

24. Goldring MB, Tsuchimochi K, Ijiri K (2006) The control of chondrogenesis. J Cell Biochem 97:33–44. doi:10.1002/jcb.20652

25. Jansen H, Meffert RH, Birkenfeld F et al (2012) Detection of vascular endothelial growth factor (VEGF) in moderate osteoarthritis in a rabbit model. Ann Anat 194:452–456. doi:10.1016/j.aanat.2012.01.006

26. Huebner JL, Johnson KA, Kraus VB, Terkeltaub RA (2009) Transglutaminase 2 is a marker of chondrocyte hypertrophy and osteoarthritis severity in the Hartley guinea pig model of knee OA. Osteoarthr Cartil 17:1056–1064. doi:10.1016/j.joca.2009.02.015

27. Pfander D, Swoboda B, Kirsch T (2001) Expression of early and late differentiation markers (proliferating cell nuclear antigen, syndecan-3, annexin VI, and alkaline phosphatase) by human osteoarthritic chondrocytes. Am J Pathol 159:1777–1783. doi:10.1016/S0002-9440(10)63024-6

28. Pulsatelli L, Addimanda O, Brusi V et al (2012) New findings in osteoarthritis pathogenesis: therapeutic implications. Ther Adv Chronic Dis 4(1):23–43. doi:10.1177/2040622312462734

29. Maldonado M, Nam J (2013) The role of changes in extracellular matrix of cartilage in the presence of inflammation on the pathology of osteoarthritis. Biomed Res Int. doi:10.1155/2013/284873

30. Brew CJ, Clegg PD, Boot-Handford RP et al (2010) Gene expression in human chondrocytes in late osteoarthritis is changed in both fibrillated and intact cartilage without evidence of generalised chondrocyte hypertrophy. Ann Rheum Dis 69:234–240. doi:10.1136/ard.2008.097139

31. Heinegård D, Saxne T (2011) The role of the cartilage matrix in osteoarthritis. Nat Rev Rheumatol 7:50–56. doi:10.1038/nrrheum.2010.198

32. Sandy JD, Verscharen C (2001) Analysis of aggrecan in human knee cartilage and synovial fluid indicates that aggrecanase (ADAMTS) activity is responsible for the catabolic turnover and loss of whole aggrecan whereas other protease activity is required for C-terminal processing in vivo. Biochem J 358:615–626

33. Stanton H, Melrose J, Little CB, Fosang AJ (2011) Proteoglycan degradation by the ADAMTS family of proteinases. Biochim Biophys Acta Mol Basis Dis 1812:1616–1629. doi:10.1016/j.bbadis.2011.08.009

34. Melrose J, Fuller ES, Roughley PJ et al (2008) Fragmentation of decorin, biglycan, lumican and keratocan is elevated in degenerate human meniscus, knee and hip articular cartilages compared with age-matched macroscopically normal and control tissues. Arthritis Res Ther 10:R79. doi:10.1186/ar2453

35. Gendron C, Kashiwagi M, Lim NH et al (2007) Proteolytic activities of human ADAMTS-5: comparative studies with ADAMTS-4. J Biol Chem 282:18294–18306. doi:10.1074/jbc.M701523200

36. Arner EC (2002) Aggrecanase-mediated cartilage degradation. Curr Opin Pharmacol 2:322–329

37. Little CB, Meeker CT, Golub SB et al (2007) Blocking aggrecanase cleavage in the aggrecan interglobular domain abrogates cartilage erosion and promotes cartilage repair. J Clin Invest 117:1627–1636. doi:10.1172/JCI30765

38. Lohmander LS, Atley LM, Pietka TA, Eyre DR (2003) The release of crosslinked peptides from type II collagen into human synovial fluid is increased soon after joint injury and in osteoarthritis. Arthritis Rheum 48:3130–3139. doi:10.1002/art.11326

39. Sandy JD, Flannery CR, Neame PJ, Lohmander LS (1992) The structure of aggrecan fragments in human synovial fluid. Evidence for the involvement in osteoarthritis of a novel proteinase which cleaves the Glu 373-Ala 374 bond of the interglobular domain. J Clin Invest 89:1512–1516. doi:10.1172/JCI115742

40. Bank RA, Soudry M, Maroudas A et al (2000) The increased swelling and instantaneous deformation of osteoarthritic cartilage is highly correlated with collagen degradation. Arthritis Rheum 43:2202–2210. doi:10.1002/1529-0131(200010)43:10<2202:AID-ANR7>3.0.CO;2-E

41. Billinghurst RC, Dahlberg L, Ionescu M et al (1997) Enhanced cleavage of type II collagen by collagenases in osteoarthritic articular cartilage. J Clin Invest 99:1534–1545. doi:10.1172/JCI119316

42. Tardif G, Pelletier JP, Dupuis M et al (1999) Collagenase 3 production by human osteoarthritic chondrocytes in response to growth factors and cytokines is a function of the physiologic state of the cells. Arthritis Rheum 42:1147–1158. doi:10.1002/1529-0131(199906)42:6<1147:AID-ANR11>3.0.CO;2-Y

43. Nagase H (1997) Activation mechanisms of matrix metalloproteinases. Biol Chem 378:151–160

44. Ra H-J, Parks WC (2007) Control of matrix metalloproteinase catalytic activity. Matrix Biol 26:587–596. doi:10.1016/j.matbio.2007.07.001

45. Dreier R, Grässel S, Fuchs S et al (2004) Pro-MMP-9 is a specific macrophage product and is activated by osteoarthritic chondrocytes via MMP-3 or a MT1-MMP/MMP-13 cascade. Exp Cell Res 297:303–312. doi:10.1016/j.yexcr.2004.02.027

46. Iliopoulos D, Malizos KN, Oikonomou P, Tsezou A (2008) Integrative microRNA and proteomic approaches identify novel osteoarthritis genes and their collaborative metabolic and inflammatory networks. PLoS ONE 3:e3740. doi:10.1371/journal.pone.0003740

47. Reboul P, Pelletier J, Tardif G, Cloutier JM, Martel-Pelletier J (1996) The new collagenase, collagenase-3, is expressed and synthesized by human chondrocytes but not by synoviocytes. J Clin Invest 97(9):2011–2019. doi:10.1172/JCI118636

48. Troeberg L, Nagase H (2012) Proteases involved in cartilage matrix degradation in osteoarthritis. Biochim Biophys Acta 1824:133–145. doi:10.1016/j.bbapap.2011.06.020

49. Yang C-C, Lin C-Y, Wang H-S, Lyu S-R (2013) Matrix metalloproteases and tissue inhibitors of metalloproteinases in medial plica and pannus-like tissue contribute to knee osteoarthritis progression. PLoS ONE 8:e79662. doi:10.1371/journal.pone.0079662

50. Goldring MB, Marcu KB (2009) Cartilage homeostasis in health and rheumatic diseases. Arthritis Res Ther 11:224. doi:10.1186/ar2592

51. Visse R, Nagase H (2003) Matrix metalloproteinases and tissue inhibitors of metalloproteinases: structure, function, and biochemistry. Circ Res 92:827–839. doi:10.1161/01.RES.0000070112.80711.3D

52. Zhang F, Yu W, Luo W et al (2014) Effect of osteopontin on TIMP-1 and TIMP-2 mRNA in chondrocytes of human knee osteoarthritis in vitro. Exp Ther Med 8:391–394

53. Mi M, Shi S, Li T et al (2012) TIMP2 deficient mice develop accelerated osteoarthritis via promotion of angiogenesis upon destabilization of the medial meniscus. Biochem Biophys Res Commun 423:366–372. doi:10.1016/j.bbrc.2012.05.132

54. Miosge N, Hartmann M, Maelicke C, Herken R (2004) Expression of collagen type I and type II in consecutive stages of human osteoarthritis. Histochem Cell Biol 122:229–236. doi:10.1007/s00418-004-0697-6

55. Brandt KD, Myers SL, Burr D, Albrecht M (1991) Osteoarthritic changes in canine articular cartilage, subchondral bone, and synovium fifty-four months after transection of the anterior cruciate ligament. Arthritis Rheum 34:1560–1570

56. Felson DT, Neogi T (2004) Osteoarthritis: is it a disease of cartilage or of bone? Arthritis Rheum 50:341–344. doi:10.1002/art.20051

57. Aigner T, Fundel K, Saas J et al (2006) Large-scale gene expression profiling reveals major pathogenetic pathways of cartilage degeneration in osteoarthritis. Arthritis Rheum 54:3533–3544. doi:10.1002/art.22174

58. Blanco FJ, Ochs RL, Schwarz H, Lotz M (1995) Chondrocyte apoptosis induced by nitric oxide. Am J Pathol 146:75–85

59. López-Armada MJ, Caramés B, Lires-Deán M et al (2006) Cytokines, tumor necrosis factor-alpha and interleukin-1beta, differentially regulate apoptosis in osteoarthritis cultured human chondrocytes. Osteoarthr Cartil 14:660–669. doi:10.1016/j.joca.2006.01.005

60. Kong D, Zheng T, Zhang M et al (2013) Static mechanical stress induces apoptosis in rat endplate chondrocytes through MAPK and mitochondria-dependent caspase activation signaling pathways. PLoS ONE 8:1–10. doi:10.1371/journal.pone.0069403

61. Zamli Z, Sharif M (2011) Chondrocyte apoptosis: a cause or consequence of osteoarthritis? Int J Rheum Dis 14:159–166. doi:10.1111/j.1756-185X.2011.01618.x

62. Tait SWG, Green DR (2010) Mitochondria and cell death: outer membrane permeabilization and beyond. Nat Rev Mol Cell Biol 11:621–632. doi:10.1038/nrm2952

63. Blanco FJ, Rego I, Ruiz-Romero C (2011) The role of mitochondria in osteoarthritis. Nat Rev Rheumatol 7:161–169. doi:10.1038/nrrheum.2010.213

64. Intekhab-Alam NY, White OB, Getting SJ et al (2013) Urocortin protects chondrocytes from NO-induced apoptosis: a future therapy for osteoarthritis? Cell Death Dis 4:e717. doi:10.1038/cddis.2013.231

65. Csaki C, Mobasheri A, Shakibaei M (2009) Synergistic chondroprotective effects of curcumin and resveratrol in human articular chondrocytes: inhibition of IL-1beta-induced NF-kappaB-mediated inflammation and apoptosis. Arthritis Res Ther 11:R165. doi:10.1186/ar2850

66. Henrotin Y, Priem F, Mobasheri A (2013) Curcumin: a new paradigm and therapeutic opportunity for the treatment of osteoarthritis: curcumin for osteoarthritis management. Springerplus 2:56. doi:10.1186/2193-1801-2-56

67. Li W, Cai L, Zhang Y et al (2015) Intra-articular resveratrol injection prevents osteoarthritis progression in a mouse model by activating SIRT1 and thereby silencing HIF-2α. J Orthop Res. doi:10.1002/jor.22859

68. Wang J, Gao J-S, Chen J-W et al (2012) Effect of resveratrol on cartilage protection and apoptosis inhibition in experimental osteoarthritis of rabbit. Rheumatol Int 32:1541–1548. doi:10.1007/s00296-010-1720-y

69. Švajger U, Jeras M (2012) Anti-inflammatory effects of resveratrol and its potential use in therapy of immune-mediated diseases. Int Rev Immunol 31:202–222. doi:10.3109/08830185.2012.665108

70. Riveiro-Naveira RR, Loureiro J, Valcárcel-Ares MN et al (2014) Anti-inflammatory effect of resveratrol as a dietary supplement in an antigen-induced arthritis rat model. Osteoarthr Cartil 22:S290. doi:10.1016/j.joca.2014.02.539

71. Pfander D, Rahmanzadeh R, Scheller EE (1999) Presence and distribution of collagen II, collagen I, fibronectin, and tenascin in rabbit normal and osteoarthritic cartilage. J Rheumatol 26:386–394

72. Veje K, Hyllested-Winge JL, Ostergaard K (2003) Topographic and zonal distribution of tenascin in human articular cartilage from femoral heads: normal versus mild and severe osteoarthritis. Osteoarthr Cartil 11:217–227

73. Hayami T, Funaki H, Yaoeda K et al (2003) Expression of the cartilage derived anti-angiogenic factor chondromodulin-I decreases in the early stage of experimental osteoarthritis. J Rheumatol 30:2207–2217
74. Henrotin Y, Pesesse L, Sanchez C (2012) Subchondral bone and osteoarthritis: biological and cellular aspects. Osteoporos Int 23(Suppl 8):S847–S851. doi:10.1007/s00198-012-2162-z
75. Pearle AD, Warren RF, Rodeo SA (2005) Basic science of articular cartilage and osteoarthritis. Clin Sports Med 24:1–12. doi:10.1016/j.csm.2004.08.007
76. Lorenz H, Richter W (2006) Osteoarthritis: cellular and molecular changes in degenerating cartilage. Prog Histochem Cytochem 40:135–163. doi:10.1016/j.proghi.2006.02.003
77. Little CB, Ghosh P, Bellenger CR (1996) Topographic variation in biglycan and decorin synthesis by articular cartilage in the early stages of osteoarthritis: an experimental study in sheep. J Orthop Res 14:433–444. doi:10.1002/jor.1100140314
78. Matyas JR, Huang D, Chung M, Adams ME (2002) Regional quantification of cartilage type II collagen and aggrecan messenger RNA in joints with early experimental osteoarthritis. Arthritis Rheum 46:1536–1543. doi:10.1002/art.10331
79. Matyas JR, Ehlers PF, Huang D, Adams ME (1999) The early molecular natural history of experimental osteoarthritis. I. Progressive discoordinate expression of aggrecan and type II procollagen messenger RNA in the articular cartilage of adult animals. Arthritis Rheum 42:993–1002. doi:10.1002/1529-0131(199905)42:5<993:AID-ANR19>3.0.CO;2-U
80. Young AA, Smith MM, Smith SM et al (2005) Regional assessment of articular cartilage gene expression and small proteoglycan metabolism in an animal model of osteoarthritis. Arthritis Res Ther 7:R852–R861. doi:10.1186/ar1756
81. Adams ME, Matyas JR, Huang D, Dourado GS (1995) Expression of proteoglycans and collagen in the hypertrophic phase of experimental osteoarthritis. J Rheumatol Suppl 43:94–97
82. Lorenz H, Wenz W, Ivancic M et al (2005) Early and stable upregulation of collagen type II, collagen type I and YKL40 expression levels in cartilage during early experimental osteoarthritis occurs independent of joint location and histological grading. Arthritis Res Ther 7:R156–R165. doi:10.1186/ar1471
83. Bluteau G, Conrozier T, Mathieu P et al (2001) Matrix metalloproteinase-1, -3, -13 and aggrecanase-1 and -2 are differentially expressed in experimental osteoarthritis. Biochim Biophys Acta 1526:147–158
84. Aigner T, Zien A, Gehrsitz A et al (2001) Anabolic and catabolic gene expression pattern analysis in normal versus osteoarthritic cartilage using complementary DNA-array technology. Arthritis Rheum 44:2777–2789
85. Le Graverand M-PH, Eggerer J, Vignon E et al (2002) Assessment of specific mRNA levels in cartilage regions in a lapine model of osteoarthritis. J Orthop Res 20:535–544. doi:10.1016/S0736-0266(01)00126-7
86. Akizuki S, Mow VC, Muller F et al (1987) Tensile properties of human knee joint cartilage. II. Correlations between weight bearing and tissue pathology and the kinetics of swelling. J Orthop Res 5:173–186. doi:10.1002/jor.1100050204
87. Lai WM, Hou JS, Mow VC (1991) A triphasic theory for the swelling and deformation behaviors of articular cartilage. J Biomech Eng 113:245–258
88. Setton LA, Mow VC, Müller FJ et al (1994) Mechanical properties of canine articular cartilage are significantly altered following transection of the anterior cruciate ligament. J Orthop Res 12:451–463. doi:10.1002/jor.1100120402
89. Goldring SR (2012) Alterations in periarticular bone and cross talk between subchondral bone and articular cartilage in osteoarthritis. Ther Adv Musculoskelet Dis 4:249–258. doi:10.1177/1759720X12437353
90. Lories RJ, Luyten FP (2011) The bone–cartilage unit in osteoarthritis. Nat Rev Rheumatol 7:43–49. doi:10.1038/nrrheum.2010.197
91. Pan J, Wang B, Li W et al (2012) Elevated cross-talk between subchondral bone and cartilage in osteoarthritic joints. Bone 51:212–217. doi:10.1016/j.bone.2011.11.030

92. Radin EL, Paul IL (1970) Does cartilage compliance reduce skeletal impact loads?. the relative force-attenuating properties of articular cartilage, synovial fluid, periarticular soft tissues and bone. Arthritis Rheum 13:139–144. doi:10.1002/art.1780130206

93. Roemer FW, Neogi T, Nevitt MC et al (2010) Subchondral bone marrow lesions are highly associated with, and predict subchondral bone attrition longitudinally: the MOST study. Osteoarthr Cartil 18:47–53. doi:10.1016/j.joca.2009.08.018

94. Lewis CW, Williamson AK, Chen AC et al (2005) Evaluation of subchondral bone mineral density associated with articular cartilage structure and integrity in healthy equine joints with different functional demands. Am J Vet Res 66:1823–1829

95. Seah S, Wheaton D, Li L et al (2012) The relationship of tibial bone perfusion to pain in knee osteoarthritis. Osteoarthr Cartil 20:1527–1533. doi:10.1016/j.joca.2012.08.025

96. Li B, Aspden RM (1997) Composition and mechanical properties of cancellous bone from the femoral head of patients with osteoporosis or osteoarthritis. J Bone Miner Res 12:641–651. doi:10.1359/jbmr.1997.12.4.641

97. Gelse K, Söder S, Eger W et al (2003) Osteophyte development—molecular characterization of differentiation stages. Osteoarthr Cartil 11:141–148

98. Van der Kraan PM, van den Berg WB (2007) Osteophytes: relevance and biology. Osteoarthr Cartil 15:237–244. doi:10.1016/j.joca.2006.11.006

99. Hashimoto S, Creighton-Achermann L, Takahashi K et al (2002) Development and regulation of osteophyte formation during experimental osteoarthritis. Osteoarthr Cartil 10:180–187. doi:10.1053/joca.2001.0505

100. Blom AB, van Lent PLEM, Holthuysen AEM et al (2004) Synovial lining macrophages mediate osteophyte formation during experimental osteoarthritis. Osteoarthr Cartil 12:627–635. doi:10.1016/j.joca.2004.03.003

101. Kishimoto K, Kitazawa R, Kurosaka M et al (2006) Expression profile of genes related to osteoclastogenesis in mouse growth plate and articular cartilage. Histochem Cell Biol 125:593–602. doi:10.1007/s00418-005-0103-z

102. Neve A, Cantatore FP, Corrado A et al (2013) In vitro and in vivo angiogenic activity of osteoarthritic and osteoporotic osteoblasts is modulated by VEGF and vitamin D3 treatment. Regul Pept 184:81–84. doi:10.1016/j.regpep.2013.03.014

103. Yuan Q, Sun L, Li J-J, An C-H (2014) Elevated VEGF levels contribute to the pathogenesis of osteoarthritis. BMC Musculoskelet Disord 15:437. doi:10.1186/1471-2474-15-437

104. Mobasheri A, Matta C, Zákány R, Musumeci G (2015) Chondrosenescence: Definition, hallmarks and potential role in the pathogenesis of osteoarthritis. Maturitas 80:237–244. doi:10.1016/j.maturitas.2014.12.003

105. Ellman MB, Yan D, Chen D, Im H (2012) Biochemical mediators involved in cartilage degradation and the induction of pain in osteoarthritis. In: Rothschild BM (ed) Princ. osteoarthritis—its defin. character, deriv. modality-related recognit. InTech, pp 367–399

106. Clarke C, Held A, Stange R et al (2015) A4.11 Syndecan-4 is an important player in regulating the WNT signalling pathway in articular cartilage. Ann Rheum Dis 74:A40–A41. doi:10.1136/annrheumdis-2015-207259.93

107. Funck-Brentano T, Bouaziz W, Marty C et al (2014) Dkk-1-mediated inhibition of Wnt signaling in bone ameliorates osteoarthritis in mice. Arthritis Rheumatol (Hoboken, NJ) 66:3028–3039. doi:10.1002/art.38799

108. Leijten JCH, Emons J, Sticht C et al (2012) Gremlin 1, frizzled-related protein, and Dkk-1 are key regulators of human articular cartilage homeostasis. Arthritis Rheum 64:3302–3312. doi:10.1002/art.34535

109. Chan BY, Fuller ES, Russell AK et al (2011) Increased chondrocyte sclerostin may protect against cartilage degradation in osteoarthritis. Osteoarthr Cartil 19:874–885. doi:10.1016/j.joca.2011.04.014

110. Papathanasiou I, Malizos KN, Tsezou A (2012) Bone morphogenetic protein-2-induced Wnt/β-catenin signaling pathway activation through enhanced low-density-lipoprotein receptor-related protein 5 catabolic activity contributes to hypertrophy in osteoarthritic chondrocytes. Arthritis Res Ther 14:R82. doi:10.1186/ar3805

111. Van Der Kraan PM, Davidson ENB, Blom A, Van Den Berg W (2009) Review TGF-beta signaling in chondrocyte terminal differentiation and osteoarthritis Modulation and integration of signaling pathways through receptor-Smads. Osteoarthr Cartil 17:1539–1545. doi:10.1016/j.joca.2009.06.008

112. Zhen G, Wen C, Jia X et al (2013) Inhibition of TGF-β signaling in mesenchymal stem cells of subchondral bone attenuates osteoarthritis. Nat Med 19:704–712. doi:10.1038/nm.3143

113. Blaney Davidson EN, Vitters EL, van der Kraan PM, van den Berg WB (2006) Expression of transforming growth factor- (TGF) and the TGF signalling molecule SMAD-2P in spontaneous and instability-induced osteoarthritis: role in cartilage degradation, chondrogenesis and osteophyte formation. Ann Rheum Dis 65:1414–1421. doi:10.1136/ard.2005.045971

114. Scharstuhl A, Glansbeek HL, van Beuningen HM et al (2002) Inhibition of endogenous TGF—during experimental osteoarthritis prevents osteophyte formation and impairs cartilage repair. J Immunol 169:507–514. doi:10.4049/jimmunol.169.1.507

115. Kingsley DM (1994) The TGF-beta superfamily: new members, new receptors, and new genetic tests of function in different organisms. Genes Dev 8:133–146

116. Kobayashi T, Lyons KM, McMahon AP, Kronenberg HM (2005) BMP signaling stimulates cellular differentiation at multiple steps during cartilage development. Proc Natl Acad Sci USA 102:18023–18027. doi:10.1073/pnas.0503617102

117. Sharma AR, Jagga S, Lee S-S, Nam J-S (2013) Interplay between cartilage and subchondral bone contributing to pathogenesis of osteoarthritis. Int J Mol Sci 14:19805–19830. doi:10.3390/ijms141019805

118. Wozney JM, Rosen V (1998) Bone morphogenetic protein and bone morphogenetic protein gene family in bone formation and repair. Clin Orthop Relat Res 346:26–37

119. Cook SD, Rueger DC (1996) Osteogenic protein-1: biology and applications. Clin Orthop Relat Res 324:29–38

120. Hayashi M, Muneta T, Ju Y-J et al (2008) Weekly intra-articular injections of bone morphogenetic protein-7 inhibits osteoarthritis progression. Arthritis Res Ther 10:R118. doi:10.1186/ar2521

121. Hunter DJ, Pike MC, Jonas BL et al (2010) Phase 1 safety and tolerability study of BMP-7 in symptomatic knee osteoarthritis. BMC Musculoskelet Disord 11:232. doi:10.1186/1471-2474-11-232

122. Blaney Davidson EN, Vitters EL, van Lent PLEM et al (2007) Elevated extracellular matrix production and degradation upon bone morphogenetic protein-2 (BMP-2) stimulation point toward a role for BMP-2 in cartilage repair and remodeling. Arthritis Res Ther 9:R102. doi:10.1186/ar2305

123. Blaney Davidson EN, Vitters EL, Bennink MB et al (2014) Inducible chondrocyte-specific overexpression of BMP2 in young mice results in severe aggravation of osteophyte formation in experimental OA without altering cartilage damage. Ann Rheum Dis 74(6):1257–1264. doi:10.1136/annrheumdis-2013-204528

124. Smeets TJM, Barg EC, Kraan MC et al (2003) Analysis of the cell infiltrate and expression of proinflammatory cytokines and matrix metalloproteinases in arthroscopic synovial biopsies: comparison with synovial samples from patients with end stage, destructive rheumatoid arthritis. Ann Rheum Dis 62:635–638

125. Smith MD, Triantafillou S, Parker A et al (1997) Synovial membrane inflammation and cytokine production in patients with early osteoarthritis. J Rheumatol 24:365–371

126. Tchetina EV, Squires G, Poole AR (2005) Increased type II collagen degradation and very early focal cartilage degeneration is associated with upregulation of chondrocyte differentiation related genes in early human articular cartilage lesions. J Rheumatol 32:876–886

127. Orita S, Koshi T, Mitsuka T et al (2011) Associations between proinflammatory cytokines in the synovial fluid and radiographic grading and pain-related scores in 47 consecutive patients with osteoarthritis of the knee. BMC Musculoskelet Disord 12:144. doi:10.1186/1471-2474-12-144

128. Furman BD, Mangiapani DS, Zeitler E et al (2014) Targeting pro-inflammatory cytokines following joint injury: acute intra-articular inhibition of interleukin-1 following knee injury prevents post-traumatic arthritis. Arthritis Res Ther 16:R134. doi:10.1186/ar4591

129. Imagawa K, de Andrés MC, Hashimoto K et al (2011) The epigenetic effect of glucosamine and a nuclear factor-kappa B (NF-kB) inhibitor on primary human chondrocytes–implications for osteoarthritis. Biochem Biophys Res Commun 405:362–367. doi:10.1016/j.bbrc.2011.01.007

130. Niederberger E, Geisslinger G (2008) The IKK-NF-kappaB pathway: a source for novel molecular drug targets in pain therapy? FASEB J 22:3432–3442. doi:10.1096/fj.08-109355

131. Rigoglou S, Papavassiliou AG (2013) The NF-kB signalling pathway in osteoarthritis. Int J Biochem Cell Biol 45:2580–2584. doi:10.1016/j.biocel.2013.08.018

132. Keifer JA, Guttridge DC, Ashburner BP, Baldwin AS (2001) Inhibition of NF-kappa B activity by thalidomide through suppression of IkappaB kinase activity. J Biol Chem 276:22382–22387. doi:10.1074/jbc.M100938200

133. Gomes WF, Lacerda ACR, Mendonça VA et al (2012) Effect of aerobic training on plasma cytokines and soluble receptors in elderly women with knee osteoarthritis, in response to acute exercise. Clin Rheumatol 31:759–766. doi:10.1007/s10067-011-1927-7

134. Messier SP, Mihalko SL, Legault C et al (2013) Effects of intensive diet and exercise on knee joint loads, inflammation, and clinical outcomes among overweight and obese adults with knee osteoarthritis: the IDEA randomized clinical trial. JAMA 310:1263–1273. doi:10.1001/jama.2013.277669

135. Martel-Pelletier J, Boileau C, Pelletier J-P, Roughley PJ (2008) Cartilage in normal and osteoarthritis conditions. Best Pract Res Clin Rheumatol 22:351–384. doi:10.1016/j.berh.2008.02.001

136. Lee AS, Ellman MB, Yan D et al (2013) A current review of molecular mechanisms regarding osteoarthritis and pain. Gene 527:440–447. doi:10.1016/j.gene.2013.05.069

137. Haseeb A, Haqqi TM (2013) Immunopathogenesis of osteoarthritis. Clin Immunol 146:185–196. doi:10.1016/j.clim.2012.12.011

138. Borzi RM, Mazzetti I, Macor S et al (1999) Flow cytometric analysis of intracellular chemokines in chondrocytes in vivo: constitutive expression and enhancement in osteoarthritis and rheumatoid arthritis. FEBS Lett 455:238–242

Chapter 5
Cartilage Tissue Engineering and Regenerative Strategies

Alain da Silva Morais, Joaquim Miguel Oliveira and Rui Luís Reis

Abstract Human adult articular cartilage is a unique avascular tissue which displays the ability to resist to repetitive compressive stress. However, this connective tissue exhibits slight capacity for intrinsic restoration and, then even injuries or lesions can lead to progressive damage and osteoarthritic joint deterioration. Therefore, the field of cartilage repair continues to expand, bridging the gap between palliative care and chondral defects reconstruction. Tissue engineering strategy, centered on three actors: cells, proteins and scaffolds, received a lot of attention in the aim to develop an articular cartilage regeneration process that will be efficient, simple, and based on global market, cost-effective. The current state of cartilage tissue engineering with respect to different cell-sources, growth factors and biomaterial scaffolds, as well as the strategies employed in the restoration and repair of damaged articular cartilage will be the focus of this book chapter.

5.1 Introduction

One of the most stimulating challenges for orthopedic surgeons is the management of articular cartilage lesions. After injury due to trauma or degenerative pathologies, the restricted intrinsic healing capability of this highly organized tissue results frequently in fibrillation and gradual tissue deterioration, leading to incapacitating joint pain, functional damage and degenerative arthritis [1]. The absence of vascularization needed to improve the entry of progenitor cells from blood or bone

A. da Silva Morais (✉) · J.M. Oliveira · R.L. Reis
3B's Research Group—Biomaterials, Biodegradables and Biomimetics,
Headquarters of the European Institute of Excellence on Tissue Engineering
and Regenerative Medicine, University of Minho, AvePark, S. Claudio de Barco,
Guimarães, Portugal
e-mail: alain.morais@dep.uminho.pt

A. da Silva Morais · J.M. Oliveira · R.L. Reis
ICVS/3B's—PT Government Associate Laboratory, Braga, Guimarães, Portugal

© Springer International Publishing AG 2017
J.M. Oliveira and R.L. Reis (eds.), *Regenerative Strategies for the Treatment of Knee Joint Disabilities*, Studies in Mechanobiology, Tissue Engineering and Biomaterials 21, DOI 10.1007/978-3-319-44785-8_5

marrow seems to be one of the principal key for the limited repair capacity of cartilage tissue.

To restore normal joint congruity and reduce supplementary joint deterioration, associated to symptomatic cartilage lesions, autologous osteochondral graft transplantation and osteotomy have been suggested. Presently, a wide variety of surgical methods has been considered [2–4]. Often, the most of these techniques are not long-term clinical solutions, except for the total joint replacement as surgical procedure for end-stage degenerative joint pathology, encouraging the development of tissue engineering and regenerative medicine (TERM) approaches to restore articular cartilage.

The goal of TERM is to regenerate tissues using preferentially an autologous cell approach, biodegradable biomaterials, and pertinent growth factors, alone or in combination to increase the efficiency of the process. In recent decades, the focus of the initial tissue engineering researches was to generate cartilage tissue in order to restore or replace the structure and function of the injured or degenerated cartilage. Despite these efforts, the development of clinically relevant and functionally correspondent cartilage tissue constructs remains vague, with some few exceptions associated to non-loading cartilage needs in restricted clinical trials [5].

Actually, it remains to be established a consensus reached on few critical components of cartilage tissue engineering approaches, such as the cell source, the biomaterial design norms, the dose and delivery mode of signaling growth factors, as well as the required stage of maturation and associated release criteria predictive of clinical potency.

Despite the many challenges, engineered cartilage tissues have been progressively recognized as biological systems offering the possibility to investigate mechanisms and processes of chondrogenesis and its regulation by physico-chemical signals or pharmacological compounds. Finally, the produced knowledge is expected to develop pioneering and effective strategies for the future of cartilage repair in patients.

5.2 Articular Cartilage

Articular Cartilage (AC) is a unique and highly specialized connective tissue, with 2–4 mm thickness, constituting the mechanical framework of the body and playing a vital role in the musculoskeletal system. This hyaline tissue, which covers the articulating ends of the bones inside the synovial joints providing frictionless and pain free movements during skeletal motion and forming a weight-bearing layer on the joint surfaces, is composed by a dense extracellular matrix (ECM) rich in water, collagen, proteoglycans and elastin fibers, with dispersed cells called chondrocytes. Its ECM composition provides strength and flexibility to the tissue [6]. However, despite the absence of blood vessels, lymphatics, and nerves, the articular cartilage is in most cases surrounded by a specific collagenous connective tissue named perichondrium, which supplies the nutrition and oxygen.

In order to develop TERM approaches for articular cartilage repair, it is important to understand the structure of the native tissue. Despite its simple appearance the articular cartilage has a complex structure and cellular organization. The tissue is divided in four zones, three non-mineralized called superficial; intermediate (middle) and radial (deep); and the mineralized zone called calcified cartilage. Inside the superficial zone, chondrocytes look flat in shape at close proximity to each other and collagen fibers aligned parallel to the articular surface. The intermediate zone is characterized by oblique in shape chondrocytes organization and a randomly dispersion of collagen fibers. Within the radial zone, spherical chondrocytes are aligned in columns and collagen fibers are perpendicularly organized and penetrate through the tidemark into the mineralized zone providing structural stability for articular cartilage on the subchondral bone [2].

5.3 Tissue Engineering and Regenerative Strategies

5.3.1 Material

The different actors of tissue engineering and their involvement in accomplishing tissue regeneration are illustrated in Fig. 5.1.

5.3.1.1 Cell Sources

Current cartilage repair options that utilize implanted cells are limited by the number of cells available for isolation and by the uncontrolled phenotypic alterations in those cells. As such, chondrocytes and stem cells have been investigated as cell sources for cartilage engineering due to their well-established ability to generate cartilage-like ECM under the appropriate culture conditions.

Chondrocytes

Articular cartilage contains 1–5 % of total tissue volume of specialized and highly differentiated cells, the chondrocytes [7]. They are the sole cells found in the lacunae of the cartilage, usually scattered individually throughout extracellular matrix of the articular cartilage primarily composed of water, proteoglycans, collagens and non-collagenous proteins [6, 8]. During growth, chondrocytes usually have circular shape, but their shape is variable depending on pathological state, age and the cartilage layer. As articular cartilage lacks vascularization, lymphatic drainage, and nervous system innervation, chondrocytes function under avascular, anaerobic

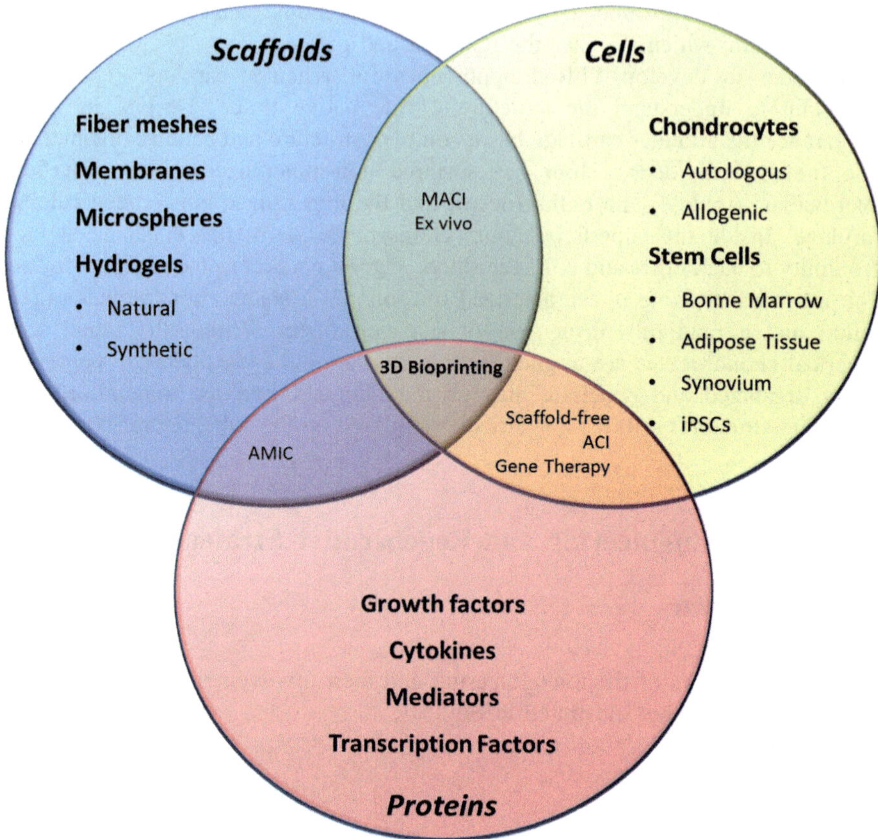

Fig. 5.1 The tissue engineering paradigm. Scaffolds, cells and signals are the key elements of tissue engineering approaches to cartilage repair. Therefore, many different scaffolds types and biomaterials, cell sources and signals are under investigation. Combination of two of these factors lead to important cartilage regeneration approaches: 1. cell-free, scaffold-based; 2. cell-seeded scaffolds and 3. scaffold-free, cell-based implants. Full combinations of these three factors drive to a most promising cartilage engineering strategy: 3D bioprinting. *AMIC* autologous matrix-induced chondrogenesis, *ACI* autologous chondrocyte implantation, *MACI* matrix-assisted autologous chondrocyte implantation, *iPSCs* induced pluripotent stem cells, *3D* three dimensional

conditions, obtaining nutrients by diffusion from synovial fluid [7, 9]. Given the accessibility of articular cartilage by arthroscopic surgery, native chondrocytes are a logical cell source for cartilage repair. However, the isolation of chondrocytes from their collagen matrixes requires overnight digestion by collagenase, which might be damaging to the cells [10]. Moreover, the fact that the chondrocytes lose their phenotype during culture and expansion process represents a discouraging point for their use. This process has been termed dedifferentiation, well known since decades, and induces several phenotypically alterations in chondrocytes. Dedifferentiation involves a decrease in the expression of chondrocyte markers, such as collagen type II, aggrecan, and the transcription factor SOX9, and an increase in the expression of fibroblastic markers, such as collagen type I and versican [11]. This is of critical importance for cartilage tissue engineering since dedifferentiated chondrocytes acquire fibroblastic phenotype and do not produce the necessary components for hyaline cartilage, which is required for the proper function of articulating joints. Extensive evidences demonstrate rapid chondrocyte dedifferentiation during culture expansion on polystyrene [12–14]. Thus, the loss of chondrocyte phenotype prior to utilization in experimentation or implantation is of huge concern for the tissue engineering field. Several studies attempt the determine the cellular pathways involved in this dedifferentiation process [11, 15, 16]. Preserving the chondrocyte phenotype with healthy hyaline tissue synthesis in vitro during expansion for autologous chondrocytes implantation (ACI) represents an ongoing challenge.

Stem Cells

Despite the fact that the use of autologous cells represented the best model to improve clinical translation, some limitations related with the quantity of primary cells available and donor site morbidity led to the use of allogeneic and xenogeneic sources [17]. Application of stem cells, especially mesenchymal stem cells, is an attractive approach to improve the regenerative processes in sites of cartilage injury compared with the implantation of differentiated cells like articular chondrocytes [18]. Stem cells, defined by their abilities for self-renewal and for differentiation into a wide variety of cell lineages, hold great potential for applications in tissue engineering and regenerative medicine [19–23]. In the context of cartilage tissue engineering, the major sources of stem cells are adult mesenchymal stem cells (MSCs) and embryonic stem cells (ESCs) [20, 24–26]. A uniform characterization of MSCs, based on a minimal set of standard criteria, has been established by the Mesenchymal and Tissue Stem Cell Committee of the International Society for Cellular Therapy. Therefore, MSCs must be plastic-adherent cells when maintained in standard culture conditions and they must express CD105, CD73, and CD90; they must lack surface expression of CD45, CD34, CD14 (CD11b), CD79α (CD19), and HLA-DR [27]. Other than surface markers and plastic adherence, MSCs must have ability a reliable potential of differentiation into cells of the mesodermal lineage (osteoblasts, adipocytes, chondrocytes and myocytes) under in vitro conditions [28, 29]. The differentiation process to a particular phenotype can be regulated by some regulatory genes which can induce progenitor cells' differentiation to a specific lineage. Even though MSCs can differentiate into a number of tissues in vitro, the resulting cell population does not mimic the targeted tissues entirely in their biochemical and biomechanical properties [30]. The fact that MSCs also displayed critical homing, trophic, and immunomodulatory activities [31–33] may favorably influence the fate and activities of unaffected cells in the surrounding cartilage upon implantation in sites of cartilage damage or injury.

MSCs can be easily isolated different human or animal tissues such as bone marrow, adipose tissue and synovium [34]. **Bone marrow-derived MSCs (BMSCs)** have been a focus of stem cell research in light of their high potential for differentiation [35]. Chondrogenesis has been usefully achieved in vitro in aggregate condition favoring the induction of the first phase, characterized by cell condensation, as well as cell-cell and cell-extracellular matrix (ECM) interactions [10]. Afterwards, cells progress into a highly proliferative phase and start to produce typical components of the cartilaginous matrix (i.e., collagen type 2 and aggrecan). The different stages of chondrogenic differentiation are regulated by signaling factors like Bone morphogenetic proteins (BMPs), Fibroblast growth factor (FGF), Transforming growth factor-β (TGF-β), Wnt, and Indian hedgehog (Ihh) that promote these processes. Different key transcription factors (SOX9, SOX5, SOX6, Slug, TRSP1, and GDF5) have a central role in the control of stem cells properties as well as in the chondrogenic status by driving the mesenchymal condensation and differentiation [36–38]. However, the relationship between age and rate remained to be clearly established, being a challenge for application in elderly patients [39]. Nevertheless, proof-of-concept for the use of BMSCs in vivo has been confirmed in different animal models of articular cartilage defects and osteoarthritic disease (rat, rabbit, pig, sheep, horse), showing improved repair of lesions related with conditions where cells were not provided [29]. **Adipose-derived MSCs** (ASCs), isolated from lipoaspirates, can be acquired using a less invasive procedure and in large amounts than BMSCs. Comparing to, ASCs are smaller, present different cell surface marker and gene expression profiles, and show enhanced rates of proliferation before senescence [40]. ASCs can differentiate into the chondrogenic, osteogenic, adipogenic, myogenic, neurogenic, and hepatogenic lineages [41, 42]. Therefore, and associated with the minimal injury to the donor site, ASCs become an attractive cell source for cartilage repair [29]. However, ASCs show reduced responses to TGF-β-induced chondrogenesis, due to their absent expression of efficient TGF-β receptor 1, but differentiation has been nevertheless established by addition of bone morphogenetic protein [43]. Therefore, due to the absence of an ideal protocol for ASCs' chondrogenic differentiation, the use of ASCs in cartilage tissue engineering remains under major restriction. **Synovial-derived MSCs**, isolated from synovial membrane via arthroscopy in a low invasive way with minimal complications at the donor site, present higher proliferative and chondrogenic properties than other MSCs especially when incubated with bone morphogenetic proteins [44–46]. Even if clinical studies using these cells have not yet been conducted, animal models studies have been performed, leading to enhanced cartilage repair by synovial-derived MSC [47–49]. Finally, **Pluripotent stem cells** (PSCs), composed by embryonic stem cells (ESCs) and induced pluripotent stem cells (iPSCs), represent a novel and potentially unlimited source of chondrocytes and tissues for therapeutic applications due to the capability of these cells to generate a large spectrum of cell types under appropriate culture conditions. The use of ESCs and iPSCs for cartilage repair has decreased due to the limitation of clinical application by ethical concerns regarding the controversies related with the high risk of inducing tumor growth and teratomas [29, 50–57].

5.3.1.2 Scaffolds

Current approaches for articular cartilage repair are centered on the use of scaffolds providing a suitable three-dimensional (3D) environment supporting the growth of a cartilaginous repair tissue. These 3D structures are often critical, both in vitro as well as in vivo, to summarizing the in vivo milieu and allowing cells to modulate their own microenvironment [58]. The ideal scaffolds for cartilage tissue engineering must to be based on the following basic requirements: porous, biocompatible, biodegradable, and appropriate for cell attachment, proliferation and differentiation. **The architectural design** of the scaffolds remains of importance; the scaffolds have to be available for vascularization, new tissue formation and remodeling so as to enable tissue integration upon implantation. Moreover, they have to present enough structure for an efficient nutrient and metabolic delivery without significantly affecting their own mechanical stability. For cartilage defect regeneration, various scaffold structure have been used from fiber meshes, membranes, hydrogels, three-dimensional porous scaffolds, and microspheres [59–62]. Regarding the mechanical properties, the scaffolds have to provide shape stability to the tissue defect. Therefore, the owned mechanical properties of the biomaterials used to create the scaffolds or their post-modification properties have to match that of the host tissue, highlighting the importance of mechanical properties of a scaffold on the seeded cells [63]. **Scaffold biocompatibility** is based on the cellular and tissue compatibility of the framework to provide support for seeded or endogenous cells to bond, grow and differentiate during both in vitro culture and in vivo implantation. As for the scaffold design, the biomaterial used to produce the framework plays an important role it will be selected regarding its compatibility with the cellular components of the engineered tissues and endogenous cells in host tissue. Furthermore, the biomaterial has also to be biodegradable upon implantation at a rate similar to that of the new matrix production by the emerging tissue [61, 64–67]. The **bioactive** property of the scaffold, essentially based on the biomaterial used to construct it, characterizes his interaction with the cellular components of the engineered tissues to assist and regulate their actions. The scaffold biomaterial is one of the key design factors to be considered in scaffold-based cartilage tissue engineering. Several biomaterials are currently used to establish scaffolds with efficient response in terms of cartilage repair engineering. Hydrogel scaffolds, easily prepared, can be made of a wide variety of biomaterials, including natural materials, which may be carbohydrate-based (e.g., alginate [68], agarose [69], chitosan [70, 71], hyaluronic acid [72, 73]), protein-based (e.g., fibrin [74, 75], collagen [76, 77], silk [78]) or some combination of two [79]. Among all the naturally derived polymers, the use of Gellan gum, a polysaccharide derived from Sphingomonas elodea, and Ulvans, the major constituents of green seaweeds cell walls, as supportive systems for tissue development presents an increasing trend in the biomedical area [80, 81]. Synthetically gel-like scaffolds, such as polyethylene glycol and its derivatives [82], poly(vinyl alcohol) (PVA) [83], and others [63, 84]. From easier fabrication, polyester-derived solid scaffolds are mostly created from synthetic biomaterials and present higher biomechanical properties to those based on hydrogels [61, 85, 86].

Some frequently used synthetic polymers in cartilage tissue engineering include poly (glycolic acid) (PGA), poly(lactic acid) (PLA), poly(lactic-co-glycolic acid) (PLGA), and poly(ethyl glycol) (PEG), between others [87–92]. Recently, scaffolds modifications are under investigation in order to include biological cues such as cell-adhesive ligands to enhance attachment [93, 94] or physical cues such as topography to influence cell morphology and alignment [95–97]. An important advantage of using scaffolds for cell delivery and implantation inside the lesion sites, is that biomaterials could prevent scarring [98, 99]. Moreover, the scaffolds may also be used as delivery system for exogenous growth factors [100–106], thus the biomaterials used need to be biocompatible with the molecules and available for encapsulation techniques in order to control the release of the bioactive molecules.

5.3.1.3 Growth Factors

The healthy articular cartilage is a low-turnover tissue with moderate degradation of the matrix by proteolytic enzymes and its renewal by the chondrocytes. These processes are controlled by various growth factors, cytokines and others mediators. The use of growth factors, actually proposed to overcome the disadvantages associated with the use of chondrocytes such as poor motility and matrix deposition, has demonstrated a potential therapeutic application in the field of tissue engineering. The anabolic processes, involving proliferation, differentiation and matrix production, are regulated by soluble growth factors including **insulin growth factor-I** (IGF-I), the main anabolic growth factor of articular cartilage [107, 108]; **bone morphogenetic proteins** (BMPs), promoting maintenance of a chondrogenic phenotype and up-regulation of cartilage matrix synthesis [109–111], the members of the sex-determining region Y-type high-mobility group box (SOX) family of transcription factors SOX9 [112], **parathyroid hormone (PTH)** [113], **parathyroid hormone related peptide** (PTHrP) [114, 115], **growth/differentiation factor 5** (GDF-5) [116] as well as **basic fibroblast growth factor** (bFGF) and **transforming growth factor-beta** (TGF-β). All these growth factors represented potent candidates showing beneficial effects on cartilage repair [117]. Basic FGF is an extensively investigated growth factor with 22 different forms with the isoform FGF-2 demonstrating high chondrogenic capacity. However, to elicit cartilage tissue regeneration, bFGF requires the association with secondary factors [111, 118, 119]. The most commonly used growth factor for in vitro chondrogenesis is TGF-β which is able to control proliferation, differentiation and maintenance of chondrogenic phenotype of differentiated cells. It also shows an important involvement in cartilage homeostasis and, as for bFGF, in combination with other growth factors leads to greater chondrogenic response than alone [107, 109, 110, 120]. Frequently used as combinatory growth factor, **platelet-derived growth factor** (PDGF) has demonstrated beneficial impact on cartilage repair process [121, 122].

5.3.2 Strategies

See Fig. 5.2.

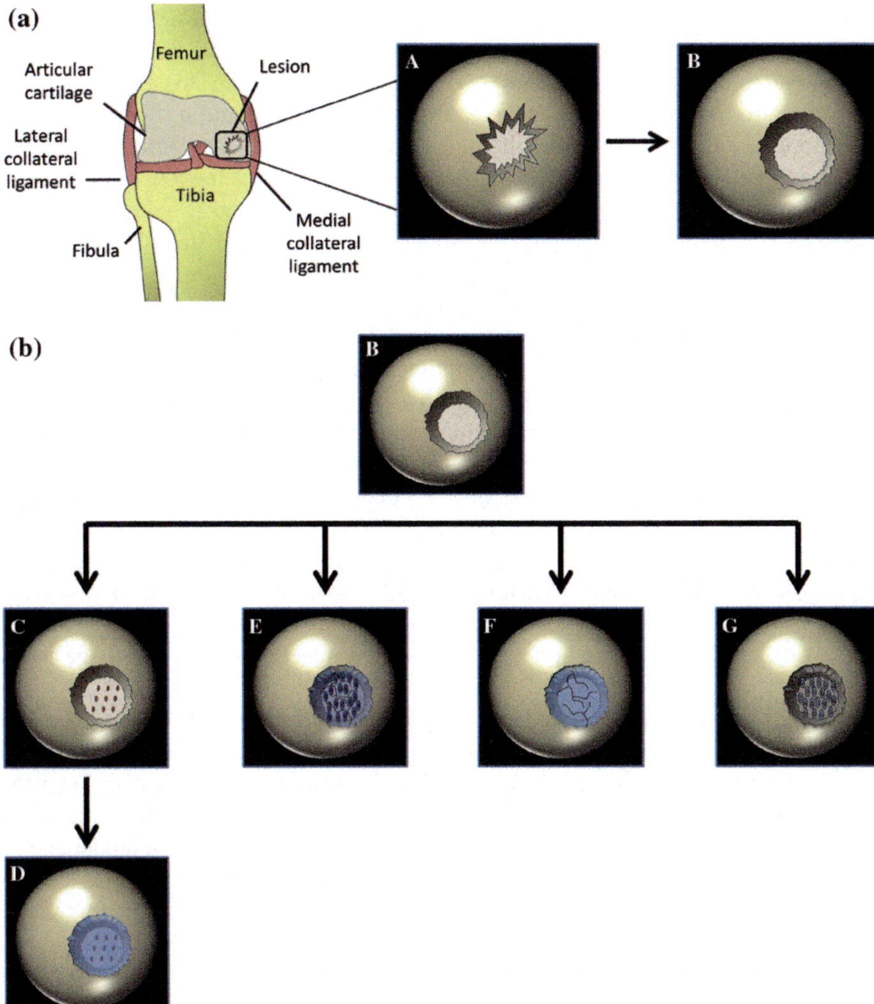

Fig. 5.2 An illustration of the cartilage repair surgical approaches. *A* Full-thickness focal chondral lesion. *B* Lesion debrided to certify healthy, stable margins for complete integration of the host tissue with the neotissue. *C* Microfracture. Channels are created to penetrate the subchondral bone, allowing MSCs to migrate from the marrow to the cartilage defect. *D* Microfracture releases blood and bone marrow MSCs, then collagen type I/III, hyaluronic acid or fibrin matrix are sutured or glued into the defect. *E* MACI uses scaffolds plus either primary articular chondrocytes or MSCs. *F* AMIC is a cell-free, scaffold-based single surgery. *G* Scaffold-free techniques include a self-assembling process. In the absence of a scaffold to interrupt cell–cell interaction, cells are able to respond to stimuli and promote integration of neotisssue with the surrounding tissue. *AMIC* autologous matrix-induced chondrogenesis, *ACI* autologous chondrocyte implantation, *MACI* matrix-assisted autologous chondrocyte implantation, *iPSCs* induced pluripotent stem cells

5.3.2.1 Current Repair Techniques

The limited ability of articular cartilage to regenerate has encouraged the development of cell-based tissue engineering techniques, such as microfracture and autologous chondrocyte implantation. The most common surgical option to treat defects of articular cartilage includes bone marrow stem cells (BMSCs) stimulation techniques that perforate the subchondral layer, creating small fractures with the aim of stimulating a wound healing response by BMSCs. This process, called microfracture, presents the advantages to be cost effective and a relatively simple procedure [123]. However, despite the fact that satisfactory outcomes of this technique have been reported [123], full recovery is not achieved in all patients [124–126]. The varying outcomes of microfracture prompted the development of autologous chondrocyte implantation (ACI). Frequently used in cell-based strategies for articular cartilage repair, ACI is based on collection, by biopsy of a full-thickness sample from low-weight-bearing region of the joint in order to provide chondrocyte population thereafter expended in vitro [127]. Since first step, ACI has evolved through different generations. Despite the cell source was patient's autologous chondrocytes harvested from non-weight-bearing cartilage surface during an index arthroscopic procedure and cultured in vitro, the method of introducing cells into cartilage defect varied into generations. While in the first generation of ACI chondrocytes are injected under periosteum patch, in the second generation, cells were injected under a prefabricated collagen membrane and thereafter sutured over the border of the cartilage defect. In third generation, autologous chondrocytes are grown in a 3D biological matrix optimized to size and inserted into the cartilage defect. The last generation, fourth, is characterized by suspension of chondrocytes into a fibrin gel then injected into the cartilage lesion site [9].

All ACI generations have been documented to significantly improve knee pain and function in several patients. Moreover, no difference was observed between first- and third-generation, between second- and third-generation, or between two different third-generation ACI systems. Third-generation ACI, which introduces the highest number of chondrocytes per defect volume, has no long-term advantage for graft quality [128]. Recently second-generation ACI leaded to higher clinical long-term outcome compared to first-generation [129]. The benefits of this procedure are the small size of the biopsy minimizing complications for the donor, the use of patient's own cells avoids eventual immune complications resulting from allogenic transplantation and, this technique include the capacity to treat lesions above 4 cm^2 [130]. Even though the suitable clinical outcomes reported for ACI [129, 131], reported complications as absence of incorporation, presence of graft hypertrophy, and loss of the implant, have encouraged further research to investigate the therapeutic potential of other scaffolds and cell sources such as MSCs [129, 132, 133]. The complexity of ACI and contraindications in varied clinical applications have motivated the development of matrix-assisted chondrocyte implantation (MACI), which uses scaffolds to provide mechanical stability and support chondrogenesis.

5.3.2.2 Scaffold-Based Strategies

Autologous chondrocyte implantation (ACI) has become a recognized technique for the restoration of full thickness cartilage defects in the joint knee. However, the absence of a scaffold material used as support to guide matrix synthesis and organization might, in part, justify the variability of results observed after treatment of patients with chondrocyte implantation techniques [3, 9]. Second generation of ACI aimed to develop a scaffold-based approach for chondrocytes delivery to the cartilage lesion site. Scaffold-based strategies presented main advantages comparing to scaffold-free methods i.e. increased control to improved fill of cartilage defect; reduced donor site problems; and shorter postoperative recovery time related with the improved graft stability. Furthermore, as chondrocytes are cultured in a tridimensional environment, they were less disposed to dedifferentiation and consequently produced a more hyaline-like cartilage [134]. Matrix-associated autologous chondrocyte transplantation/implantation (MACT/MACI) is a third-generation ACI product and consists in an initial arthroscopic harvesting of healthy cartilage followed by isolation and expansion of autologous chondrocytes ex vivo which were seeded on a porcine collagen I/III membrane, specifically engineered to endorse chondrocyte infiltration on the side fronting the subchondral bone and low-friction surface facing the joint cavity on the other, prior implantation into debrided defect. After implantation, the matrix is secured using fibrin glue, which support cell migration and proliferation [135]. This combined tissue-engineering product (TEP) is the most common scaffold-plus-cell-based cartilage repair technique used as Advanced Therapy Medicinal Product (ATMP) in clinical practice. Then, successful patient outcomes involved pain-free and return to a normally active lifestyle. Some factors were proposed to influence improved MACI outcomes: (1) successful cell culture, (2) efficient surgical procedure, (3) patient collaboration in all aspects of the pre- and post-surgery program, and (4) timely progression postoperative rehabilitation [136]. Although promising clinical outcomes and repair tissue have been shown with MACI, associated to safety profile and reduced periosteal hypertrophy compared to ACI procedures [135–140], the superiority of MACI over existing procedures remains unproven. Rather than the use of porcine collagen I/III for scaffolds/membrane production, different natural polymers have been investigated, some of them based on animal-free formulations. Scaffold based on hyaluronic acid, a natural polymer also known as hyaluronan, represented an additional means for chondrocytes implantation, at the lesion site, within a 3D biodegradable environment. Alone or in combination with fibrin, polyethylene glycol, tricalcium phosphate-collagen, or poly-ε-caprolactone, hyaluronic-acid-based MACI presented better clinical outcomes than microfracture [5, 72, 141–145]. Animal-free protocols for hyaluronic-acid production are presently under investigation [141]. Actually, several studies are evaluating the effect of the anionic microbial heteropolysaccharide Gellan Gum (GG) and its methacrylated form (GGMA) on cartilage repair. Up to now, GG and GGMA hydrogels provided great in vitro and in vivo outcomes for cartilage and intervertebral disc engineering [74, 80, 146, 147]. Despite the technical advantages of implantation of 3D cell-based constructs on defect sites, there is a time-delay until the neotissue formation, resulting in an immature and vulnerable implant. Therefore to improve the efficiency of this tissue engineering approach, as well as graft maturation and biomechanical integrity, cells can be cultured in vitro before implantation on 3D matrices in presence of exogenous stimuli. After seeded into scaffolds, the resulting cellular 3D construct can be cultured in physiological-like conditions. Then, the use of bioreactors represented an important tool for this purpose. Bioreactors are devices to culture tissue through delivery of a well-regulated, well-established, and mechanically active environment, in order to improve neotissue formation, based on cellularity and molecular composition, by biomechanical modulation. Novel bioreactor tissue engineering approaches aiming for articular cartilage repair remained under investigation [148]. Maturation of the neotissue after long-term ex vivo 3D culture increased mechanical robustness and functionality of the implants compared to MACI and demonstrated promising outcomes in preclinical studies [149, 150]. Due to the fact that MACI-based techniques are cellular, the Food and Drug Administration (FDA) considered it as a medical and a biological device, resulting in long and expensive regulatory approval. Therefore, there is an enthusiastic interest in cell-free material-based products for cartilage regeneration and repair. Autologous matrix-induced chondrogenesis (AMIC) is a cell-free technique performed through a mini-arthrotomy, which exposes and cleans the defect site, then a microfracture released both blood and bone marrow containing MSCs, and finally a mixed collagen types I/III matrix is sutured or glued into the cartilage defect [151]. The implanted collagen matrix aimed to stabilize the resultant blood clot, promoting early mechanical stability and cartilage regeneration. This procedure is providing two major advantages; (1) single surgery without the need of cartilage harvesting that could lead to donor site morbidity, and (2) cost-effective without in vitro cell expansion [151–153].

5.3.2.3 Scaffold-Free Approaches

Actually, associated with the tissue engineering pattern of cells, signals, and scaffolds, several approaches based on the lack of scaffold have appeared as suitable modalities to engineer functional tissues. Scaffold-free or scaffoldless approaches for tissue engineering employ a high cell density to create cartilage neotissue, called neocartilage. Scaffold-free tissue engineering refers to any platform that does not require cell seeding or adherence within a three-dimensional material and several approaches have been employed successfully for articular cartilage, meniscus, and temporomandibular joint and intervertebral discs. In case of clinically relevant scaffoldless tissue engineering, cell sourcing, cell proliferation, cell differentiation, stimuli for tissue-specific extracellular matrix production, as well as tissue organization and maturation have to be considered in order to form neocartilage. Lacking a scaffold to disturb the cell–cell signaling and stress shielding, the neotissue microenvironment might be more bioactive, improving the response to stimulation and integration with surrounding tissue. Scaffold-free approaches is divided in two distinct categories, self-assembly and self-organization [154, 155]. Self-assembly refers to a process in which order spontaneously results from disorder without the use of external input, and self-organization, by contrast, refers to a process in which order appears when external energy or forces are input into the system [154]. Even if many scaffold free methods exist, the three commonly approaches included pellet culture, aggregate culture, and the self-assembling process. Pellet culture has presented promising results about chondrocyte phenotype, nevertheless the limited sizes and forms of the neocartilage have limited the translatability of this approach. By contrast, the aggregate culture promoted chondrogenic development and the self-assembling process recapitulated the cartilage development [156]. In scaffold-free culture, only cell secreting ECM proteins contributed to neotissue properties, mimicking cartilage development. Then, exogenous stimuli, such as enzymes, growth factors, and mechanical stimulation, are used to improve matrix formation and maturation to reproduce native tissue structure–function as demonstrated in ex vivo experiments where the stimulation of articular cartilage self-assembly by hypoxia, growth factors or crosslinking agents promoted the development of healthy neotissue with functional properties as native cartilage [157–159].

5.3.2.4 3D Bioprinting

Three dimensional (3D) bioprinting technology provided the necessary capabilities to create viable, 3D tissue constructs. Bioprinting or direct cell printing represented an extension of tissue engineering, and used manufacturing technologies, including laser-based writing and extrusion-based deposition [160–162]. However, the high cost of these systems is also a concern for basic tissue engineering research. By contrast, thermal inkjet printing is usually water based in order to minimize the clogging of the printhead [163]. Organ/tissue printing using inkjet technology is evolving into a practical method to fabricate viable 3D tissue constructs [164] through the capacity of precisely placing cells, biomaterials scaffolds, and biological factors in a layer-by-layer approach up to the required 3D locations with digital control [160, 165]. Then, inkjet approach represented one of the most promising additive manufacturing applications in the field of tissue engineering and regenerative medicine [166].

5.3.2.5 Gene Therapy

The concept of using gene transfer approaches for cartilage repair was based on the transfer of genes encoding for therapeutic factors into the defect site. Gene therapy aimed to treat human diseases using gene transfer procedures resulting into a temporarily and spatially defined delivery of therapeutic molecules to sites of damage [64, 167]. The transfer of genetic material into cells needed generally a viral (process named transduction) or non-viral vector (process named transfection) to allow transport through cell membranes into cytosol, and ultimately into nucleus. Once in nuclear localization, the foreign material either integrates in the host genome or stays under episomal forms allowing only for transient transgene expression. Therefore, gene therapy could be used to surpass the difficulties related with the restricted delivery of growth factors such as their half-life and the needed for repeated administration of the proteins resulting in the dispersion of the material trough the desired area [2]. Several gene of interest heave been investigated from sequences coding for growth (TGF-β, BMPs, IGF-I, FGF-2) and transcription factors (SOX5, SOX6, SOX9), to signaling molecules (hedgehogs, PTHrP) in chondrocytes and MSCs models [2, 18, 110, 168–171]. The production of therapeutically significant concentrations of transgene products is obtained only when a sufficient number of cells is expressing them [172]. Many studies involving gene delivery for cartilage regeneration appear to prefer viral rather than non-viral vectors. The reason is simply due to the superior ability of viruses to transfect cells, with transduction efficiencies of around 80–90 %, by contrast with the transfection of 40–50 % the cell population by non-viral vectors. As viral vectors, adenovirus, retrovirus, lentivirus, non-pathogenic human adeno-associated virus (AAV), recombinant AAV (rAAV) as well as self-complementary (double strained) AAV (scAAV) vectors have been investigated in several studies [111, 171, 173–179]. Although most gene therapy protocols in clinical trials actually employ viral vectors due to their high transfection efficiency, their disadvantages, such as immunogenicity, potential infectivity, complicated production, and oncogenic effects, may prevent their further use in the clinic [118]. Non-viral methods, based on the chemical complexion of DNA to different macromolecules including cationic lipids and liposomes, polymers and nanoparticles, avoid the risk of getting replication competence inherent to viral vectors. The non-viral vectors have been used in several studies due to the fact that they can be repeatedly administered, have the capacity to deliver large therapeutic genes, are easy to produce on a large scale, are efficient cell/tissue targeting, do not elicit immune responses in the host organism, and have unrestricted plasmid size [180–184]. Actually, clinical trials focused on the treatment of articular cartilage damage diseases are ongoing [171].

References

1. Huey DJ, Hu JC, Athanasiou KA (2012) Unlike bone, cartilage regeneration remains elusive. Science 338:917–921. doi:10.1126/science.1222454
2. Matsiko A, Levingstone T, O'Brien F (2013) Advanced strategies for articular cartilage defect repair. Materials (Basel) 6:637–668. doi:10.3390/ma6020637
3. Moran CJ, Pascual-Garrido C, Chubinskaya S et al (2014) Restoration of articular cartilage. J Bone Joint Surg Am 96:336–344. doi:10.2106/JBJS.L.01329
4. Johnstone B, Alini M, Cucchiarini M (2013) Tissue engineering for articular cartilage repair—the state of the art. Eur Cell Mater 25:248–267
5. Makris EA, Gomoll AH, Malizos KN et al (2014) Repair and tissue engineering techniques for articular cartilage. Nat Rev Rheumatol. doi:10.1038/nrrheum.2014.157

6. Wilusz RE, Sanchez-adams J, Guilak F (2014) The structure and function of the pericellular matrix of articular cartilage. Matrix Biol 39:25–32. doi:10.1016/j.matbio.2014.08.009

7. Nukavarapu SP, Dorcemus DL (2013) Osteochondral tissue engineering: current strategies and challenges. Biotechnol Adv 31:706–721. doi:10.1016/j.biotechadv.2012.11.004

8. Hsueh M, Önnerfjord P, Byers V (2014) Biomarkers and proteomic analysis of osteoarthritis. Matrix Biol 39:56–66. doi:10.1016/j.matbio.2014.08.012

9. Dewan AK, Gibson MA, Elisseeff JH, Trice ME (2014) Evolution of autologous chondrocyte repair and comparison to other cartilage repair techniques. Biomed Res Int 2014:11. doi:10.1155/2014/272481

10. Leyh M, Seitz A, Dürselen L et al (2014) Subchondral bone influences chondrogenic differentiation and collagen production of human bone marrow-derived mesenchymal stem cells and articular chondrocytes. Arthritis Res Ther 16:1–18. doi:10.1186/s13075-014-0453-9

11. Hong E, Reddi AH (2013) Dedifferentiation and redifferentiation of articular chondrocytes from surface and middle zones: changes in microRNAs-221/-222, -140, and -143/145 expression. Tissue Eng Part A 19:1015–1022. doi:10.1089/ten.TEA.2012.0055

12. Hubka KM, Dahlin RL, Meretoja VV et al (2014) Enhancing chondrogenic phenotype for cartilage tissue engineering: monoculture and coculture of articular chondrocytes and mesenchymal stem cells. Tissue Eng Part B Rev 20:641–654. doi:10.1089/ten.TEB.2014.0034

13. Li S, Sengers BG, Oreffo RO, Tare RS (2015) Chondrogenic potential of human articular chondrocytes and skeletal stem cells: a comparative study. J Biomater Appl 29:824–836. doi:10.1177/0885328214548604

14. Rosenzweig DH, Matmati M, Khayat G et al (2012) Culture of primary bovine chondrocytes on a continuously expanding surface inhibits dedifferentiation. Tissue Eng Part A 18:120803081750003. doi:10.1089/ten.tea.2012.0215

15. Rosenzweig DH, Ou SJ, Quinn TM (2013) P38 mitogen-activated protein kinase promotes dedifferentiation of primary articular chondrocytes in monolayer culture. J Cell Mol Med 17:508–517. doi:10.1111/jcmm.12034

16. Ma B, Leijten JCH, Wu L et al (2013) Gene expression profiling of dedifferentiated human articular chondrocytes in monolayer culture. Osteoarthr Cartil 21:599–603. doi:10.1016/j.joca.2013.01.014

17. DuRaine GD, Brown WE, Hu JC, Athanasiou KA (2014) Emergence of scaffold-free approaches for tissue engineering musculoskeletal cartilages. Ann Biomed Eng. doi:10.1007/s10439-014-1161-y

18. Cucchiarini M, Venkatesan JK, Ekici M et al (2012) Human mesenchymal stem cells overexpressing therapeutic genes: from basic science to clinical applications for articular cartilage repair. Biomed Mater Eng 22:197–208. doi:10.3233/BME-2012-0709

19. Trappmann B, Gautrot JE, Connelly JT et al (2012) Extracellular-matrix tethering regulates stem-cell fate. Nat Mater 11:642–649. doi:10.1038/nmat3339

20. Paschos NK, Brown WE, Eswaramoorthy R et al (2014) Advances in tissue engineering through stem cell-based co-culture. J Tissue Eng Regen Med 4:524–531. doi:10.1002/term.1870

21. Park H, Jung S, Yang K et al (2014) Biomaterials paper-based bioactive scaffolds for stem cell-mediated bone tissue engineering. Biomaterials 35:9811–9823. doi:10.1016/j.biomaterials.2014.09.002

22. Chen W, Villa-Diaz LG, Sun Y et al (2012) Nanotopography influences adhesion, spreading, and self-renewal of human embryonic stem cells. ACS Nano 6:4094–4103. doi:10.1021/nn3004923

23. Kamei K-I, Hirai Y, Yoshioka M et al (2013) Phenotypic and transcriptional modulation of human pluripotent stem cells induced by nano/microfabrication materials. Adv Healthc Mater 2:287–291. doi:10.1002/adhm.201200283

24. Baghaban Eslaminejad M, Malakooty Poor E (2014) Mesenchymal stem cells as a potent cell source for articular cartilage regeneration. World J Stem Cells 6:344–354. doi:10.4252/wjsc.v6.i3.344

25. Toh WS, Foldager CB, Pei M, Hui JHP (2014) Advances in mesenchymal stem cell-based strategies for cartilage repair and regeneration. Stem Cell Rev 10:686–696. doi:10.1007/s12015-014-9526-z

26. Brown PT, Handorf AM, Jeon WB, Li W-J (2013) Stem cell-based tissue engineering approaches for musculoskeletal regeneration. Curr Pharm Des 19:3429–3445. doi:10.1016/j.biotechadv.2011.08.021.Secreted

27. Fernández Vallone VB, Romaniuk MA, Choi H et al (2013) Mesenchymal stem cells and their use in therapy: what has been achieved? Differentiation 85:1–10. doi:10.1016/j.diff.2012.08.004

28. Lee T, Jang J, Kang S et al (2014) Mesenchymal stem cell-conditioned medium enhances embryonic stem cells and human induced pluripotent stem cells by mesodermal lineage induction. Tissue Eng Part A 20:1306–1313. doi:10.1089/ten.tea.2013.0265

29. Orth P, Rey-Rico A (2014) Current perspectives in stem cell research for knee cartilage repair. Stem Cells 7:1–17

30. Patel DM, Shah J, Srivastava AS (2013) Therapeutic potential of mesenchymal stem cells in regenerative medicine. Stem Cells Int. doi:10.1155/2013/496218

31. Ma S, Xie N, Li W et al (2014) Immunobiology of mesenchymal stem cells. Cell Death Differ 21:216–225. doi:10.1038/cdd.2013.158

32. Wei X, Yang X, Han Z et al (2013) Mesenchymal stem cells: a new trend for cell therapy. Acta Pharmacol Sin 34:747–754. doi:10.1038/aps.2013.50

33. Figueroa FE, Carrión F, Villanueva S, Khoury M (2012) Mesenchymal stem cell treatment for autoimmune diseases: a critical review. Biol Res 45:269–277. doi:10.4067/S0716-97602012000300008

34. Kean TJ, Lin P, Caplan AI, Dennis JE (2013) MSCs: delivery routes and engraftment, cell-targeting strategies, and immune modulation. Stem Cells Int. doi:10.1155/2013/732742

35. Mabuchi Y, Houlihan DD, Akazawa C et al (2013) Prospective isolation of murine and human bone marrow mesenchymal stem cells based on surface markers. Stem Cells Int 2013:507301. doi:10.1155/2013/507301

36. Wang S, Chang Q, Kong X, Wang C (2015) The chondrogenic induction potential for bone marrow-derived stem cells between autologous platelet-rich plasma and common chondrogenic induction agents: a preliminary comparative study. Stem Cells Int 2015:1–7, Article ID 589124. doi: 10.1155/2015/589124

37. Torreggiani E, Lisignoli G, Manferdini C et al (2012) Role of slug transcription factor in human mesenchymal stem cells. J Cell Mol Med 16:740–751. doi:10.1111/j.1582-4934.2011.01352.x

38. Bosetti M, Boccafoschi F, Leigheb M et al (2012) Chondrogenic induction of human mesenchymal stem cells using combined growth factors for cartilage tissue engineering. J Tissue Eng Regen Med 6:205–213. doi:10.1002/term.416

39. Lynch K, Pei M (2014) Age associated communication between cells and matrix: a potential impact on stem cell-based tissue regeneration strategies. Organogenesis. doi:10.4161/15476278.2014.970089

40. Orbay H, Tobita M, Mizuno H (2012) Mesenchymal stem cells isolated from adipose and other tissues: basic biological properties and clinical applications. Stem Cells Int. doi:10.1155/2012/461718

41. Zuk P (2013) Adipose-derived stem cells in tissue regeneration: a review. ISRN Stem Cells 2013:1–35. doi:10.1155/2013/713959

42. Li X, Yuan J, Li W et al (2014) Direct differentiation of homogeneous human adipose stem cells into functional hepatocytes by mimicking liver embryogenesis. J Cell Physiol 229:801–812. doi:10.1002/jcp.24501

43. Sun H, Liu Y, Jiang T et al (2014) Chondrogenic differentiation and three dimensional chondrogenesis of human adipose-derived stem cells induced by engineered cartilage-derived conditional media. Tissue Eng Regen Med 11:59–66. doi:10.1007/s13770-013-1120-y

44. De Sousa E, Casado PL, Neto VM et al (2014) Synovial fluid and synovial membrane mesenchymal stem cells: latest discoveries and therapeutic perspectives. Stem Cell Res Ther 5:1–6
45. Gupta PK, Das AK, Chullikana A, Majumdar AS (2012) Mesenchymal stem cells for cartilage repair in osteoarthritis. Stem Cell Res Ther 3:25. doi:10.1186/scrt116
46. Campbell D, Pei M (2012) Surface markers for chondrogenic determination: a highlight of synovium-derived stem cells. Cells 1:1107–1120. doi:10.3390/cells1041107
47. Nakamura T, Sekiya I, Muneta T et al (2012) Arthroscopic, histological and MRI analyses of cartilage repair after a minimally invasive method of transplantation of allogeneic synovial mesenchymal stromal cells into cartilage defects in pigs. Cytotherapy 14:327–338. doi:10.3109/14653249.2011.638912
48. Lee JC, Min HJ, Park HJ et al (2013) Synovial membrane-derived mesenchymal stem cells supported by platelet-rich plasma can repair osteochondral defects in a rabbit model. Arthrosc J Arthrosc Relat Surg 29:1034–1046. doi:10.1016/j.arthro.2013.02.026
49. Lee J-C, Lee SY, Min HJ et al (2012) Synovium-derived mesenchymal stem cells encapsulated in a novel injectable gel can repair osteochondral defects in a rabbit model. Tissue Eng Part A 18:2173–2186. doi:10.1089/ten.tea.2011.0643
50. Gong SP, Kim B, Kwon HS et al (2014) The co-injection of somatic cells with embryonic stem cells affects teratoma formation and the properties of teratoma-derived stem cell-like cells. PLoS ONE 9:1–9. doi:10.1371/journal.pone.0105975
51. Lee M-O, Moon SH, Jeong H-C et al (2013) Inhibition of pluripotent stem cell-derived teratoma formation by small molecules. Proc Natl Acad Sci USA 110:E3281–E3290. doi:10.1073/pnas.1303669110
52. Cheng A, Kapacee Z, Peng J et al (2014) Cartilage repair using human embryonic stem cell-derived chondroprogenitors. Stem Cells Trans Med Publ 3:1–8. doi:10.5966/sctm.2014-0101
53. Craft AM, Ahmed N, Rockel JS et al (2013) Specification of chondrocytes and cartilage tissues from embryonic stem cells. Development 140:2597–2610. doi:10.1242/dev.087890
54. Tsumaki N, Okada M, Yamashita A (2014) IPS cell technologies and cartilage regeneration. Bone. doi:10.1016/j.bone.2014.07.011
55. Stromps J-P, Paul NE, Rath B et al (2014) Chondrogenic differentiation of human adipose-derived stem cells: a new path in articular cartilage defect management? Biomed Res Int 2014:740926. doi:10.1155/2014/740926
56. Fisher MC (2012) The potential of human embryonic stem cells for articular cartilage repair and osteoarthritis treatment. Rheumatol Curr Res. doi:10.4172/2161-1149.S3-004
57. Diekman BO, Christoforou N, Willard VP et al (2012) Cartilage tissue engineering using differentiated and purified induced pluripotent stem cells. Proc Natl Acad Sci. doi:10.1073/pnas.1210422109
58. Irion VH, Flanigan DC (2013) New and emerging techniques in cartilage repair: other scaffold-based cartilage treatment options. Oper Tech Sports Med 21:125–137. doi:10.1053/j.otsm.2013.03.001
59. Liu M, Yu X, Huang F et al (2013) Tissue engineering stratified scaffolds for articular cartilage and subchondral bone defects repair. Orthopedics 36:868–873. doi:10.3928/01477447-20131021-10
60. Salgado AJ, Oliveira JM, Martins A et al (2013) Tissue engineering and regenerative medicine: past, present, and future. Int Rev Neurobiol. doi:10.1016/B978-0-12-410499-0.00001-0
61. Cao Z, Dou C, Dong S (2014) Scaffolding biomaterials for cartilage regeneration. J Nanomater 2014:1–8. doi:10.1155/2014/489128
62. Rodrigues MT, Lee SJ, Gomes ME et al (2012) Bilayered constructs aimed at osteochondral strategies: the influence of medium supplements in the osteogenic and chondrogenic differentiation of amniotic fluid-derived stem cells. Acta Biomater 8:2795–2806. doi:10.1016/j.actbio.2012.04.013

63. Izadifar Z, Chen X, Kulyk W (2012) Strategic design and fabrication of engineered scaffolds for articular cartilage repair. J Funct Biomater 3:799–838. doi:10.3390/jfb3040799

64. Demoor M, Ollitrault D, Gomez-Leduc T et al (2014) Cartilage tissue engineering: molecular control of chondrocyte differentiation for proper cartilage matrix reconstruction. Biochim Biophys Acta 1840:2414–2440. doi:10.1016/j.bbagen.2014.02.030

65. Zhang Z, Gupte M, Ma P (2013) Biomaterials and stem cells for tissue engineering. Expert Opin Biol 13:527–540. doi:10.1517/14712598.2013.756468.Biomaterials

66. Griffin M, Butler P, Seifalian A, Szarko M (2013) Update into articular cartilage tissue engineering. OapublishinglondonCom 1:1–6

67. Musumeci G, Castrogiovanni P, Leonardi R et al (2014) New perspectives for articular cartilage repair treatment through tissue engineering: a contemporary review. World J Orthop 5:80–88. doi:10.5312/wjo.v5.i2.80

68. Sun J, Tan H (2013) Alginate-based biomaterials for regenerative medicine applications. Materials (Basel) 6:1285–1309. doi:10.3390/ma6041285

69. Goldman SM, Barabino GA (2014) Cultivation of agarose-based microfluidic hydrogel promotes the development of large, full-thickness, tissue-engineered articular cartilage constructs. J Tissue Eng Regen Med. doi:10.1002/term.1954

70. Martins EAN, Michelacci YM, Baccarin RYA et al (2014) Evaluation of chitosan-GP hydrogel biocompatibility in osteochondral defects: an experimental approach. BMC Vet Res 10:1

71. Whu SW, Hung K-C, Hsieh K-H et al (2013) In vitro and in vivo evaluation of chitosan–gelatin scaffolds for cartilage tissue engineering. Mater Sci Eng C Mater Biol Appl 33:2855–2863. doi:10.1016/j.msec.2013.03.003

72. Unterman SA, Gibson M, Lee JH et al (2012) Hyaluronic acid-binding scaffold for articular cartilage repair. Tissue Eng Part A 18:120814114305007. doi:10.1089/ten.tea.2011.0711

73. Levett PA, Hutmacher DW, Malda J, Klein TJ (2014) Hyaluronic acid enhances the mechanical properties of tissue-engineered cartilage constructs. PLoS ONE 9:e113216. doi:10.1371/journal.pone.0113216

74. Ahearne M, Kelly DJ (2013) A comparison of fibrin, agarose and gellan gum hydrogels as carriers of stem cells and growth factor delivery microspheres for cartilage regeneration. Biomed Mater 8:035004. doi:10.1088/1748-6041/8/3/035004

75. Chung JY, Song M, Ha C-W et al (2014) Comparison of articular cartilage repair with different hydrogel-human umbilical cord blood-derived mesenchymal stem cell composites in a rat model. Stem Cell Res Ther 5:39. doi:10.1186/scrt427

76. Mastbergen SC, Saris DB, Lafeber FP (2013) Functional articular cartilage repair: here, near, or is the best approach not yet clear? Nat Rev Rheumatol 9:277–290. doi:10.1038/nrrheum.2013.29

77. Freymann U, Petersen W, Kaps C (2013) Cartilage regeneration revisited: entering of new one-step procedures for chondral cartilage repair. OapublishinglondonCom 1:1–6

78. Yodmuang S, Mcnamara SL, Nover AB et al (2014) Silk microfiber-reinforced silk hydrogel composites for functional cartilage tissue repair. Acta Biomater. doi:10.1016/j.actbio.2014.09.032

79. Snyder TN, Madhavan K, Intrator M et al (2014) A fibrin/hyaluronic acid hydrogel for the delivery of mesenchymal stem cells and potential for articular cartilage repair. J Biol Eng 8:10. doi:10.1186/1754-1611-8-10

80. Pereira DR, Canadas RF, Silva-Correia J et al (2013) Gellan gum-based hydrogel bilayered scaffolds for osteochondral tissue engineering. Key Eng Mater 587:255–260. doi:10.4028/www.scientific.net/KEM.587.255

81. Popa EG, Reis RL, Gomes ME (2014) Seaweed polysaccharide-based hydrogels used for the regeneration of articular cartilage. Crit Rev Biotechnol 8551:1–14. doi:10.3109/07388551.2014.889079

82. Sharma B, Fermanian S, Gibson M et al (2013) Human cartilage repair with a photoreactive adhesive-hydrogel composite. Sci Transl Med 5:167ra6. doi:10.1126/scitranslmed.3004838

83. Norton AB, Hancocks RD, Grover LM (2014) Poly (vinyl alcohol) modification of low acyl gellan hydrogels for applications in tissue regeneration. Food Hydrocoll 42:373–377. doi:10.1016/j.foodhyd.2014.05.001

84. Balakrishnan B, Joshi N, Banerjee R (2013) Borate aided Schiff's base formation yields in situ gelling hydrogels for cartilage regeneration. J Mater Chem B 1:5564. doi:10.1039/c3tb21056a

85. Chen J-L, Duan L, Zhu W et al (2014) Extracellular matrix production in vitro in cartilage tissue engineering. J Transl Med 12:88. doi:10.1186/1479-5876-12-88

86. Nooeaid P, Salih V, Beier JP, Boccaccini AR (2012) Osteochondral tissue engineering: scaffolds, stem cells and applications. J Cell Mol Med 16:2247–2270. doi:10.1111/j.1582-4934.2012.01571.x

87. Alves da Silva ML, Costa-Pinto AR, Martins A et al (2013) Conditioned medium as a strategy for human stem cells chondrogenic differentiation. J Tissue Eng Regen Med 4:524–531. doi:10.1002/term.1812

88. Fernandes-Silva S, Moreira-Silva J, Silva TH et al (2013) Porous hydrogels from shark skin collagen crosslinked under dense carbon dioxide atmosphere. Macromol Biosci 13:1621–1631. doi:10.1002/mabi.201300228

89. Chen C-H, Shyu VB-H, Chen J-P, Lee M-Y (2014) Selective laser sintered poly-ε-caprolactone scaffold hybridized with collagen hydrogel for cartilage tissue engineering. Biofabrication 6:015004. doi:10.1088/1758-5082/6/1/015004

90. Yan LP, Oliveira JM, Oliveira AL et al (2012) Macro/microporous silk fibroin scaffolds with potential for articular cartilage and meniscus tissue engineering applications. Acta Biomater 8:289–301. doi:10.1016/j.actbio.2011.09.037

91. Yan L-P, Silva-Correia J, Oliveira MB et al (2015) Bilayered silk/silk-nanoCaP scaffolds for osteochondral tissue engineering: in vitro and in vivo assessment of biological performance. Acta Biomater. doi:10.1016/j.actbio.2014.10.021

92. Yan L, Oliveira JM, Oliveira AL, Reis RL (2014) In vitro evaluation of the biological performance of macro/micro-porous silk fibroin and silk-nano calcium phosphate scaffolds. J Biomed Mater Res Part B. doi:10.1002/jbm.b.33267

93. Ferris CJ, Stevens LR, Gilmore KJ et al (2014) Peptide modification of purified gellan gum. J Mater Chem B. doi:10.1039/c4tb01727g

94. Ferris CJ, Gilmore KJ, Wallace GG, Panhuis M et al (2013) Modified gellan gum hydrogels for tissue engineering applications. Soft Matter 9:3705. doi:10.1039/c3sm27389j

95. Beachley V, Hepfer RG, Katsanevakis E et al (2014) Precisely assembled nanofiber arrays as a platform to engineer aligned cell sheets for biofabrication. Bioengineering 1:114–133. doi:10.3390/bioengineering1030114

96. Bourget J, Guillemette M, Veres T et al (2013) Alignment of cells and extracellular matrix within tissue-engineered substitutes. Adv Biomater Sci Biomed Appl Ref. doi:10.5772/54142

97. Mashhadikhan M, Soleimani M, Parivar K, Yaghmaei P (2015) ADSCs on PLLA/PCL hybrid nanoscaffold and gelatin modification: cytocompatibility and mechanical properties. Avicenna J Med Biotechnol 7:32–38

98. Venugopal JR, Prabhakaran MP, Mukherjee S et al (2012) Biomaterial strategies for alleviation of myocardial infarction. J R Soc Interface 9:1–19. doi:10.1098/rsif.2011.0301

99. Markeson D, Pleat JM, Sharpe JR et al (2013) Scarring, stem cells, scaffolds and skin repair. J Tissue Eng Regen Med 4:524–531. doi:10.1002/term.1841

100. Zeng W, Rong M, Hu X et al (2014) Incorporation of chitosan microspheres into collagen–chitosan scaffolds for the controlled release of nerve growth factor. PLoS ONE. doi:10.1371/journal.pone.0101300

101. Thomopoulos S, Sakiyama-Elbert S, Silva M et al (2014) Polymer nanofiber scaffold for a heparin/fibrin based growth factor delivery system

102. Blackwood KA, Bock N, Dargaville TR, Ann Woodruff M (2012) Scaffolds for growth factor delivery as applied to bone tissue engineering. Int J Polym Sci. doi:10.1155/2012/174942

103. García Cruz DM, Sardinha V, Escobar Ivirico JL et al (2013) Gelatin microparticles aggregates as three-dimensional scaffolding system in cartilage engineering. J Mater Sci Mater Med 24:503–513. doi:10.1007/s10856-012-4818-9

104. Zhang W, Zhu C, Ye D et al (2014) Porous silk scaffolds for delivery of growth factors and stem cells to enhance bone regeneration. PLoS ONE 9:1–9. doi:10.1371/journal.pone. 0102371

105. Almeida HV, Liu Y, Cunniffe GM et al (2014) Controlled release of transforming growth factor-β3 from cartilage-extra-cellular-matrix-derived scaffolds to promote chondrogenesis of human-joint-tissue-derived stem cells. Acta Biomater. doi:10.1016/j.actbio.2014.05.030

106. Santo VE, Gomes M, Mano J, Reis RL (2012) Controlled release strategies for bone, cartilage and osteochondral engineering—part II: challenges on the evolution from single to multiple bioactive factor delivery. Tissue Eng Part B Rev 19:327–352. doi:10.1089/ten.TEB. 2012.0727

107. Jonitz A, Lochner K, Tischer T et al (2012) TGF-b1 and IGF-1 influence the re-differentiation capacity of human chondrocytes in 3D pellet cultures in relation to different oxygen concentrations. Int J Mol Med 30:666–672. doi:10.3892/ijmm.2012.1042

108. Loffredo FS, Pancoast JR, Cai L et al (2014) Targeted delivery to cartilage is critical for in vivo efficacy of insulin-like growth factor 1 in a rat model of osteoarthritis. Arthritis Rheumatol (Hoboken, NJ) 66:1247–1255. doi:10.1002/art.38357

109. Reyes R, Delgado A, Solis R et al (2013) Cartilage repair by local delivery of transforming growth factor-β1 or bone morphogenetic protein-2 from a novel, segmented polyurethane/polylactic-co-glycolic bilayered scaffold. J Biomed Mater Res A. 102:1–11. doi:10.1002/jbma.34769

110. Lu C-H, Yeh T-S, Fang Y-HD et al (2014) Regenerating cartilages by engineered ASCs: Prolonged TGF-(beta)3/BMP-6 expression improved articular cartilage formation and restored zonal structure. Mol Ther 22:186–195. doi:10.1038/mt.2013.165

111. Li X, Su G, Wang J et al (2013) Exogenous bFGF promotes articular cartilage repair via up-regulation of multiple growth factors. Osteoarthr Cartil 21:1567–1575. doi:10.1016/j.joca. 2013.06.006

112. Liao J, Hu N, Zhou N et al (2014) Sox9 potentiates BMP2-induced chondrogenic differentiation and inhibits BMP2-induced osteogenic differentiation. PLoS ONE 9:e89025. doi:10.1371/journal.pone.0089025

113. Zhang Y, Kumagai K, Saito T (2014) Effect of parathyroid hormone on early chondrogenic differentiation from mesenchymal stem cells. J Orthop Surg Res 9:1–7. doi:10.1186/s13018-014-0068-5

114. Zhang W, Chen J, Zhang S, Ouyang HW (2012) Inhibitory function of parathyroid hormone-related protein on chondrocyte hypertrophy: the implication for articular cartilage repair. Arthritis Res Ther 14:221. doi:10.1186/ar4025

115. Wu XC, Huang B, Wang J et al (2013) Collagen-targeting parathyroid hormone-related peptide promotes collagen binding and in vitro chondrogenesis in bone marrow-derived MSCs. Int J Mol Med 31:430–436. doi:10.3892/ijmm.2012.1219

116. Murphy MK, Huey DJ, Hu JC, Athanasiou KA (2014) TGF-β1, GDF-5, and BMP-2 stimulation induces chondrogenesis in expanded human articular chondrocytes and marrow-derived stromal cells. Stem Cells 00:00. doi:10.1002/stem.1890

117. Mariani E, Pulsatelli L, Facchini A (2014) Signaling pathways in cartilage repair. Int J Mol Sci 15:8667–8698. doi:10.3390/ijms15058667

118. Gurusinghe S, Strappe P (2014) Gene modification of mesenchymal stem cells and articular chondrocytes to enhance chondrogenesis. Biomed Res Int. doi:10.1155/2014/369528

119. Croutze R, Jomha N, Uludag H, Adesida A (2013) Matrix forming characteristics of inner and outer human meniscus cells on 3D collagen scaffolds under normal and low oxygen tensions. BMC Musculoskelet Disord 14:353. doi:10.1186/1471-2474-14-353

120. McNary S, Athanasiou K, Reddi AH (2013) Transforming growth factor beta-induced superficial zone protein accumulation in the surface zone of articular cartilage is dependent on the cytoskeleton. Tissue Eng Part A. doi:10.1089/ten.TEA.2013.0043

121. Montaseri A, Busch F, Mobasheri A et al (2011) IGF-1 and PDGF-bb suppress IL-1β-induced cartilage degradation through down-regulation of NF-κB signaling: Involvement of Src/PI-3k/AKT pathway. PLoS ONE. doi:10.1371/journal.pone.0028663

122. Lee JM, Kim B-S, Lee H, Im G-I (2012) In vivo tracking of mesechymal stem cells using fluorescent nanoparticles in an osteochondral repair model. Mol Ther 20:1434–1442. doi:10.1038/mt.2012.60

123. Montoya F, Martínez F, García-Robles M et al (2013) Clinical and experimental approaches to knee cartilage lesion repair and mesenchymal stem cell chondrocyte differentiation. Biol Res 46:441–451. doi:10.4067/S0716-97602013000400015

124. Salzmann GM, Sah B, Südkamp NP, Niemeyer P (2013) Clinical outcome following the first-line, single lesion microfracture at the knee joint. Arch Orthop Trauma Surg 133:303–310. doi:10.1007/s00402-012-1660-y

125. Xu X, Shi D, Shen Y et al (2015) Full-thickness cartilage defects are repaired via a microfracture technique and intra-articular injection of the small molecule compound kartogenin. Arthritis Res Ther. doi:10.1186/s13075-015-0537-1

126. Goyal D, Keyhani S, Lee EH, Hui JHP (2013) Evidence-based status of microfracture technique: a systematic review of Level I and II studies. Arthrosc J Arthrosc Relat Surg 29:1579–1588. doi:10.1016/j.arthro.2013.05.027

127. Mobasheri A, Kalamegam G, Musumeci G, Batt ME (2014) Chondrocyte and mesenchymal stem cell-based therapies for cartilage repair in osteoarthritis and related orthopaedic conditions. Maturitas 78:188–198. doi:10.1016/j.maturitas.2014.04.017

128. Pestka JM, Bode G, Salzmann G et al (2013) Clinical outcomes after cell-seeded autologous chondrocyte implantation of the knee: when can success or failure be predicted? Am J Sports Med 42:208–215. doi:10.1177/0363546513507768

129. Niemeyer P, Porichis S, Steinwachs M et al (2013) Long-term outcomes after first-generation autologous chondrocyte implantation for cartilage defects of the knee. Am J Sports Med. doi:10.1177/0363546513506593

130. Minas T, Von Keudell A, Bryant T, Gomoll AH (2014) The John Insall Award: a minimum 10-year outcome study of autologous chondrocyte implantation knee. Clin Orthop Relat Res 472:41–51. doi:10.1007/s11999-013-3146-9

131. Trinh TQ, Harris JD, Siston RA, Flanigan DC (2013) Improved outcomes with combined autologous chondrocyte implantation and patellofemoral osteotomy versus isolated autologous chondrocyte implantation. Arthrosc J Arthrosc Relat Surg 29:566–574. doi:10.1016/j.arthro.2012.10.008

132. Filardo G, Kon E, Di MA et al (2012) Second-generation arthroscopic autologous chondrocyte implantation for the treatment of degenerative cartilage lesions. Knee Surg Sport Traumatol Arthrosc 20:1704–1713. doi:10.1007/s00167-011-1732-5

133. Kreuz PC, Müller S, von Keudell A et al (2013) Influence of sex on the outcome of autologous chondrocyte implantation in chondral defects of the knee. Am J Sports Med 41:1541–1548. doi:10.1177/0363546513489262

134. Caron MMJ, Emans PJ, Coolsen MME et al (2012) Redifferentiation of dedifferentiated human articular chondrocytes: comparison of 2D and 3D cultures. Osteoarthr Cartil 20:1170–1178. doi:10.1016/j.joca.2012.06.016

135. Ebert JR, Smith A, Edwards PK et al (2013) Factors predictive of outcome 5 years after matrix-induced autologous chondrocyte implantation in the tibiofemoral joint. Am J Sports Med 41:1245–1254. doi:10.1177/0363546513484696

136. Ebert JR, Fallon M, Zheng MH et al (2012) A randomized trial comparing accelerated and traditional approaches to postoperative weightbearing rehabilitation after matrix-induced autologous chondrocyte implantation: findings at 5 years. Am J Sports Med 40:1527–1537. doi:10.1177/0363546512445167

137. Ebert JR, Smith A, Fallon M et al (2014) Correlation between clinical and radiological outcomes after matrix-induced autologous chondrocyte implantation in the femoral condyles. Am J Sports Med 42:1857–1864. doi:10.1177/0363546514534942

138. Marlovits S, Aldrian S, Wondrasch B et al (2012) Clinical and radiological outcomes 5 years after matrix-induced autologous chondrocyte implantation in patients with symptomatic, traumatic chondral defects. Am J Sports Med 40:2273–2280. doi:10.1177/0363546512457008

139. Edwards PK, Ackland TR, Ebert JR (2013) Accelerated weightbearing rehabilitation after matrix-induced autologous chondrocyte implantation in the tibiofemoral joint: early clinical and radiological outcomes. Am J Sports Med 41:2314–2324. doi:10.1177/0363546513495637

140. Saris D, Price A, Widuchowski W et al (2014) Matrix-applied characterized autologous cultured chondrocytes versus microfracture: two-year follow-up of a prospective randomized trial. Am J Sports Med 42:1384–1394. doi:10.1177/0363546514528093

141. Boeriu CG, Springer J, Kooy FK et al (2013) Production methods for hyaluronan. Int J Carbohydr Chem 2013:1–14. doi:10.1155/2013/624967

142. Saw K-Y, Anz A, Siew-Yoke Jee C et al (2013) Articular cartilage regeneration with autologous peripheral blood stem cells versus hyaluronic acid: a randomized controlled trial. Arthroscopy 29:684–694. doi:10.1016/j.arthro.2012.12.008

143. Jang JD, Moon YS, Kim YS et al (2013) Novel repair technique for articular cartilage defect using a fibrin and hyaluronic acid mixture. Tissue Eng Regen Med 10:1–9. doi:10.1007/s13770-013-0361-0

144. Frith JE, Menzies DJ, Cameron AR et al (2014) Effects of bound versus soluble pentosan polysulphate in PEG/HA-based hydrogels tailored for intervertebral disc regeneration. Biomaterials 35:1150–1162. doi:10.1016/j.biomaterials.2013.10.056

145. Meng F, He A, Zhang Z et al (2014) Chondrogenic differentiation of ATDC5 and hMSCs could be induced by a novel scaffold-tricalcium phosphate–collagen–hyaluronan without any exogenous growth factors in vitro. J Biomed Mater Res Part A 102:2725–2735. doi:10.1002/jbm.a.34948

146. Silva-Correia J, Correia SI, Oliveira JM, Reis RL (2013) Tissue engineering strategies applied in the regeneration of the human intervertebral disk. Biotechnol Adv 31:1514–1531. doi:10.1016/j.biotechadv.2013.07.010

147. Tsaryk R, Silva-Correia J, Oliveira JM et al (2014) Biological performance of cell-encapsulated methacrylated gellan gum-based hydrogels for nucleus pulposus regeneration. J Tissue Eng Regen Med. doi:10.1002/term.1959

148. Emans PJ, Peterson L (2014) Developing insights in cartilage repair. © Springer, London. doi:10.1007/978-1-4471-5385-6

149. Jeon JE, Schrobback K, Hutmacher DW, Klein TJ (2012) Dynamic compression improves biosynthesis of human zonal chondrocytes from osteoarthritis patients. Osteoarthr Cartil 20:906–915. doi:10.1016/j.joca.2012.04.019

150. Tatsumura M, Sakane M, Ochiai N, Mizuno S (2014) Off-loading of cyclic hydrostatic pressure promotes production of extracellular matrix by chondrocytes. Cells Tissues Organs. doi:10.1159/000360156

151. Sadlik B, Wiewiorski M (2014) Implantation of a collagen matrix for an AMIC repair during dry arthroscopy. Knee Surg Sport Traumatol Arthrosc. doi:10.1007/s00167-014-3062-x

152. Dhollander A, Moens K, van der Mass J et al (2014) Treatment of patellofemoral cartilage defects in the knee by autologous matrix-induced chondrogenesis (AMIC). Acta Orthop Belg 80:251–259

153. Piontek T, Ciemniewska-Gorzela K, Szulc A et al (2012) All-arthroscopic AMIC procedure for repair of cartilage defects of the knee. Knee Surg Sport Traumatol Arthrosc 20:922–925. doi:10.1007/s00167-011-1657-z

154. Athanasiou KA, Eswaramoorthy R, Hadidi P, Hu JC (2013) Self-organization and the self-assembling process in tissue engineering. Annu Rev Biomed Eng 15:115–136. doi:10.1146/annurev-bioeng-071812-152423

155. Huey DJ, Hu JC, Athanasiou KA (2012) Unlike bone, cartilage regeneration remains elusive. Science (80-) 338:917–921. doi:10.1126/science.1222454

156. Athanasiou KA, Responte DJ, Brown WE, Hu JC (2015) Harnessing biomechanics to develop cartilage regeneration strategies. J Biomech Eng 137:020901. doi:10.1115/1.4028825

157. Makris EA, MacBarb RF, Paschos NK et al (2014) Combined use of chondroitinase-ABC, TGF-β1, and collagen crosslinking agent lysyl oxidase to engineer functional neotissues for fibrocartilage repair. Biomaterials 35:6787–6796. doi:10.1016/j.biomaterials.2014.04.083

158. Makris EA, Hu JC, Athanasiou KA (2013) Hypoxia-induced collagen crosslinking as a mechanism for enhancing mechanical properties of engineered articular cartilage. Osteoarthr Cartil 21:634–641. doi:10.1016/j.joca.2013.01.007

159. Makris EA, MacBarb RF, Responte DJ et al (2013) A copper sulfate and hydroxylysine treatment regimen for enhancing collagen cross-linking and biomechanical properties in engineered neocartilage. FASEB J 27:2421–2430. doi:10.1096/fj.12-224030

160. Murphy S, Atala A (2014) 3D bioprinting of tissues and organs. Nat Biotechnol 32(8):494–539

161. Ozbolat IT, Yu Y (2013) Bioprinting toward organ fabrication: challenges and future trends. IEEE Trans Biomed Eng 60:691–699. doi:10.1109/TBME.2013.2243912

162. Conese M (2014) Bioprinting: a further step to effective regenerative medicine and tissue engineering. Adv Genet Eng 2:2–5. doi:10.4172/2169-0111.1000e112

163. Gao G, Yonezawa T, Hubbell K, Dai GCX (2015) Inkjet-bioprinted acrylated peptides and PEG hydrogel with human mesenchymal stem cells promote robust bone and cartilage formation with minimal printhead clogging. Biotechnol J. doi:10.1002/biot.201400635. Submitted

164. Xu T, Binder KW, Albanna MZ et al (2013) Hybrid printing of mechanically and biologically improved constructs for cartilage tissue engineering applications. Biofabrication 5:015001. doi:10.1088/1758-5082/5/1/015001

165. Cui X, Boland T, D'Lima D, Lotz M (2012) Thermal inkjet printing in tissue engineering and regenerative medicine. Recent Patents Drug 6:149–155

166. Cui X, Gao G, Yonezawa T, Dai G (2014) Human cartilage tissue fabrication using three-dimensional inkjet printing technology. J Vis Exp. doi:10.3791/51294

167. Santhagunam A, Madeira C, Cabral JMS (2012) Genetically engineered stem cell-based strategies for articular cartilage regeneration. Biotechnol Appl Biochem 59:121–131. doi:10.1002/bab.1016

168. Lorda-Diez CI, Montero JA, Garcia-Porrero JA, Hurle JM (2014) Divergent differentiation of skeletal progenitors into cartilage and tendon: lessons from the embryonic limb. ACS Chem Biol 9:72–79. doi:10.1021/cb400713

169. Lu C-H, Lin K-J, Chiu H-Y et al (2012) Improved chondrogenesis and engineered cartilage formation from TGF-β3-expressing adipose-derived stem cells cultured in the rotating-shaft bioreactor. Tissue Eng Part A 18:2114–2124. doi:10.1089/ten.tea.2012.0010

170. Madry H, Kaul G, Zurakowski D et al (2013) Cartilage constructs engineered from chondrocytes overexpressing IGF-I improve the repair of osteochondral defects in a rabbit model. Eur Cells Mater 25:229–247

171. Hu Y-C (2014) Gene Therapy for Cartilage and Bone Tissue Engineering. Gene Ther Cartil Bone Tissue Eng 2:1–15. doi:10.1007/978-3-642-53923-7

172. Madeira C, Santhagunam A, Salgueiro JB, Cabral JMS (2015) Advanced cell therapies for articular cartilage regeneration. Trends Biotechnol 33:35–42. doi:10.1016/j.tibtech.2014.11.003

173. Brunger JM, Huynh NPT, Guenther CM et al (2014) Scaffold-mediated lentiviral transduction for functional tissue engineering of cartilage. Proc Natl Acad Sci USA 111:E798–E806. doi:10.1073/pnas.1321744111

174. Cucchiarini M, Madry H (2014) Overexpression of human IGF-I via direct rAAV-mediated gene transfer improves the early repair of articular cartilage defects in vivo. Gene Ther 21:1–9. doi:10.1038/gt.2014.58

175. Cucchiarini M, Orth P, Madry H (2013) Direct rAAV SOX9 administration for durable articular cartilage repair with delayed terminal differentiation and hypertrophy in vivo. J Mol Med 91:625–636. doi:10.1007/s00109-012-0978-9

176. Goodrich LR, Phillips JN, McIlwraith CW et al (2013) Optimization of scAAVIL-1ra in vitro and in vivo to deliver high levels of therapeutic protein for treatment of osteoarthritis. Mol Ther Nucleic Acids 2:e70. doi:10.1038/mtna.2012.61

177. Ha C-W, Noh MJ, Choi KB, Lee KH (2012) Initial phase I safety of retrovirally transduced human chondrocytes expressing transforming growth factor-beta-1 in degenerative arthritis patients. Cytotherapy 14:247–256. doi:10.3109/14653249.2011.629645

178. Li X, Ellman MB, Kroin JS et al (2012) Species-specific biological effects of FGF-2 in articular cartilage: implication for distinct roles within the FGF receptor family. J Cell Biochem 113:2532–2542. doi:10.1002/jcb.24129

179. Watson R, Broome T, Levings P et al (2013) scAAV-mediated gene transfer of Interleukin 1-receptor antagonist to synovium and articular cartilage in large mammalian joints. Gene Ther 20:670–677. doi:10.1038/gt.2012.81.scAAV-Mediated

180. Gascón AR, del Pozo-Rodríguez A, Solinís MÁ (2014) Non-viral delivery systems in gene therapy. Gene Ther Tools Potential Appl. doi:10.5772/52704

181. He CX, Zhang TY, Miao PH et al (2012) TGF-β1 gene-engineered mesenchymal stem cells induce rat cartilage regeneration using nonviral gene vector. Biotechnol Appl Biochem 59:163–169. doi:10.1002/bab.1001

182. Oliveira PH, Mairhofer J (2013) Marker-free plasmids for biotechnological applications—implications and perspectives. Trends Biotechnol 31:539–547. doi:10.1016/j.tibtech.2013.06.001

183. Li P, Wei X, Guan Y et al (2014) MicroRNA-1 regulates chondrocyte phenotype by repressing histone deacetylase 4 during growth plate development. FASEB J 28:3930–3941. doi:10.1096/fj.13-249318

184. Qi B, Yu A, Zhu S et al (2013) Chitosan/poly(vinyl alcohol) hydrogel combined with Ad-hTGF-β1 transfected mesenchymal stem cells to repair rabbit articular cartilage defects. Exp Biol Med (Maywood) 238:23–30. doi:10.1258/ebm.2012.012223

Chapter 6
Advances in Biomaterials for the Treatment of Articular Cartilage Defects

Cristiana Gonçalves, Hajer Radhouani, Joaquim Miguel Oliveira and Rui Luís Reis

Abstract The management of cartilage defects is one of the most challenging problems for public and medical communities. The complete repairing of the damaged cartilage is a complex procedure, since articular cartilage is characterized by a poor vascularization (absence of blood vessels and nerve source), which limits the capacity to repair itself. Cartilage tissue engineering and regenerative medicine are relatively novel areas of research and may hold the key to the successful treatment of cartilage diseases and disorders. Materials such as natural and synthetic biomaterials have been explored to recreate the microarchitecture of articular cartilage through multilayered biomimetic scaffolds. In this chapter, an overview is given of the natural and synthetic biomaterials used on cartilage repair, describing the procedures to obtain these biomaterials, their chemical structure, their modifications to enhance their properties, and also their medical applications.

6.1 Introduction

Extracellular matrix of articular cartilage (ECM) is predominantly composed of the network type II collagen (COL II) and an interlocking mesh of fibrous proteins and proteoglycans (PGs), hyaluronic acid (HA), and chondroitin sulphate (CS) [1]. Complex interactions between individual macromolecules synergistically provide

Cristiana Gonçalves and Hajer Radhouani have contributed equally to this chapter.

C. Gonçalves (✉) · H. Radhouani · J.M. Oliveira · R.L. Reis
3B's Research Group—Biomaterials, Biodegradables and Biomimetics,
Headquarters of the European Institute of Excellence on Tissue Engineering
and Regenerative Medicine, University of Minho, Avepark, Parque da Ciência e
Tecnologia, Zona Industrial da Gandra, 4805-017 Barco, Guimarães, Portugal
e-mail: cristiana.goncalves@dep.uminho.pt

C. Gonçalves · H. Radhouani · J.M. Oliveira · R.L. Reis
ICVS/3B's—PT Government Associate Laboratory, Braga, Guimarães, Portugal

© Springer International Publishing AG 2017
J.M. Oliveira and R.L. Reis (eds.), *Regenerative Strategies for the Treatment of Knee Joint Disabilities*, Studies in Mechanobiology, Tissue Engineering and Biomaterials 21, DOI 10.1007/978-3-319-44785-8_6

an advanced lubrication mechanism to provide low friction and wear protection at both low and high loads and sliding velocities that must last over a lifetime [2].

Studies on tissue engineering have revealed promising approaches to regenerate damaged articular cartilage and to contribute on normal body functions recover [3].

When biomaterials are considered for cartilage tissue repair, their development, as medical devices or pharmaceutical products, is made considering the knowledge of the anatomical and structural complexity of articular cartilage [4]. The increasing knowledge in the biomaterial field, combined with the progress in new technology, such as cellular, molecular biology and biochemistry, provide a unique and unprecedented toolbox to create biomaterial with interesting and specific properties [4].

Besides being biocompatible, the ideal biomaterial for cartilage regeneration should be bioactive, biomimetic, biodegradable, bio-responsive, highly porous, suitable for cell attachment, proliferation and differentiation, osteoconductive, non-cytotoxic, flexible and elastic, and non-antigenic [5]. Furthermore, recent efforts were also made to find a minimally invasive procedure, using these novel biomaterials with such characteristics and properties [6].

To date, an extensive range of natural and synthetic biomaterials have been explored as therapy for cartilage repair. Concerning natural polymers, that are found in nature and can be isolated, have been investigated as bioactive scaffolds for cartilage engineering such as alginate, agarose, fibrin, HA, collagen, gelatin, chitosan, chondroitin sulphate, and cellulose [7]. It is important to highlight that usually the natural biomaterials acquire most of the above characteristics described since they are found in biological systems and are well adapted in their environments [8]. The main advantage of these natural polymers is their interaction capability with cells and cellular enzymes, but also to be well adjusted and/or degraded when space for growing tissue is required [9]. When comparing the natural biomaterials with synthetic ones, such as poly(ethylene glycol) (PEG), poly (lactic-*co*-glycolic acid) (PLGA), poly-L-lactic acid (PLLA), polycaprolactone (PCL), poly(*N*-isopropylacrylamide) (poly NiPAAm), and polycarbonate they are not so bioactive (Fig. 6.1).

Although, these synthetic materials provide more control over chemical, physical and mechanical properties making them more interesting and potential biomaterials. Moreover, innovations and advances in several sciences have been improving their biocompatibility [10]. In fact, these studies encompass elements of medicine, biology, chemistry, tissue engineering, and materials science [11].

Interestingly, it has been shown that any natural or man-made material can be a biomaterial, as long as it serves the stated medical and surgical purposes. Defined as materials working under biological constraint, they are present in numerous therapeutic strategies. It is been shown that this field represents 2–3 % of the overall health expenses in developed countries. World expenses could be estimated at 100 billion dollars. The availability of these biomaterials in considerable quantities for

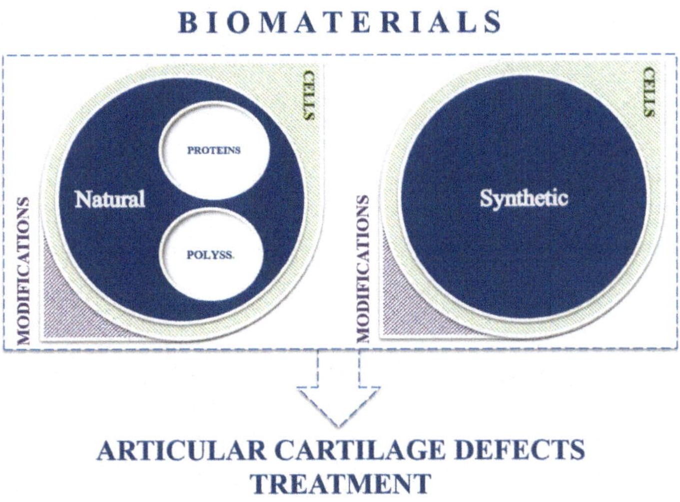

Fig. 6.1 Biomaterials application on articular cartilage treatment

the industry, their extraction, isolation, and purification methodologies to get standardized products are several issues to consider in their applications [11].

In the following section, it is mate an outline of current trends in the development of biomaterials for cartilage repair.

6.2 Biomaterials Used for Cartilage Tissue Engineering

Articular cartilage tissue shows unique combinations of nonlinear tensile and compressive properties due to hierarchically arranged collagen fibrils, proteoglycans, and proteins [12, 13]. The structural and mechanical properties of such natural tissue can be imitated by engineering bio-nanocomposites made from polymer and nanoparticles [13].

The natural biomaterials development is not a recent scientific area. It has existed as a dynamic area with numerous progresses made throughout centuries. The use of natural products as a biomaterial is presently undergoing a renaissance in the biomedical field [11]. One of the most important points in natural biomaterials development is the selection of a starting material, which will somehow mimic a naturally occurring one. New biomimetic material is one of those advances, where restoration of an organ's function is assumed to be obtained if the tissues themselves are imitated [14].

Over the last few decades, a large number of literature reviews have covered several biomaterials that have been considered for articular cartilage repair and regeneration [4, 12, 15–18]. Contemporary materials are either synthetically made, naturally-occurring, or combinations of the two [19], many of which have been optimized to not only provide 3D scaffold as suitable support for cell migration and proliferation but also to provide their differentiated phenotype [20, 21].

6.2.1 Natural Polymers

6.2.1.1 Proteins

Collagen

Collagen, the most abundant protein in mammals, is found in the extracellular matrices of many connective tissues in mammals, making up about 25–35 % of the whole-body protein content [14]. It is mostly found in fibrous tissues such as tendons, ligaments and skin, but is also found in other tissues such as cartilage, bone, and intervertebral disc. It makes up 1–2 % of muscle tissue, and accounts for 6 % of strong, tendinous muscle-weight [8, 22]. Collagen in the body is generally synthesized by fibroblasts, which originate from pluripotential adventitial cells or reticulum cells. To date, more than 29 types of collagen have been found and described. Over 90 % of the collagen in the body is of type I and is identified in bones, skins, tendons, vascular, ligatures, and organs [13]. Collagen type I participates in the maintenance of the many tissue integrity via its interactions which cell surfaces, other ECM molecules, and growth and differentiation factors. Though, in the human formation of scar tissue, as a result of age or injury, there is an change in the abundance of types I and III collagen, and also their proportion to one another [23].

Collagen, due to its unique structural characteristics, has been fabricated in large quantities. It has been shown to carry interesting structural, physical, chemical and immunological properties, and to be biodegradable, biocompatible, non-cytotoxic, with an ability to support cellular growth, and can be processed, through chemical and biochemical modifications, into a variety of forms including cross-linked films, steps, sheets, beads, meshes, fibres, and sponges [14, 24].

Bovine skin, pig skin and chicken waste are perhaps the major source of collagen. However, collagen obtained from pig bones cannot be freely used, due to religious and ethnic constraints; and can face regulatory and quality control difficulties and can contain biological contaminants and poisons, such as BSE (Mad Cow Disease), transmissible spongiform encephalopathy (TSE) and foot-and-mouth disease (FMD). This situation engenders the increasing interest to alternative collagen sources. Many reports have been showed to extract collagen from marine sources and have used to screen their potential industrial applications. In fact, marine organisms have been renowned as interesting and promising alternative sources, due to their availability, lack of disease risk, and high collagen profits [25, 26].

Collagen molecules are structural macromolecule of the ECM that contains in their structure one or several fields that have a characteristic triple-helix. The

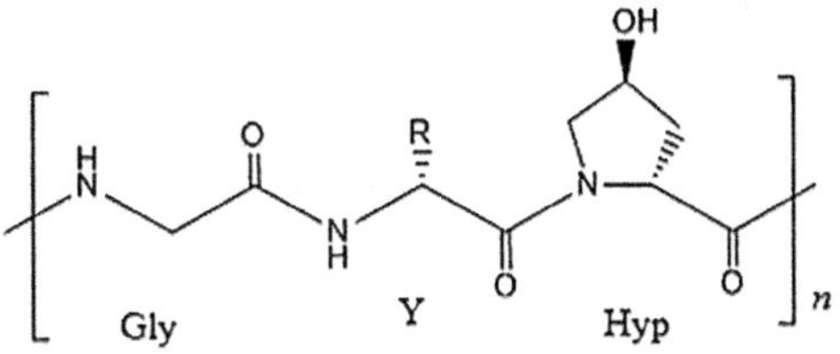

Fig. 6.2 Example of a repeating sequence of collagen

tropocollagen or "collagen molecule", rod-shape molecule, is a subunit of larger collagen aggregates such as fibrils. The length of each subunit is ~ 300 nm, and the diameter of the triple helix is ~ 1.5 nm. A distinctive feature of collagen is the regular arrangement of amino acids in each of the three chains of these collagen subunits. The primary structure of collagen (with its high content of proline and hydroxyproline and with every third amino acid being glycine) shows a strong sequence homology across genus and adjacent family line. Figure 6.2 depicts the common sequence follows the pattern Gly-Pro-Y or Gly-X-Hyp where X and Y may be any of numerous other amino acid residues [14].

The three main methods of collagen extraction produce are neutral salt-solubilized collagen, acid-solubilized collagen and pepsin-solubilized collagen (PSC) [27]. PSC extraction method of different sources have been reported such as collagen from the skin of brownstripe red snapper [28], albacore tuna and silver-line grunt skin [29], black drum and sheepshead seabream skin [30], bone and scale of black drum and sheepshead seabream [31] and fish waste material [32]. Different extraction procedures were performed and optimized using changed parameters such as low temperatures 4–6 °C with a high-speed centrifuge 30,000 rpm [29]. Some other procedures use higher temperature (up to 20 °C) and use lower speed centrifugation 10,000 rpm [31]. It has been shown that the extraction procedure using high temperatures could be more convenient for a pilot scale. It is important to highlight that the biochemical properties and thermal stability of collagen extracted could be different function to the extraction method.

Collagen production is quite simple, and is achieved in water resulting in a diversity of matrix systems, such as meshes, hydrogels, scaffolds, injectable solutions and dispersions, among others [14].

Many efforts have been made to develop biomimetically collagen-based macromolecules with variable physicochemical properties and macromolecular arrangement with respect to fibre, film, nano- and micro-particles or also porous scaffold formations with potential use in tissue engineering [26, 33].

In the late nineteenth century, collagen was first employed as a biomaterial in medical surgery century. Over the past few years, this biomaterial has already considered significant usage in clinical medicine such as injectable collagen for the augmentation of tissue defects, haemostasis, burn and wound dressings, hernia repair, bioprostetic heart valves, vascular grafts, a drug–delivery system, ocular surfaces, and nerve regeneration [14]. Collagen type I is most usually used in medical devices [34].

Many reports have demonstrated that a combination of collagens (such as type I and type II) with chondrocytes and stem cells facilitated cartilage tissue growth in vitro and in vivo [3, 35–37].

Silk Fibroin

Silk (SF) is a biopolymer essentially composed of two proteins: hydrophobic fibroin and hydrophilic sericin. This biopolymer is a natural polymeric biomaterial with remarkable oxygen and water vapour permeability, mechanical properties, biocompatibility and biodegradability [38]. The use of silk for this material production is not novel, since it has been widely used as a textile fibre for a very long time. Nevertheless, silk biomaterials have received an increasing attention as a source of promising technologies capable of producing on-demand therapeutic materials as scaffolds for tissue engineering [39, 40]. Moreover, raw silk is composed by silk fibroin (75–83 wt%), glue-like coating sericine (17–25 wt%), wax and fats (1.5 %), colorants and mineral components (1 %). These rates depend on the species, origin and culture conditions [41]. Silk fibroin heavy chains (350 kDa) comprise crystalline regions with an amino acid sequence (Fig. 6.3), and non-repetitive amorphous regions. SF is also known to have three crystal structures denoted as silk I, silk II and silk III [42, 43].

SF, the structural protein of fiber [44], can be easily extracted and at low cost from the cocoons of both wild (e.g. *Antheraea pernyi*) and domesticated (e.g. *Bombyx mori*) silkworms, spiders, wasps, honeybees, and ants [45, 46]. Usually, SF is extracted from the silk filaments, produced by *Bombyx mori* silkworms. The raw silk fibers (cocoons) are degummed with a Na_2CO_3 (0.1 % w/v) solution (around 100 °C) and then rinsed with distilled water (repeated thwice) to extract the highly immunogenic sericin protein (the other protein secreted by the silk worm) and then air-dried. SF thus extracted is dissolved in a ternary solvent containing $CaCl_2$, CH_3CH_2OH and H_2O in a 1:2:8 molar ratio, at 70 °C for 1 h, with continuous stirring. The final regenerated SF solution is obtained following dialysis against distilled water and filtration at room temperature [41, 46, 47].

Besides the traditional use of silk in the textile industry, due to its excellent lustre, softness, hygroscopicity and mechanical strength [42], SF fibres have also been used as surgical sutures without any serious side effects. Thus SF, one of the most abundant naturally found proteins, is useful material not only in the textile field but also in biomedical applications [42]. Moreover, due to its excellent

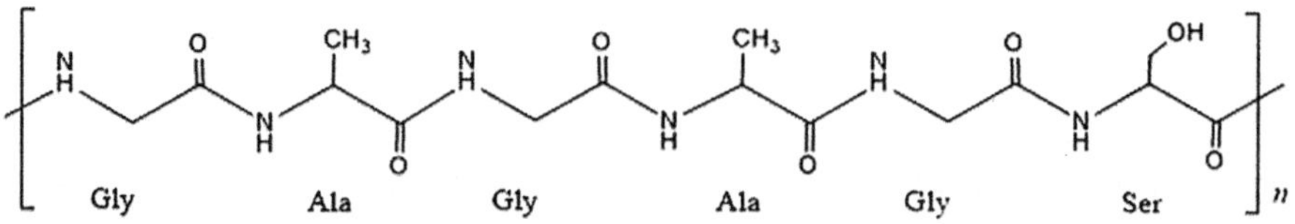

Gly   Ala   Gly   Ala   Gly   Ser

Fig. 6.3 Example of a repeating sequence of a crystalline region of SF protein heavy chains

biocompatibility, biodegradability and mechanical strength, SF has high oxygen/water permeability and minimal inflammatory reactions in vivo, being considered to be one of the best biomaterial for skeletal tissue regeneration [43, 46, 48]. It can be processed to obtain films, nonwoven nets, sponges, and hydrogels able to sustain cell adhesion, proliferation, and ECM production both in vivo and in vitro while also possessing a low inflammatory potential [49].

The protein-based natural polymer SF can be processed into various carries such as regenerated fibers, particles, membranes, films, hydrogels and porous matrix for different applications [50, 51]. Silk fibroin derived from the silk-worm *Bombyx mori* has proved to be a promising candidate as a scaffolding material for use in tissue engineering, namely for cartilage and meniscus regeneration [51]. Saha et al. [48] found that silk fibroin scaffolds of *Antheraea mylitta* are more chondroinductive while those of *Bombyx mori* are more osteoinductive, indicating the potential of using a multi-layered combination of these, for osteochondral defect regeneration with good integration, the stem cell-silk biomaterial interaction, which is exploited in the context of bone or cartilage.

Despite their potential, because of low hardness and ease of deformation under physical stress, application of pure and unmodified SF materials in bone regeneration is challenging [46]. For instance, SF/hyaluronic acid 3D matrices [49], as well as, genipin-cross-linked chitosan/SF sponges [52], silk fibroin-chondroitin sulfate-sodium alginate (SF-CHS-SA) porous hybrid scaffolds [53], and others, were considered as cartilage engineering strategies, presenting promising results.

6.2.1.2 Polysaccharides

Alginate

Alginate is natural polysaccharide, extensively used in several industries like textile, agri-food, paper, cosmetic, biomedical and pharmaceutical [54]. The main industrial applications of alginates are linked to their ability to crosslink with ions, retain water, gelling, viscosifying and stabilizing properties [55].

Furthermore, the biological function of alginate in algae is usually believed to be as a structure-forming component. The intercellular alginate gel matrix gives the plants both mechanical strength and flexibility [33, 56]. Alginates are presented in both salt and acid forms. Concerning the salt form, it is a significant cell wall component in all brown seaweeds, constituting up to 40–47 % of the dry weight of algal biomass. For the acid form, it is a linear polyuronic acid and mentioned as alginic acid [55], Fig. 6.4.

Concerning the isolation, alginate is produced by *Pseudomonas* species and *Azotobacter* bacteria such as *Azotobacter chroococcum* and *A. vinelandii* [57]. Alginate is also naturally found in the cell walls of brown seaweeds [58].

Alginates are linear binary copolymers of $(1 \rightarrow 4)$ linked β-D Mannuronic acid (M) and α-L-glucuronic acid (G) residues of different compositions and sequence [59].

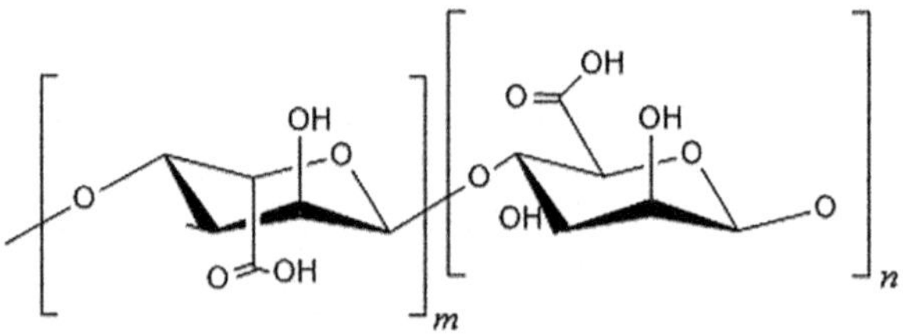

Fig. 6.4 Example of a
repeating sequence of alginate
chains

Haug et al. report, for the first time, informations about the sequential structure of alginates came [60–62]. By partial acidic hydrolysis and fractionation, they were able to separate alginate into three fractions of widely differing composition [56]. On partial acid hydrolysis, in which two fractions contains almost homoplolymeric molecules of G and M, respectively, and the third fraction of alginates contains of approximately equal proportions of both monomers and was shown to contain a large number of MG dimer residues [63]. The monomer sequence (M and G) can vary among algal species and among different tissues of the same species. The M/G ratio and the block structure have a significant effect on the physicochemical and rheological properties of alginates. These biopolymers with more industrial significance usually show high G content [56, 59].

The first step of alginate extraction from algal material is an acidification step. In fact, alginate is insoluble within the algae with a counter-ion composition, which is determined by the ion exchange equilibrium with seawater. The acidification step will also permit the elimination of contaminant glycans, like laminaran and fucan. Concerning the second step, it consists to brought the alginic acid into solution by neutralisation with alkali as NaOH or Na_2CO_3 to form the water-soluble sodium alginate. After extensive separation procedures such as sifting, flotation, centrifugation and filtration to remove algal particles, the soluble sodium alginate is precipitated directly by alcohol, by mineral acid or calcium chloride, converted to the sodium form if is necessary and ultimately dried and milled.

The used algal raw material and also the algae's age will determine the colour of extracted alginate. This alginate pigmentation could be controlled by a bleaching procedure. Alginate's extraction procedure from brown algae has been methodically optimized in order to develop economically and industrially sustainable products, with controlled properties as to envisage different therapeutic applications [33, 56].

The most important feature of alginate, from both the industrial and the biotechnological point of view, is connected with its capability to efficiently bind divalent cations, such as Ca^{2+}, Sr^{2+}, and Ba^{2+} to name a few, leading finally to hydrogel formation [64]. It is important to highlight that alginate, which possess a number of free hydroxyl and carboxyl groups distributed along the backbone represent a perfect candidate for chemical functionalization. By forming alginate derivatives through chemical functionalization, the several properties such as solubility, hydrophobicity and physicochemical and biological characteristics may be changed. Several studies have been reported the alginate chemical modification of hydroxyl groups through different techniques such as oxidation [65], reductive-amination of oxidized alginate

[66], sulphation [67], copolymerization or cyclodextrin-linked alginate [68], and also chemical modification of carboxyl groups using other techniques as esterification [69], Ugi reaction [70] or amidation [71, 72].

Alginate has been extensively used for many biomedical applications due to its excellent biocompatibility, low toxicity, and the mild gelation conditions required to form cross-linked structures [73]. It could form scaffolds through the use of ionic cross-linking, allowing for encapsulation of cells. Several researches have evaluated alginate scaffolds as a platform for generating cartilage [74–78]. Both bone marrow-derived mesenchymal stem cells and adipose-derived adult stem cells have been shown to survive and differentiate in chondrocytes in these studies [75].

Chitosan

Chitosan (CS) is a naturally occurring biodegradable polysaccharide, derived from chitin, which is copolymer of β(1–4) linked D-glucosamine and *N*-acetyl D-gluco-samine residues [79–82]. Chitin is a polysaccharide, a homopolymer comprised of 2-acetamido-2-deoxy-b-D-glucopyranose units [41], Fig. 6.5.

The linear polysaccharide chitosan could be easily bought since it is commercially available. CS can be obtained from chitin by the deacetylation process [41], being possible to obtain different degree of deacetylation [83]. It could also be reacetylated to improve the solubility of the chitosan.

CS is soluble in dilute acids, where it carries a strong positive charge because of protonation of the amino group [80]. This amino polysaccharide can been employed as a physical cross-linking component in the scaffold construction [84]. This polymer is generally inert in vivo and has favourable degradation kinetics and mimics the glycosaminoglycan component [79]. Chitosan itself is a very interesting material for biomedical applications [41], but since it has poor mechanical properties and very high swelling ability, which causes it to get easily deformed, it is generally improved by blending with other polymers [38].

The blending of CS with SF is reported in numerous studies [38, 43, 52, 79]. They are two preeminent natural materials for the design of tissue-engineered scaffolds due to their excellent biocompatibility, degradability, excellently engineered structures and also mechanical characteristics [85]. These SFCS scaffolds have strong interactions between the synthetic and biological components occur, mainly due to hydrogen bonds, and can mimic the in vivo extracellular matrix,

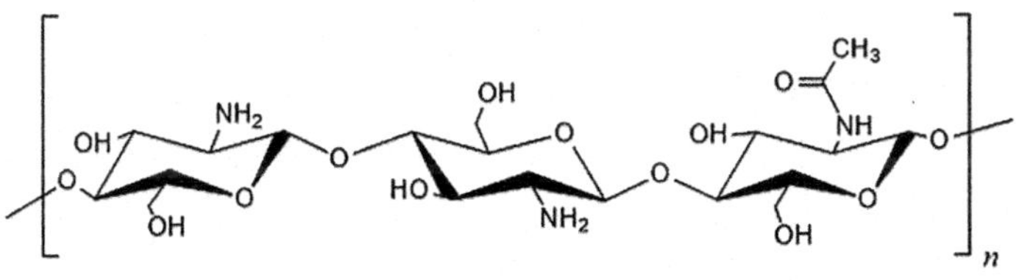

Fig. 6.5 Example of a repeating sequence of chitosan chains

having biological, structural and mechanical properties that can be adjusted to meet specific clinical needs producing promising outcomes both in vitro and in vivo in repairing abdominal wall defects, healing skin wounds and regenerating bone and tracheal cartilage [79, 86]. The intra-articular injection of alginate-chitosan beads dispersed in an hydrogel was studied with interesting results supporting this new biphasic biomaterial for treatment of ostearthritis in human [87].

The potentiality of a polyelectrolyte poly(lactide-*co*-glycolide) (PLGA) with CS complex scaffold to repair a cartilage defect in vivo, by a tissue engineering approach, was studied and was found that it provided a favourable environment for the maintenance of cellular proliferation, migration and chondrogenic differentiation, achieving successful hyaline cartilage regeneration, repairing a full thickness articular cartilage defect over a period of 12 weeks [84].

Hybrid chitosan/blood implants have also been studied to heal cartilage defects [88, 89]. It was found that chitosan implant treatment suppresses fibrocartilage scar tissue formation, and promotes bone remodelling, which allows more blood vessel migration and woven bone repair towards the cartilage lesion area [90].

Hyaluronic Acid

Hyaluronic acid (also called hyaluronan or hyaluronate, HA) is a linear, hydrophilic, polyanionic polysaccharide. It is found naturally as high molecular weight glycosaminoglycan in extracellular matrix, connective tissues, epithelial tissues, neural tissues, synovial fluid, vitreous humor, and umbilical cord [8, 91].

Meyer and Palmer, in 1934, isolated for the first time from the vitreous body [92]. Nowadays, HA is produced through bacterial fermentation of *streptococcus* species or extracted from rooster combs, umbilical cords, synovial fluids, skin or vitreous humour for commercial purposes. The usual sources for its industrial production are bovine vitreous humour, bovine synovial liquid, and rooster crest, with an increasing interest to bacterial cultivations [93].

HA is composed of repeating disaccharide units of β-1, 4-glucuronic acid and β-1, 3-*N*-acetyl-D-glucosamine (Fig. 6.6) with a molecular weight up to 6 million Daltons. This polysaccharide is the only GAG that is not synthesized in the Golgi apparatus and its synthesis does not necessitate attachment to a core protein. HA catabolism is achieved either by hyaluronidase-mediated enzymatic degradation or by chemical mechanisms to generate fragments of various sizes [94].

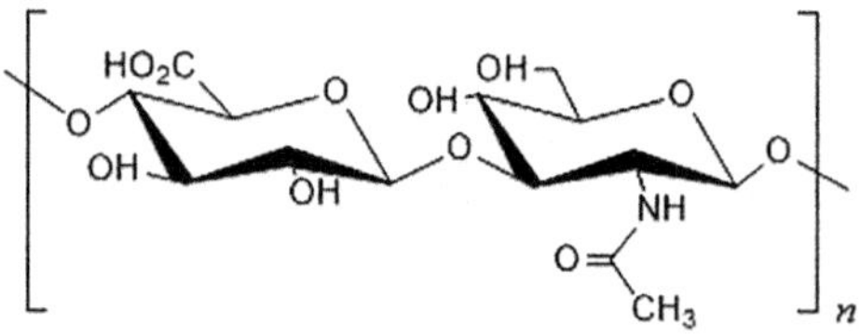

Fig. 6.6 Example of a repeating sequence of hyaluronic acid chains

Extraction procedures from animal tissues were optimized to remove protein and to minimize the inevitable degradation of hyaluronic acid. Over the years, extraction protocols have been developed and optimized but are still limited to low yields, due to the intrinsic low hyaluronic acid concentration of hyaluronic acid in the tissue, and from high polydispersity of polysaccharide products due to both the uncontrolled degradation during extraction and thie natural hyaluronic acid polydispersity [95]. Because of the HA products could from animal origins, exist a potential risk of contamination with proteins and viruses, but this can be reduced by using tissues from healthy animals and also by an extensive purification procedure. However, this statement raised the interest in the biotechnological hyaluronic acid production.

Concerning the microbial HA production on industrial scale was firstly achieved in 1980s by Shiseido [96]. Optimization of cultivation conditions and also culture media along with strain improvement were performed to raise hyaluronic acid yield and also its quality. However, this technical is also limited due to mass transfer limitation causing by high viscosity of the fermentation broth. It is important to highlight that bacterial fermentation produces hyaluronic acid with high molecular weight and purity, but risk of contamination with bacterial endotoxins, proteins, nucleic acids, and heavy metals occurs [95].

Techniques used for derivatization of the HA include sulphation, carbodiimude-mediated modification, esterification, hydrazide modification, cross-linking with polyfunctional epoxides, divinyl sulphone and glutaraldehyde, among others. Hyaluronic acid products derived from esterification are more rigid and stable, hydrophobic, and less susceptible by enzyme [97]. Concerning, HA derivatized using carbomiide-mediated modification allows covalent binding of the carboxyl group with an amine of another. Depending on the pH condition, carbodiimide reactions are sensitive and frequently result in the formation of acylurea, an unreactive intermediate from the carboxyl group [98]. Furthermore, yields of derivarized hyaluronic acid are also affected by the reaction conditions of carbodiimide-meditate modification. Furthermore, HA-derived from sulphation can occur by reacting the hydroxyl group with sulphur groups per dissacharide. The cross-linking approach to engineer hyaluronic acid's properties implicates either a one-step procedure of exposing the hyaluronic acid to a cross-linker, or a two-step procedure of synthesizing a highly reactive hyaluronic acid derivative followed subsequently by cross-linking [8, 97].

HA is characterized by its 3D structure, which has a significant porosity, surface and space area. This represents an advantage for growth of cell adhesion, extracellular matrix deposition, the transport of gases and nutrients, and metabolic product release, and offers a good interface of material-cell function. This polymer plays an important role in joint lubrication, nutrition and preserving cartilage properties [97, 99, 100]. In fact, HA protect the surface of articular cartilage helping in the control of water balance. Several reports showed that HA also can maintain normal growth of cartilage cells and promote the integration of transplanted chondrocytes and damaged cartilage [91, 101–104].

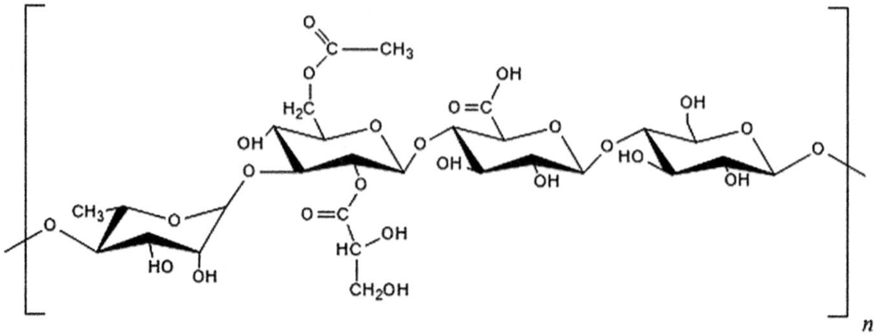

Fig. 6.7 Example of a repeating sequence of high acyl gellan gum chains

Gellan Gum

Gellan gum (GG) is a linear anionic deacetylated exopolysaccharide produced in high yield by aerobic fermentation of the non-pathogenic bacteria *Sphingomonas* (formerly *Pseudomonas*) elodea [105–109]. This is one of the most extensively studied and described member of bacterial polysaccharides.

GG is a linear, anionic exopolysaccharide, with a tetrasaccharide repeating unit of one α-L-rhamnose, one β-D-glucuronic acid and two β-D-glucoses [105, 107], and two acyl groups, acetate and glycerate bound to glucose residue adjacent to glucuronic acid [110], Fig. 6.7. The approximate composition is glucose 60 %, rhamnose 20 % and glucuronic acid 20 % [110].

GG is produced by a pure culture fermentation of carbohydrates by *Pseudomonas elodea*, purified by recovery with isopropyl alcohol, dried and milled. The fermentation process efficiency is dependent on many factors such as media composition, pH, temperature, agitation rate and available oxygen [110]. GG can be processed into hydrogels that are resistant to heat and acid. A temperature of at least 70 °C is needed with subsequent cooling (to room temperature) to conformation changes of the polymer chains, inducing coil-to-helix transition [105, 106, 111]. The acetyl groups in native gellan gum can be removed by alkaline treatment to produce deacetylated gellan gum [110]. Native gellan gives soft, easily deformable gels, while the deacetylated one forms rigid and brittle gels [105]. GG can also be clarified by filtration of hot deacetylated gellan gum, so it can be used as an agar substitute [110].

GG has widespread applications in food and cosmetics mainly as a multi-functional gelling, stabilizing and suspending agent, and has received both US FDA and EU (E418) approval for these purposes [108]. GG has been utilized in the field of tissue engineering, mostly as a material for cartilage reconstruction, due to its gelation and rheological properties [105, 108], as well as resembles extracellular matrix glycosaminoglycan composition [111]. Oliveira et al. [109] results demonstrate the adequacy of gellan gum hydrogels processed by simple methods for noninvasive injectable applications toward the formation of a functional cartilage tissue-engineered construct and report the preliminary response of a living organism to the subcutaneous implantation of the gellan gum hydrogels. However, some problems are associated to gellan gel to qualify as genuinely injectable cell vehicle for practical use. For instance, its gelation temperature (Tg) is too high (>42 °C), when compared to physiological temperature (approx. 37.5 °C). Metacrilated GG hydrogels have highly tunable physical and mechanical properties without affecting their biocompatibility, being applicable for a wide range of tissue engineering applications [112]. The complex gel of gellan gum and carboxymethyl chitosan is also a promising material for cartilage tissue engineering [113].

Chondroitin Sulphate

Chondroitin sulphate (CS) is acid mucopolysaccharide and extensively distributed in human, other mammals and invertebrates, as well as some bacteria [114]. It is an important component of the extracellular matrix (ECM). CS is the most frequent glucose amino glycan (GAG) in the aggrecan molecule of the cartilage. Due to the negative charge of CS, it is responsible for the water retention of the cartilage, which is essential for pressure resistance [115], Fig. 6.8.

CS can be extracted from different tissue of various natural sources such as terrestrial (bovine, chicken, porcine…) and also marine species (whale [116], shark [117, 118], squid [119, 120], salmon [121], king crab [122] and sea cucumber [123] being extracted by proteolytic digestion and further purified by precipitation with organic solvents, chromatography or enzymatic degradation of contaminants) [124]. Several researches reported the extraction of cartilage from sea cucumber and also marine invertebrates like cnidaria, molluscs and polychaeta [125, 126]. However, the most successful commercial products of CS, by market volume and profits, are those related with cartilage regeneration, anti-inflammatory activity and also osteoarthritis (OA) [121].

Chondroitin sulphate, one of the natural glycosaminglycans, is a polymeric carbohydrate comprising repeating disaccharide units of glucuronic acid and *N*-acetyl-galactosamine linked by β-(1 → 3) glycosidic bonds and sulphated in different carbon positions (CS no sulphated is CS-O). Moreover, CS has a molecular mass of 20–60 kDa and contains free carboxylic acid groups available for carbodiimide-mediated modification [127]. Chain size is known to vary with

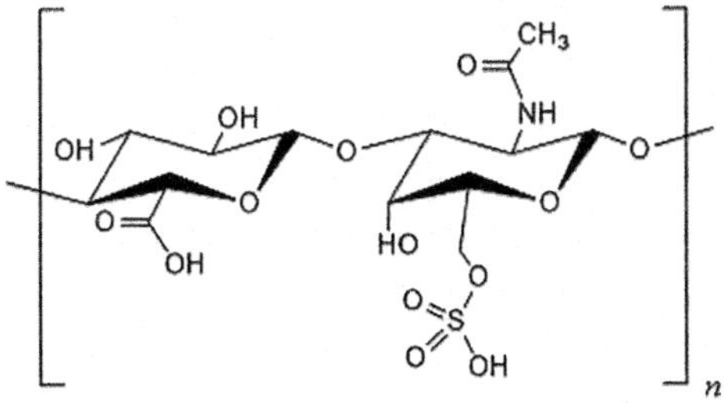

Fig. 6.8 Example of a repeating sequence of chondroitin sulphate chains

source, and available data suggest that tracheal CS is 20–25 kDa, while shark CS is larger at 50–80 kDa. However, any population of CS chains, even from a single tissue source, is heterogeneous with respect to size. Concentration and composition of CS depends on the organism and tissue, thus, CS from terrestrial and marine sources contains diverse chain lengths and oversulphated disaccharides [114].

From an industrial viewpoint, the processes of CS isolation from cartilage have been defined for many years and generally include four steps: (1) chemical hydrolysis of cartilage; (2) breakdown of proteoglycan core; (3) elimination of proteins and CS recovery; (4) purification of CS [114]. The two first steps are mostly conducted by means of alkaline hydrolysis at high concentration of NaOH, urea, cysteine or guanidine HCl, subsequently combined with selective precipitation of GAG using cationic quaternary ammonium chemicals, non-ionic detergents, potassium thiocyanate or alcoholic solutions, podeproteinization with tri-chloroacetic acid and finally purification with gel filtration and/or ion-exchange and size-exclusion chromatography. Recently, Murado research [93] showed the optimization of a rapid and economical extraction procedure obtaining a purified CS product (99 %) from skate cartilage wastes with a high concentration 35–45 g L^{-1} [114].

Types of application for CS-based products, and so their market prices, depends on the purity and quality of GAG in the commercial products. Consequently, it is mainly important to develop low-cost and high yield extraction procedures maintaining the quality and purity of CS. Clinical applications request highly concentrated and pure CS in comparison with other applications [114]. CS controls the symptoms of OA as inflammation and pain; it could impede OA progression, could change the progression of this disease by stimulating the anabolic processes involved in new cartilage formation. Moreover, it has been found that CS prevents enzymes such as hyaluronidase presented in the patient synovial at high concentration [115]. CS also participates in the increasing of the hyaluronic acid production by synovial cells, which subsequently improves the viscosity and the synovial fluid levels.

Cellulose

Cellulose is a semi-crystalline homopolymer of glucose, being the most widely spread natural polymer worldwide. It is found in all plants and is the main constituent of the cell wall green plants. Cellulose is degradable by enzymes like cellulases; the biocompatibility of cellulose and its derivatives is well established [17]. This polysaccharide is easily processable, versatile, and exist in a wide range of forms and shapes, e.g. as membrane sponges, microspheres and non-woven, woven or knitted textiles; their mechanical properties with those of hard and soft tissue has been also reported [8, 128].

Cellulose is produced by all plant matter and is the main constituent of the cell wall green plants. It is also secreted by some species of bacteria and prokaryotes

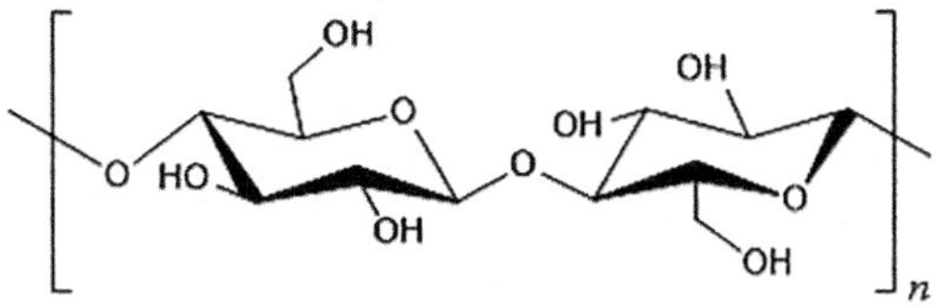

Fig. 6.9 Example of a repeating sequence of cellulose chains

such as acetobacter, rhizobium, agrobacterium, as well as fresh water and marine algae [8].

Cellulose with the formula $(C_6H_{10}O_5)_n$ is one of the many beta-glucan compounds, it is a polycarbohydrate composed of a series of cellubiose units, formed by two anhydroglucose subunits, Fig. 6.9. This polysaccharide has the unique characteristic that it cannot be synthesized from or hydrolyzed into monosaccharaides. This exclusive property is a result of the structure of the cellulose and its intricate hydrogen bond network that inhibits this polysaccharide from melting or dissolving in common solvents. This complex is also responsible for the high mechanical properties of cellulose that are important for its function in nature, as well as making it suitable for composite reinforcement applications [129].

Cellulose, by being the most abundant renewable and biodegradable polymer, represents a promising attractive candidate to be converted to derivatives. Purification and isolation of cellulose comprises several steps including a pulping process, partial hydrolysis, dissolution, re-precipitation, and extraction with organic solvents. Usually, purification and isolation processes almost engender the degradation of cellulose and also permit the cellulose to undergo oxidation by reaction with both acids and bases [8].

It has been found through in vitro approach that the use of a cellulose polymer permitted the proliferation of chondrocytes and showed an interesting biocompatibility [128]. Furthermore, it has been shown that an injectable product based on hydroxypropylmethylcellulose hydrogel, for the 3D culture of chondrocytes, may be used for articular cartilage repair [17].

Agarose

Agarose is a polysaccharide polymer material, derived from seaweed, composed by residues of L and D-galactose and is isolated from Chinese algae [17], Fig. 6.10. Agarose gel is obtained through changing temperatures, and a cross-linked alginate

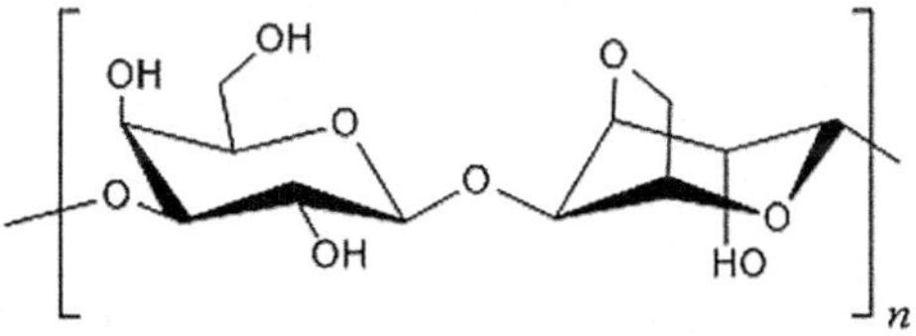

Fig. 6.10 Example of a repeating sequence of agarose chains

matrix can be formed via ionic bonding in the presence of Ca^{2+} [3]. Agarose hydrogels are well-characterized and have been extensively studied for chondrocyte culture and cartilage tissue engineering, being determined by the effects of factors such as mechanical loading and cell-seeding density on cartilage tissue formation [130–132]. The feasibility of using high-concentration agarose for applying in vitro compression to chondrocytes, as a model for understanding how chondrocytes respond to in vivo loading, was recently demonstrated [132]. Moreover, agarose implantation containing chondrocytes or mesenchymal stem cells (chondrocyte- or MSC-laden hydrogels) in osteochondral defects, which ensure physiologically relevant mechanical properties, allows the formation of a repaired tissue containing collagens and proteoglycans. Despite its promising performance, agarose has been poorly studied in vivo due to its inhibitory effect and inability to degrade in vivo, which prevents graft integration with the host tissue [17, 130, 131]. Furhermore, cartilage tissue formation within an agarose hydrogel is complicated by an inability to customize agarose scaffold structure and composition [130].

6.2.2 Synthetic Polymers

In the last few decades, a variety of synthetic polymer material applications have significantly been developed. Due to the ease processability and chemical modification, high biocompatibility, high versatility, important mechanical properties, and controllable biodegradability, synthetic polymers presently stimulate increasing interest from researchers who are investigating their potential for tissue engineering applications [3]. In the field of articular *cartilage tissue engineering*, the most popular synthetic polymers are poly-lactic acid (PLA, which is present in both L and D forms), poly-glycolic acid (PGA), and their copolymer poly-lactic-*co*-glycolic acid (PLGA). These FDA approved biodegradable polymers can be fabricated into 3D matrices [3]. Some examples of synthetic biomaterials will be discussed in the following topics.

6.2.2.1 Poly(ethylene glycol)

Poly(ethylene glycol) (PEG), first appeared in scientific literature during the 1980s [133], is a synthetic polymer that has been used significantly in biomedical application due to its excellent biocompatibility [134]. PEG is the most commonly applied non-ionic hydrophilic polymer (Fig. 6.11) with stealth behaviour and has

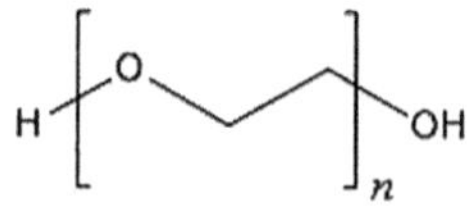

Fig. 6.11 Repeating sequence of poly(ethylene glycol)

been most widely used polymer for antifouling applications, such that it has often been termed the "gold standard" of antifouling polymers [135, 136]. PEGylation approach contributes to (i) rising the aqueous solubility and stability, (ii) decreasing immunogenicity, (iii) reducing intermolecular aggregation and (iv) prolonging the systemic circulation time of a compound [137].

The PEG, a polyether compound, its molar mass used in diverse medical and pharmaceutical applications ranges from 400 Da to about 50 kDa. PEG with a molar mass of 20–50 kDa is mostly used for the conjugation of low-molar-mass drugs as small molecules, oligonucleotides, and siRNA. PEGs with lower molar masses of 1–5 kDa are frequently applied for the conjugation of larger drugs, such as antibodies or nanoparticule systems [135].

PEG macromonomers may be polymerized via a variety of mechanisms such as anionic, cationic, ring opening metathesis, or free-radical polymerization. Most commonly these are functionalized with acrylate and methacrylate groups at the chain ends [138], although fumarate [139] and other derivatives capable of free radical polymerization have been described [134, 140].

The scaffolds made from PEG can also be chemically modified to contain bioactive molecules, including peptides and heparin. In combination with stem cells and/or growth factor, these scaffolds have been evaluated for their suitability and showed to be a promise candidate as therapy for the cartilage tissue regeneration [141, 142].

6.2.2.2 Poly(lactic-*co*-glycolic acid)

Poly(lactic-co-glycolic acid) (PLGA) is a biodegradable polymer approved by the US Food and Drug Administration (FDA) as a safe biomaterial for clinical applications [143]. PLGA is a polypeptide (Fig. 6.12) with no antigenicity or immunogenicity, exhibiting good biological and physico-chemical properties (non-toxic, hydrophilic and biodegradable) [84]. This polymer is one of the most typical biomaterials for bone/cartilage tissue engineering. However, conventional PLGA does not mimic important properties of biological tissue, including cartilage tissue [144], and that is why it is commonly studied as a component of a complex, with other compounds.

A microdevice of PLGA–P188–PLGA matrix was developed with an efficient easy-to-handle and represents an injectable tool for cartilage repair [145].

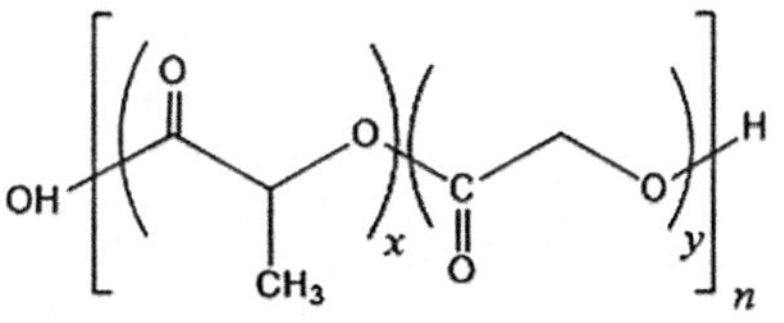

Fig. 6.12 Repeating unit of poly(lactic-co-glycolic acid)

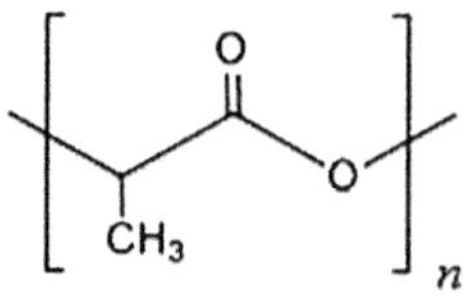

Fig. 6.13 Repeating unit of
polylactic acid

PLGA/chitosan polyelectrolyte complex also possesses great potential as a scaffold for cartilage tissue engineering [84]. It was found that is feasible to repair articular cartilage using a single-layer porous PLGA scaffold without exogenous cells and that cartilage regeneration can be improved further by the application of short-term continuous passive motion [146]. Moreover, Eswaramoorthy et al. [143] showed that sustained release of PTH(1–34) from PLGA microspheres suppresses osteoarthritis progression in rats.

6.2.2.3 Poly-L-Lactic Acid

In 1954, French chemists first synthesized Poly-L-lactic acid (PLLA), a polymer of lactic acid monomers derived from corn dextrose fermentation [147]. PLLA is a biocompatible, biodegradable, synthetic, and immunologically inert polymer device derived from the alpha-hydroxy-acid family, and it has a long history of safe use in numerous therapeutic applications [148]. The main disadvantages of PLLA limiting its applications are its poor chemical modifiability and mechanical ductility, slow degradation profile, and poor hydrophilicity [149].

Each PLLA molecule is relatively heavy (140 kDa) and irregularly crystalline shaped (2–50 μm in diameter) (Fig. 6.13), structures that contribute to its slow physiologic absorption. Pure L-lactic acid or D-lactic acid (the two isomers of lactic acid), or mixtures of both components are needed for the synthesis of PLA [150]. Furthermore, the half-life of PLLA is estimated at 31 days, with total absorption occurring by 18 months. PLLA are metabolized along a similar metabolic pathway as lactate/pyruvate and it is aliphatic polyester, which has the capability to degrade into water and carbon dioxide by simple hydrolysis of ester bonds [147, 151].

PLLA has a longer than 30-years history of safe use in the medical field. These synthetic materials are reported and used alone or mixed with other matrices for cartilage tissue engineering. Several forms of these polymers, from the fine fibrillary layer to the sponge, have been developed [17].

6.2.2.4 Polycaprolactone

Polycaprolactone (PCL) is a FDA approved biodegradable polymer [152], Fig. 6.14. It is, hydrophobic, semi-crystalline polymer; and its crystallinity tends to decrease with increasing molecular weight [153]. Moreover, its good solubility, low

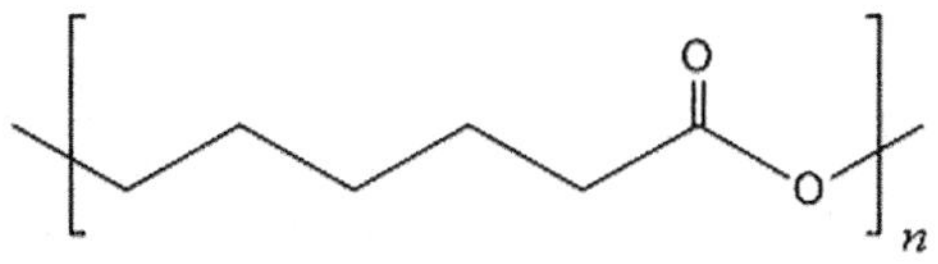

Fig. 6.14 Repeating unit of
polycaprolactone

melting point (59–64 °C) and excellent blend-compatibility have made PCL a
continuous research focus in the biomedical engineering area [153]. PCL has a
lower degradation profile with respect to other polyesters such as poly-lactic and
poly-glycolic acids [152].

PCL is able to support the adhesion and growth of different cell types (i.e.
skeletal muscle cells, smooth muscle cells, fibroblasts, chondrocytes, endothelial
cells, human mesenchymal stem cells) [152]. Due to its viscoelastic properties, it
has been used as the tissue-engineering scaffold for various organs or tissues,
including bone, cartilage, tendon and ligament, blood vessel, skin and nerve [153].
For instance, polycaprolactone scaffolds manufactured by selective laser sintering
(SLS) surface modified through immersion coating with collagen are feasible
scaffolds for cartilage tissue engineering in craniofacial reconstruction [154]. It is
usually used with other compounds, for instance, it was found that MWNTs/PCL
composite scaffolds have the potential for bone tissue engineering and the relatively
low concentration of MWNTs (0.5 wt%) is preferred [153].

6.2.2.5 Poly(*N*-Isopropylacrylamide)

Poly(*N*-isopropylacrylamide), Fig. 6.15, is the most relevant and attractive thermoresponsive polymer for biomedical applications, because it possesses a sharp
phase transition or a lower critical solution temperature (LCST) around 32 °C in
aqueous solutions. Below the LCST, the polymer is in the hydrated state and
therefore it is soluble in cold water [106, 155]. The LCST can be removed by
copolymerizing Poly(NIPAAm) with other more hydrophilic or hydrophobic
monomers [156, 157].

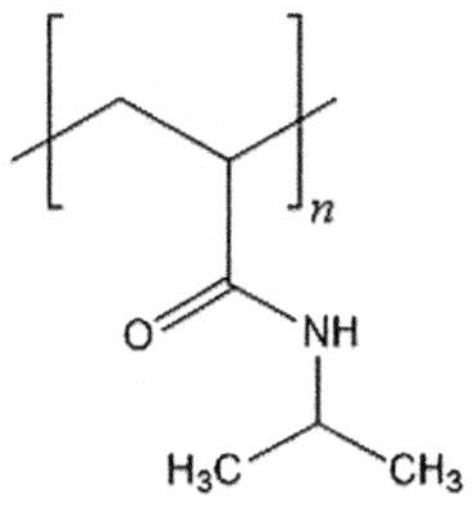

Fig. 6.15 Repeating unit of
poly(N-isopropylacrylamide)

Though, as temperature is increased above its LCST, the polymer becomes hydrophobic by exposing the hydrophobic isopropyl groups to the water interface and so insoluble, experiencing a collapse into gel form [158].

It has been shown that this transition may involve the breakage of intermolecular hydrogen bonds with the water molecules, which are substituted by intramolecular hydrogen bonds amongst the dehydrated amide groups. The hydrophobic methyl groups on poly(NIPAAm) then breakdown and permit for entanglements of the polymer chains, resulting in a gel [158].

Poly(NIPAAm) can be synthesized from N-isopropylacrylamide, which is commercially available. NIPAAm can be polymerized through different methods; those polymers with high molecular weight (Mw) Poly(NIPAAm) was achieved via radical polymerization. Polymers produced in this way have the disadvantage of a high polydispersity index. Narrowly dispersed Poly(NIPAAm) with the desired molecular weights can be obtained in a controlled manner via nitroxide mediated polymerization (NMP), anionic polymerization, reversible addition–fragmentation chain transfer (RAFT) and atom transfer radical polymerization (ATRP) [159].

Several researches reported the modifications of poly(NIPAAm) which showed a great interest to tailor the LCST of poly(NIPAAm)-based systems. For example, N-isopropylacrylamide (NIPAAm) was copolymerized with N-acryloxysuccinimide (NASI) via free radical polymerization. For example, for in vivo applications Poly (NIPAAm-co-NASI) was modified to obtain Poly(NIPAAm-co-cysteamine) through a nucleophilic attack on the carbonyl group of the NASI by the amine group of the cysteamine in order to create in situ physically and chemically cross-linking hydrogel [160].

Moreover, it has been found that the addition of hydrophobic monomer is soluble inside the surfactant micelles, while the hydrophilic monomer is soluble in the aqueous continuous medium. Furthermore, block copolymers based on PEO–PPO sequences are a family of commercially available triblock copolymers, which reveal a sol-gel transition below or close to the physiological temperature, a gel-sol transition around 50 °C and a LCST [161].

Poly(NIPAAm) is one of the most intensively studied synthetic polymers for use in controlled drug delivery, cell-sheet engineering, as a biosensor or in tissue engineering [159].

6.2.3 Others

Other biomaterials for cartilage repair could be referred, such as ceramic polymers (hydroxyapatite and zirconia). In fact, the working time slot of this chapter comprises the referred compounds as the most important ones. However, it is vital to highlight that novel biomaterials emerge promptly while others lose interest at the same rate, and, in the meantime another group of materials are still being continuously and intensely studied.

6.3 Final Remarks

This chapter looks at the most recent research in natural and synthetic biomaterials used in cartilage tissue engineering. The purpose of this section was to update the available information, on the procedures to obtain these biomaterials and also their chemical structures, their modifications and also their applications on cartilage repair, when applicable.

It is important to highlight that these materials such as native biological materials and synthetic polymeric materials are characterized by their advantages and disadvantages. The overcoming disadvantages could be observed through the use of either physical or biochemical modifications of the biomaterial in order to improve its biological and mechanical properties. These required properties include for example biocompatibility, suitable microstructure, desired mechanical strength and degradation rate as well as the capability to support cell residence and permit retention of metabolic functions.

To conclude, tissue engineering, which is an interdisciplinary and multidisciplinary research areas, is currently considered one of the most exciting and challenging fields in biology; and is rising exponentially over time through better technology and research. In fact, the development of novel biomaterials in cartilage tissue regeneration is an extremely active and challenging area of research, due to the complex physiology of cartilage tissue and its poor healing ability. Ideal biomaterials for cartilage tissue engineering should have several interesting properties and characteristics, which actively participate to cartilage repair and regeneration.

More exciting findings lie ahead. We believe undoubtedly that further investigations with a multidisciplinary approach are imperative to develop useful novel polymers for cartilage tissue engineering.

References

1. Gao Y, Liu S, Huang J, Guo W, Chen J, Zhang L, Zhao B, Peng J, Wang A, Wang Y, Xu W, Lu S, Yuan M, Guo Q (2014) The ECM-cell interaction of cartilage extracellular matrix on chondrocytes. Biomed Res Int 2014:648459. doi:10.1155/2014/648459
2. Gaharwar AK, Schexnailder PJ, Schmidt G (2011) Nanocomposite polymer biomaterials for tissue repair of bone and cartilage: a material science perspective. In: Taylor and Francis Group L (ed) Nanobiomaterials handbook. UK, p 20
3. Zhang L, Hu J, Athanasiou KA (2009) The role of tissue engineering in articular cartilage repair and regeneration. Crit Rev Biomed Eng 37(1–2):1–57
4. Johnstone B, Alini M, Cucchiarini M, Dodge GR, Eglin D, Guilak F, Madry H, Mata A, Mauck RL, Semino CE, Stoddart MJ (2013) Tissue engineering for articular cartilage repair—the state of the art. Eur Cells Mater 25:248–267
5. Khaled EG, Saleh M, Hindocha S, Griffin M, Khan WS (2011) Tissue engineering for bone production—stem cells, gene therapy and scaffolds. Open Orthop J 5(Suppl 2):289–295. doi:10.2174/1874325001105010289
6. Lee KB, Wang VT, Chan YH, Hui JH (2012) A novel, minimally-invasive technique of cartilage repair in the human knee using arthroscopic microfracture and injections of

mesenchymal stem cells and hyaluronic acid–a prospective comparative study on safety and short-term efficacy. Ann Acad Med Singapore 41(11):511–517

7. Chung C, Burdick JA (2008) Engineering cartilage tissue. Adv Drug Deliv Rev 60(2):243–262. doi:10.1016/j.addr.2007.08.027

8. Ong KL, Lovald S, Black J (2015) Orthopaedic biomaterials in research and practice, 2nd edn. CRC Press, Boca Raton

9. Sannino A, Demitri C, Madaghiele M (2009) Biodegradable cellulose-based hydrogels: design and applications. Materials 2(2):353–373. doi:10.3390/Ma2020353

10. Kon E, Verdonk P, Condello V, Delcogliano M, Dhollander A, Filardo G, Pignotti E, Marcacci M (2009) Matrix-assisted autologous chondrocyte transplantation for the repair of cartilage defects of the knee: systematic clinical data review and study quality analysis. Am J Sports Med 37(Suppl 1):156S–166S. doi:10.1177/0363546509351649

11. Ige OO, Umoru LE, Aribo S (2012) Natural products: a minefield of biomaterials. ISRN Mater Sci 20:1–20

12. Doulabi AH, Mequanint K, Mohammadi H (2014) Blends and nanocomposite biomaterials for articular cartilage tissue engineering. Materials 7:5327–5355

13. Gaharwar AK, Sant S, Hancock MJ, Hacking SA (2013) Nanocomposite polymer: biomaterials for tissue repair of bone and cartilage: a material science perspective. In: Gaharwar AK, Sant S, Hancock MJ, Hacking SA (eds) Nanomaterials in tissue engineering: fabrication and applications. Woodhead Publishing, Cambridge, p 468

14. Gorgieva S, Kokol V (2011) Collagen- vs. gelatine-based biomaterials and their biocompatibility: review and perspectives. In: Pignatello R (ed) Biomaterials applications for nanomedicine. INTECH Open Access Publisher. doi:10.5772/24118

15. Temenoff JS, Mikos AG (2000) Review: tissue engineering for regeneration of articular cartilage. Biomaterials 21(5):431–440

16. Bonzani IC, George JH, Stevens MM (2006) Novel materials for bone and cartilage regeneration. Curr Opin Chem Biol 10(6):568–575. doi:10.1016/j.cbpa.2006.09.009

17. Vinatier C, Bouffi C, Merceron C, Gordeladze J, Brondello JM, Jorgensen C, Weiss P, Guicheux J, Noel D (2009) Cartilage tissue engineering: towards a biomaterial-assisted mesenchymal stem cell therapy. Curr Stem Cell Res Ther 4(4):318–329

18. Cao Z, Dou C, Dong S (2014) Scaffolding biomaterials for cartilage regeneration. J Nanomater 2014:1–8

19. Hollinger JO (2011) An introduction to biomaterials, 2nd edn. The Biomedical Engineering Series. CRC Press, Boca Raton

20. Kloxin AM, Kasko AM, Salinas CN, Anseth KS (2009) Photodegradable hydrogels for dynamic tuning of physical and chemical properties. Science 324(5923):59–63. doi:10.1126/science.1169494

21. Ou KL, Hosseinkhani H (2014) Development of 3D in vitro technology for medical applications. Int J Mol Sci 15(10):17938–17962. doi:10.3390/ijms151017938

22. Quereshi S, Mhaske A, Raut D, Singh R, Mani A, Patel J (2010) Extraction and partial characterization of collagen from different animal skins. Recent Res Sci Technol 2(9):28–31

23. Cheng W, Yan-hua R, Fang-gang N, Guo-an Z (2011) The content and ratio of type I and III collagen in skin differ with age and injury. Afr J Biotechnol 10(13):2524–2529

24. Misra A (2010) Parenteral delivery of peptides and proteins. In: Misra A (ed) Challenges in delivery of therapeutic genomics and proteomics. Elsevier, Burlington

25. Abedin MZ, Karim AA, Ahmed F, Latiff AA, Gan CY, Che Ghazali F, Islam Sarker MZ (2013) Isolation and characterization of pepsin-solubilized collagen from the integument of sea cucumber (*Stichopus vastus*). J Sci Food Agric 93(5):1083–1088. doi:10.1002/jsfa.5854

26. Potaros T, Raksakulthai N, Runglerdkreangkrai J, Worawattanamateekul W (2009) Characteristics of collagen from Nile Tilapia (*Oreochromis niloticus*) skin isolated by two different methods. Kasetsart J 43:584–593

27. Aberoumand A (2012) Comparative study between different methods of collagen extraction from fish and its properties. World Appl Sci J 16(3):316–319

28. Jongjareonrak A (2006) Characterization and functional properties of collagen and gelatin from Bigeye Snapper (*Priacanthus macracanthus*) and Brownstripe Red Snapper (*Lutjanus vitta*) Skins., Prince of Songkla University
29. Noitup P, Garnjanagoonchorn W, Morrissey MT (2005) Fish skin type I collagen characteristic comparison of albacore tuna (*Thunnus alalunga*) and silver-line grunt (*Pomadasys kaakan*). J Aquat Food Prod Technol 14(1):17–27
30. Li H, Liu BL, Gao LZ, Chen HL (2004) Studies on bullfrog skin collagen. Food Chem 84 (1):65–69. doi:10.1016/s0308-8146(03)00167-5
31. Ogawa M, Moody MW, Portier RJ, Bell J, Schexnayder MA, Losso JN (2003) Biochemical properties of black drum and sheepshead seabream skin collagen. J Agric Food Chem 51 (27):8088–8092. doi:10.1021/jf034350r
32. Nagai T, Suzuki N (2000) Isolation of collagen from fish waste material—skin, bone and fins. Food Chem 68(3):277–281. doi:10.1016/S0308-8146(99)00188-0
33. Silva TH, Alves A, Ferreira BM, Oliveira JM, Reys LL, Ferreira RJF, Sousa RA, Silva SS, Mano JF, Reis RL (2012) Materials of marine origin: a review on polymers and ceramics of biomedical interest. Int Mater Rev 57(5):276–306
34. Burke KE, Naughton G, Waldo E, Cassai N (1983) Bovine collagen implant: histologic chronology in pig dermis. J Dermatol Surg Oncol 9(11):889–895
35. Parenteau-Bareil R, Gauvin R, Berthod F (2010) Collagen-based biomaterials for tissue engineering applications. Materials 3(3):1863–1887. doi:10.3390/Ma3031863
36. Yudoh K, Karasawa R (2012) A novel biomaterial for cartilage repair generated by self-assembly: creation of a self-organized articular cartilage-like tissue. J Biomater Nanobiotechnol 3:125–129
37. Kock L, van Donkelaar CC, Ito K (2012) Tissue engineering of functional articular cartilage: the current status. Cell Tissue Res 347(3):613–627. doi:10.1007/s00441-011-1243-1
38. Bhardwaj N, Kundu SC (2011) Silk fibroin protein and chitosan polyelectrolyte complex porous scaffolds for tissue engineering applications. Carbohydr Polym 85(2):325–333. doi:10.1016/j.carbpol.2011.02.027
39. Jin J, Wang J, Huang J, Huang F, Fu J, Yang X, Miao Z (2014) Transplantation of human placenta-derived mesenchymal stem cells in a silk fibroin/hydroxyapatite scaffold improves bone repair in rabbits. J Biosci Bioeng 118(5):593–598. doi:10.1016/j.jbiosc.2014.05.001
40. Tabatabai AP, Kaplan DL, Blair DL (2015) Rheology of reconstituted silk fibroin protein gels: the epitome of extreme mechanics. Soft Matter 11(4):756–761. doi:10.1039/c4sm02079k
41. Sionkowska A, Planecka A, Lewandowska K, Michalska M (2014) The influence of UV-irradiation on thermal and mechanical properties of chitosan and silk fibroin mixtures. J Photochem Photobiol B 140:301–305. doi:10.1016/j.jphotobiol.2014.08.017
42. Hashimoto T, Taniguchi Y, Kameda T, Tamada Y, Kurosu H (2015) Changes in the properties and protein structure of silk fibroin molecules in autoclaved fabrics. Polym Degrad Stab 112:20–26. doi:10.1016/j.polymdegradstab.2014.12.007
43. Lai GJ, Shalumon KT, Chen SH, Chen JP (2014) Composite chitosan/silk fibroin nanofibers for modulation of osteogenic differentiation and proliferation of human mesenchymal stem cells. Carbohydr Polym 111:288–297. doi:10.1016/j.carbpol.2014.04.094
44. Lin L, Hao R, Xiong W, Zhong J (2015) Quantitative analyses of the effect of silk fibroin/nano-hydroxyapatite composites on osteogenic differentiation of MG-63 human osteosarcoma cells. J Biosci Bioeng 119(5):591–595. doi:10.1016/j.jbiosc.2014.10.009
45. Freddi G (2014) Silk fibroin microfiber and nanofiber scaffolds for tissue engineering and regeneration. In: Kundu S (ed) Silk biomaterials for tissue engineering and regenerative medicine. Woodhead, Cambridge (and imprint of Elsevier), pp 157–190. doi:10.1533/9780857097064.1.157
46. Jin SH, Kweon H, Park JB, Kim CH (2014) The effects of tetracycline-loaded silk fibroin membrane on proliferation and osteogenic potential of mesenchymal stem cells. J Surg Res 192(2):e1–e9. doi:10.1016/j.jss.2014.08.054

47. Gong X, Liu H, Ding X, Liu M, Li X, Zheng L, Jia X, Zhou G, Zou Y, Li J, Huang X, Fan Y (2014) Physiological pulsatile flow culture conditions to generate functional endothelium on a sulfated silk fibroin nanofibrous scaffold. Biomaterials 35(17):4782–4791. doi:10.1016/j.biomaterials.2014.02.050

48. Saha S, Kundu B, Kirkham J, Wood D, Kundu SC, Yang XB (2013) Osteochondral tissue engineering in vivo a comparative study using layered silk fibroin scaffolds from mulberry and nonmulberry silkworms. PLoS ONE 8(11):e80004. doi:10.1371/journal.pone.0080004.t001

49. Foss C, Merzari E, Migliaresi C, Motta A (2013) Silk fibroin/hyaluronic acid 3D matrices for cartilage tissue engineering. Biomacromolecules 14(1):38–47. doi:10.1021/bm301174x

50. Zhang Y-Q, Shen W-D, Xiang R-L, Zhuge L-J, Gao W-J, Wang W-B (2007) Formation of silk fibroin nanoparticles in water-miscible organic solvent and their characterization. J Nanopart Res 9:885–900. doi:10.1007/s11051-006-9162-x

51. Yan LP, Oliveira JM, Oliveira AL, Caridade SG, Mano JF, Reis RL (2012) Macro/microporous silk fibroin scaffolds with potential for articular cartilage and meniscus tissue engineering applications. Acta Biomater 8(1):289–301. doi:10.1016/j.actbio.2011.09.037

52. Silva SS, Gomes ME, Motta A, Mano J, Rodrigues TT, Reis RL, Pinheiro AFM, Migliaresi AC (2008) Novel genipin-cross-linked chitosan silk fibroin sponges for cartilage engineering strategies novel genipin-cross. Biomacromolecules 9:2764–2774. doi:10.1021/bm800874q

53. Naeimi M, Fathi M, Rafienia M, Bonakdar S (2014) Silk fibroin-chondroitin sulfate-alginate porous scaffolds structural properties and in vitro studies. J Appl Polym Sci. doi:10.1002/app.41048

54. Vauchel P, Le Roux K, Kaas R, Arhaliass A, Baron R, Legrand J (2009) Kinetics modeling of alginate alkaline extraction from *Laminaria digitata*. Bioresour Technol 100(3):1291–1296. doi:10.1016/j.biortech.2008.03.005

55. Nalamothu N, Potluri A, Muppalla MB (2014) Review on marine alginates and its applications. Indo Am J Pharm Res 4(10):4006–4015

56. Draget KI, Smidsrød O, Skjåk-Bræk G (2005) Alginates from algae. In: Biopolymers Online. Wiley Online Library

57. Vu B, Chen M, Crawford RJ, Ivanova EP (2009) Bacterial extracellular polysaccharides involved in biofilm formation. Molecules 14(7):2535–2554. doi:10.3390/molecules14072535

58. Kloareg B, Quatrano RS (1988) Structure of the cell-walls of marine-algae and ecophysiological functions of the matrix polysaccharides. Oceanogr Mar Biol 26:259–315

59. Fertah M, Belfkira A, Em D, Taourirte M, Brouillette F (2015) Extraction and characterization of sodium alginate from Moroccan *Laminaria digitata* brown seaweed. Arab J Chem 8(1):1–142

60. Haug A (1964) Composition and properties of alginates. Norwegian Institute of Technology, Trondheim

61. Haug A, Larsen B (1966) A study on the constitution of alginic acid by partial acid hydrolysis. Proc Int Seaweed Symp 5:271–277

62. Haug A, Larsen B, Smidsrød O (1966) A study of the constitution of alginic acid by partial hydrolysis. Acta Chem Scand 20:183–190

63. Andersen T, Strand BL, Formo K, Alsberg E, Christensen BE (2012) Alginates as biomaterials in tissue engineering. J Carbohydr Chem 37:227–258

64. Rehm AHB (2009) Alginates: biology and applications. Springer, London

65. Boontheekul T, Kong HJ, Mooney DJ (2005) Controlling alginate gel degradation utilizing partial oxidation and bimodal molecular weight distribution. Biomaterials 26(15):2455–2465. doi:10.1016/j.biomaterials.2004.06.044

66. Li C, Ni C, Xiong C, Li Q (2009) Preparation and drug release of hydrophobically modified alginate. Chemistry 1:93–96

67. Alban S, Schauerte A, Franz G (2002) Anticoagulant sulfated polysaccharides: Part I. Synthesis and structure-activity relationships of new pullulan sulfates. Carbohydr Polym 47(3):267–276. doi:10.1016/S0144-8617(01)00178-3

68. Pluemsab W, Sakairi N, Furuike T (2005) Synthesis and inclusion property of alpha-cyclodextrin-linked alginate. Polymer 46(23):9778–9783. doi:10.1016/J.Polymer.08.005
69. Pelletier S, Hubert P, Payan E, Marchal P, Choplin L, Dellacherie E (2001) Amphiphilic derivatives of sodium alginate and hyaluronate for cartilage repair: rheological properties. J Biomed Mater Res 54(1):102–108
70. Bu HT, Kjoniksen AL, Elgsaeter A, Nystrom B (2006) Interaction of unmodified and hydrophobically modified alginate with sodium dodecyl sulfate in dilute aqueous solution—calorimetric, rheological, and turbidity studies. Colloid Surf A 278(1–3):166–174. doi:10.1016/J.Colsurfa.12.016
71. Yang J-S, Xie Y-J, He W (2010) Research progress on chemical modification of alginate: a review. Carbohydr Polym 84(1):33–39
72. Galant C, Kjoniksen AL, Nguyen GT, Knudsen KD, Nystrom B (2006) Altering associations in aqueous solutions of a hydrophobically modified alginate in the presence of beta-cyclodextrin monomers. J Phys Chem B 110(1):190–195. doi:10.1021/jp0518759
73. Park H, Lee KY (2014) Cartilage regeneration using biodegradable oxidized alginate/hyaluronate hydrogels. J Biomed Mater Res Part A 102(12):4519–4525. doi:10.1002/jbm.a.35126
74. Awad HA, Wickham MQ, Leddy HA, Gimble JM, Guilak F (2004) Chondrogenic differentiation of adipose-derived adult stem cells in agarose, alginate, and gelatin scaffolds. Biomaterials 25(16):3211–3222. doi:10.1016/j.biomaterials.2003.10.045
75. Willerth SM, Sakiyama-Elbert SE (2008) Combining stem cells and biomaterial scaffolds for constructing tissues and cell delivery. In: StemBook. Cambridge (MA)
76. Hannouche D, Terai H, Fuchs JR, Terada S, Zand S, Nasseri BA, Petite H, Sedel L, Vacanti JP (2007) Engineering of implantable cartilaginous structures from bone marrow-derived mesenchymal stem cells. Tissue Eng 13(1):87–99. doi:10.1089/ten.2006.0067
77. Wayne JS, McDowell CL, Shields KJ, Tuan RS (2005) In vivo response of polylactic acid-alginate scaffolds and bone marrow-derived cells for cartilage tissue engineering. Tissue Eng 11(5–6):953–963. doi:10.1089/ten.2005.11.953
78. Jin X, Sun Y, Zhang K, Wang J, Shi T, Ju X, Lou S (2007) Ectopic neocartilage formation from predifferentiated human adipose derived stem cells induced by adenoviral-mediated transfer of hTGF beta2. Biomaterials 28(19):2994–3003. doi:10.1016/j.biomaterials.2007.03.002
79. Dunne LW, Iyyanki T, Hubenak J, Mathur AB (2014) Characterization of dielectrophoresis-aligned nanofibrous silk fibroin-chitosan scaffold and its interactions with endothelial cells for tissue engineering applications. Acta Biomater 10(8):3630–3640. doi:10.1016/j.actbio.2014.05.005
80. Bhardwaj N, Nguyen QT, Chen AC, Kaplan DL, Sah RL, Kundu SC (2011) Potential of 3-D tissue constructs engineered from bovine chondrocytes/silk fibroin-chitosan for in vitro cartilage tissue engineering. Biomaterials 32(25):5773–5781. doi:10.1016/j.biomaterials.2011.04.061
81. Fang J, Zhang Y, Yan S, Liu Z, He S, Cui L, Yin J (2014) Poly(L-glutamic acid)/chitosan polyelectrolyte complex porous microspheres as cell microcarriers for cartilage regeneration. Acta Biomater 10(1):276–288. doi:10.1016/j.actbio.2013.09.002
82. Baran ET, Tuzlakoglu K, Mano JF, Reis RL (2012) Enzymatic degradation behavior and cytocompatibility of silk fibroin-starch-chitosan conjugate membranes. Mater Sci Eng C 32 (6):1314–1322. doi:10.1016/j.msec.2012.02.015
83. Fei Liu X, Lin Guan Y, Zhi Yang D, Li Z, De Yao K (2001) Antibacterial action of chitosan and carboxymethylated chitosan. J Appl Polym Sci 79(7):1324–1335
84. Zhang K, Zhang Y, Yan S, Gong L, Wang J, Chen X, Cui L, Yin J (2013) Repair of an articular cartilage defect using adipose-derived stem cells loaded on a polyelectrolyte complex scaffold based on poly(l-glutamic acid) and chitosan. Acta Biomater 9(7):7276–7288. doi:10.1016/j.actbio.2013.03.025
85. Mirahmadi F, Tafazzoli-Shadpour M, Shokrgozar MA, Bonakdar S (2013) Enhanced mechanical properties of thermosensitive chitosan hydrogel by silk fibers for cartilage tissue engineering. Mater Sci Eng C 33(8):4786–4794. doi:10.1016/j.msec.2013.07.043

86. Sionkowska A, Płanecka A (2013) Surface properties of thin films based on the mixtures of chitosan and silk fibroin. J Mol Liq 186:157–162. doi:10.1016/j.molliq.2013.07.008

87. Oprenyeszk F, Chausson M, Maquet V, Dubuc JE, Henrotin Y (2013) Protective effect of a new biomaterial against the development of experimental osteoarthritis lesions in rabbit: a pilot study evaluating the intra-articular injection of alginate-chitosan beads dispersed in an hydrogel. Osteoarthr Cartil 21(8):1099–1107. doi:10.1016/j.joca.2013.04.017

88. Lafantaisie-Favreau CH, Guzman-Morales J, Sun J, Chen G, Harris A, Smith TD, Carli A, Henderson J, Stanish WD, Hoemann CD (2013) Subchondral pre-solidified chitosan/blood implants elicit reproducible early osteochondral wound-repair responses including neutrophil and stromal cell chemotaxis, bone resorption and repair, enhanced repair tissue integration and delayed matrix deposition. BMC Musculoskelet Disord 14:27. doi:10.1186/1471-2474-14-27

89. Bell AD, Lascau-Coman V, Sun J, Chen G, Lowerison MW, Hurtig MB, Hoemann CD (2012) Bone-induced chondroinduction in sheep jamshidi biopsy defects with and without treatment by subchondral chitosan-blood implant: 1-day, 3-week, and 3-month repair. Cartilage 4(2):131–143. doi:10.1177/1947603512463227

90. Mathieu C, Chevrier A, Lascau-Coman V, Rivard GE, Hoemann CD (2013) Stereological analysis of subchondral angiogenesis induced by chitosan and coagulation factors in microdrilled articular cartilage defects. Osteoarthr Cartil 21(6):849–859. doi:10.1016/j.joca.2013.03.012

91. Zhao F, He W, Yan Y, Zhang H, Zhang G, Tian D, Gao H (2014) The application of polysaccharide biocomposites to repair cartilage defects. Int J Polym Sci 2014:9

92. Meyer K, Palmer JW (1934) The polysaccharide of the vitreous humor. J Biol Chem 107:629–634

93. Murado MA, Montemayor MI, Cabo ML, Vazquez JA, Gonzalez MP (2012) Optimization of extraction and purification process of hyaluronic acid from fish eyeball. Food Bioprod Process 90(C3):491–498. doi:10.1016/J.Fbp.11.002

94. Shen B, Wei A, Bhargav D, Kishen T, Diwan AD (2010) Hyaluronan: its potential application in intervertebral disc regeneration. Orthop Res Rev 2:17–26

95. Boeriu CG, Springer J, Kooy FK, van den Broek LAM, Eggink G (2013) Production methods for hyaluronan. Int J Carbohydr Chem 2013:14

96. Liu L, Liu Y, Li J, Du G, Chen J (2011) Microbial production of hyaluronic acid: current state, challenges, and perspectives. Microb Cell Fact 10:99. doi:10.1186/1475-2859-10-99

97. Collins MN, Birkinshaw C (2013) Hyaluronic acid based scaffolds for tissue engineering—a review. Carbohydr Polym 92(2):1262–1279. doi:10.1016/j.carbpol.2012.10.028

98. Collins MN, Birkinshaw C (2008) Physical properties of crosslinked hyaluronic acid hydrogels. J Mater Sci Mater Med 19(11):3335–3343. doi:10.1007/s10856-008-3476-4

99. Kang JY, Chung CW, Sung JH, Park BS, Choi JY, Lee SJ, Choi BC, Shim CK, Chung SJ, Kim DD (2009) Novel porous matrix of hyaluronic acid for the three-dimensional culture of chondrocytes. Int J Pharm 369(1–2):114–120. doi:10.1016/j.ijpharm.2008.11.008

100. Kim IL, Mauck RL, Burdick JA (2011) Hydrogel design for cartilage tissue engineering: a case study with hyaluronic acid. Biomaterials 32(34):8771–8782. doi:10.1016/j.biomaterials.2011.08.073

101. Julovi SM, Ito H, Nishitani K, Jackson CJ, Nakamura T (2011) Hyaluronan inhibits matrix metalloproteinase-13 in human arthritic chondrocytes via CD44 and P38. J Orthop Res 29(2):258–264. doi:10.1002/jor.21216

102. Laroui H, Grossin L, Leonard M, Stoltz JF, Gillet P, Netter P, Dellacherie E (2007) Hyaluronate-covered nanoparticles for the therapeutic targeting of cartilage. Biomacromolecules 8(12):3879–3885. doi:10.1021/bm700836y

103. Gomoll AH (2009) Serum levels of hyaluronic acid and chondroitin sulfate as a non-invasive method to evaluate healing after cartilage repair procedures. Arthritis Res Ther 11(4):118. doi:10.1186/ar2730

104. Unterman SA, Gibson M, Lee JH, Crist J, Chansakul T, Yang EC, Elisseeff JH (2012) Hyaluronic acid-binding scaffold for articular cartilage repair. Tissue Eng Part A 18(23–24):2497–2506. doi:10.1089/ten.TEA.2011.0711
105. Osmalek T, Froelich A, Tasarek S (2014) Application of gellan gum in pharmacy and medicine. Int J Pharm 466(1–2):328–340. doi:10.1016/j.ijpharm.2014.03.038
106. da Silva RMP, Lopez-Perez PM, Elvira C, Mano JF, Roman JS, Reis RL (2008) Poly (N-Isopropylacrylamide) surface-grafted chitosan membranes as a new substrate for cell sheet engineering and manipulation. Biotechnol Bioeng 101(6):1321–1331. doi:10.1002/Bit.22004
107. Novac O, Lisa G, Profire L, Tuchilus C, Popa MI (2014) Antibacterial quaternized gellan gum based particles for controlled release of ciprofloxacin with potential dermal applications. Mater Sci Eng 35:291–299. doi:10.1016/j.msec.2013.11.016
108. Kang D, Zhang F, Zhang H (2015) Fabrication of stable aqueous dispersions of graphene using gellan gum as a reducing and stabilizing agent and its nanohybrids. Mater Chem Phys 149–150:129–139. doi:10.1016/j.matchemphys.2014.09.055
109. Oliveira JT, Martins L, Picciochi R, Malafaya PB, Sousa RA, Neves NM, Mano JF, Reis RL (2010) Gellan gum: a new biomaterial for cartilage tissue engineering applications. J Biomed Mater Res Part A 93(3):852–863. doi:10.1002/jbm.a.32574
110. Prajapati VD, Jani GK, Zala BS, Khutliwala TA (2013) An insight into the emerging exopolysaccharide gellan gum as a novel polymer. Carbohydr Polym 93(2):670–678. doi:10.1016/j.carbpol.2013.01.030
111. da Silva LP, Cerqueira MT, Sousa RA, Reis RL, Correlo VM, Marques AP (2014) Engineering cell-adhesive gellan gum spongy-like hydrogels for regenerative medicine purposes. Acta Biomater 10(11):4787–4797. doi:10.1016/j.actbio.2014.07.009
112. Coutinho DF, Sant SV, Shin H, Oliveira JT, Gomes ME, Neves NM, Khademhosseini A, Reis RL (2010) Modified Gellan Gum hydrogels with tunable physical and mechanical properties. Biomaterials 31(29):7494–7502. doi:10.1016/j.biomaterials.2010.06.035
113. Tang Y, Sun J, Fan H, Zhang X (2012) An improved complex gel of modified gellan gum and carboxymethyl chitosan for chondrocytes encapsulation. Carbohydr Polym 88(1):46–53. doi:10.1016/j.carbpol.2011.11.058
114. Shi YG, Meng YC, Li JR, Chen J, Liu YH, Bai X (2014) Chondroitin sulfate: extraction, purification, microbial and chemical synthesis. J Chem Technol Biot 89(10):1445–1465. doi:10.1002/Jctb.4454
115. Jerosch J (2011) Effects of glucosamine and chondroitin sulfate on cartilage metabolism in OA: outlook on other nutrient partners especially omega-3 fatty acids. Int J Rheumatol 2011:17
116. Michelacci YM, Dietrich CP (1976) Structure of chondroitin sulfates. Analyses of the products formed from chondroitin sulfates A and C by the action of the chondroitinases C and AC from Flavobacterium heparinum. Biochim Biophys Acta 451(2):436–443
117. Seno N, Meyer K (1963) Comparative biochemistry of skin; the mucopolysaccharides of shark skin. Biochim Biophys Acta 78:258–264
118. Chen JS, Chang CM, Wu JC, Wang SM (2000) Shark cartilage extract interferes with cell adhesion and induces reorganization of focal adhesions in cultured endothelial cells. J Cell Biochem 78(3):417–428
119. Srinivasan SR, Radhakrishinamurthy B, Dalferes ER Jr, Berenson GS (1969) Glycosaminoglycans from squid skin. Comp Biochem Physiol 28(1):169–176
120. Karamanos NK, Aletras AJ, Antonopoulos CA, Tsegenidis T, Tsiganos CP, Vynios DH (1988) Extraction and fractionation of proteoglycans from squid skin. Biochim Biophys Acta 966(1):36–43
121. Vazquez JA, Rodriguez-Amado I, Montemayor MI, Fraguas J, Gonzalez Mdel P, Murado MA (2013) Chondroitin sulfate, hyaluronic acid and chitin/chitosan production using marine waste sources: characteristics, applications and eco-friendly processes: a

review. Mar Drugs 11(3):747–774. doi:10.3390/md11030747

122. Sugahara K, Tanaka Y, Yamada S, Seno N, Kitagawa H, Haslam SM, Morris HR, Dell A (1996) Novel sulfated oligosaccharides containing 3-O-sulfated glucuronic acid from king crab cartilage chondroitin sulfate K. Unexpected degradation by chondroitinase ABC. J Biol Chem 271(43):26745–26754

123. Borsig L, Wang LC, Cavalcante MCM, Cardilo-Reis L, Ferreira PL, Mourao PAS, Esko JD, Pavao MSG (2007) Selectin blocking activity of a fucosylated chondroitin sulfate glycosaminoglycan from sea cucumber—effect on tumor metastasis and neutrophil recruitment. J Biol Chem 282(20):14984–14991. doi:10.1074/Jbc.M610560200

124. Mano JF (2013) Biomimetic approaches for biomaterials development. Wiley, New York

125. Kim S-K (2015) Springer handbook of marine biotechnology. Springer, London

126. Cole AG, Hall BK (2004) The nature and significance of invertebrate cartilages revisited: distribution and histology of cartilage and cartilage-like tissues within the Metazoa. Zoology (Jena) 107(4):261–273. doi:10.1016/j.zool.2004.05.001

127. Lai JY, Li YT, Cho CH, Yu TC (2012) Nanoscale modification of porous gelatin scaffolds with chondroitin sulfate for corneal stromal tissue engineering. Int J Nanomed 7:1101–1114. doi:10.2147/IJN.S28753

128. Muller FA, Muller L, Hofmann I, Greil P, Wenzel MM, Staudenmaier R (2006) Cellulose-based scaffold materials for cartilage tissue engineering. Biomaterials 27 (21):3955–3963. doi:10.1016/j.biomaterials.2006.02.031

129. Orellana J (2013) Advanced biomaterials from renewable resources: an investigation on cellulose nanocrystal composites and CO_2 extraction of rendered materials. Clemson University, Clemson

130. Yodmuang S, McNamara SL, Nover AB, Mandal BB, Agarwal M, Kelly TA, Chao PH, Hung C, Kaplan DL, Vunjak-Novakovic G (2015) Silk microfiber-reinforced silk hydrogel composites for functional cartilage tissue repair. Acta Biomater 11:27–36. doi:10.1016/j. actbio.2014.09.032

131. Khanarian NT, Haney NM, Burga RA, Lu HH (2012) A functional agarose-hydroxyapatite scaffold for osteochondral interface regeneration. Biomaterials 33(21):5247–5258. doi:10. 1016/j.biomaterials.2012.03.076

132. Zignego DL, Jutila AA, Gelbke MK, Gannon DM, June RK (2014) The mechanical microenvironment of high concentration agarose for applying deformation to primary chondrocytes. J Biomech 47(9):2143–2148. doi:10.1016/j.jbiomech.2013.10.051

133. Masson P, Beinert G, Franta E, Rempp P (1982) Synthesis of polyethylene oxide macromers. Polym Bull 7(1):17–22

134. Dai X, Chen X, Yang L, Foster S, Coury AJ, Jozefiak TH (2011) Free radical polymerization of poly(ethylene glycol) diacrylate macromers: impact of macromer hydrophobicity and initiator chemistry on polymerization efficiency. Acta Biomater 7(5):1965–1972. doi:10.1016/j.actbio.2011.01.005

135. Knop K, Hoogenboom R, Fischer D, Schubert US (2010) Poly(ethylene glycol) in drug delivery: pros and cons as well as potential alternatives. Angew Chem Int Ed Engl 49 (36):6288–6308. doi:10.1002/anie.200902672

136. Konradi R, Acikgoz C, Textor M (2012) Polyoxazolines for nonfouling surface coatings–a direct comparison to the gold standard PEG. Macromol Rapid Commun 33(19):1663–1676. doi:10.1002/marc.201200422

137. Sah H, Thoma LA, Desu HR, Sah E, Wood GC (2013) Concepts and practices used to develop functional PLGA-based nanoparticulate systems. Int J Nanomed 8:747–765. doi:10. 2147/IJN.S40579

138. Bencherif SA, Srinivasan A, Sheehan JA, Walker LM, Gayathri C, Gil R, Hollinger JO, Matyjaszewski K, Washburn NR (2009) End-group effects on the properties of PEG-co-PGA hydrogels. Acta Biomater 5(6):1872–1883. doi:10.1016/j.actbio.2009.02.030

139. Kinard LA, Kasper FK, Mikos AG (2012) Synthesis of oligo(poly(ethylene glycol) fumarate). Nat Protoc 7(6):1219–1227. doi:10.1038/nprot.2012.055

140. Micic M, Jeremic M, Radotic K, Leblanc RM (2000) A comparative study of enzymatically and photochemically polymerized artificial lignin supramolecular structures using environmental scanning electron microscopy. J Colloid Interface Sci 231(1):190–194. doi:10.1006/jcis.2000.7136

141. Hutanu D, Frishberg MD, Guo L, Darie CC (2014) Recent applications of polyethylene glycols (PEGs) and PEG derivatives. Mod Chem Appl 2(12):2

142. Ferretti M, Marra KG, Kobayashi K, Defail AJ, Chu CR (2006) Controlled in vivo degradation of genipin crosslinked polyethylene glycol hydrogels within osteochondral defects. Tissue Eng 12(9):2657–2663. doi:10.1089/ten.2006.12.2657

143. Eswaramoorthy R, Chang CC, Wu SC, Wang GJ, Chang JK, Ho ML (2012) Sustained release of PTH(1-34) from PLGA microspheres suppresses osteoarthritis progression in rats. Acta Biomater 8(6):2254–2262. doi:10.1016/j.actbio.2012.03.015

144. Zhu Y, Wan Y, Zhang J, Yin D, Cheng W (2014) Manufacture of layered collagen/chitosan-polycaprolactone scaffolds with biomimetic microarchitecture. Colloids Surf B 113:352–360. doi:10.1016/j.colsurfb.2013.09.028

145. Morille M, Van-Thanh T, Garric X, Cayon J, Coudane J, Noel D, Venier-Julienne MC, Montero-Menei CN (2013) New PLGA-P188-PLGA matrix enhances TGF-beta3 release from pharmacologically active microcarriers and promotes chondrogenesis of mesenchymal stem cells. J Control Release 170(1):99–110. doi:10.1016/j.jconrel.2013.04.017

146. Chang NJ, Lin CC, Li CF, Wang DA, Issariyaku N, Yeh ML (2012) The combined effects of continuous passive motion treatment and acellular PLGA implants on osteochondral regeneration in the rabbit. Biomaterials 33(11):3153–3163. doi:10.1016/j.biomaterials.2011.12.054

147. Rotunda AM, Narins RS (2006) Poly-L-lactic acid: a new dimension in soft tissue augmentation. Dermatol Ther 19(3):151–158. doi:10.1111/j.1529-8019.2006.00069.x

148. Narins RS, Baumann L, Brandt FS, Fagien S, Glazer S, Lowe NJ, Monheit GD, Rendon MI, Rohrich RJ, Werschler WP (2010) A randomized study of the efficacy and safety of injectable poly-ʟ-lactic acid versus human-based collagen implant in the treatment of nasolabial fold wrinkles. J Am Acad Dermatol 62(3):448–462

149. Ramos AR (2014) Applications of PLA-poly(lactic acid) in tissue engineering and delivery systems. Dissertation, Biomedical Materials and Devices from University of Aveiro (Portugal).

150. Lai WC (2011) Thermal behavior and crystal structure of poly(L-lactic acid) with 1,3:2,4-dibenzylidene-D-sorbitol. J Phys Chem B 115(38):11029–11037. doi:10.1021/jp2037312

151. Hyun MY, Lee Y, No YA, Yoo KH, Kim MN, Hong CK, Chang SE, Won CH, Kim BJ (2015) Efficacy and safety of injection with poly-L-lactic acid compared with hyaluronic acid for correction of nasolabial fold: a randomized, evaluator-blinded, comparative study. Clin Exp Dermatol 40(2):129–135. doi:10.1111/ced.12499

152. Danesin R, Brun P, Roso M, Delaunay F, Samouillan V, Brunelli K, Iucci G, Ghezzo F, Modesti M, Castagliuolo I, Dettin M (2012) Self-assembling peptide-enriched electrospun polycaprolactone scaffolds promote the h-osteoblast adhesion and modulate differentiation-associated gene expression. Bone 51(5):851–859. doi:10.1016/j.bone.2012.08.119

153. Pan L, Pei X, He R, Wan Q, Wang J (2012) Multiwall carbon nanotubes/polycaprolactone composites for bone tissue engineering application. Colloids Surf B 93:226–234. doi:10.1016/j.colsurfb.2012.01.011

154. Chen C-H, Liu J, Chua C-K, Chou S-M, Shyu V, Chen J-P (2014) Cartilage tissue engineering with silk fibroin scaffolds fabricated by indirect additive manufacturing technology. Materials 7(3):2104–2119. doi:10.3390/ma7032104

155. Constantin M, Cristea M, Ascenzi P, Fundueanu G (2011) Lower critical solution temperature versus volume phase transition temperature in thermoresponsive drug delivery systems. Express Polym Lett 5(10):839–848
156. Shen ZY, Terao K, Maki Y, Dobashi T, Ma GH, Yamamoto T (2006) Synthesis and phase behavior of aqueous poly(N-isopropylacrylamide-co-acrylamide), poly(N-isopropylacrylamide-co-N, N-dimethylacrylamide) and poly (N-isopropylacrylamide-co-2-hydroxyethyl methacrylate). Colloid Polym Sci 284(9):1001–1007. doi:10.1007/S00396-005-1442-Y
157. Schilli CM, Zhang MF, Rizzardo E, Thang SH, Chong YK, Edwards K, Karlsson G, Muller AHE (2004) A new double-responsive block copolymer synthesized via RAFT polymerization: poly(N-isopropylacrylamide)-block-poly(acrylic acid). Macromolecules 37 (21):7861–7866. doi:10.1021/Ma035838w
158. Bearat HH, Lee BH, Valdez J, Vernon BL (2011) Synthesis, characterization and properties of a physically and chemically gelling polymer system using poly(NIPAAm-co-HEMA-acrylate) and poly(NIPAAm-co-cysteamine). J Biomater Sci 22:1299–1318
159. Spasojevic M, Vorenkamp J, Jansen MRPACS, de Vos P, Schouten AJ (2014) Synthesis and phase behavior of poly(N-isopropylacrylamide)-b-poly(L-lysine hydrochloride) and poly (N-Isopropylacrylamide-co-acrylamide)-b-poly(L-lysine hydrochloride). Materials 7 (7):5305–5326. doi:10.3390/Ma7075305
160. Robb SA, Lee BH, McLemore R, Vernon BL (2007) Simultaneously physically and chemically gelling polymer system utilizing a poly(NIPAAm-co-cysteamine)-based copolymer. Biomacromolecules 8(7):2294–2300. doi:10.1021/bm070267r
161. Aguilar MR, Elvira C, Gallardo A, Vázquez B, Román JS (2007) Smart polymers and their applications as biomaterials. In: Ashammakhi N, Reis R, Chiellini E (eds) Topics in tissue engineering, vol 3

Chapter 7
Fundamentals on Osteochondral Tissue Engineering

Viviana Ribeiro, Sandra Pina, Joaquim Miguel Oliveira
and Rui Luís Reis

Abstract The repair and regeneration of osteochondral (OC) defects has been increasing owing the high number of diseases, trauma and injuries. Although current clinical options are effective for the treatment of the OC lesions, new therapeutic options are necessary for the complete regeneration of the damaged articular cartilage which has a limited healing capacity. OC tissue engineering has been proposing advanced tools and technologies involving structured scaffolds, bioactive molecules, and cells for the repair and regeneration of the bone and cartilage tissues, as well as their interface. Multi-phased or stratified scaffolds with distinct bone and cartilage sections have been designed for OC repair. Diverse forms, as porous scaffolds, fibres, and hydrogels are the most commonly strategies used for OC tissue engineering. This chapter presents the current treatment and biomimetic strategies for OC tissue engineering. Structure and properties of the OC tissue are also briefly described.

7.1 Introduction

Osteochondral (OC) tissue engineering has the potential of producing grafts with tailor-made mechanical properties and topology of the graft essential for the repair/regeneration of OC defects. The purpose is to culture cells on the scaffolds over a period of time to be further implanted in vivo (Fig. 7.1). OC defects are

---

Viviana Ribeiro and Sandra Pina have contributed equally to this work.

V. Ribeiro (✉) · S. Pina · J.M. Oliveira · R.L. Reis
3B's Research Group—Biomaterials, Biodegradables and Biomimetics,
Headquarters of the European Institute of Excellence on Tissue
Engineering and Regenerative Medicine, University of Minho,
AvePark, Zona Industrial da Gandra, 4805-017 Barco, Guimarães, Portugal
e-mail: viviana.ribeiro@dep.uminho.pt

S. Pina
e-mail: sandra.pina@dep.uminho.pt

V. Ribeiro · S. Pina · J.M. Oliveira · R.L. Reis
ICVS/3B's—PT Government Associate Laboratory, Braga, Guimarães, Portugal

© Springer International Publishing AG 2017
J.M. Oliveira and R.L. Reis (eds.), *Regenerative Strategies for the Treatment of Knee Joint Disabilities*, Studies in Mechanobiology, Tissue Engineering and Biomaterials 21, DOI 10.1007/978-3-319-44785-8_7

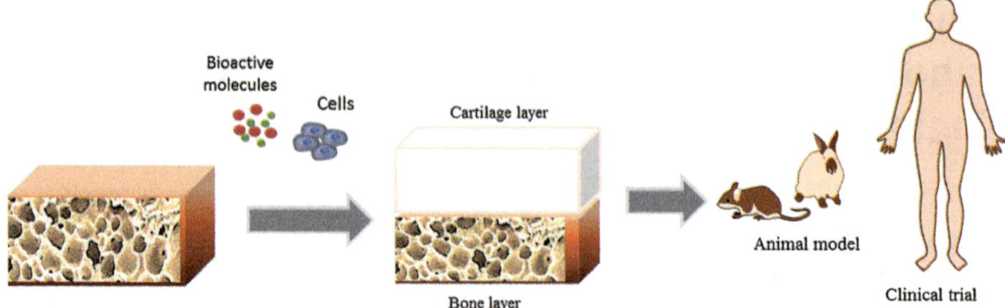

Fig. 7.1 Osteochondral tissue engineering strategy

lesions of the articular cartilage and underlying subchondral bone often derived from trauma related injuries or osteoarthritis, causing joint pain and deformity, impaired function, limited range of motion and stiffness [1].

Currently, OC treatment strategies include nonsurgical treatment with immobilization and use of non-steroidal anti-inflammatory drugs, and surgical treatment such as osteotomy, abrasion arthroplasty, microfracture, implantation of autografts, and autologous chondrocyte implantation [2–5]. These treatments are well established and effective for reducing pain improving the patient's quality of life. However, the articular cartilage has a limited regeneration capacity demanding new therapeutic options for complete healing of the OC lesions. Thus, specific structured biomimetic scaffolds have emerged as promising alternatives for the repair and regeneration of subchondral bone and cartilage considering their different architectures [6]. Cartilage can be distinguished into four distinct zones, defined by a particular composition and organization of cells and extracellular matrix (ECM) that significantly influence the mechanical properties of each area [7]. Bone is a dynamic and complex tissue composed of water and an ECM consisting of an organic matrix (collagen and non-collagenous proteins) and hydroxyapatite (HAp) crystals that provide stiffness and structural support to the body [8, 9]. It is a highly vascularised tissue with a remarkable intrinsic ability to remodel and spontaneously regenerate [10]. However, the true challenge lies in the complexity of the tissue interface, making it challenging for the design and fabrication of scaffolds [1, 9].

Different strategies have been proposed for OC repair and regeneration namely the development of different scaffold types: (i) single scaffolds for the bone and the cartilage parts combined at the time of implantation, (ii) single scaffold for the bone part, (iii) single homogeneous scaffold for both components, and (iv) single heterogeneous composite scaffolds [8, 11]. The latter, also known as bilayered scaffolds, have attracted singular attention and some of them have already been used in preclinical animal studies and clinical trials [12]. These types of scaffolds have unique composition and organization, structural strength and specific biological properties. They engineer cartilage using biopolymers (e.g., proteins, polysaccharides, glycosaminoglycans) and synthetic polymers, and bone engineering is carried out by combining polymers with bioactive/resorbable inorganic materials. Polymers have high strength and design flexibility and natural polymers also hold significant similarities with the ECM, chemical versatility, and good biological performance without toxicity or immunological reactions [13]. Conversely, bioresorbable materials, such as calcium phosphates (CaPs) (e.g., β-tricalcium phosphate, HAp) and bioactive glasses have favourable osteoconductivity, resorbability and biocompatibility [14, 15].

Usefulness of bilayered scaffolds, however, is limited by the lack of stability and scarce tissue integration. To overcome these limitations, adherent multi-phased or stratified scaffolds with distinct subchondral bone and cartilage compartments have been employed. This was based on the use of different biomaterials and entailed stratifications of mineral content, ECM components and porosity. These studies propose the incorporation of an interface band with homogeneous intermediate characteristics, a strategy that has brought forth promising results.

Herein, it is updated current treatments and strategies for OC repair and regeneration, with special focus on the scaffolds design. OC tissue structure and properties are briefly presented, in order to better understand the ideal constructs to be developed.

7.2 Osteochondral Tissue: Structure and Properties

Before designing scaffolds for OC tissue engineering, it is essential to consider the functional environment where the scaffolds will be implanted, as well as the anatomy of the surrounding tissues involved in the regenerative process. The natural OC tissue consists of two main tissues, articular cartilage and subchondral bone, connected by a stable interface. A description of each tissue is issued as follows.

7.2.1 Articular Cartilage

Cartilage is a flexible and supportive tissue found in several areas of the human body. The three main types of cartilage are the fibrocartilage, elastic cartilage and hyaline cartilage. Hyaline cartilage is a dense connective tissue that provides smooth surfaces for joint motion, being found in the articular cartilage of all human body joints [1, 16, 17]. Biochemically, articular cartilage tissue is comprised of a solid phase of about 15–32 % and a fluid phase of about 68–85 %. Collagen, chondrocytes, proteoglycans and a few more minor proteins are the main constituents of the solid phase. The fluid phase is mainly composed of water [17–19]. Each component of cartilage holds a specific role to maintain its supportive nature. Unlike highly vascularised tissues like bone, articular cartilage is avascular and has a low cell density, which leads to a poor self-repair capacity [7]. In all areas of the tissue, collagen fibrils form a dense and highly interconnected matrix that host aggrecan, a highly glycosylated molecule with net negative charge. It is bound in large amounts to glycosaminoglycans (GAG's) chains, including hyaluronic acid, chondroitin and keratin sulphates, to form proteoglycan aggregates responsible for raising the osmolarity of the tissue. The resulting swelling is countered by the resistance of the collagen matrix, generating a large internal pressure that gives to cartilage its unique mechanical properties [16, 20]. Type II collagen is the main containing of articular cartilage, however, types V, VI, IX and XI also play an important role in the intermolecular interactions and as modulators of type II collagen activity [19]. The entire articular cartilage contains the same basic components; however their different proportions and arrangement in ECM composition allowed defining the cartilage tissue as a multi-layered structure composed of four distinct zones: superficial (tangential), middle (transitional), deep (radical) and calcified cartilage zones (Fig. 7.2) [8, 17, 19]. The superficial zone constitutes the upper 10–20 % of the articular cartilage and is composed of thin collagen fibrils parallel to the articular surface. Cell density is at its highest point, unlike aggrecan that is found at the lowest concentrations. The flattened chondrocytes found in this zone also secrete specialized proteins that maintain the upper region of the articular surface lubricated and prevents the wear and friction of the tissue [18, 21–23]. The next zone is the middle zone, which consists of the following 40–60 % of the articular cartilage. It is composed by thicker collagen fibrils with a more random organization (less parallel) [24]. Chondrocytes are presented in a lower number with a round morphology, however, it is noted an abundance of proteoglycans. The properties presented by the middle zone lead to high compressive modulus that

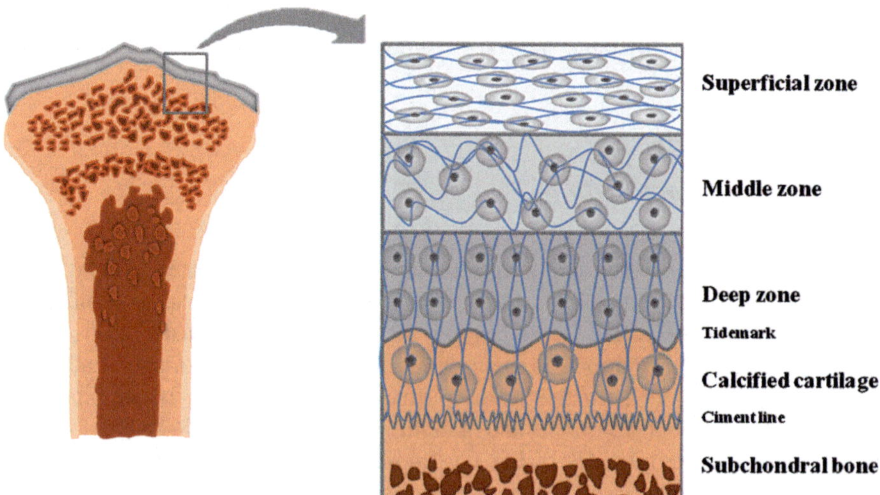

Fig. 7.2 Schematic diagram of a long bone cross section showing the normal articular cartilage divided in *superficial, middle, deep* and *calcified cartilage* zones with a *tidemark* between the deep zone and the calcified region, followed by the underling subchondral bone with a *cement line* at the interface

facilitates the recovery from impacts sustained by the articular surface [16, 18, 21, 25]. The remaining thickness of the articular cartilage is taken up by the deep zone and the calcified cartilage zone [21]. In the deep zone, both cells and collagen fibrils are perpendicularly orientated to the articular surface, with the chondrocytes grouped in vertical columns or clusters. Collagen fibrils diameter is maximal in this region, but the collagen content is also the lowest [24]. This zone also has a high compressive modulus and content of proteoglycans, but chondrocytes are presented in its lowest number and with a larger size [16, 18, 25]. In the bottom of the deep zone is a thin wavy line known as "tidemark". This mark represents a calcification front by dividing the deep zone and the last small region, the calcified zone [26]. The calcified cartilage zone constitutes the interface between the flexible cartilage and the rigid subchondral bone [1]. This region contains hypertrophic circular chondrocytes surrounded by type X collagen that is believed to assist the mineralization process between the cartilage and the underlying bone [18, 27, 28]. Structurally, the collagen fibrils from the deep zone traverse through this area anchoring in the subchondral bone. The existence of interdigitations ("cement line") between the calcified cartilage and the underlying bone provides a tighter adherence at the interface, favouring the load bearing and force distribution [7, 29–31]. The wavy tidemark and vertical orientation of the collagen fibrils, combined with the interdigitations existing at the interface, may reduce the stress concentrated at this area and allow for a better integration within the bone tissue [9].

7.2.2 Subchondral Bone

Bone is a dynamic and complex tissue with a high ability for self-repair and remodelling. The main roles of bone tissue are to provide mechanical support to the body and mineral homeostasis [32, 33]. Briefly, bone tissue can be arranged in two architectural forms: compact or cortical bone and trabecular or spongy bone. Compact bone is a dense structural tissue that comprises 80 % of an adult human skeleton, with a low porosity of about 5–10 % [32–34]. The components of compact bone are organized into repeated cylindrical units called osteons or haversian systems, composed of a central (haversian) canal surrounded by concentric lamellae rings of calcified extracellular matrix [35]. Trabecular bone presents a higher porosity of 50–90 % and comprises about 20 % of the total skeleton. It is arranged in a sponge-like form, characterized by plates or struts of various sizes called trabeculae. Another important function of trabecular bone is to host the bone marrow that contains high proportions of mesenchymal and hematopoietic stem cells [32–34]. The elaboration, maintenance and resorption of bone tissue results from the interaction of three cell types originating from the bone marrow reservoir: osteoblasts (bone formation), osteocytes (calcification of the matrix and calcium homeostasis) and osteoclasts (bone resorption) [36, 37]. The cooperation of these cells makes possible the constant production and degradation of bone tissue, but there is a need for a balance between bone formation and resorption that is controlled by signalling molecules and biochemical forces. Quantitatively, cells do not represent a high volume of the bone tissue, which is mainly composed of small crystals of an inorganic phase of HAp and some amorphous calcium phosphate compounds (65–70 % of the matrix), and an organic phase composed of collagen, glycoproteins, proteoglycans, sialoproteins and bone "gla" proteins (25–30 % of the total matrix) [38]. Bone tissue formation process (osteogenesis) is characterized by relatively high cell content with numerous osteoprogenitor cells and growth factors. Similarly, the bone healing process after a fracture event also mobilize a high number of cells and growth factors, however at this stage the inflammatory response is also initiated [39]. Subchondral bone comprises the latter region of the OC tissue, composed of two bone types: immediately below the calcified cartilage lies compact bone (subchondral bone plate) followed by trabecular bone (supporting trabecular bone) [31]. The subchondral bone plate consists of a solid mineralized mass of bone with low porosity and vascularity that separates the articular cartilage from the bone marrow. Depending upon the joint, the subchondral bone plate varies in thickness, ranging from 0.2 to 0.4 mm in humans. It is composed of an abundant matrix of collagen fibrils, mainly type I collagen, and minerals (e.g. calcium and phosphate), able to adapt its structure for acting on load bearing. This functional adaptation of the subchondral bone plate results in a preferential direction of the collagen fibrils at microstructural level [9, 17]. The trabeculae of the supporting trabecular bone are highly vascularised and present a predominant perpendicular orientation to the joint surface. These trabeculae are themselves crossed perpendicularly by finer trabeculae that continue from the lamellar sheets of the subchondral bone plate [31]. The two main functions of subchondral bone are to absorb and maintain joint shape. Although articular cartilage provides some compressive strength to the joint, it is the subchondral bone that sustains most of the strength since it has a larger area and a lower modulus of elasticity [40]. In fact, there is an increased stiffness gradient among the non-calcified cartilage, calcified cartilage, subchondral bone plate and supporting trabecular bone, which may be one of the causes of the cartilage damage caused by mechanical stresses [31]. While more research has been directed to the damages on articular cartilage resulting from certain OC defects [41–44], it is now recognized the integral role of subchondral bone in the pathogenesis of these defects [45]. The loss of articular cartilage associated with some OC defects, at some point leads to the exposure of the underlying subchondral bone to the surrounding forces, which is signalized by the occurrence of pain generated by the nerve endings existing at this tissue [46]. Therefore, changes in subchondral bone may act as an early marker for more complicated OC defects.

Summarizing, the natural joint is a complex composite that gradually changes the material combinations/properties, volume fractions and anisotropy along its structure. It can be modelled as a biphasic system composed of a permeable cartilage phase and an underlying porous but rigid subchondral bone region. Given the complexity of the biology and mechanisms of OC tissue, the challenges to engineer this tissue include developing a low friction surface that integrates well with the surrounding cartilage tissue and ensuring a proper implant integration maintaining the mechanical properties of the tissue.

7.3 Strategies for Osteochondral Tissue Engineering

OC tissue engineering involve bioactive and biodegradable structures with controlled pore structure and mechanical stability, and capable of directing cell-matrix and cell–cell interactions, thus mimicking the natural ECM. Scaffolding materials such as porous structures (single and bilayered), hydrogels and fibrous networks are the prospective candidates for OC defects repair/regeneration which promote growth of both bone and cartilage layers in a single integrated scaffold. The following section presents a brief description of OC tissue engineering approaches focusing on bilayered, fibrous and hydrogel scaffolds, design and properties.

7.3.1 Bilayered Scaffolds

Scaffolds are designed to act as temporary support structures to the surrounding bone tissue mimicking ECM, with advantageous characteristics, including: (i) porous structure that promotes cell-biomaterial interactions, cell adhesion, growth and migration, (ii) interconnected pores to facilitate transport of mass, nutrients, and regulatory factors to allow cell survival, proliferation, and differentiation, (iii) adequate mechanical properties as tensile strength and elasticity, (iv) controlled degradation, (v) synthesis of new bone formation with homogeneous distribution to avoid necrosis, and (vi) minimal degree of inflammation or toxicity in vivo [47]. Besides, scaffolds have desirable characteristics for cell transfer into a defect site and to restrict cell loss, instead of simple injection of cells to the defects.

Several conventional methods have been used to produce scaffolds with controlled pore size and porosity, such as foam replica, particulate-leaching/solvent casting, freeze-drying, phase separation, gas foaming, and rapid prototyping [48–52]. The challenge is the production of scaffolds ensuring a good compatibility between the phases while keeping the porous structure and the mechanical properties. It has been reported that a pore size larger than 300 μm and porosity of about 50–90 % is acclaimed for an enhanced osteogenesis, while a pore size in the range 90–120 μm are recommended for chondrogenesis [53, 54]. Besides, it is important to achieve a homogeneous structure between the polymer matrix and the bioresorbable fillers.

Bilayered scaffolds have been constructed aiming OC tissue engineering applications, consisting of both osteogenic and chondrogenic regions, manufactured in a single integrated implant, thus simultaneously promoting regeneration of bone and cartilage with different properties and biological requirements [49, 55–57]. Oliveira

Fig. 7.3 Chitosan/HAp bilayered scaffold showing two different layers, respectively for cartilage (*upper layer*) and bone (*bottom layer*). Reprinted with permission [49]. Copyright 2006, Elsevier

et al. [49] developed an integrated macroporous bilayered scaffold made of HAp and chitosan combining sintering and freeze-drying techniques, with two distinct layers, respectively for cartilage and bone (Fig. 7.3). The bone layer was constituted of porous HAp and produced through foam replica method, while the cartilage layer was composed of chitosan poured on the top of the HAp scaffold, previously prepared, followed by freeze-drying technique. The scaffolds presented porosities of 59.3 ± 1.7 % for HAp layer and 74.6 ± 1.2 % for chitosan layer. The compression modulus (E) for the HAp and chitosan was determined to be 153 ± 12 and 2.9 ± 0.4 MPa, respectively. Moreover, it was shown through in vitro cell studies that both layers provided an adequate support for cell attachment, proliferation and differentiation into osteoblasts and chondrocytes, respectively. More recently, Yao et al. [57] prepared a 45S5 Bioglass®/chitosan-polycaprolactone (CS-PCL) bilayered scaffold by a combination of foam replication and freeze-drying methods aiming OC tissue engineering applications. The bone layer was composed by a bioglass-based scaffold coated with CS-PCL blend, while a PCL-CS foam fabricated by freeze-drying was used in the cartilage layer. The scaffolds showed a porosity of 81 % with favourable mechanical strength. In vitro studies indicated that the scaffolds supported the adhesion and growth of MG-63 cells with favourable proliferation behavior after 3 weeks of incubation. Wang et al. [56] fabricated a bilayered scaffold using articular cartilage extracellular matrix and HAp, which involved a porous oriented upper layer and a dense mineralized lower layer using the "liquid-phase cosynthesis" technique [58] (Fig. 7.4). Mean pore

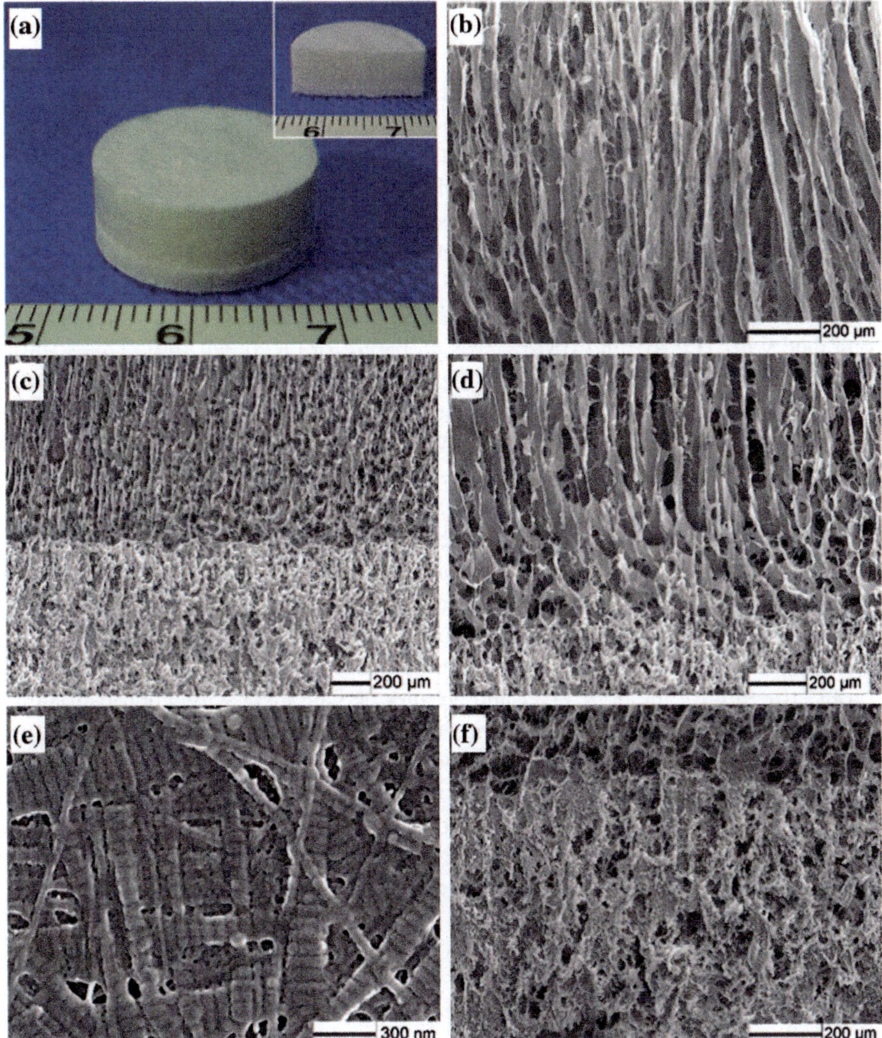

Fig. 7.4 Image of the articular cartilage ECM/HAp bilayered scaffold prototype (**a**) and respective scanning electron microstructures: *upper layer* (**b**), interface region (**c**, **d**), and nanofibrous ECM orientation in the *upper layer* of the scaffold wall at a high magnification (**e**) and in the *lower layer* (**f**). Reprinted with permission [56]. Copyright 2014, BioMed Central

sizes of 21.2 ± 3.1 μm and 128.2 ± 20.3 μm, and porosity of 92.6 ± 6 % and 44 ± 3 %, respectively for the upper and lower layers were obtained. In vitro culture performed using rabbit chondrocytes seeded on the bilayered scaffolds showed that the chondrocytes were well-distributed in the non-mineralized component (upper layer) and do not penetrated into the mineralized component (lower layer).

Growth factors have been used for tissue engineering due to their beneficial effect in the cell growth and tissue formation. The construction of scaffolds using this strategy has been endeavoured. A biomimetic and multi-phasic scaffold design, spatially controlled and localized gene delivery system and multi-lineage differentiation of a single stem cell population were combined to design a bilayered gene-activated OC scaffold [59]. The scaffold consisted of plasmid transforming growth factor-β1 (TGF-β1)-activated chitosan-gelatin for the cartilage layer and plasmid bone morphogenetic protein-2 (BMP-2)-activated HAp/chitosan-gelatin for the subchondral bone layer. This scaffold could induce mesenchymal stem cells (MSCs) to differentiate into chondrocytes and osteoblasts, respectively for the cartilage and subchondral bone, using a rabbit knee OC defect model. Results showed cell proliferation, high expression of TGF-β1 protein and BMP-2 protein. In another study, a scaffold made of poly D,L-lactic-co-glycolic acid (PLGA) microspheres combined with TGF-β1 and BMP-2 was prepared for OC repair [60]. This scaffold consisted of PLGA microspheres and TGF-β1 for the chondrogenic layer and PLGA microspheres and BMP-2 with or without HAp for the osteogenic layer. In vivo results showed complete bone ingrowth, with an overlying cartilage layer with high glycosaminoglycan content, appropriate thickness, and integration with the surrounding cartilage and underlying bone, after implantation in rabbit knees.

Recently, Ding et al. [61] took a different approach by preparing an integrated trilayered scaffold composed of silk fibroin and HAp by combining paraffin-sphere leaching with a modified temperature gradient-guided thermal-induced phase separation (TIPS) technique for OC tissue engineering. The scaffold is constituted by three layers: longitudinally oriented microtubular structure for the cartilage layer, a porous structure for the bone layer and an intermediate layer with a dense structure. Live/dead tests indicated good biocompatibility of the scaffolds for supporting the growth, proliferation, and infiltration of adipose-derived stem cells, which could be induced to differentiate toward chondrocytes or osteoblasts in the presence of chondrogenic- or osteogenic-induced culture medium, respectively for the cartilage and bone layers. Besides, the intermediate layer could play an isolating role for preventing the mixture of the cells within the cartilage and bone layers.

7.3.2 Fibrous Scaffolds

Fibrous scaffolds, particularly nanoscale fibrous, have been employed for the repair and regeneration of natural tissues, as they are capable of mimicking the network of

ECM. These types of scaffolds have high porosities and homogeneous pore distribution, which facilitates cell adhesion, proliferation, and differentiation [62].

Fibrous scaffolds can be obtained through electrospinning, phase separation, and molecular self-assembly. Electrospinning is the most commonly used technique since it can produce fibres in the order of nanometers, with interconnected pore structure and higher mechanical strength [63].

Composite scaffolds using fibres have been used as a strategy for developing OC tissue repair [64]. Yunos et al. [64] developed a bilayered scaffold using a matrix of electrospun poly-DL-lactide (PDLLA) fibres combined with 45S5 Bioglass®-based foam for OC tissue replacement (Fig. 7.5). PDLLA fibres were produced through electrospinning with diameters between ∼100 nm and ∼0.2 μm and good integration onto the scaffolds (Fig. 7.5b). In vitro tests using chondrocyte cells showed increased cell attachment and proliferation on the scaffolds.

By its turn, Zhang et al. [65] fabricated a bilayered collagen/microporous poly-L-lactide (PLLA) nanofibers scaffold for focal OC defect regeneration. It was shown that the scaffolds promoted osteogenic differentiation of bone marrow stem cells (BMSCs) in vitro, and improved bone tissue formation after implantation in a rabbit. Filová et al. [66] evaluated the effect of a hyaluronate/ type I collagen/fibrin composite scaffold containing polyvinyl alcohol (PVA) nanofibers with liposomes and functionalized with basic fibroblast growth factor (bFGF) and insulin on the

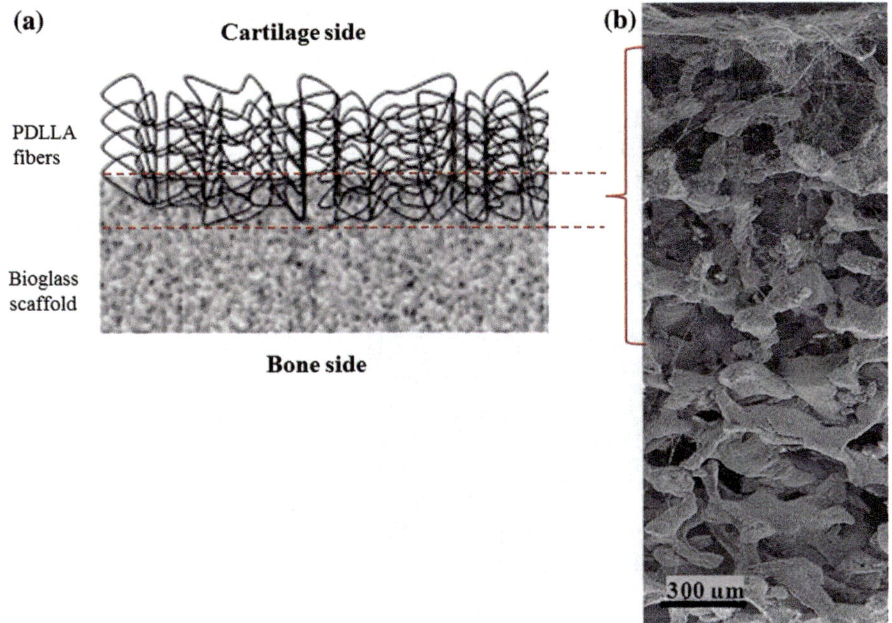

Fig. 7.5 Schematic diagram of the bilayered composite scaffolds (**a**) and scanning electron micrograph of the scaffolds cross-section (**b**). Adapted with permission [64]. Copyright 2011, SAGE Publications

regeneration of OC defects. It was shown that the nanofibers improved the composite scaffolds mechanical properties and MSC viability in vitro. Besides, OC regeneration towards hyaline cartilage and/or fibrocartilage was enhanced after 12 weeks implantation in miniature pigs.

7.3.3 Hydrogel Scaffolds

Hydrogels comprise a three-dimensional network highly hydrated, hydrophilic cross-linked that mimic the natural ECM, including cartilage [67]. Hydrogels have been employed as scaffolding materials for tissue engineering owed their swollen network structure, diffusion of nutrients and by-products from cells metabolism, and biocompatibility. In addition, hydrogels are able to encapsulate cells, biomolecules and growth factors, for a controlled delivery after implantation. Techniques such as, solvent casting and particulate leaching, freeze-drying, phase separation, gas foaming, solvent evaporation, and blending with non-cross-linkable linear polymers have been used to produce the porosity in hydrogels [68].

Hydrogel biphasic scaffolds combining different materials have attracted great deal of attention for OC repair/regeneration. Pereira et al. [69] produced biphasic hydrogel scaffolds consisting of low acyl gellan gum (LAGG) and HAp for the bone layer and LAGG for the cartilage layer, with total cohesion of the whole structure (Fig. 7.6). Rodrigues et al. [70] reported the development of a bilayered

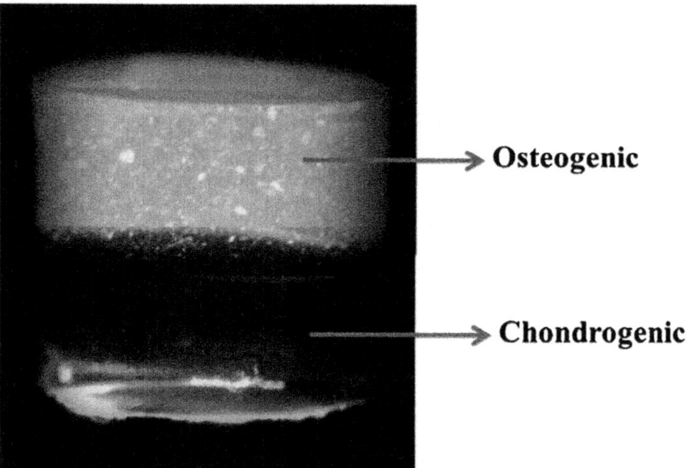

Fig. 7.6 Macroscopic appearance of the LAGG/LAGG-HAp hydrogels. Reprinted with permission [69]. Copyright 2014, Trans Tech Publications

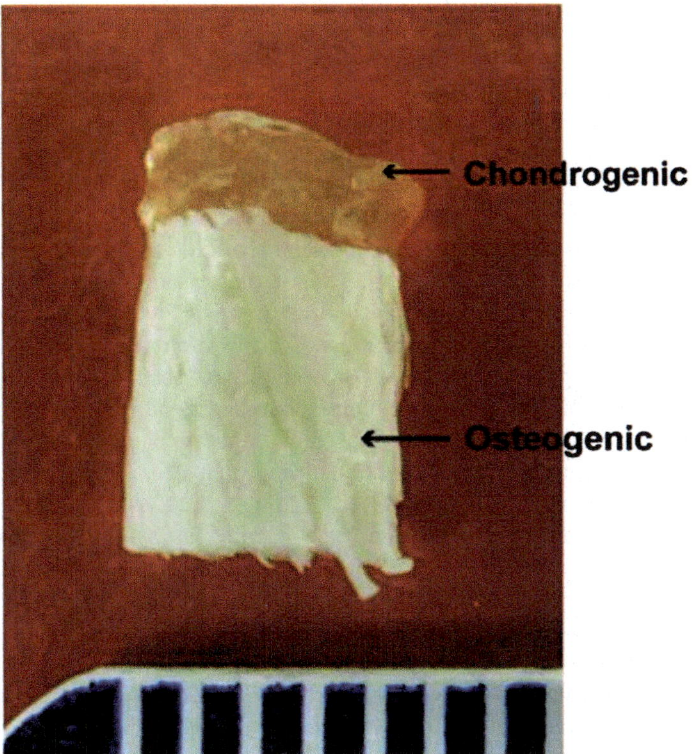

Fig. 7.7 Image of agarose-SPCL bilayered scaffold. Reprinted with permission [70]. Copyright 2012, Elsevier

scaffold which combines a starch/polycaprolactone (SPCL) scaffold for osteogenesis and hydrogel of agarose for chondrogenesis and containing pre-differentiated amniotic fluid-derived stem cells (Fig. 7.7).

Biodegradable composite hydrogels encapsulating growth factors have been also investigated. Bilayered OC constructs mimicking the hierarchical OC structure were produced using oligo polyethylene glycol fumarate hydrogel containing gelatin microparticles and stem cells, via a two-step cross-linking process [71, 72]. Results of live/dead assay showed that cell viability was maintained in osteogenic and chondrogenic layers and the incorporation of TGF-β3 significantly stimulated chondrogenic differentiation of the stem cells in the chondrogenic layer [71]. Besides, after 12 weeks of implantation in a rabbit full-thickness OC defect, it was observed a significant improvement in the cartilage morphology when comparing with a blank hydrogel [72].

7.4 Concluding Remarks

Functional repair and regeneration of OC tissue remains a challenge, and none of the existing treatment procedures gives the reliable good outcome. OC tissue engineering has the potential to develop novel strategies for the repair and regeneration of OC lesions comprising subchondral bone, and cartilage, as well as the interface between them, which involve physicochemical and mechanical properties of materials, cell types and interaction between scaffolds and native tissues. These strategies are influenced by the complexity of the cartilage and underlying subchondral bone compositions, mechanical structure, and functional properties.

Stratified scaffolds with distinct bone and cartilage phases in a single structure have been proposing as one of the most optimal OC scaffolds, characterized by differential microporosity, chemical composition, and good mechanical properties, aiming ECM deposition and adequate cell differentiation and proliferation, and for the regeneration of new tissue. Current emerging trends for OC tissue engineering focused on the production of different scaffolding structures e.g., bilayered structures, fibrous, and hydrogels, using diverse natural and synthetic polymers, and inorganic materials, and resulting composites, using several fabrication methods. Inorganic materials are optimal candidates for the bone part due to their inherent biocompatibility, while the polymers are suitable for the cartilage part because they confer adequate mechanical strength and toughness to the structures.

Another important factor to be considered in OC tissue engineering is the combination of the scaffolds with specific cells for chondrogenic and osteogenic potential, and growth factors in order to regulate cell differentiation and also to improve the mechanical properties after implantation. Under this perspective, in vivo studies in animal models must be considered to evaluate integration, osteogenesis, and vascularization of the scaffolds, and long term stability to translate into the clinical.

Acknowledgments The research leading to this work has received funding from the European Union's Seventh Framework Program (FP7/2007-2013) under grant Agreement No REGPOT-CT2012-316331-POLARIS, and from QREN (ON.2—NORTE-01-0124-FEDER-000016) cofinanced by North Portugal Regional Operational Program (ON.2—O Novo Norte), under the National Strategic Reference Framework (NSRF), through the European Regional Development Fund (ERDF). Thanks are also due to the Portuguese Foundation for Science and Technology (FCT) and FSE/POCH (Fundo Social Europeu através do Programa Operacional do Capital Humano), PD/59/2013, for the project PEst-C/SAU/LA0026/201, for the fellowship grants of Sandra Pina (SFRH/BPD/108763/2015) and Viviana Ribeiro (PD/BD/113806/2015), and for the distinction attributed to J.M. Oliveira under the Investigator FCT program (IF/00423/2012).

References

1. Nukavarapu SP, Dorcemus DL (2013) Osteochondral tissue engineering: current strategies and challenges. Biotechnol Adv 31(5):706–721
2. Xing LZ, Jiang YQ, Gui JC, Lu YM, Gao F, Xu Y, Xu Y (2013) Microfracture combined with osteochondral paste implantation was more effective than microfracture alone for full-thickness cartilage repair. Knee Surg Sport Traumatol Arthrosc 21(8):1770–1776

3. Kim YS, Park EH, Lee HJ, Koh YG, Lee JW (2012) Clinical comparison of the osteochondral autograft transfer system and subchondral drilling in osteochondral defects of the first metatarsal head. Am J Sport Med 40(8):1824–1833

4. de Girolamo L, Quaglia A, Bait C, Cervellin M, Prospero E, Volpi P (2012) Modified autologous matrix-induced chondrogenesis (AMIC) for the treatment of a large osteochondral defect in a varus knee: a case report. Knee Surg Sport Traumatol Arthrosc 20(11):2287–2290

5. Miska M, Wiewiorski M, Valderrabano V (2012) Reconstruction of a large osteochondral lesion of the distal tibia with an iliac crest graft and autologous matrix-induced chondrogenesis (AMIC): a case report. J Foot Ankle Surg 51(5):680–683

6. Puppi D, Chiellini F, Piras AM, Chiellini E (2010) Polymeric materials for bone and cartilage repair. Prog Polym Sci 35(4):403–440

7. Harley BA, Lynn AK, Wissner-Gross Z, Bonfield W, Yannas IV, Gibson LJ (2010) Design of a multiphase osteochondral scaffold III: fabrication of layered scaffolds with continuous interfaces. J Biomed Mater Res A 92A(3):1078–1093

8. Nooeaid P, Salih V, Beier JP, Boccaccini AR (2012) Osteochondral tissue engineering: scaffolds, stem cells and applications. J Cell Mol Med 16(10):2247–2270

9. Yang PJ, Temenoff JS (2009) Engineering orthopedic tissue interface. Tissue Eng Part B 15 (2):127–141

10. Shrivats AR, McDermott MC, Hollinger JO (2014) Bone tissue engineering: state of the union. Drug Discov Today 19(6):781–786

11. Martin I, Miot S, Barbero A, Jakob M, Wendt D (2007) Osteochondral tissue engineering. J Biomech 40(4):750–765

12. Yan L, Oliveira J, Oliveira A, Reis R (2014) Silk fibroin/nano-CaP bilayered scaffolds for osteochondral tissue engineering. Key Eng Mater 587:245

13. Mano J, Silva G, Azevedo H, Malafaya P, Sousa R, Silva S, Reis R (2007) Natural origin biodegradable systems in tissue engineering and regenerative medicine: present status and some moving trends. J R Soc Interface 4:999–1030

14. Bohner M (2000) Calcium orthophosphates in medicine: from ceramics to calcium phosphate cements. Injury-Int J Care Injured 31:37–47

15. Dorozhkin S (2009) Calcium orthophosphates in nature. Biol Med Mater 2:399–498

16. Verbruggen G, Wang J, Elewaut D, Veys EM (2004) Biology of articular cartilage. In: Proceedings of the 5th symposium of the international cartilage repair society—ICRS, pp 21–25

17. Swieszkowski W, Tuan BHS, Kurzydlowski KJ, Hutmacher DW (2007) Repair and regeneration of osteochondral defects in the articular joints. Biomol Eng 24(5):489–495

18. Poole AR, Kojima T, Yasuda T, Mwale F, Kobayashi M, Laverty S (2001) Composition and structure of articular cartilage—a template for tissue repair. Clin Orthop Relat Res 391:S26–S33

19. García-Carvajal ZY, Garciadiego-Cázares D, Parra-Cid C, Aguilar-Gaytán R, Velasquillo C, Ibarra C, Carmona JSC (2013) Cartilage tissue engineering: the role of extracellular matrix (ECM) and novel strategies. Regen Med Tissue Eng, 365–397. InTech, Croatia

20. Athanasiou KA, Zhu CF, Wang X, Agrawal CM (2000) Effects of aging and dietary restriction on the structural integrity of rat articular cartilage. Ann Biomed Eng 28(2):143–149

21. Pearle AD, Warren RF, Rodeo SA (2005) Basic science of articular cartilage and osteoarthritis. Clin Sport Med 24(1):1

22. Khalafi A, Schmid TM, Neu C, Reddi AH (2007) Increased accumulation of superficial zone protein (SZP) in articular cartilage in response to bone morphogenetic protein-7 and growth factors. J Orthop Res 25(3):293–303

23. Neu CP, Khalafi A, Komvopoulos K, Schmid TM, Reddi AH (2007) Mechanotransduction of bovine articular cartilage superficial zone protein by transforming growth factor beta signaling. Arthritis Rheum 56(11):3706–3714

24. Korhonen RK, Julkunen P, Wilson W, Herzog W (2008) Importance of collagen orientation and depth-dependent fixed charge densities of cartilage on mechanical behavior of chondrocytes. J Biomech Eng-T ASME 130(2):021003

25. Huber M, Trattnig S, Lintner F (2000) Anatomy, biochemistry, and physiology of articular cartilage. Invest Radiol 35(10):573–580

26. Lyons TJ, Stoddart RW, McClure SF, McClure J (2005) The tidemark of the chondro-osseous junction of the normal human knee joint. J Mol Histol 36(3):207–215

27. Eyre D (2002) Collagen of articular cartilage. Arthritis Res 4(1):30–35

28. Hyc A, Osiecka-Iwan A, Jozwiak J, Moskalewski S (2001) The morphology and selected biological properties of articular cartilage. Ortop Traumatol Rehabil 3(2):151–162

29. Zizak I, Roschger P, Paris O, Misof BM, Berzlanovich A, Bernstorff S, Amenitsch H, Klaushofer K, Fratzl P (2003) Characteristics of mineral particles in the human bone/cartilage interface. J Struct Biol 141(3):208–217

30. Lu HH, Subramony SD, Boushell MK, Zhang XZ (2010) Tissue engineering strategies for the regeneration of orthopedic interfaces. Ann Biomed Eng 38(6):2142–2154

31. Madry H, van Dijk CN, Mueller-Gerbl M (2010) The basic science of the subchondral bone. Knee Surg Sport Traumatol Arthrosc 18(4):419–433

32. Costa-Pinto AR, Reis RL, Neves NM (2011) Scaffolds based bone tissue engineering: the role of chitosan. Tissue Eng Part B-Rev 17(5):331–347

33. Cordonnier T, Sohier J, Rosset P, Layrolle P (2011) Biomimetic materials for bone tissue engineering—state of the art and future trends. Adv Eng Mater 13(5):B135–B150

34. Salgado AJ, Coutinho OP, Reis RL (2004) Bone tissue engineering: State of the art and future trends. Macromol Biosci 4(8):743–765

35. Hutmacher DW, Schantz JT, Lam CX, Tan KC, Lim TC (2007) State of the art and future directions of scaffold-based bone engineering from a biomaterials perspective. J Tissue Eng Regen Med 1(4):245–260

36. Ducy P, Schinke T, Karsenty G (2000) The osteoblast: a sophisticated fibroblast under central surveillance. Science 289(5484):1501–1504

37. Mackie EJ (2003) Osteoblasts: novel roles in orchestration of skeletal architecture. Int J Biochem Cell Biol 35(9):1301–1305

38. Sommerfeldt DW, Rubin CT (2001) Biology of bone and how it orchestrates the form and function of the skeleton. Eur Spine J 10:S86–S95

39. Sundelacruz S, Kaplan DL (2009) Stem cell- and scaffold-based tissue engineering approaches to osteochondral regenerative medicine. Semin Cell Dev Biol 20(6):646–655. doi:10.1016/j.semcdb.2009.03.017

40. Kawcak CE, McIlwraith CW, Norrdin RW, Park RD, James SP (2001) The role of subchondral bone in joint disease: a review. Equine Vet J 33(2):120–126

41. Mahmoudifar N, Doran PM (2005) Tissue engineering of human cartilage and osteochondral composites using recirculation bioreactors. Biomaterials 26(34):7012–7024

42. Dahlin RL, Kinard LA, Lam J, Needham CJ, Lu S, Kasper FK, Mikos AG (2014) Articular chondrocytes and mesenchymal stem cells seeded on biodegradable scaffolds for the repair of cartilage in a rat osteochondral defect model. Biomaterials 35(26):7460–7469

43. Kon E, Filardo G, Perdisa F, Venieri G, Marcacci M (2014) Acellular matrix–based cartilage regeneration techniques for osteochondral repair. Oper Tech Orthop 24(1):14–18

44. Sherwood JK, Riley SL, Palazzolo R, Brown SC, Monkhouse DC, Coates M, Griffith LG, Landeen LK, Ratcliffe A (2002) A three-dimensional osteochondral composite scaffold for articular cartilage repair. Biomaterials 23(24):4739–4751

45. Burr DB (2004) Anatomy and physiology of the mineralized tissues: role in the pathogenesis of osteoarthrosis. Osteoarth Cartil 12:20–30

46. Moisio K, Eckstein F, Chmiel JS, Guermazi A, Prasad P, Almagor O, Song J, Dunlop D, Hudelmaier M, Kothari A, Sharma L (2009) Denuded subchondral bone and knee pain in persons with knee osteoarthritis. Arthritis Rheum 60(12):3703–3710

47. Thorrez L, Shansky J, Wang L, Fast L, VandenDriessche T, Chuah M, Mooney D, Vandenburgh H (2008) Growth, differentiation, transplantation and survival of human skeletal myofibers on biodegradable scaffolds. Biomaterials 29(1):75–84

48. Hou Q, Grijpma D, Feijen J (2003) Porous polymeric structures for tissue engineering prepared by a coagulation, compression moulding and salt leaching technique. Biomaterials 24:1937–1947
49. Oliveira JM, Rodrigues MT, Silva SS, Malafaya PB, Gomes ME, Viegas CA, Dias IR, Azevedo JT, Mano JF, Reis RL (2006) Novel hydroxyapatite/chitosan bilayered scaffold for osteochondral tissue-engineering applications: Scaffold design and its performance when seeded with goat bone marrow stromal cells. Biomaterials 27:6123–6137
50. Dehghani F, Annabi N (2011) Engineering porous scaffolds using gas-based techniques. Curr Opin Biotechnol 22:661–666
51. vande Witte P, Dijkstra P, vanden Berg J, Feijen J (1996) Phase separation processes in polymer solutions in relation to membrane formation. J Membr Sci 117:1–31
52. Woodfield TBF, Guggenheim M, Von Rechenberg B, Riesle J, Van Blitterswijk CA, Wedler V (2009) Rapid prototyping of anatomically shaped, tissue-engineered implants for restoring congruent articulating surfaces in small joints. Cell Prolif 42(4):485–497
53. Karageorgiou V, Kaplan D (2005) Porosity of 3D biomaterial scaffolds and osteogenesis. Biomaterials 26:5474–5491
54. Kim K, Yeatts A, Dean D (2010) Stereolithographic bone scaffold design parameters: osteogenic differentiation and signal expression. Tissue Eng Part B Rev 16:523–539
55. Yan LP, Silva-Correia J, Correia C, Caridade SG, Fernandes EM, Sousa RA, Mano JF, Oliveira JM, Oliveira AL, Reis RL (2013) Bioactive macro/micro porous silk fibroin/nano-sized calcium phosphate scaffolds with potential for bone-tissue-engineering applications. Nanomedicine (Lond) 8(3):359–378
56. Wang Y, Meng H, Yuan X, Peng J, Guo Q, Lu S, Wang A (2014) Fabrication and in vitro evaluation of an articular cartilage extracellular matrix-hydroxyapatite bilayered scaffold with low permeability for interface tissue engineering. Biomed Eng Online 13:80
57. Yao Q, Nooeaid P, Detsch R, Roether JA, Dong Y, Goudouri OM, Schubert DW, Boccaccini AR (2014) Bioglass((R))/chitosan-polycaprolactone bilayered composite scaffolds intended for osteochondral tissue engineering. J Biomed Mater Res A 102(12):4510–4518
58. Harley B, Lynn A, Wissner-Gross Z, Bonfield W, Yannas I, Gibson L (2010) Design of a multiphase osteochondral scaffold III: fabrication of layered scaffolds with continuous interfaces. J Biomed Mater Res A 92:1078–1093
59. Chen J, Chen H, Li P, Diao H, Zhu S, Dong L, Wang R, Guo T, Zhao J, Zhang J (2011) Simultaneous regeneration of articular cartilage and subchondral bone in vivo using MSCs induced by a spatially controlled gene delivery system in bilayered integrated scaffolds. Biomaterials 32(21):4793–4805
60. Mohan N, Dormer NH, Caldwell KL, Key VH, Berkland CJ, Detamore MS (2011) Continuous gradients of material composition and growth factors for effective regeneration of the osteochondral interface. Tissue Eng Part A 17(21–22):2845–2855
61. Ding X, Zhu M, Xu B, Zhang J, Zhao Y, Ji S, Wang L, Wang L, Li X, Kong D, Ma X, Yang Q (2014) Integrated trilayered silk fibroin scaffold for osteochondral differentiation of adipose-derived stem cells. ACS Appl Mater Interfaces 6(19):16696–16705
62. Ng R, Zang R, Yang K, Liu N, Yang S (2012) Three-dimensional fibrous scaffolds with microstructures and nanotextures for tissue engineering. RSC Adv 2:10110–10124
63. Hong J, Madihally S (2011) Next generation of electrosprayed fibers for tissue regeneration. Tissue Eng Part B Rev 17:125–142
64. Yunos DM, Ahmad Z, Salih V, Boccaccini AR (2013) Stratified scaffolds for osteochondral tissue engineering applications: electrospun PDLLA nanofibre coated Bioglass(R)-derived foams. J Biomater Appl 27(5):537–551
65. Zhang S, Chen L, Jiang Y, Cai Y, Xu G, Tong T, Zhang W, Wang L, Ji J, Shi P, Ouyang HW (2013) Bi-layer collagen/microporous electrospun nanofiber scaffold improves the osteochondral regeneration. Acta Biomater 9(7):7236–7247
66. Filová E, Rampichová M, Litvinec A, Držík M, Míčková A, Buzgo M, Košťáková E, Martinová L, Usvald D, Prosecká E, Uhlík J, Motlík J, Vajner L, Amler E (2013) A cell-free

nanofiber composite scaffold regenerated osteochondral defects in miniature pigs. Int J Pharm 447(1–2):139–149

67. Slaughter BV, Khurshid SS, Fisher OZ, Khademhosseini A, Peppas NA (2009) Hydrogels in regenerative medicine. Adv Mater 21(32–33):3307–3329

68. Annabi N, Nichol J, Zhong X, Ji C, Koshy S, Khademhosseini A, Dehghani F (2010) Controlling the porosity and microarchitecture of hydrogels for tissue engineering. Tissue Eng Part B Rev 16:371–383

69. Pereira D, Canadas R, Silva-Correia J, Marques A, Reis R, Oliveira J (2014) Gellan gum-based hydrogel bilayered scaffolds for osteochondral tissue engineering. Key Eng Mater 587:255–260

70. Rodrigues MT, Lee SJ, Gomes ME, Reis RL, Atala A, Yoo JJ (2012) Bilayered constructs aimed at osteochondral strategies: the influence of medium supplements in the osteogenic and chondrogenic differentiation of amniotic fluid-derived stem cells. Acta Biomater 8(7):2795–2806

71. Guo X, Liao J, Park H, Saraf A, Raphael RM, Tabata Y, Kasper FK, Mikos AG (2010) The effects of TGF-β3 and preculture period of osteogenic cells on the chondrogenic differentiation of rabbit marrow mesenchymal stem cells encapsulated in a bilayered hydrogel composite. Acta Biomater 6(8):2920–2931

72. Kim K, Lam J, Lu S, Spicer PP, Lueckgen A, Tabata Y, Wong ME, Jansen JA, Mikos AG, Kasper FK (2013) Osteochondral tissue regeneration using a bilayered composite hydrogel with modulating dual growth factor release kinetics in a rabbit model. J Controll Release 168 (2):166–178

Chapter 8
Pre-clinical and Clinical Management of Osteochondral Lesions

Sandra Pina, Viviana Ribeiro, Joaquim Miguel Oliveira
and Rui Luís Reis

Abstract The majority of osteochondral (OC) lesions occur after injury or trauma of both bone and the overlying cartilage, and symptoms are pain and disability, leading to the risk of inducing osteoarthritis. These lesions are currently repaired by non-surgical and surgical methods or by advanced tissue engineering strategies, which require a proof of efficacy and safety for regulatory approval for human application. Pre-clinical studies using animal models have been the support of OC repair and regeneration with successful clinical outcomes. Small animal models as mice and rabbits, and large animal models as sheep, goats and horses, have been most commonly used according with the outcome goals. Small animals are recommended as a proof of concept, while large animals are endorsed for truly translational research in order to get the regulatory approval for clinical use in humans. An up-to-date of the in vivo studies using different animal models and ongoing clinical trials for the repair and regeneration of OC lesions are presented. Commercialised products for OC repair are also indicated.

Sandra Pina and Viviana Ribeiro have contributed equally to this work.

S. Pina (✉) · V. Ribeiro · J.M. Oliveira · R.L. Reis
3B's Research Group—Biomaterials, Biodegradables and Biomimetics,
Headquarters of the European Institute of Excellence on Tissue Engineering
and Regenerative Medicine, University of Minho, Avepark—Parque de Ciência e
Tecnologia, Zona Industrial da Gandra, 4805-017 Barco GMR, Portugal
e-mail: sandra.pina@dep.uminho.pt

V. Ribeiro
e-mail: viviana.ribeiro@dep.uminho.pt

S. Pina · V. Ribeiro · J.M. Oliveira · R.L. Reis
ICVS/3B's—PT Government Associate Laboratory, Braga, Guimarães, Portugal

© Springer International Publishing AG 2017
J.M. Oliveira and R.L. Reis (eds.), *Regenerative Strategies for the Treatment
of Knee Joint Disabilities*, Studies in Mechanobiology, Tissue Engineering
and Biomaterials 21, DOI 10.1007/978-3-319-44785-8_8

8.1 Introduction

Osteochondral (OC) lesions may have numerous causes, varying from natural degradation to trauma related lesions that involve the degeneration of articular cartilage, subchondral bone and bone-cartilage interface. Articular cartilage and subchondral bone form the load-bearing system allowing joint motion, stability and uniform distribution of high acting loads [1]. Articular cartilage increases joint congruence and protects the subchondral bone from high stresses, reducing the friction movements within the joint and the normal contact pressure at the edge of bones [1, 2]. Considering the avascular nature of cartilage tissue, in case of injury its response does not follow the typical tissue remodeling cascade events. Cartilage lesions are therefore usually irreversible [3, 4]. Subchondral bone lesions are commonly associated with articular cartilage defects. In the past two decades promising results have been achieved for the treatment of isolated articular cartilage defects [5–8]. However, deeper OC defects are related to an extension of the cartilage defect into the underlying subchondral bone. Other possible scenarios could be that the subchondral bone becomes less dense and unable to support cartilage in transmitting loads to the cancellous and cortical bone, resulting in the fracture of the articular cartilage [1]. In this sense, basic scientific and technical issues that continue to complicate the treatment of OC lesions must be considered for the development of new strategies.

Osteoarthritis (OA) is the major cause of natural OC defects (Fig. 8.1). It is a common disease causing joint paint, deformity and functional disability and is expected that at least 40 % of patients aged 65 years and older may suffer of OA in large joints [9, 10]. Since 2008, it was estimated that over 39 million people in Europe and over than 20 million Americans suffer of OA, expecting that this number will double in 2020 [11]. Nowadays, OA can be treated using non-surgical (pharmacologic and/or physical treatments) and surgical (joint replacement and osteotomy) therapies, depending on the severity of the damage [12–14]. These treatments are well established and demonstrated to be effective for reducing pain

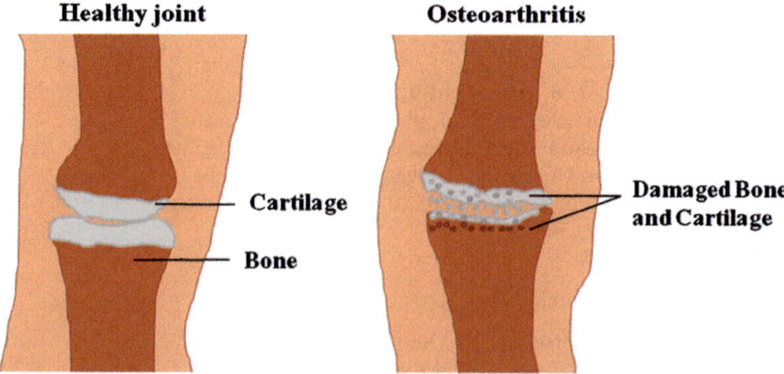

Fig. 8.1 Illustrative image of a knee healthy joint and joint with osteoarthritis

improving the patient's quality of life, but are not regenerative thus limiting the complete healing of the articular cartilage [1, 15, 16].

Surgeons and researchers have been testing different treatments for the repair and regeneration of OC defects. The most common methods are joint debridement, bone marrow stimulation technique (microfracture) and the use of OC allografts. OC autograft implantation (mosaicplasty) and cell-based therapy, as autologous chondrocytes implantation (ACI), are other possibilities [2, 17–21]. Joint debridement method eliminates debris from the joint space or surface and is applied only as a palliative treatment. In clinical treatment, it is usually associated with the microfracture technique that consists in penetrating the subchondral bone and releasing bone marrow progenitor cells into the defect [22, 23]. Mosaicplasty technique removes cylindrical plugs of the hyaline cartilage with subchondral bone from an unaffected area and implants the removed tissue into the chondral defect. Perpendicular edges of normal cartilage are previously prepared to create a "mosaic" pattern [24]. For OC defects >2.5 cm^2, OC allografts and cells with chondrogenic potential are usually the chosen treatments. ACI technique consist of isolate chondrocytes from a small cartilage piece in a low-weight-bearing area of the knee joint, and after 2–3 weeks of expansion in vitro, cells are implanted into the chondral defect [25]. Surgical approaches using OC allografts are often indicated as rescue treatments for post-traumatic lesions related to articular fractures in which there is a significant bone loss [24]. Although these treatments are effective, significant disadvantages exist, such as the lack of regenerative hyaline cartilage tissue, immune rejection and limited available tissues [26, 27]. Tissue engineering strategies emerged as promising alternatives to overcome such problems, mainly involving biodegradable scaffolds, tissue forming cells and growth factors. The initially proposed OC scaffolds were made out of a single phase structure, including different scaffolds for bone and cartilage layers or single homogeneous scaffolds for both components [28]. Recently, it was found that a successful tissue-engineered approach for the treatment of OC lesions involves the development of bilayered or stratified scaffolds, with distinct bone and cartilage parts, holding great biocompatibility, biodegradability, and biomechanical features able to establish an adequate environment for cell distribution, proliferation and differentiation [27, 29–32]. Different cell-based tissue engineering strategies have also been studied, in which OC scaffolds are combined with appropriate cells, progenitor cells (mesenchymal stem cells) or tissue specific cells (osteocytes and chondrocytes), and growth factors in order to generate and deposit new extracellular matrix (ECM) [33–36].

Several tissue-engineered structures have been developed for OC repair and regeneration and tested preclinically, but only some them, have been investigated through clinical trials. In this context, this chapter provides an up-to-date of pre-clinical animal models studies in OC tissue engineering approaches applying different scaffolding strategies. Ongoing and complete clinical trials with high relevance and the related marketed tissue-engineered products towards OC tissue regeneration are also presented.

8.2 Pre-clinical Studies for Osteochondral Regeneration

As abovementioned, OC defects can be conventionally treated by clinical/surgical methods or by advanced tissue engineering strategies [2, 28, 37–40]. As any tissue engineering approach, the development of OC constructs requires the evaluation of its performance on several pre-clinical studies prior to the evaluation in humans. There is a need to evaluate the structures response under physiologic conditions that better simulate human clinical pathologies. Recently reported animal pre-clinical studies for the treatment of OC lesions using different scaffolds strategies are summarized in Table 8.1. The success of a pre-clinical study is determined by an appropriate selection of the animal model, such as mice, rabbit, and sheep, chosen based on the final functional application of the tissue-engineered construct [41]. Small-animal models (e.g. mice and rabbit) are usually used as proof-of-concept, mainly because they can reproduce fast, which enable the study of the function of particular genes during a reasonable period of time, and they are quite inexpensive [42]. However, the absence of the joint biochemical and biomechanical environment in small animals can significantly prejudice the results, therefore considering large animals more appropriate for modelling human OC defects, thus used for final preclinical evaluation. Though, few published pre-clinical using large animal models for OC repair/regeneration are available. A typical image section stained with hematoxylin-eosin of gelatin/β-tricalcium phosphate (TCP) scaffolds after

Table 8.1 Recent pre-clinical studies for OC repair using different animal models

Scaffold	Animal model	Cells/growth factors incorporation	Results	Reference
Alginate/PCL triphasic scaffolds	Rats	Chondrocytes, BMP-7	Limited bone formation after 12 weeks of implantation	[44]
Poly(ester-urethane)/ HAp scaffolds	Rabbit	n.a.	Tissue repair after 12 weeks implantation	[45]
Platelet-rich fibrin glue/HAp biphasic scaffolds	Rabbit	MSCs	OC defects were resurfaced with more hyaline-like at 8 weeks post-implantation	[46]
PEG hydrogel/β-TCP biphasic scaffolds	Rabbit	n.a.	Tidemark formation at 52 weeks post-implantation. Formation of the repaired subchondral bone from 16 to 52 weeks in a "flow like" manner from surrounding bone to the defect center gradually	[47]

(continued)

Table 8.1 (continued)

Scaffold	Animal model	Cells/growth factors incorporation	Results	Reference
PLLA nanofibers/collagen bilayered scaffolds	Rabbit	MSCs	Rapid induction of subchondral bone appearance, and better cartilage formation after 12 weeks implantation	[48]
Oligo (PEG fumarate)/gelatin composites	Rabbit	IGF-1 and TGF-β3	Significant improvement in cartilage morphology at 12 weeks post-implantation	[49]
Collagen/HAp-silk bilayered scaffolds	Rabbit	Administration of parathyroid hormone-related protein (PTHrP)	Optimal time for administering PTHrP with the scaffold is 4–6 weeks post-injury for OC defect repair	[50]
HAp/chitosan-gelatin scaffolds	Rabbit	MSCs, TGF-β1, and BMP-2	Good restoration of OC defect and excellent integration with the native OC tissue after 12 weeks of implantation	[51]
Coralline aragonite and hyaluronic acid biphasic scaffolds	Goat	n.a.	Nearly completely biodegradation and replacement by newly OC formation tissue by 6 months after implantation	[52]
PLGA scaffolds	Sheep	Autologous chondrocytes	Significant improvement of the OC lesion after 20 weeks implantation	[53]
Allogenous bone/collagen biphasic scaffold	Sheep	Autologous chondrocytes	Full integration of the allogenous bone cand detection of conntinuous chondral layer after 6 months implantation	[54]
Microporous β-TCP scaffolds	Sheep	Autologous chondrocytes	Degradation and subchondral bone formation after 52 weeks of implantation. Healthy and biomechanical stable cartilage were not reached after 1 year	[55]
Collagen Type-I/HAp scaffolds	Pig	Autologous chondrocytes	Formation of a reparative tissue with high cellularity after 3 months of implantation	[42]

(continued)

Table 8.1 (continued)

Scaffold	Animal model	Cells/growth factors incorporation	Results	Reference
β-TCP/collagen Type-I and Type-III scaffolds	Mini-pig	BMP-2, 3, 4, 6, 7, and TGF-β1, 2, 3	Increasing new bone formation at 12 weeks post-implantation, and almost fully degradation after 52 weeks, and defect restoration	[56]
Gelatin/β-TCP bilayered sponge scaffolds	Horse	MSCs, chondrocytes, BMP-2, and platelet rich plasma	OC regeneration after 4 months of implantation	[43]

n.a. not applicable, *PCL* polycaprolactone, *BMP* bone morphogenetic protein, *HAp* hydroxyapatite, *MSCs* mesenchymal stem cells, *TGF* transforming growth factor, *IGF* insulin-like growth factor, *BMSCs* bone marrow-derived stem cells, *PLLA* poly-L-lactide, *PLGA* Poly-lactic-co-glycolic acid, *PEG* poly(ethylene glycol), *TCP* tricalcium phosphate

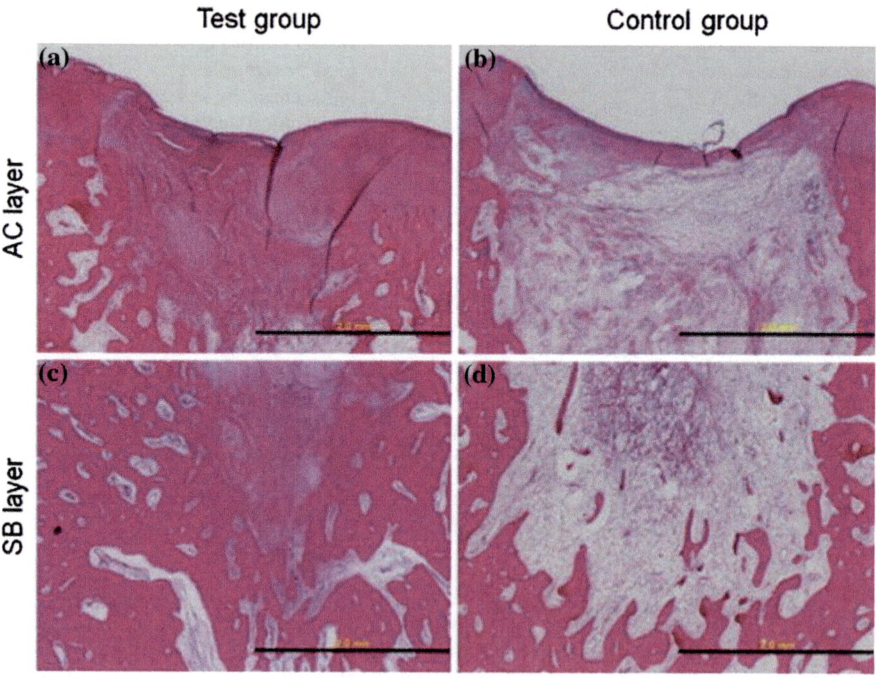

Fig. 8.2 Microscopy images of histological hematoxylin-eosin stained sections of gelatin/β-TCP scaffolds, with (test group) or without (control group) MSCs and growth factors, after implantation in horses for 4 months. *AC* articular cartilage, *SB* subchondral bone. *Scale bar* 2 mm. Reprinted with permission from [43]. Copyright 2013, Elsevier

implantation in horses talus defect for 4 months is presented in Fig. 8.2, combined with or without cells and growth factors, respectively, Test group and Control group [43]. It can be seen that histological scores were significantly superior in the test group that in the control group. Moreover, no remaining implant material, any chronic inflammations, and bone regeneration was observed.

8.3 Clinical Trials for Osteochondral Regeneration

Although most of the pre-clinical approaches are yet to be translated into clinical trials, the outcomes reveal several tissue engineered OC grafts promising for future clinical applications. For example, Kon et al. [57] performed a short-term clinical trial in thirteen human patients treated with nanostructured and multilayered scaffolds for OC defects of the knee joint. The clinical observation period provided promising preliminary results for the constructs attachment rate, stability and healing process, encouraging a clinical study with longer follow-up to confirm the high potential of the proposed OC scaffold. Williams et al. [58] developed a synthetic resorbable biphasic device made predominantly from poly lactide-co-glycolide (PLGA) copolymer, calcium sulphate and PGA, revealing favourable clinical results and a good safety profile. In fact, this clinically tested OC device has already been approved by the European Medicines Agency (EMEA) for the treatment of accurate focal articular cartilage or OC defects. Looking at the increasing progress in this area, US Food and Drug Administration (FDA) revealed that tissue engineered OC grafts would be expected to be seen in the market within the next 5–10 years. Considering all OC tissue engineering strategies for clinical application, ongoing and completed clinical trials (with no reported results yet) for OC repair and regeneration using scaffolds or cell therapies, or even scaffolds combined with cells pre-cultures in vitro are listed in Table 8.2.

Table 8.2 Overview of ongoing and complete clinical trials using strategies for OC regeneration. Information obtained from https://clinicaltrials.gov/

NCT number	Denomination	Procedure	Patients age	Follow-up	Period time
NCT00891501	The use of autologous bone marrow MSCs in the treatment of articular cartilage defects	Bone marrow MSCs aspiration and implantation	15–55 years	n.d.	2006–2014
NCT00560664	Comparison of ACI versus mosaicoplasty	Autologous chondrocytes transplantation and mosaicoplasty	18–50 years	2 years	2007–2013

(continued)

Table 8.2 (continued)

NCT number	Denomination	Procedure	Patients age	Follow-up	Period time
NCT00945399	Comparison of microfracture treatment and CARTIPATCH® chondrocyte graft treatment in femoral condyle lesions	ACI and microfracture	18–45 years	18 months	2008–2011
NCT00793104	Evaluation of the CR plug (Allograft) for the treatment of a cartilage Injury in the knee	Placement of allograft CR plug in primary injury site	≥ 18 years	2 years	2008–2012
NCT00821873	Evaluation of the CR plug for repair of defects created at the harvest site from an autograft in the knee	CR Plug implantation in the harvest site	18–55 years	2 years	2008–2012
NCT01409447	Repair of articular OCD	Biphasic OC composite implantation	18–60 years	1 year	2009–2011
NCT00984594	Evaluation of a composite cancellous and demineralized bone plug (CR-plug) for repair of knee OCDs	Autograft implantation in the primary defect site; CR-Plug implantation in the harvest site	18–55 years	2 years	2009–2012
NCT01183637	Evaluation of an acellular OC graft for cartilage lesions	Microfacture	≥ 21 years	2 years	2010–2014
NCT01159899	Transplantation of BM stem cells stimulated by proteins scaffold to heal defects articular cartilage of the knee	Transplantation of bone marrow stem cells activated in knee arthrosis	30–75 years	1 year	2010–2014
NCT01209390	A prospective, post-marketing registry on the use of chondroMimetic for the repair of OCDs	Chondromimetic	18–65 years	3 years	2010–2016
NCT01473199	BioPoly RS knee registry study for cartilage defect replacement	BioPoly RS partial resurfacing knee implantation	≥ 21 years	5 years	2011
NCT01290991	A study to evaluate the safety of Augment™ bone graft	Augment bone graft	18–40 years	1 year	2011–2012
NCT01410136	Chondrofix OC allograft prospective study	Allogeneic OC grafting	18–70 years	2 years	2011–2014
NCT01477008	BiPhasic cartilage repair implant	Marrow stimulation	Up to 54 years	1 year	2011–2014

(continued)

Table 8.2 (continued)

NCT number	Denomination	Procedure	Patients age	Follow-up	Period time
NCT01282034	Study for the treatment of knee chondral and OC lesions	Marrow stimulation—drilling or Microfractures	18–60 years	2 years	2011–2015
NCT01471236	Evaluation of the Agili-C biphasic implant in the knee joint	Agili-C Bi-phasic implantation and mini-arthrotomy or arthroscopy	18–55 years	2 years	2011–2017
NCT01347892	DeNovo NT Ankle LDC study	DeNovo NT natural tissue grafting	≥ 18 years	5 years	2011–2019
NCT01747681	Results at 10–14 years after microfracture in the knee	Microfracture	18–80 years	10 years	2012–2013
NCT01554878	Observational study on the treatment of knee OC lesions of grade III–IV	Knee surgery	30–60 years	1 year	2012–2014
NCT01920373 (cancelled)	Platelet-rich plasma versus corticosteroid Injection as treatment for degenerative pathology of the temporo-mandibular joint	Corticosteroid and platelet rich plasma injection	n.d.	6 months	2013
NCT01799876	Use of cell therapy to enhance arthroscopic knee cartilage surgery	Autologous cell and standard microfracture arthroscopic surgery	18–68 years	1 year	2013–2015
NCT02005861	"One-step" bone marrow mononuclear cell transplantation in talar OC Lesions	Bone marrow derived cells transplantation on collagen scaffold	15–50 years	2 years	2013–2016
NCT02011295	Bone marrow aspirate concentrate supplementation for OC lesions	Ankle arthroscopy with debridement and microfracture	18–95 years	2 years	2013–2017

n.d. not defined, *OC* osteochondral, *MSCs* mesenchymal stem cells, *ACI* autologous chondrocyte implantation, *OCDs* osteochondral defects

8.4 Marketed Products for Osteochondral Regeneration

The commercialization process of the scaffolds for implantation involves multiple stages of R&D replications before reaching the final approval from the government. R&D stages ensure safety and efficacy of the implants, which involve the production of medical grade scaffolds followed by animal testing under regulatory

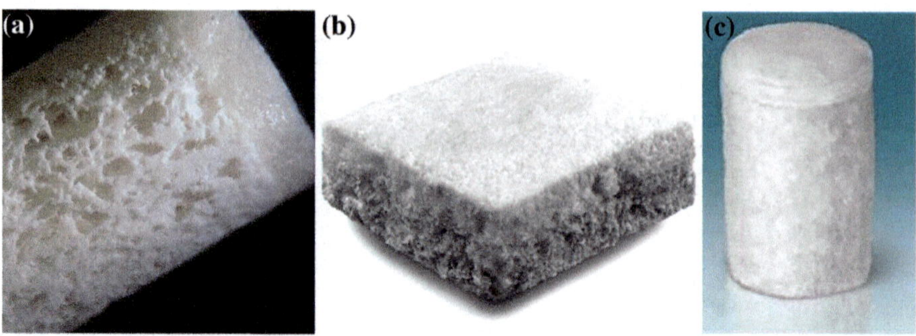

Fig. 8.3 Commercial bi/multilayered scaffolds for OC repair/regeneration: **a** TruFit® (reprinted with permission from [59]), **b** Maioregen® (reprinted with permission from [60]), and **c** CRD technology (reprinted with permission from [61])

approved conditions. Some of the products present in Table 8.2 are already being commercialized. However, only three scaffolds for OC applications are reported in the literature, namely: (i) *TruFit*® (Smith ans Nephew, USA) that is a bilayered 75:25 PLGA-PGA/calcium-sulfate copolymer porous and resorbable scaffold (Fig. 8.3a) [59]; (ii) *Maioregen*® (Finceramic, Italy) is a trilayered porous scaffold composed of collagen and Mg enriched HAp (Fig. 8.3b) [60]; and (iii) *Cartilage Repair Device* (Kensey Nash Corp.) is a biphasic bioresorbable scaffold consisting of a unique bovine collagen Type-I matrix and β-TCP, intended to be implanted at the site of a focal articular cartilage lesion or OC in the knee (Fig. 8.3c) [61]. In addition, *ChondroMimetic* is an off-the-shelf bilayered implant based on collagen and calcium phosphate for the treatment of small chondral and subchondral defects. It will be marketed as a procedure pack with the collagen implant preloaded in an accurate, easy to use arthroscopic delivery device. The literature also describes *Agili-C*™ as an off-the-shelf, aragonite-based cell-free implant that reproducibly regenerates hyaline cartilage and its underlying subchondral bone. Another available medical product to repair full-thickness OC lesions to a new level of convenience is *Chondrofix*® *Allograft*, the first off-the-shelf OC allograft [62]. It is composed of donated human decellularized hyaline cartilage and cancellous bone, and possesses relevant mechanical properties that are comparable to unprocessed OC tissue.

8.5 Conclusions

Over the last few years, great progress has been made to validate tissue engineering strategies in pre-clinical studies and clinical trials on the regeneration of OC defects. Foremost strategies used in pre-clinical studies involve the development of bi/multilayered scaffolds, alone or incorporating stem cells and/or growth factors. The chondrogenic and osteogenic repair capacities of OC defects have been evaluated subcutaneously in mice and rabbit models, although large animal models (e.g., sheep or horse) closely resemble the human tissue compared to smaller animal models, being therefore crucial for a final in vivo testing prior to clinical application in controlled clinical trials. Animal studies are used to assess biological responses, degradation and durability time, and dose response of the implanted materials, as a proof-of-concept.

Concerning commercialized OC scaffolds, only three have been documented in the literature, plus three off-the-shelf scaffolds. Among these, the three-layered scaffold (Maioregen) is showing the best follow-up outcomes, with encouraging results for the treatment of complex OC lesions and larger articular defects.

Despite significant advances in OC scaffolding, further investigations relating to clinical surveys are demanded for consistent long-term efficacy, as many challenges remain to be determined.

Acknowledgments The research leading to this work has received funding from the European Union's Seventh Framework Program (FP7/2007-2013) under Grant Agreement No REGPOT-CT2012-316331-POLARIS, and from QREN (ON.2—NORTE-01-0124-FEDER-000016) cofinanced by North Portugal Regional Operational Program (ON.2—O Novo Norte), under the National Strategic Reference Framework (NSRF), through the European Regional Development Fund (ERDF). Thanks are also due to the Portuguese Foundation for Science and Technology (FCT) and FSE/POCH (Fundo Social Europeu através do Programa Operacional do Capital Humano), PD/59/2013, for the project PEst-C/SAU/LA0026/201, for the fellowship grants of Sandra Pina (SFRH/BPD/108763/2015) and Viviana Ribeiro (PD/BD/113806/2015), and for the distinction attributed to J.M. Oliveira under the Investigator FCT program (IF/00423/2012).

References

1. Swieszkowski W, Tuan BHS, Kurzydlowski KJ, Hutmacher DW (2007) Repair and regeneration of osteochondral defects in the articular joints. Biomol Eng 24(5):489–495. doi:10.1016/j.bioeng.2007.07.014
2. Panseri S, Russo A, Cunha C, Bondi A, Di Martino A, Patella S, Kon E (2012) Osteochondral tissue engineering approaches for articular cartilage and subchondral bone regeneration. Knee Surg Sport Traumatol Arthrosc 20(6):1182–1191. doi:10.1007/s00167-011-1655-1
3. Pearle AD, Warren RF, Rodeo SA (2005) Basic science of articular cartilage and osteoarthritis. Clin Sport Med 24(1):1–12. doi:10.1016/j.csm.2004.08.007
4. Liu M, Yu X, Huang FG, Cen SQ, Zhong G, Xiang Z (2013) Tissue engineering stratified scaffolds for articular cartilage and subchondral bone defects repair. Orthopedics 36(11):868–873. doi:10.3928/01477447-20131021-10
5. Chu CR, Coutts RD, Yoshioka M, Harwood FL, Monosov AZ, Amiel D (1995) Articular cartilage repair using allogeneic perichondrocyteseeded biodegradable porous polylactic acid (PLA): a tissue-engineering study. J Biomed Mater Res 29(9):1147–1154
6. Sherwood JK, Riley SL, Palazzolo R, Brown SC, Monkhouse DC, Coates M, Griffith LG, Landeen LK, Ratcliffe A (2002) A three-dimensional osteochondral composite scaffold for articular cartilage repair. Biomaterials 23(24):4739–4751
7. Yan LP, Wang YJ, Ren L, Wu G, Caridade SG, Fan JB, Wang LY, Ji PH, Oliveira JM, Oliveira JT (2010) Genipin-cross-linked collagen/chitosan biomimetic scaffolds for articular cartilage tissue engineering applications. J Biomed Mater Res Part A 95(2):465–475

8. Yan L-P, Oliveira JM, Oliveira AL, Caridade SG, Mano JF, Reis RL (2012) Macro/microporous silk fibroin scaffolds with potential for articular cartilage and meniscus tissue engineering applications. Acta Biomater 8(1):289–301

9. Shimomura K, Moriguchi Y, Murawski CD, Yoshikawa H, Nakamura N (2014) Osteochondral tissue engineering with biphasic scaffold: current strategies and techniques. Tissue Eng Part B Rev 20(5):468–476

10. Moyer R, Ratneswaran A, Beier F, Birmingham T (2014) Osteoarthritis year in review 2014: mechanics–basic and clinical studies in osteoarthritis. Osteoarth Cartil 22(12):1989–2002

11. Csaki C, Schneider P, Shakibaei M (2008) Mesenchymal stem cells as a potential pool for cartilage tissue engineering. Ann Anat-Anatomischer Anzeiger 190(5):395–412

12. Zhang W, Moskowitz R, Nuki G, Abramson S, Altman R, Arden N, Bierma-Zeinstra S, Brandt K, Croft P, Doherty M (2007) OARSI recommendations for the management of hip and knee osteoarthritis. Part I: critical appraisal of existing treatment guidelines and systematic review of current research evidence. Osteoarth Cartil 15(9):981–1000

13. Zhang W, Moskowitz R, Nuki G, Abramson S, Altman R, Arden N, Bierma-Zeinstra S, Brandt K, Croft P, Doherty M (2008) OARSI recommendations for the management of hip and knee osteoarthritis. Part II: OARSI evidence-based, expert consensus guidelines. Osteoarth Cartil 16(2):137–162

14. Zhang W, Nuki G, Moskowitz R, Abramson S, Altman R, Arden N, Bierma-Zeinstra S, Brandt K, Croft P, Doherty M (2010) OARSI recommendations for the management of hip and knee osteoarthritis. Part III: changes in evidence following systematic cumulative update of research published through January 2009. Osteoarth Cartil 18(4):476–499

15. Redman S, Oldfield S, Archer C (2005) Current strategies for articular cartilage repair. Eur Cell Mater 9(23–32):23–32

16. Shimomura K, Ando W, Tateishi K, Nansai R, Fujie H, Hart DA, Kohda H, Kita K, Kanamoto T, Mae T (2010) The influence of skeletal maturity on allogenic synovial mesenchymal stem cell-based repair of cartilage in a large animal model. Biomaterials 31 (31):8004–8011

17. Xing L, Jiang Y, Gui J, Lu Y, Gao F, Xu Y, Xu Y (2013) Microfracture combined with osteochondral paste implantation was more effective than microfracture alone for full-thickness cartilage repair. Knee Surg Sports Traumatol Arthrosc 21(8):1770–1776

18. Kim YS, Park EH, Lee HJ, Koh YG, Lee JW (2012) Clinical comparison of the osteochondral autograft transfer system and subchondral drilling in osteochondral defects of the first metatarsal head. Am J Sports Med 40(8):1824–1833

19. De Girolamo L, Quaglia A, Bait C, Cervellin M, Prospero E, Volpi P (2012) Modified autologous matrix-induced chondrogenesis (AMIC) for the treatment of a large osteochondral defect in a varus knee: a case report. Knee Surg Sports Traumatol Arthrosc 20(11):2287–2290

20. Miska M, Wiewiorski M, Valderrabano V (2012) Reconstruction of a large osteochondral lesion of the distal tibia with an iliac crest graft and autologous matrix-induced chondrogenesis (AMIC): a case report. J Foot Ankle Surg 51(5):680–683

21. Christensen BB, Foldager CB, Hansen OM, Kristiansen AA, Le DQS, Nielsen AD, Nygaard JV, Bünger CE, Lind M (2012) A novel nano-structured porous polycaprolactone scaffold improves hyaline cartilage repair in a rabbit model compared to a collagen type I/III scaffold: in vitro and in vivo studies. Knee Surg Sports Traumatol Arthrosc 20(6):1192–1204

22. Jackson RW, Dieterichs C (2003) The results of arthroscopic lavage and debridement of osteoarthritic knees based on the severity of degeneration. Arthrosc J Arthrosc Relat Surg 19 (1):13–20

23. Shannon F, Devitt A, Poynton A, Fitzpatrick P, Walsh M (2001) Short-term benefit of arthroscopic washout in degenerative arthritis of the knee. Int Orthop 25(4):242–245

24. Hangody L, Kish G, Karpati Z, Szerb I, Udvarhelyi I (1997) Arthroscopic autogenous osteochondral mosaicplasty for the treatment of femoral condylar articular defects A preliminary report. Knee Surg Sports Traumatol Arthrosc 5(4):262–267

25. Minas T, Peterson L (2012) Autologous chondrocyte transplantation. Oper Tech Sports Med 20(1):72–86

26. Getgood AM, Kew SJ, Brooks R, Aberman H, Simon T, Lynn AK, Rushton N (2012) Evaluation of early-stage osteochondral defect repair using a biphasic scaffold based on a collagen–glycosaminoglycan biopolymer in a caprine model. Knee 19(4):422–430

27. Siclari A, Mascaro G, Gentili C, Cancedda R, Boux E (2012) A cell-free scaffold-based cartilage repair provides improved function hyaline-like repair at one year. Clin Orthop Relat Res® 470(3):910–919

28. Martin I, Miot S, Barbero A, Jakob M, Wendt D (2007) Osteochondral tissue engineering. J Biomech 40(4):750–765. doi:10.1016/j.jbiomech.2006.03.008

29. Sotoudeh A, Jahanshahi A, Takhtfooladi MA, Bazazan A, Ganjali A, Harati MP (2013) Study on nano-structured hydroxyapatite/zirconia stabilized yttria on healing of articular cartilage defect in rabbit. Acta Cir Bras 28(5):340–345

30. Jiang J, Tang A, Ateshian GA, Guo XE, Hung CT, Lu HH (2010) Bioactive stratified polymer ceramic-hydrogel scaffold for integrative osteochondral repair. Ann Biomed Eng 38(6):2183–2196

31. Nejadnik H, Daldrup-Link HE (2012) Engineering stem cells for treatment of osteochondral defects. Skeletal Radiol 41(1):1–4

32. Keeney M, Pandit A (2009) The osteochondral junction and its repair via bi-phasic tissue engineering scaffolds. Tissue Eng Part B Rev 15(1):55–73

33. Wang X, Wenk E, Zhang X, Meinel L, Vunjak-Novakovic G, Kaplan DL (2009) Growth factor gradients via microsphere delivery in biopolymer scaffolds for osteochondral tissue engineering. J Controll Release 134(2):81–90

34. Im GI, Lee JH (2010) Repair of osteochondral defects with adipose stem cells and a dual growth factor-releasing scaffold in rabbits. J Biomed Mater Res B Appl Biomater 92(2):552–560

35. Mohan N, Dormer NH, Caldwell KL, Key VH, Berkland CJ, Detamore MS (2011) Continuous gradients of material composition and growth factors for effective regeneration of the osteochondral interface. Tissue Eng Part A 17(21–22):2845–2855

36. Schütz K, Despang F, Lode A, Gelinsky M (2016) Cell-laden biphasic scaffolds with anisotropic structure for the regeneration of osteochondral tissue. J Tissue Eng Regen Med 10 (5):404–417

37. Harley BA, Lynn AK, Wissner-Gross Z, Bonfield W, Yannas IV, Gibson LJ (2010) Design of a multiphase osteochondral scaffold III: Fabrication of layered scaffolds with continuous interfaces. J Biomed Mater Res A 92A(3):1078–1093. doi:10.1002/Jbm.A.32387

38. Grayson WL, Chao PHG, Marolt D, Kaplan DL, Vunjak-Novakovic G (2008) Engineering custom-designed osteochondral tissue grafts. Trends Biotechnol 26(4):181–189. doi:10.1016/j.tibtech.2007.12.009

39. Nukavarapu SP, Dorcemus DL (2013) Osteochondral tissue engineering: Current strategies and challenges. Biotechnol Adv 31(5):706–721. doi:10.1016/j.biotechadv.2012.11.004

40. Nooeaid P, Salih V, Beier JP, Boccaccini AR (2012) Osteochondral tissue engineering: scaffolds, stem cells and applications. J Cell Mol Med 16(10):2247–2270. doi:10.1111/j.1582-4934.2012.01571.x

41. Goldstein SA (2002) Tissue engineering—functional assessment and clinical outcome. Ann N Y Acad Sci 961:183–192

42. Sosio C, Di Giancamillo A, Deponti D, Gervaso F, Scalera F, Melato M, Campagnol M, Boschetti F, Nonis A, Domeneghini C, Sannino A, Peretti GM (2015) Osteochondral repair by a novel interconnecting collagen-hydroxyapatite substitute: a large-animal study. (1937-335X (Electronic))

43. J-P Seo, Tanabe T, Tsuzuki N, Haneda S, Yamada K, Furuoka H, Tabata Y, Sasaki N (2013) Effects of bilayer gelatin/β-tricalcium phosphate sponges loaded with mesenchymal stem cells, chondrocytes, bone morphogenetic protein-2, and platelet rich plasma on osteochondral defects of the talus in horses. Res Vet Sci 95(3):1210–1216. doi:10.1016/j.rvsc.2013.08.016

44. Jeon JE, Vaquette C, Theodoropoulos C, Klein TJ, Hutmacher DW (2014) Multiphasic construct studied in an ectopic osteochondral defect model, vol 11. vol 95. doi:10.1098/rsif.2014.0184

45. Dresing I, Zeiter S, Auer J, Alini M, Eglin D (2014) Evaluation of a press-fit osteochondral poly(ester-urethane) scaffold in a rabbit defect model. J Mater Sci Mater Med 25(7):1691–1700. doi:10.1007/s10856-014-5192-6

46. Jang KM, Lee JH, Park CM, Song HR, Wang JH (2014) Xenotransplantation of human mesenchymal stem cells for repair of osteochondral defects in rabbits using osteochondral biphasic composite constructs. Knee Surg Sport Traumatol Arthrosc 22(6):1434–1444. doi:10.1007/s00167-013-2426-y

47. Zhang WJ, Lian Q, Li DC, Wang KZ, Hao DJ, Bian WG, He JK, Jin ZM (2014) Cartilage repair and subchondral bone migration using 3D printing osteochondral composites: a one-year-period study in rabbit trochlea. Biomed Res Int. doi:10.1155/2014/746138

48. Zhang SF, Chen LK, Jiang YZ, Cai YZ, Xu GW, Tong T, Zhang W, Wang LL, Ji JF, Shi PH, Ouyang HW (2013) Bi-layer collagen/microporous electrospun nanofiber scaffold improves the osteochondral regeneration. Acta Biomater 9(7):7236–7247. doi:10.1016/j.actbio.2013.04.003

49. Kim K, Lam J, Lu S, Spicer PP, Lueckgen A, Tabata Y, Wong ME, Jansen JA, Mikos AG, Kasper FK (2013) Osteochondral tissue regeneration using a bilayered composite hydrogel with modulating dual growth factor release kinetics in a rabbit model. J Control Release 168 (2):166–178. doi:10.1016/j.jconrel.2013.03.013

50. Zhang W, Chen JL, Tao JD, Hu CC, Chen LK, Zhao HS, Xu GW, Heng BC, Ouyang HW (2013) The promotion of osteochondral repair by combined intra-articular injection of parathyroid hormone-related protein and implantation of a bi-layer collagen-silk scaffold. Biomaterials 34(25):6046–6057. doi:10.1016/j.biomaterials.2013.04.055

51. Chen JN, Chen HA, Li P, Diao HJ, Zhu SY, Dong L, Wang R, Guo T, Zhao JN, Zhang JF (2011) Simultaneous regeneration of articular cartilage and subchondral bone in vivo using MSCs induced by a spatially controlled gene delivery system in bilayered integrated scaffolds. Biomaterials 32(21):4793–4805. doi:10.1016/j.biomaterials.2011.03.041

52. Kon E, Filardo G, Robinson D, Eisman JA, Levy A, Zaslav K, Shani J, Altschuler N (2015) Osteochondral regeneration using a novel aragonite-hyaluronate bi-phasic scaffold in a goat model. (1433-7347 (Electronic))

53. Fonseca C, Caminal M, Peris D, Barrachina J, Fabregas PJ, Garcia F, Cairo JJ, Godia F, Pla A, Vives J (2014) An arthroscopic approach for the treatment of osteochondral focal defects with cell-free and cell-loaded PLGA scaffolds in sheep. Cytotechnology 66(2):345–354. doi:10.1007/s10616-013-9581-3

54. Schleicher I, Lips KS, Sommer U, Schappat I, Martin AP, Szalay G, Hartmann S, Schnettler R (2013) Biphasic scaffolds for repair of deep osteochondral defects in a sheep model. J Surg Res 183(1):184–192

55. Bernstein A, Niemeyer P, Salzmann G, Südkamp NP, Hube R, Klehm J, Menzel M, von Eisenhart-Rothe R, Bohner M, Görz L, Mayr HO (2013) Microporous calcium phosphate ceramics as tissue engineering scaffolds for the repair of osteochondral defects: Histological results. Acta Biomater 9(7):7490–7505. doi:10.1016/j.actbio.2013.03.021

56. Gotterbarm T, Breusch SJ, Jung M, Streich N, Wiltfang J, Berardi Vilei S, Richter W, Nitsche T (2014) Complete subchondral bone defect regeneration with a tricalcium phosphate collagen implant and osteoinductive growth factors: a randomized controlled study in Gottingen minipigs. (1552-4981 (Electronic))

57. Kon E, Delcogliano M, Filardo G, Pressato D, Busacca M, Grigolo B, Desando G, Marcacci M (2010) A novel nano-composite multi-layered biomaterial for treatment of osteochondral lesions: technique note and an early stability pilot clinical trial. Injury 41 (7):693–701. doi:10.1016/j.injury.2009.11.014

58. Williams RJ, Gamradt SC (2008) Articular cartilage repair using a resorbable matrix scaffold. Instr Course Lect 57:563–571

59. Melton JTK, Wilson AJ, Chapman-Sheath P, Cossey AJ (2010) TruFit CB® bone plug: chondral repair, scaffold design, surgical technique and early experiences. Expert Rev Med Devices 7(3):333–341. doi:10.1586/erd.10.15
60. Kon E, Delcogliano M, Filardo G, Pressato D, Busacca M, Grigolo B, Desando G, Marcacci M (2010) A novel nano-composite multi-layered biomaterial for treatment of osteochondral lesions: technique note and an early stability pilot clinical trial. Injury 41:693–701
61. Ostrovsky G (2010) Bioresorbable, acellular, biphasic scaffold gets EU approval for knee cartilage repair. medGadget. Accessed 25 Nov 2014
62. Gomoll AH (2013) Osteochondral allograft transplantation using the chondrofix implant. Oper Tech Sports Med 21(2):90–94. doi:10.1053/j.otsm.2013.03.002

Chapter 9
Rapid Prototyping for the Engineering of Osteochondral Tissues

Alessandra Marrella, Marta Cavo and Silvia Scaglione

Abstract The reconstruction of complex joints represents one of the major challenges in Tissue Engineering, whose aim is to realize bioactive 3D grafts interacting with the articular environment while providing structural and mechanical functionality. Due to the complex hierarchical structure and the co-existence of several architectural organizations of natural articular tissue, a series of chemical-physical-biological features have to be carefully controlled and defined for a best tuning of the mechanical and functional properties of osteochondral tissues. However, the control over scaffold architecture using conventional manufacturing techniques is highly process dependent rather than design dependent. As a result, in the last years Rapid Prototyping (RP) techniques are proposed as promising alternative for 3D porous scaffolds fabrication, opening the possibility to realize engineered grafts with defined and reproducible complex internal structures, for an enhanced cellular response in vivo; moreover, implantable personalized articular tissue materials may be created individually for each patient according to the orthopedic requirements. In this Chapter, we will expose the major RP-based techniques, among them laser-, nozzled- and printed-based RP methods, with particular reference to the most cogent works in the field of joints repair.

9.1 Introduction

Trauma and disease of joints frequently involve structural damage to the articular cartilage surface and the underlying subchondral bone, accompanied by pain, reduced mobility and high socio-economic costs [1, 2]. Current approaches to

A. Marrella · M. Cavo · S. Scaglione (✉)
CNR—National Research Council of Italy, IEIIT Institute, Genoa, Italy
e-mail: scaglione@mailzimbra.ge.ieiit.cnr.it

A. Marrella
Department of Experimental Medicine, University of Genoa, Genoa, Italy

M. Cavo
Department of Informatics, Bioengineering, Robotics and Systems Engineering, University of Genoa, Genoa, Italy

© Springer International Publishing AG 2017
J.M. Oliveira and R.L. Reis (eds.), *Regenerative Strategies for the Treatment of Knee Joint Disabilities*, Studies in Mechanobiology, Tissue Engineering and Biomaterials 21, DOI 10.1007/978-3-319-44785-8_9

osteo-chondral (OC) treatment are usually aimed at treating the medical condition rather than curing it, consequently giving unpredictable results: removal and/or damage of this anatomical structure leads to degenerative changes of the articular cartilage, ultimately to osteoarthritis [3–5]. Hence, the reconstruction of complex joints represents a major challenge in Tissue Engineering (TE).

To hypothesise innovative and improved designs for osteochondral scaffolds, a precise definition of cartilage and bone structural characteristics is needed, exhibiting the cartilage/bone native tissues intrinsically distinct structural and biochemical features across a spatial volume, which reflect their functional environment. In particular, in the articular site, bone and the cartilage tissues work in a synchronized biofunctional manner keeping their specific properties and roles. Moreover, the low OC tissue vascularization hinders the tissue healing and union with original tissue [6–8]; although integration in the vertical direction, where cartilage joins with subchondral bone tissue, can be quite successful, it is much more challenging in the lateral direction, due to the aforementioned low healing rate. In addition, the phenotype and characteristics of newly formed tissue are significantly affected by the biological and mechanical cues to which they are exposed, where risk of formation of fibrous, hypertrophic tissue is high [9].

Among current approaches to OC treatment, some TE based solutions have been proposed to help overcoming the limitations of standard procedures. The promise of tissue engineering is rooted in the fact that engineered grafts can interact with their environment and remodel themselves while providing structural and mechanical functionality and integration with the host [10]. In particular, to engineer such constructs, the cartilage and bone components can either be generated independently in vitro and combined together (e.g. sutured or glued), or fabricated as a single composite graft.

Initially, several studies were carried out by realizing *single-phase* scaffolds supporting chondrocytes and bone cells attachment [11, 12], besides promoting the proper loading function.

For a better engineering of scaffolds functional properties resembling the hierarchy and anatomical shape of the osteochondral tissue, the clinical attention has shifted towards integrated multi-layered scaffolds, matching together different layers separately manufactured with morphological and chemical features specially designed to mimic the different tissues [6, 13–22].

Fabricating a functional osteochondral scaffold requires simultaneous consideration of the appropriate features of the two compartments (i.e. osseous and cartilaginous) and of the continuous interface. The macro- and microstructure of the scaffold may in fact significantly influence cell migration and viability, besides matrix deposition and blood vessel invasion, if required [23]. A large fraction of pore volume represents a key requirement for effective nutrient and metabolite mass transport, required for cell functions and for in vivo neo-tissue deposition. Moreover, the presence of pores accelerates the biodegradability of the graft, ultimately favouring a complete conversion of the scaffold into mature tissue [24]. Likewise, the chemical composition is responsible for selective cellular differentiation [25, 26].

Starting from these considerations, osteochondral prototypes should be properly realized recreating the most performing chemical-physical cues for an enhanced and selective cellular differentiation and matrix deposition. However, when realized with conventional techniques, including solvent casting and particulate leaching [26, 27], gas foaming [28, 29], phase separation [30, 31] and electro-spinning [32, 33], a precisely control of pores size and geometry, interconnectivity and spatial distribution cannot be provided [34]. The control over scaffold architecture using these conventional techniques, in fact, is highly process dependent rather than design dependent. Moreover, when realized as multi-phase OC grafts, these scaffolds still present some drawbacks as the possible structural delamination of the different layers in vivo [20, 35].

As a result, Rapid Prototyping (RP) is seen to be a viable alternative for achieving extensive and detailed control over scaffold architecture. In particular, some composite materials have been finely designed and manufactured through RP in order to realize 3D functionalized OC scaffolds [36].

Since different aspects must be finely tuned, such as the materials ability to induce bone and chondral tissue differentiation and the local and global mechanical proprieties of the constructs, RP techniques allow the production of scaffolds previously modeled and designed at the micro-macro scale [3, 37, 38].

This great advantage opens the possibility to realize both scaffolds with well-defined and reproducible complex internal structures—affecting cell survival, signaling, growth, propagation and reorganization, tissue growth and the preservation of native cellular phenotypes [39–41]—and customized implantable materials created individually for each patient.

9.2 Rapid Prototyping Technology for Osteochondral Tissue Engineering Applications

Rapid Prototyping is a method of quickly creating a scale model of a part or a finished product, using Computer-Aided Design (CAD) software.

In the Tissue Engineering field, a scaffold can be designed by taking directly anatomical information of the patient's target defect (e.g. CAT scan, MRI images, X-rays data) to obtain a custom-tailored implant. Starting from these files, the images can be imported in proper management platform and converted in a 3D model through dedicated software, including segmentation and visualization tools. The segmented regions are further processed with CAD software, saved in a STL (Surface Tessellation Language) file format and finally sent to the 3D printing machine for the production. The realization of such customized implants opens the possibility of creating joint replacements of the proper size for each articular defect and for each specific requirement, greatly improving surgical techniques thanks to an adequate match between individual needs and anatomical substitutes [42].

Worldwide, a lot of different RP techniques exist; in general, they can be classified according to their working principles: *laser-based, nozzled-based* and *printed-based* [43].

Although a wide range of RP technologies have been used to realize scaffold for bone [44–47] and cartilage [48–52] regeneration respectively, in the following sections we will consider RP techniques adopted to achieve osteochondral engineered constructs for the simultaneous regeneration and integration of bone and chondral tissues.

9.2.1 Laser-Based Techniques

This techniques family involves the use of a specific light that is progressively deposited on a photo-cross-linkable precursor preset pattern.

Stereo-litography (SL) is considered the first available RP technique, developed in 1986 by Charles Hull, co-founder of 3D Systems, Inc. [53]. It is a multilayer technology, which exploits the curing and following solidification of a liquid resin bath to form an object through the selective cure reaction of a polymer. The typical set-up is composed by a resin tank, a laser source, a controller for the light beam movement and a platform allowing the movement along the z-axis. After curing a 2D layer, the subsequent layers are built up moving the fabrication platform stepwise in z-direction; this allows highest accuracy and resolution, comparing with the other RP techniques.

Micro-stereolitography shares the same principles of SL, but in different dimensions. In μSL, the UV laser beam is focused to 1–2 μm to solidify a thin layer of few micron in thickness [54].

Recently, digital light projection is emerging as a method of resin illuminating [55–57]: a Digital Mirror Device (DMD) can be rotated, two-dimensional pixel-pattern is projected onto the transparent plate, and a 2D layer of resin can be cured at once. This approach, based on a top-down working principle, allows to reduce the building times, as they only depend on the layer thickness and on the required exposure time, but not on layers size.

As a RP technology, SL allows the total control on the scaffold geometry and on its morphological features with a large freedom of design, as required to mimic the hierarchical osteochondral tissue. Despite these advantages, a drawback of this technique is a limited choice of usable biomaterials, which have to be photocurable, besides showing all the required features for tissue engineering applications, such as bioactivity and biodegradability [58]. Also the resin viscosity is a bound parameter: typically, the highest resin viscosity that can be employed in SL is approximately about 5 Pa s [59].

In the last years, some works have been carried out in developing biodegradable macromers and resins [60]. Different biodegradable macromers have been recently applied in SL to create biodegradable resins based on different polymers, such as poly(propylene fumarate) (PPF) [47, 61], trimethylene carbonate (TMC), ε-caprolactone (CL) [49, 62] or d,l-lactide (DLLA) [63].

In 2010, Melchels et al. realized a PDLLA-based resin scaffold with a morphological gradient structure for osteochondral applications in which defined but not uniform porosity was desired [64]. For an enhanced emulation of the functional articular microenvironment, a pores size and volume fraction gradient was designed to gradually decrease from 70 % at the mid-section to 30 % at the bottom end of the structure (Fig. 9.1), finally showing that SL fabrication is a powerful tool for realizing complex tissue engineering scaffolds. An accurate mathematical modeling may be also provided to properly design the scaffold architecture in the RP software model [65].

Also polymer-ceramic composite scaffolds can be realized suspending ceramic particles in the photocurable resin to obtain stiffer composite and stronger structures if compared with the polymeric ones [66–69].

The composite resin processing is more difficult, since its viscosity can significantly increase upon addition of the powder. In fact, the maximum possible ceramic contents have been reported to be about 50 wt% [70]. Furthermore, the ceramic particles size should be smaller than the layer thickness in the building process to prepare the objects accurately.

All-ceramic objects have been also realized thanks to a particular SL type, properly called ceramic-stereolithography (CLS) [59, 71–73]. With this technique, 3D ceramic objects can be realized using photocurable ceramic suspensions.

In 2012, Bian et al. realized a novel bilayer scaffold by combining CSL technique with gel casting. Thanks to this innovative technology, a highly porous ceramic scaffold was realized starting from a 3D model inspired to the bone tissue and bone-cartilage interface. The bony layer was fabricated through the use of CSL technique, obtaining a beta-tricalcium phosphate scaffold designed with the

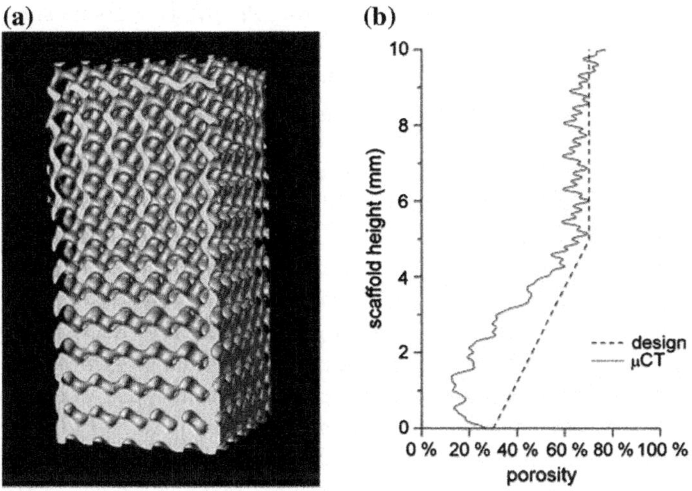

Fig. 9.1 PDLLA scaffolds with a gradient in porosity and pore size. **a** μCT visualisation. **b** Change in the average porosity with scaffold height (*solid line*) in comparison with the designed porosity (*dotted line*) [64]

following proprieties: 700–900 µm pore size, 200–500 µm interconnected pore size, 50–65 % porosity.

The chondral layer was produced pouring a collagen solution on the surface of the ceramic scaffold in vacuum conditions, crosslinking and then freeze-drying. A suitable integration between the two phases was obtained by physical locking and also biological analysis performed culturing bone marrow stromal cells (MSCs) on the specimens showed good results. This work revealed the high versatility of scaffold fabrication by RP, allowing to obtain a composite graft for the osteochondral tissue, more performing in terms of structural and mechanical proprieties, indeed, compared to the traditional fabrication process, presented a great advantage in controllability and rate of production.

Moreover, the shearing resistance of the final constructs indicated that gel casting combined RP technology might be a promising way to avoid delamination in vivo, typical of the bilayer scaffolds [74].

Also hydrogel, combined always with ceramic grafts, were used to mimic the bone and chondral niches, as in 2014, when Zhang et al. realized Poly(ethylene glycol)(PEG)/β-TCP scaffolds by SL technology [75]. The β-TCP ceramic scaffold was fabricated by gel casting process, while for the chondral layer PEG hydrogel, widely used in cartilage tissue engineering [36, 76, 77], was processed with SL. Anatomy shaped CAD models were used as inputs to a custom-made SL machine and PEG diacrylate (PEGDA) was directly cured on β-TCP scaffolds to fabricate biphasic hydrogel-ceramic osteochondral composites. In particular, the bone phase was realized with the following morphological proprieties: 700–900 µm pore size, 200–500 µm interconnected pore size and 50–65 % porosity [74, 78]. The resultant biphasic structure was tested in vivo in rabbit trochlea critical size defect of skeletal mature rabbit model and animals were sacrificed after different months up to 1 year to evaluate the evolution of the healing process. PEG/β-TCP composites fabricated by stereolitography were revealed to be a feasible strategy for osteochondral tissue engineering applications, since the histological results showed that hyaline-like cartilage formed along with white smooth surface and invisible margin at 24 weeks postoperatively, while typical tidemark formation at 1 year. The repaired subchondral bone formed from 16 to 52 weeks in a "flow like" manner from surrounding bone to the center of the defect [75].

Later, the same group optimized this biphasic scaffold evaluating the most suitable hydrogel microstructure to enhance the interfacial integration between bone and chondral tissues. A series of models were designed changing the different pore area percentages (in the range of 0–60 %) at the interface, while pore size about 0.4 mm in diameter and 1 mm in depth were arrayed in a quadrilateral pattern; the length (L) between every two pore centers was automatically calculated by an optimization function, according to different pore area percentages.

Interfacial shear tests were performed to determine the effect of the interface different structures on the integration between the two phases. The interfacial shear strength increased gradually to 0.34 MPa from 0 to 30 % pore area percentage groups, but no significant differences were observed among 30/40/50/60 % pore area groups, which might be due to the biomechanical property of applied PEG hydrogel. Although still below the level of native osteochondral integration [79], 30 % pore area group showed three-fold improvement of interfacial shear strength respect to 0 % pore area where none interfacial structures was designed. Authors showed also more than 50-fold improvement compared with that of traditional integration (5.91 ± 0.59 kPa). Increasing the area pore percentage, the compressive modulus of ceramic component rapidly decreased, anyway the substrate compressive modulus of 30 % pore area percentage group was 18.4 MPa, which is comparable to that of human cancellous bone [80]. So, in this study thanks to the high degree of controllability and reproducibility of the controlled geometrical patterns ensured by RP techniques (Fig. 9.2), it was been possible to perform a deep evaluation of the most suitable geometric features to improve the proprieties of the complex osteochondral interface [81].

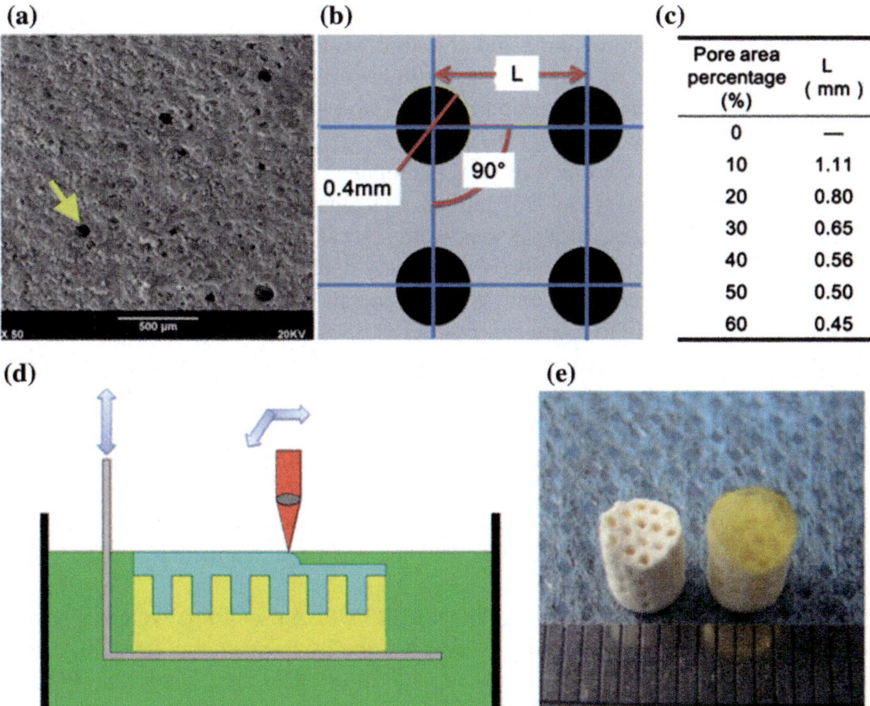

Pore area percentage (%)	L (mm)
0	—
10	1.11
20	0.80
30	0.65
40	0.56
50	0.50
60	0.45

Fig. 9.2 Osteochondral interface biomimetic design and osteochondral scaffold fabrication. **a** Microstructure located in osteochondral interface. **b** Interface microstructure design, **c** detailed data for osteochondral interface design, **d** the schematic of integration of chondral phase and osseous part via stereolithography, **e** fabricated ceramic scaffold and PEG/β-TCP scaffold [81]

9.2.2 Nozzled-Based Techniques

One of the most famous RP techniques belonging to this family is the Fused Deposition Modeling (FDM) [82]. Briefly, thin filaments are melted and guided to form layer by layer a 3D object through the use of a guided robotic device. In this way, scaffolds are built in additive manner and their micro-macro architecture can be controlled both in horizontal and vertical direction by a numerical control mechanism [83].

Thermoplastic materials leave the extruder in a liquid form and immediately solidify. This method does not require any solvent and offers great flexibility in material processing [84]. The deposition path and the parameters of each layer can be designated depending on the used material, the fabrication conditions, the applications of the designed part and the preferences of the designer [85].

Following this approach, different hybrid scaffolds may be realized. In 2007 Swieszkowski designed biphasic scaffolds composed by specifically compartments for inducing the regeneration of bone and chondral tissues respectively: the first class of scaffolds was composed of fibrin for the cartilage phase, and PCL for the subchondral one, while the second prototypes comprised PCL to promote cartilage regeneration and PCL-TCP for the bony one (Fig. 9.3). The design process of the geometry of the bi-phasic scaffold was performed by CAD computers systems, while the prediction of the mechanical and fluid-dynamic conditions was pursued by using a Computer Aided Engineering (CAE) system. Both phases were separately fabricated, seeded with bone marrow-derived mesenchymal cells (BMSCs) and finally cultured in chondrogenic and osteogenic media for cartilage and bone regeneration, respectively. Then, the two phases were integrated into a single construct (scaffold) by using fibrin glue and then were implanted in critical size defects created in the medial condyle of the rabbit model. The quantification of bone regeneration through micro CT analysis demonstrated the potential of PCL/PCL-TCP in promoting bone healing respect fibrin/PCL; moreover, the evaluation of the cartilage phase revealed PCL excelling when compared to fibrin, due to its fast degradability [86].

FDM technology may be also employed to engineer articular grafts. In particular, PGA/PLA PCL/HA biphasic scaffolds have been realized for goat femoral head articular cartilage repair [87].

Polylactic acid-coated polyglycolic acid (PGA/PLA) scaffolds were realized for cartilage regeneration, while poly-ε-caprolactone/hydroxyapatite (PCL/HA) scaffolds were designed and used for the regeneration of femoral head. The PCL/HA scaffold was fabricated by FDM according to the 3D data achieved from goat proximal femoral condyle (not containing cartilage layer) to form a 3D hemispherical scaffold to fit with a load-bearing cylindrical one, which was equal to the anatomic size of goat proximal femoral condyle (Fig. 9.4). The PCL/HA scaffold was designed to contain regular 3D interconnecting microchannels (200–400 μm in pore size) with a porosity of 54.6 ± 1.2 %. PGA/PLA scaffold was compressed into the shape of the articular surface with a thickness of about 1.2 mm, which was similar to the thickness of the goat femoral head articular cartilage.

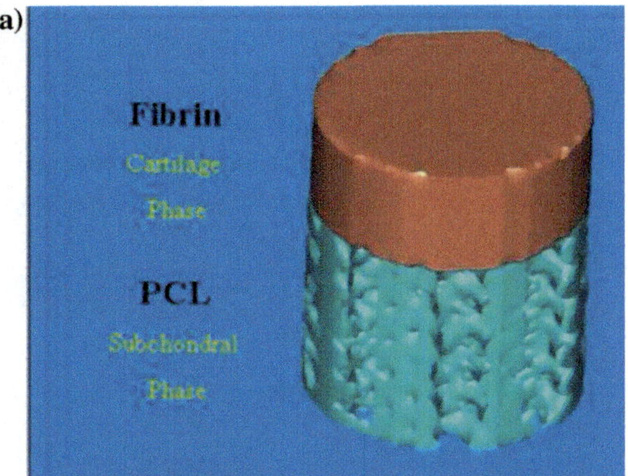

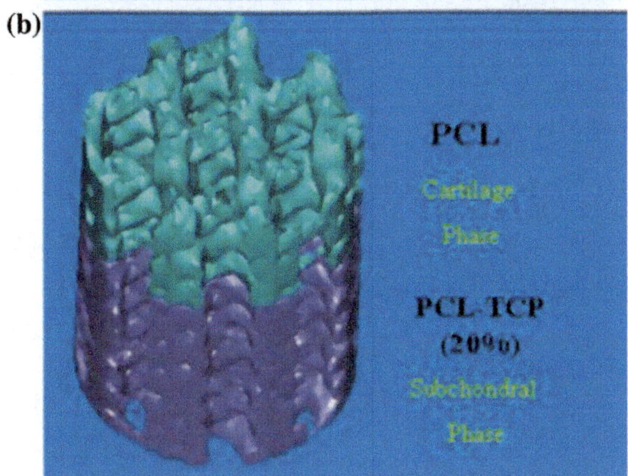

Fig. 9.3 **a** Fibrin + PCL scaffold and **b** PCL + PCL-TCP scaffold [86]

Scaffold-cells constructs were implanted into the dorsum subcutaneously of athymic nude mice for 10 weeks. Homogenous cartilage was regenerated on the surface layer of regenerated femoral heads and typical trabecular bony tissue was formed in the bony layer. Moreover, chondral and bony layers showed a satisfactory integration (Fig. 9.5), showing a typical osteochondral interface where cartilage tissue, immature calcified tissue, transitional trabecular bone and hypertrophic chondrocytes were observed. These structures were present only at the interface area showing the tissue specific regeneration induced by different microenvironments. These results confirmed that biphasic scaffolds may be properly designed and realized, through RP techniques, supporting the regeneration of newly formed tissue with similar histological structures and biophysical properties compared to the native ones, finally revealing this manufacturing technique a valid technological alternative for biological joint reconstruction [87].

Another nozzle-based RP technique used to realize osteochondral scaffolds is the multi-nozzle low temperature deposition (M-LDM), a system to extrude both

                                                                          A. Marrella et al.

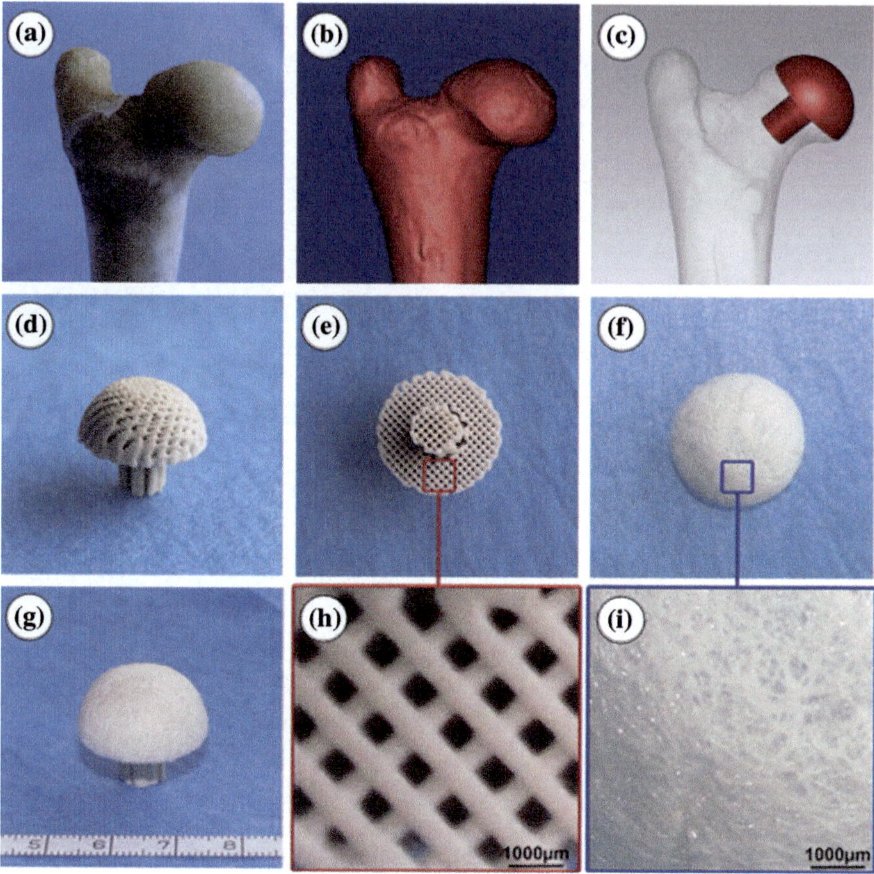

Fig. 9.4 The morphological data of the goat femoral head (**a**) the 3D data are reconstructed and designed using the CAD software (**b**, **c**). PCL/HA scaffold (**d**, **e**) fabricated by FDM is designed to be with an intramedullary stem and interconnecting microchannels (**h**). The PGA/PLA scaffold (**f**) with small pore size (**i**). The well-matched PGA/PLA PCL/HA biphasic scaffold (**g**) [87]

natural derived and synthetic polymers [88–90]. In brief, the system is based on a disposable syringe, which allows the deposition of a wide range of polymers. The syringe is fixed in a copper cylinder in which an electric resistance is set to keep a initial temperature of the extruded materials [44]. The syringe deposits filaments layer by layer within a low temperature refrigerator. Nozzle of the first material extrudes a filament along the plane x–y, which is rapidly frozen. When all the filaments of the first materials are extruded, nozzles can move and a second material can be extruded. As soon as the first layer is completed, the platform move along the z-axis and the second layer is manufactured in the same way as above. 3D scaffolds with different profiles, heterogeneous materials and highly interconnected gradient porous structures can be fabricated by repeating this layer-by-layer process [90] (Figs. 9.6, 9.7).

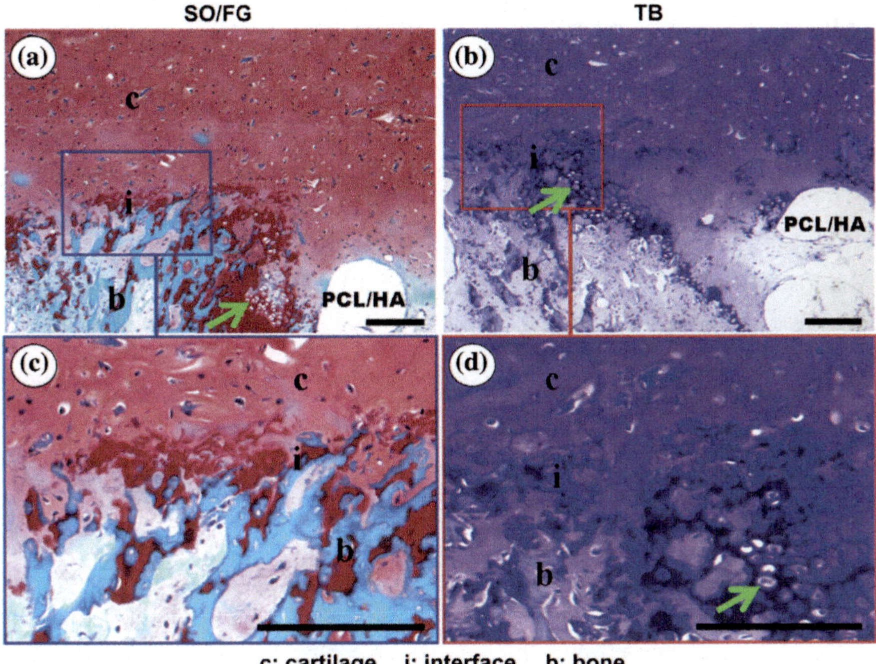

Fig. 9.5 Histological examination of the interface between regenerated cartilage and subchondral bone at different magnitudes. *Scale bar* 200 μm [87]

In 2009, Liu et al. adopted this technique to realize scaffold composed of three parts, each with different features and materials for bone, cartilage and interface respectively. Briefly, a model was realized through computer-aided design (CAD) and was divided into three regions and used to drive the M-LDM system. The model consisted in two highly porous layers with larger pores for bone and cartilage repair and compact separate layer with low porosity and smaller pores for the interface. PLGA-TCP slurry and PLGA-NaCl slurries were prepared and extruded into a low temperature room to fabricate the osteochondral scaffold as defined in the model. The gradient scaffold was cross-linked and combined with collagen sponges. Even if PLGA is a widely used polymer for cartilage and bone tissue engineering [19, 91], its surface chemistry does not promote cell adhesion [92], so collagen sponges were introduced to promote cell adhesion. On the other hand, TCP was added into bony layer to improve the biocompatibility, the mechanical properties and the osteoinductive capacity in vivo [93, 94]. These OC scaffolds combined with progenitor cells were implanted the knee joints of 5 New Zealand white rabbits. At 6 weeks post-operation, the samples contained cartilage-like tissues and some bone-like tissues in the bone region, revealing this system promising to fabricate scaffolds with heterogeneous materials and gradient hierarchical porous structures useful to mimic the osteochondral tissue [90].

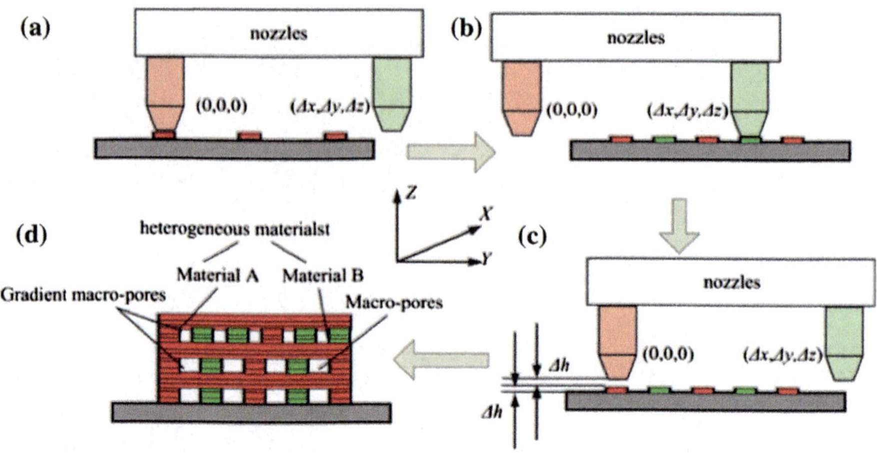

Fig. 9.6 Representation of the fabrication process via M-LDM system [90]

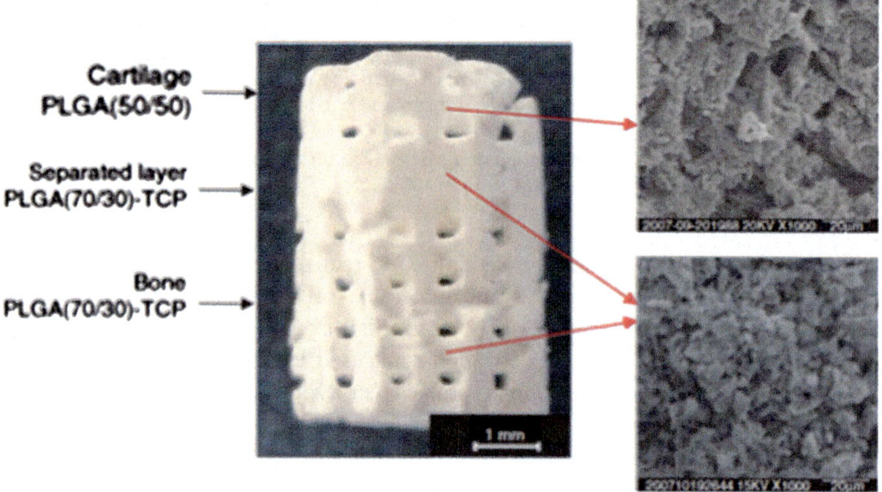

Fig. 9.7 PLGA-TCP graded composite scaffold [90]

9.2.3 Printer-Based Techniques

The 3D-Printing (3DP) technology was developed at the Massachusetts Institute of
Technology [95–97]. Basically, the bonding method of inkjet printing technology is
a layered fabrication process, working as following: a defined layer of material is
deposited on a building platform; successive layers are printed on a freshly laid
layer of powder until the whole model is completed. When the porous model is
completed, the unbounded powder is removed, and the graft must be strengthened

by a conventional pre-sintering process [97–99]. An advantage of 3DP is that it can be performed in an ambient environment [58].

In 2002 Sherwood et al. developed a heterogeneous, osteochondral scaffold using the three-dimensional printing process. The material composition, porosity and macro-architecture were varied throughout the scaffold structure, in order to reproduce the hierarchical complexity of the OC tissue. In particular, the upper cartilage region was composed of d,l-PLGA/l-PLA, with macroscopic staggered channels to facilitate homogenous cell seeding. A similar pores conformation was highly adopted in cartilage substitutes, with the aim to limit pores interconnection, finally reducing the in vivo vascularization and a non-isotropic mechanics [21, 100]. On the other hand, the lower bone portion was made by a l-PLGA/TCP composite and designed to maximize bone ingrowth while maintaining a strong enough mechanical support. The bone layer was 55 % porous, while the cartilage layer was 90 %, to reproduce the native mechanical properties of the two tissues (Fig. 9.8).

Thanks to the different porosity percentage, the tensile strength of the bone layer was similar to the fresh cancellous human bone one, making these grafts suitable for in vivo applications, including full joint replacement [19]. The transition region between these two sections was made by a materials and porosity gradient in order to prevent delamination. The in vitro cell seeding showed that chondrocytes preferentially seeded into the cartilage portion of the device, while cell attachment to the bone region was minimal.

In 2005 Leukers et al. used hydroxyapatite (HA) granulates to fabricate porous ceramic structures with designed internal architecture. In particular, a spray-dried HA-granulate containing polymeric additives was used to improve bonding and flowability, while a water soluble polymer blend was used as binder. The granule

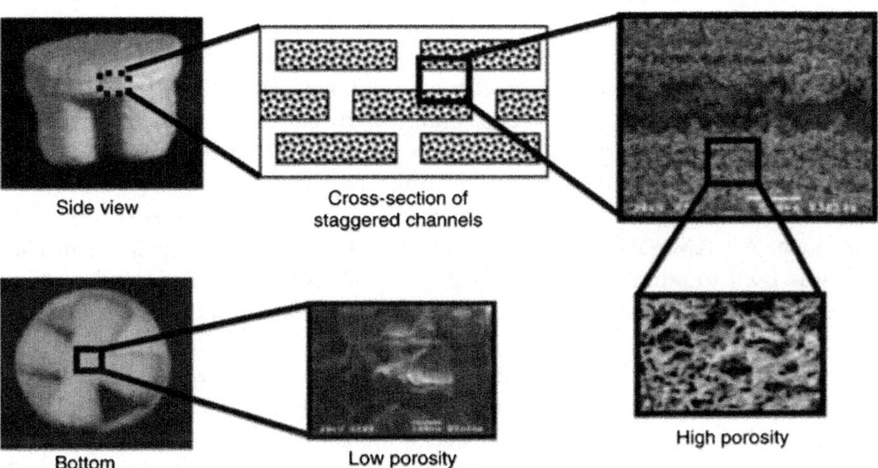

Fig. 9.8 Osteochondral scaffold obtained by 3D printing process: the upper-cartilage region is made of d,l-PLGA/l-PLA, while the bony layer is made of l-PLGA/TCP [19]

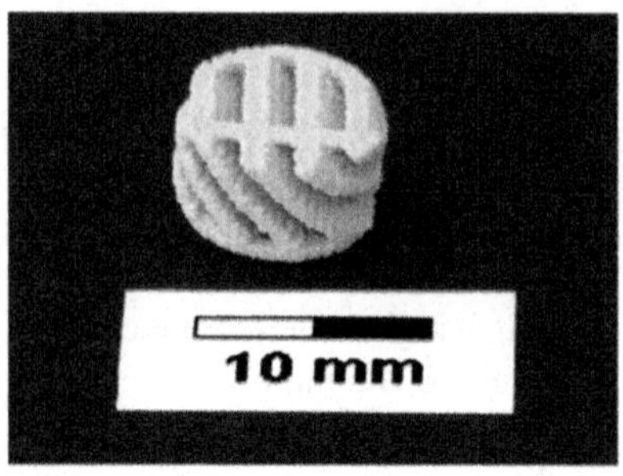

Fig. 9.9 Hydroxyapatite porous scaffold with designed internal architecture made up of walls standing in 45° to the x-axis [98]

structure remained after sintering and this has leaded to obtain parts with high microporosity in addition to the macro one. In the same study, the internal structure of the scaffold was investigated in order to facilitate the seeding process and to enhance cell attachment. Eventually, a scaffold design made up of walls standing in 45° to the x-axis was chosen, enabling cell proliferation into the inside of the structure without clogging (Fig. 9.9).

9.3 Bioprinting

In the conventional scaffold-based approach, cells and growth factor distribution inside a porous scaffold cannot be controlled; as a result, it is very difficult to predict and facilitate the invasion of native tissues in the constructs [101]. Considering these limitations, studies of new approaches called *bioprinting* and *biofabrication* have been carried out [102, 103]. In this techniques, biological materials including living cells, proteins and various biomaterials, are fabricated directly using several 3D fabrication methods.

Cell-leaden structures can be able to mimic the anatomical cell arrangement of the native tissues or organs, better than conventional scaffolds.

Tissues printed by 3D Fiber deposition (3DF) are based on a layer by layer deposition of cell-laden hydrogel and such structures could be very successful to replicate the intricate cell/matrix natural structures [104].

In 2012, Fedorovich et al. realized alginate cell-leaden hydrogel using a 3DF technique. A CAM/CAD model was input in a BioScaffolder dispensing machine that was used to build 3D constructs through the controlled motion of a syringe dispenser. To develop an osteochondral construct, human chondrocytes and human osteogenic progenitors, combined with osteo-conductive calcium phosphate particles, were combined to the hydrogels. First of all, this work revealed the feasibility of a 3DF to fabricate porous heterogeneously cell laden scaffolds. Both after 3 weeks culture in vitro and after 6 weeks implanted subcutaneously in immunodeficient mice in vivo, different tissues formation was observed (Figs. 9.10, 9.11).

Hydrogels are mostly used to produce cell-printed structures, but they exhibit poor mechanical proprieties [104, 106]. For this reason the attention has shifted

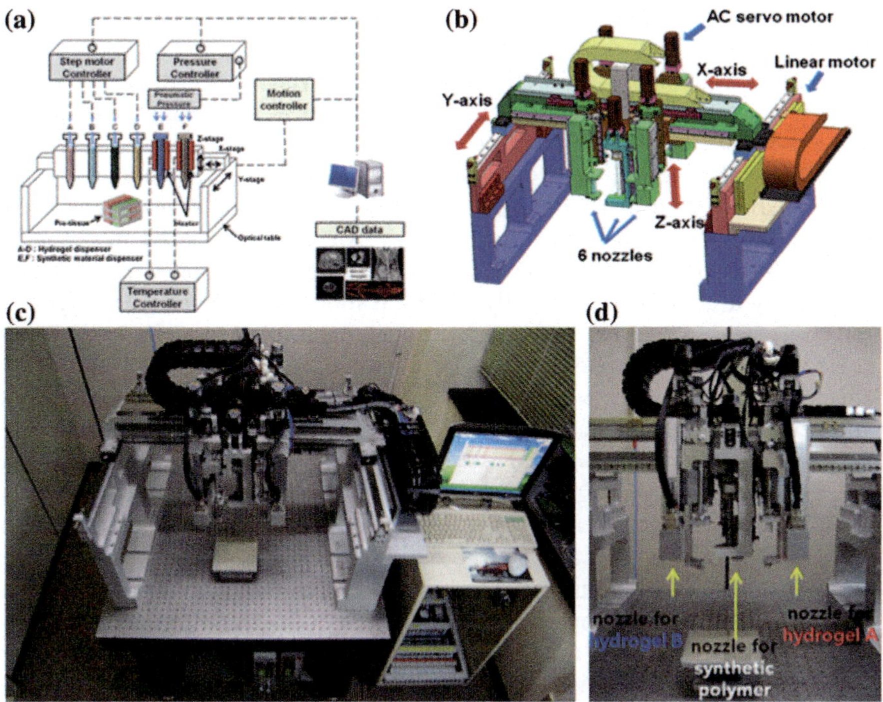

Fig. 9.10 A schematic diagram (**a**) and CAD modeling of the MtoBS (**b**). A front view of MtoBS (**c**) and its dispensing parts (**d**) [105]

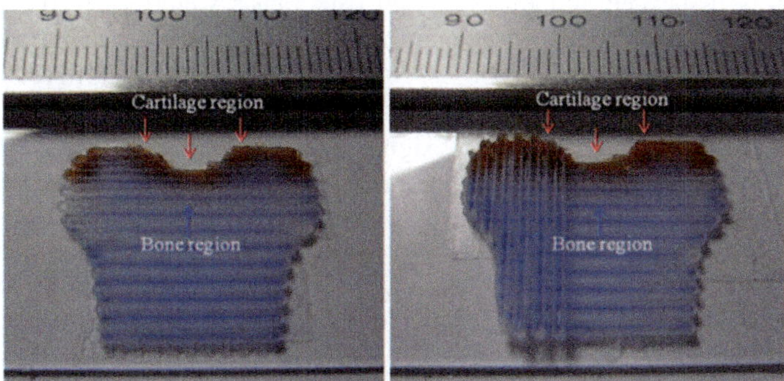

Fig. 9.11 A conceptual 3D osteochondral structure made up of PCL and two different alginates. Cartilage and bone regions are filled with *red* stained alginate and *blue* stained alginate, respectively [105]

towards the production of composite hydrogel-synthetic polymers printed materials for enhancing the mechanical performances; for example, poly(D, L-lactic-coglycolic acid) (PLGA) and polycaprolactone (PCL) were used [107]. By the way, the use of thermoplastic materials is very difficult in inkjet cell printing technology, because the high temperature required to melt the polymers are incompatible with living cells. Therefore, a new type of dispensing system is required to dispense separately both synthetic polymers and natural hydrogels to realize composite cell-laden constructs.

In 2012, Shim et al. developed a multi-head tissue/organ building system (MtoBS) having six dispensing heads to extrude a wide range of biomaterials. The thermoplastic biomaterials were dispensed by two head connected to a heating system used to melt the polymers, then they were extruded by pneumatic pressure regulated by a dispenser and rapidly cooled at room temperature. The other four heads were used to dispense cell-laden hydrogel; the heating system was obviously excluded from the hydrogel heads because natural material-based hydrogel is damaged by temperatures >37 °C. The motion of the plunger connected to the hydrogel was governed by a step motor.

In this work a synthetic polymer as Polycaprolactone (PCL) was used to create a framework to for enhancing the mechanical stability of the bioprinted construct, then two different alginate solutions were infused into the framework to create a bioprinted osteochondral graft. The porous PCL framework was designed with CAD/CAM with a line width of 250 μm and pore size of 750 μm and the height of each layer was set to 100 μm. Then the pores were filled with red and blue alginate solutions: a 1 mm width at the end portion of the construct (corresponding to the cartilage region) was filled with red-stained alginate.

The remaining construct (corresponding to the bone region) was filled with blue stained alginate. This scaffold was used as a control, then cell-laden structure consisting of a PCL framework simplified with only five pores, chondrocytes and osteoblasts were produced using the procedure above.

Separately dispensed chondrocytes and osteoblasts were clearly observed by a florescence microscope. The dispensing process was revealed to be safe for the cells: the viability of the cells (>90 %) after one day wasn't significantly different from that of unprinted cells. Dispensed chondrocytes and osteoblasts remained viable for at least 7 days, with no significant decrease in viability. Therefore, cell-laden structures may be manufactured using MtoBS, within cells viable and even proliferating for at least 1 week.

Moreover, different kinds of cells encapsulated in the alginate hydrogel could be accurately printed into the pores of a preformed polymeric frameworks. So MtoBS could be a promising system for regeneration of heterogeneous tissues, overcoming the problems related to the materials processing.

9.4 Conclusions

In this Chapter, different Rapid Prototyping techniques already adopted to realize osteochondral scaffolds for tissue engineering applications have been widely described. Although with their peculiarities, all these fabrication methods enable to have full control over the design, fabrication and production of the scaffolds, providing a systematic approach for investigating cell-matrix interactions. In particular, custom-designed implantable materials could be engineered by extrapolating defects shape and size from anatomical images of patients, such as CAT or X-rays data, towards a concept of personalized medicine.

A wide set of both synthetic and natural materials have been studied and adopted to realize functional OC grafts; synthetic polymers still represent the most challenging choice, based on their thermo properties which allow a easy manufacturing. Moreover, the recent milestones achieved by using either natural polymers or ceramics have widely boosted the application of RP technique in tissue engineering and regenerative medicine fields, and in particular for the production of such living tissues where complex chemical and morphological features are required.

Further development and advances in RP require the design of new bioactive materials, optimal scaffold design and the input of enhanced knowledge of cell physiology, including functionally efficient cell seeding and vascularization.

Additionally, indirect RP methods, coupled with conventional pore-forming techniques, further expand the range of materials that can be used in tissue engineering. In particular, fiber-based 3D structures mimicking the complex extracellular matrix microenvironment, hardly realized through standard RP techniques, could represent the next challenge, reachable combining RP methods with others, such as electro-spinning, self-assembling of phase inversion.

References

1. Nesic D, Whiteside R, Brittberg M, Wendt D, Martin I, Mainil-Varlet P (2006) Cartilage tissue engineering for degenerative joint disease. Adv Drug Deliv Rev 58(2):300–322
2. Temenoff JS, Mikos AG (2000) Review: tissue engineering for regeneration of articular cartilage. Biomaterials 21(5):431–440
3. Martin I, Miot S, Barbero A, Jakob M, Wendt D (2007) Osteochondral tissue engineering. J Biomech 40(4):750–765. doi:10.1016/j.jbiomech.2006.03.008
4. Buckwalter JA, Mankin HJ (1997) Articular cartilage: degeneration and osteoarthritis, repair, regeneration, and transplantation. Instr Course Lect 47:487–504
5. O'Driscoll SW (1998) Current concepts review-the healing and regeneration of articular cartilage*. J Bone Joint Surg 80(12):1795–1812
6. Oliveira JM, Rodrigues MT, Silva SS, Malafaya PB, Gomes ME, Viegas CA, Dias IR, Azevedo JT, Mano JF, Reis RL (2006) Novel hydroxyapatite/chitosan bilayered scaffold for osteochondral tissue-engineering applications: scaffold design and its performance when seeded with goat bone marrow stromal cells. Biomaterials 27(36):6123–6137
7. Buckwalter JA (1998) Articular cartilage: injuries and potential for healing. J Orthop Sports Phys Ther 28(4):192–202
8. Redman SN, Oldfield SF, Archer CW (2005) Current strategies for articular cartilage repair. Eur Cell Mater 9(23–32):23–32
9. Hunziker EB (2002) Articular cartilage repair: basic science and clinical progress. A review of the current status and prospects. Osteoarth Cartil 10(6):432–463
10. Grayson WL, Chao P-HG, Marolt D, Kaplan DL, Vunjak-Novakovic G (2008) Engineering custom-designed osteochondral tissue grafts. Trends Biotechnol 26(4):181–189

11. Coburn JM, Gibson M, Monagle S, Patterson Z, Elisseeff JH (2012) Bioinspired nanofibers support chondrogenesis for articular cartilage repair. Proc Natl Acad Sci 109(25):10012–10017. doi:10.1073/pnas.1121605109

12. Malda J, Woodfield TB, van der Vloodt F, Wilson C, Martens DE, Tramper J, van Blitterswijk CA, Riesle J (2005) The effect of PEGT/PBT scaffold architecture on the composition of tissue engineered cartilage. Biomaterials 26(1):63–72. doi:10.1016/j.biomaterials.2004.02.046

13. Gao J, Dennis JE, Solchaga LA, Awadallah AS, Goldberg VM, Caplan AI (2001) Tissue-engineered fabrication of an osteochondral composite graft using rat bone marrow-derived mesenchymal stem cells. Tissue Eng 7(4):363–371. doi:10.1089/107632701152436427

14. Holland TA, Bodde EW, Baggett LS, Tabata Y, Mikos AG, Jansen JA (2005) Osteochondral repair in the rabbit model utilizing bilayered, degradable oligo(poly(ethylene glycol) fumarate) hydrogel scaffolds. J Biomed Mater Res A 75(1):156–167. doi:10.1002/jbm.a.30379

15. Jiang J, Tang A, Ateshian GA, Guo XE, Hung CT, Lu HH (2010) Bioactive stratified polymer ceramic-hydrogel scaffold for integrative osteochondral repair. Ann Biomed Eng 38(6):2183–2196. doi:10.1007/s10439-010-0038-y

16. Kon E, Delcogliano M, Filardo G, Fini M, Giavaresi G, Francioli S, Martin I, Pressato D, Arcangeli E, Quarto R, Sandri M, Marcacci M (2010) Orderly osteochondral regeneration in a sheep model using a novel nano-composite multilayered biomaterial. J Orthop Res 28(1):116–124. doi:10.1002/jor.20958

17. Mano JF, Reis RL (2007) Osteochondral defects: present situation and tissue engineering approaches. J Tissue Eng Regen Med 1(4):261–273. doi:10.1002/term.37

18. Schaefer D, Martin I, Jundt G, Seidel J, Heberer M, Grodzinsky A, Bergin I, Vunjak-Novakovic G, Freed LE (2002) Tissue-engineered composites for the repair of large osteochondral defects. Arthritis Rheum 46(9):2524–2534. doi:10.1002/art.10493

19. Sherwood JK, Riley SL, Palazzolo R, Brown SC, Monkhouse DC, Coates M, Griffith LG, Landeen LK, Ratcliffe A (2002) A three-dimensional osteochondral composite scaffold for articular cartilage repair. Biomaterials 23(24):4739–4751

20. Tampieri A, Sandri M, Landi E, Pressato D, Francioli S, Quarto R, Martin I (2008) Design of graded biomimetic osteochondral composite scaffolds. Biomaterials 29(26):3539–3546. doi:10.1016/j.biomaterials.2008.05.008

21. Giannoni P, Lazzarini E, Ceseracciu L, Barone AC, Quarto R, Scaglione S (2012) Design and characterization of a tissue-engineered bilayer scaffold for osteochondral tissue repair. J Tissue Eng Regen Med. doi:10.1002/term.1651

22. Nukavarapu SP, Dorcemus DL (2013) Osteochondral tissue engineering: Current strategies and challenges. Biotechnol Adv 31(5):706–721. doi:10.1016/j.biotechadv.2012.11.004

23. Karageorgiou V, Kaplan D (2005) Porosity of 3D biomaterial scaffolds and osteogenesis. Biomaterials 26(27):5474–5491

24. Bohner M, Loosli Y, Baroud G, Lacroix D (2011) Commentary: deciphering the link between architecture and biological response of a bone graft substitute. Acta Biomater 7(2):478–484

25. Lynn AK, Best SM, Cameron RE, Harley BA, Yannas IV, Gibson LJ, Bonfield W (2010) Design of a multiphase osteochondral scaffold. I. Control of chemical composition. J Biomed Mater Res A 92(3):1057–1065

26. Scaglione S, Lazzarini E, Ilengo C, Quarto R (2010) A composite material model for improved bone formation. J Tissue Eng Regen Med 4(7):505–513

27. Suh SW, Shin JY, Kim J, Kim J, Beak CH, Kim D-I, Kim H, Jeon SS, Choo I-W (2002) Effect of different particles on cell proliferation in polymer scaffolds using a solvent-casting and particulate leaching technique. ASAIO J 48(5):460–464

28. Yoon JJ, Park TG (2001) Degradation behaviors of biodegradable macroporous scaffolds prepared by gas foaming of effervescent salts. J Biomed Mater Res 55(3):401–408

29. Nam YS, Yoon JJ, Park TG (2000) A novel fabrication method of macroporous biodegradable polymer scaffolds using gas foaming salt as a porogen additive. J Biomed Mater Res 53(1):1–7
30. Nam YS, Park TG (1999) Porous biodegradable polymeric scaffolds prepared by thermally induced phase separation. J Biomed Mater Res 47(1):8–17
31. Do Kim H, Bae EH, Kwon IC, Pal RR, Do Nam J, Lee DS (2004) Effect of PEG–PLLA diblock copolymer on macroporous PLLA scaffolds by thermally induced phase separation. Biomaterials 25(12):2319–2329
32. Yoshimoto H, Shin YM, Terai H, Vacanti JP (2003) A biodegradable nanofiber scaffold by electrospinning and its potential for bone tissue engineering. Biomaterials 24(12):2077–2082. doi:10.1016/S0142-9612(02)00635-X
33. Polini A, Pisignano D, Parodi M, Quarto R, Scaglione S (2011) Osteoinduction of human mesenchymal stem cells by bioactive composite scaffolds without supplemental osteogenic growth factors. PLoS One 6(10):e26211
34. Yeong W-Y, Chua C-K, Leong K-F, Chandrasekaran M (2004) Rapid prototyping in tissue engineering: challenges and potential. Trends Biotechnol 22(12):643–652
35. Kon E, Delcogliano M, Filardo G, Altadonna G, Marcacci M (2009) Novel nano-composite multi-layered biomaterial for the treatment of multifocal degenerative cartilage lesions. Knee Surg Sports Traumatol Arthrosc 17(11):1312–1315. doi:10.1007/s00167-009-0819-8
36. Sargeant TD, Desai AP, Banerjee S, Agawu A, Stopek JB (2012) An in situ forming collagen–PEG hydrogel for tissue regeneration. Acta Biomater 8(1):124–132. doi:10.1016/j.actbio.2011.07.028
37. Alhadlaq A, Mao JJ (2005) Tissue-engineered osteochondral constructs in the shape of an articular condyle. J Bone Joint Surg Am. doi:10.2106/jbjs.d.02104
38. Alhadlaq A, Elisseeff JH, Hong L, Williams CG, Caplan AI, Sharma B, Kopher RA, Tomkoria S, Lennon DP, Lopez A, Mao JJ (2004) Adult stem cell driven genesis of human-shaped articular condyle. Ann Biomed Eng 32(7):911–923
39. Mooney DJ, Cima L, Langer R, Johnson L, Hansen LK, Ingber DE, Vacant JP (1991) Principles of tissue engineering and reconstruction using polymer-cell constructs. In: MRS proceedings. Cambridge University Press, p 345
40. Chen CS, Mrksich M, Huang S, Whitesides GM, Ingber DE (1997) Geometric control of cell life and death. Science 276(5317):1425–1428
41. Leong KF, Cheah CM, Chua CK (2003) Solid freeform fabrication of three-dimensional scaffolds for engineering replacement tissues and organs. Biomaterials 24(13):2363–2378
42. Rengier F, Mehndiratta A, von Tengg-Kobligk H, Zechmann CM, Unterhinninghofen R, Kauczor HU, Giesel FL (2010) 3D printing based on imaging data: review of medical applications. Int J CARS 5(4):335–341. doi:10.1007/s11548-010-0476-x
43. Billiet T, Vandenhaute M, Schelfhout J, Van Vlierberghe S, Dubruel P (2012) A review of trends and limitations in hydrogel-rapid prototyping for tissue engineering. Biomaterials 33 (26):6020–6041. doi:10.1016/j.biomaterials.2012.04.050
44. Schantz J-T, Brandwood A, Hutmacher DW, Khor HL, Bittner K (2005) Osteogenic differentiation of mesenchymal progenitor cells in computer designed fibrin-polymer-ceramic scaffolds manufactured by fused deposition modeling. J Mater Sci Mater Med 16(9):807–819
45. Tellis B, Szivek J, Bliss C, Margolis D, Vaidyanathan R, Calvert P (2008) Trabecular scaffolds created using micro CT guided fused deposition modeling. Mater Sci Eng C 28 (1):171–178
46. Kim K, Yeatts A, Dean D, Fisher JP (2010) Stereolithographic bone scaffold design parameters: osteogenic differentiation and signal expression. Tissue Eng Part B Rev 16 (5):523–539. doi:10.1089/ten.TEB.2010.0171
47. Cooke MN, Fisher JP, Dean D, Rimnac C, Mikos AG (2003) Use of stereolithography to manufacture critical-sized 3D biodegradable scaffolds for bone ingrowth. J Biomed Mater Res B Appl Biomater 64(2):65–69. doi:10.1002/jbm.b.10485

48. Woodfield TB, Malda J, de Wijn J, Peters F, Riesle J, van Blitterswijk CA (2004) Design of porous scaffolds for cartilage tissue engineering using a three-dimensional fiber-deposition technique. Biomaterials 25(18):4149–4161. doi:10.1016/j.biomaterials.2003.10.056

49. Lee SJ, Kang HW, Park JK, Rhie JW, Hahn SK, Cho DW (2008) Application of microstereolithography in the development of three-dimensional cartilage regeneration scaffolds. Biomed Microdevices 10(2):233–241. doi:10.1007/s10544-007-9129-4

50. Yen H-J, Tseng C-S, S-h Hsu, Tsai C-L (2009) Evaluation of chondrocyte growth in the highly porous scaffolds made by fused deposition manufacturing (FDM) filled with type II collagen. Biomed Microdevices 11(3):615–624. doi:10.1007/s10544-008-9271-7

51. Chen C-H, Chen J-P, Lee M-Y (2011) Effects of gelatin modification on rapid prototyping PCL scaffolds for cartilage engineering. J Mech Med Biol 11(05):993–1002. doi:10.1142/S0219519411004848

52. Schuller-Ravoo S, Teixeira SM, Feijen J, Grijpma DW, Poot AA (2013) Flexible and elastic scaffolds for cartilage tissue engineering prepared by stereolithography using poly (trimethylene carbonate)-based resins. Macromol Biosci 13(12):1711–1719. doi:10.1002/mabi.201300399

53. Jacobs PF (1992) Rapid prototyping & manufacturing: fundamentals of stereolithography. The Society of Manufacturing Engineers, Dearborn.

54. Zhang X, Jiang X, Sun C (1999) Micro-stereolithography of polymeric and ceramic microstructures. Sens Actuators A 77(2):149–156

55. Lu Y, Mapili G, Suhali G, Chen S, Roy K (2006) A digital micro-mirror device-based system for the microfabrication of complex, spatially patterned tissue engineering scaffolds. J Biomed Mater Res Part A 77(2):396–405. doi:10.1002/jbm.a.30601

56. Melchels FP, Feijen J, Grijpma DW (2009) A poly(D, L-lactide) resin for the preparation of tissue engineering scaffolds by stereolithography. Biomaterials 30(23–24):3801–3809. doi:10.1016/j.biomaterials.2009.03.055

57. Choi J-W, Wicker R, Lee S-H, Choi K-H, Ha C-S, Chung I (2009) Fabrication of 3D biocompatible/biodegradable micro-scaffolds using dynamic mask projection microstereolithography. J Mater Process Technol 209(15–16):5494–5503. doi:10.1016/j.jmatprotec.2009.05.004

58. Hutmacher DW, Sittinger M, Risbud MV (2004) Scaffold-based tissue engineering: rationale for computer-aided design and solid free-form fabrication systems. Trends Biotechnol 22 (7):354–362. doi:10.1016/j.tibtech.2004.05.005

59. Hinczewski C, Corbel S, Chartier T (1998) Ceramic suspensions suitable for stereolithography. J Eur Ceram Soc 18(6):583–590. doi:10.1016/S0955-2219(97)00186-6

60. Matsuda T, Mizutani M, Arnold SC (2000) Molecular design of photocurable liquid biodegradable copolymers. 1. Synthesis and photocuring characteristics. Macromolecules 33 (3):795–800

61. Lee JW, Lan PX, Kim B, Lim G, Cho D-W (2007) 3D scaffold fabrication with PPF/DEF using micro-stereolithography. Microelectron Eng 84(5–8):1702–1705. doi:10.1016/j.mee.2007.01.267

62. Matsuda T, Mizutani M (2002) Liquid acrylate-endcapped biodegradable poly (epsilon-caprolactone-co-trimethylene carbonate). II. Computer-aided stereolithographic microarchitectural surface photoconstructs. J Biomed Mater Res 62(3):395–403. doi:10.1002/jbm.10295

63. Jansen J, Melchels FP, Grijpma DW, Feijen J (2008) Fumaric acid monoethyl ester-functionalized poly (D, L-lactide)/N-vinyl-2-pyrrolidone resins for the preparation of tissue engineering scaffolds by stereolithography. Biomacromolecules 10(2):214–220

64. Melchels FPW, Bertoldi K, Gabbrielli R, Velders AH, Feijen J, Grijpma DW (2010) Mathematically defined tissue engineering scaffold architectures prepared by stereolithography. Biomaterials 31(27):6909–6916. doi:10.1016/j.biomaterials.2010.05.068

65. Gabbrielli R, Turner I, Bowen CR (2008) Development of modelling methods for materials to be used as bone substitutes. Key Eng Mater 361:903–906

66. Yoon JJ, Park TG (2001) Degradation behaviors of biodegradable macroporous scaffolds prepared by gas foaming of effervescent salts. J Biomed Mater Res 55(3):401-408
67. Nam YS, Yoon JJ, Park TG (2000) A novel fabrication method of macroporous biodegradable polymer scaffolds using gas foaming salt as a porogen additive. J Biomed Mater Res Appl Biomater 53(1):1–7
68. Nam YS, Park TG (1999) Porous biodegradable polymeric scaffolds prepared by thermally induced phase separation. Biomed Mater Res 47(1):8–17
69. Kim SB, Kim YJ, Yoon TL, Park SA, Cho IH, Kim EJ, Kim IA, Shin, JW (2004) The characteristics of a hydroxyapatite–chitosan–PMMA bone cement. Biomaterials 25 (26):5715–5723
70. Yoshimoto H, Shin YM, Terai H, Vacanti JP (2003) A biodegradable nanofiber scaffold by electrospinning and its potential for bone tissue engineering. Biomaterials 24(12):2077–2082
71. Brady GA, Halloran JW (1997) Stereolithography of ceramic suspensions. Rapid Prototyp J 3(2):61–65. doi:10.1108/13552549710176680
72. Licciulli A, Corcione CE, Greco A, Amicarelli V, Maffezzoli A (2004) Laser stereolithography of ZrO_2 toughened Al_2O_3. J Eur Ceram Soc 24(15–16):3769–3777. doi:10.1016/j.jeurceramsoc.2003.12.024
73. Chu TM, Orton DG, Hollister SJ, Feinberg SE, Halloran JW (2002) Mechanical and in vivo performance of hydroxyapatite implants with controlled architectures. Biomaterials 23 (5):1283–1293
74. Bian W, Li D, Lian Q, Li X, Zhang W, Wang K, Jin Z (2012) Fabrication of a bio-inspired beta-Tricalcium phosphate/collagen scaffold based on ceramic stereolithography and gel casting for osteochondral tissue engineering. Rapid Prototyp J 18(1):68–80. doi:10.1108/ 13552541211193511
75. Zhang W, Lian Q, Li D, Wang K, Hao D, Bian W, He J, Jin Z (2014) Cartilage repair and subchondral bone migration using 3D printing osteochondral composites: a one-year-period study in rabbit trochlea. BioMed Res Int 2014:16. doi:10.1155/2014/746138
76. Ingavle GC, Frei AW, Gehrke SH, Detamore MS (2013) Incorporation of aggrecan in interpenetrating network hydrogels to improve cellular performance for cartilage tissue engineering. Tissue Eng Part A 19(11–12):1349–1359. doi:10.1089/ten.tea.2012.0160
77. Nguyen QT, Hwang Y, Chen AC, Varghese S, Sah RL (2012) Cartilage-like mechanical properties of poly (ethylene glycol)-diacrylate hydrogels. Biomaterials 33(28):6682–6690. doi:10.1016/j.biomaterials.2012.06.005
78. Li X, Li D, Wang L, Lu B, Wang Z (2008) Osteoblast cell response to β-tricalcium phosphate scaffolds with controlled architecture in flow perfusion culture system. J Mater Sci Mater Med 19(7):2691–2697. doi:10.1007/s10856-008-3391-8
79. Chen A, Klisch S, Bae W, Temple M, McGowan K, Gratz K, Schumacher B, Sah R (2004) Mechanical characterization of native and tissue-engineered cartilage. In: De Ceuninck F, Sabatini M, Pastoureau P (eds) Cartilage and osteoarthritis, vol 101. Methods in molecular medicine. Humana Press, New York, pp 157–190. doi:10.1385/1-59259-821-8:157
80. Goulet RW, Goldstein SA, Ciarelli MJ, Kuhn JL, Brown MB, Feldkamp LA (1994) The relationship between the structural and orthogonal compressive properties of trabecular bone. J Biomech 27(4):375–389. doi:10.1016/0021-9290(94)90014-0
81. Zhang W, Lian Q, Li D, Wang K, Hao D, Bian W, Jin Z (2015) The effect of interface microstructure on interfacial shear strength for osteochondral scaffolds based on biomimetic design and 3D printing. Mater Sci Eng C 46:10–15. doi:10.1016/j.msec.2014.09.042
82. Walters B (1991) Fast, precise, safe prototypes with FDM. In: 39th Annual technical meeting, p 1991
83. Hoque M, Hutmacher D, Feng W, Li S, Huang M-H, Vert M, Wong Y (2005) Fabrication using a rapid prototyping system and in vitro characterization of PEG-PCL-PLA scaffolds for tissue engineering. J Biomater Sci Polym Ed 16(12):1595–1610
84. Melchels FPW, Domingos MAN, Klein TJ, Malda J, Bartolo PJ, Hutmacher DW (2012) Additive manufacturing of tissues and organs. Prog Polym Sci 37(8):1079–1104. doi:10. 1016/j.progpolymsci.2011.11.007

85. Zein I, Hutmacher DW, Tan KC, Teoh SH (2002) Fused deposition modeling of novel scaffold architectures for tissue engineering applications. Biomaterials 23(4):1169–1185. doi:10.1016/S0142-9612(01)00232-0
86. Swieszkowski W, Tuan BH, Kurzydlowski KJ, Hutmacher DW (2007) Repair and regeneration of osteochondral defects in the articular joints. Biomol Eng 24(5):489–495. doi:10.1016/j.bioeng.2007.07.014
87. Ding C, Qiao Z, Jiang W, Li H, Wei J, Zhou G, Dai K (2013) Regeneration of a goat femoral head using a tissue-specific, biphasic scaffold fabricated with CAD/CAM technology. Biomaterials 34(28):6706–6716. doi:10.1016/j.biomaterials.2013.05.038
88. Liu L, Xiong Z, Yan Y, Hu Y, Zhang R, Wang S (2007) Porous morphology, porosity, mechanical properties of poly (α-hydroxy acid)–tricalcium phosphate composite scaffolds fabricated by low-temperature deposition. J Biomed Mater Res Part A 82(3):618–629
89. Xiong Z, Yan Y, Wang S, Zhang R, Zhang C (2002) Fabrication of porous scaffolds for bone tissue engineering via low-temperature deposition. Scr Mater 46(11):771–776
90. Liu L, Xiong Z, Yan Y, Zhang R, Wang X, Jin L (2009) Multinozzle low-temperature deposition system for construction of gradient tissue engineering scaffolds. J Biomed Mater Res B Appl Biomater 88(1):254–263. doi:10.1002/jbm.b.31176
91. Park TG (1995) Degradation of poly (lactic-co-glycolic acid) microspheres: effect of copolymer composition. Biomaterials 16(15):1123–1130
92. Lee SJ, Khang G, Lee YM, Lee HB (2002) Interaction of human chondrocytes and NIH/3T3 fibroblasts on chloric acid-treated biodegradable polymer surfaces. J Biomater Sci Polym Ed 13(2):197–212
93. Yuan H, De Bruijn JD, Zhang X, Van Blitterswijk CA, De Groot K (2001) Use of an osteoinductive biomaterial as a bone morphogenetic protein carrier. J Mater Sci Mater Med 12(9):761–766
94. Qu S, Leng Y, Guo X, Cheng J, Chen W, Yang Z, Zhang X (2002) Histological and ultrastructural analysis of heterotopic osteogenesis in porous calcium phosphate ceramics. J Mater Sci Lett 21(2):153–155
95. Bose S, Roy M, Bandyopadhyay A (2012) Recent advances in bone tissue engineering scaffolds. Trends Biotechnol 30(10):546–554. doi:10.1016/j.tibtech.2012.07.005
96. Hutmacher DW (2000) Scaffolds in tissue engineering bone and cartilage. Biomaterials 21 (24):2529–2543
97. Lam CXF, Mo XM, Teoh S-H, Hutmacher DW (2002) Scaffold development using 3D printing with a starch-based polymer. Mater Sci Eng C 20(1):49–56
98. Leukers B, Gülkan H, Irsen SH, Milz S, Tille C, Schieker M, Seitz H (2005) Hydroxyapatite scaffolds for bone tissue engineering made by 3D printing. J Mater Sci Mater Med 16 (12):1121–1124
99. Santos ARC, Almeida HA, Bártolo PJ (2013) Additive manufacturing techniques for scaffold-based cartilage tissue engineering: a review on various additive manufacturing technologies in generating scaffolds for cartilage tissue engineering. Virtual Phys Prototyp 8 (3):175–186
100. Scaglione S, Ceseracciu L, Aiello M, Coluccino L, Ferrazzo F, Giannoni P, Quarto R (2014) A novel scaffold geometry for chondral applications: theoretical model and in vivo validation. Biotechnol Bioeng 111(10):2107–2119. doi:10.1002/bit.25255
101. Nakamura M, Iwanaga S, Henmi C, Arai K, Nishiyama Y (2010) Biomatrices and biomaterials for future developments of bioprinting and biofabrication. Biofabrication 2 (1):014110
102. Mironov V, Boland T, Trusk T, Forgacs G, Markwald RR (2003) Organ printing: computer-aided jet-based 3D tissue engineering. Trends Biotechnol 21(4):157–161
103. Boland T, Xu T, Damon B, Cui X (2006) Application of inkjet printing to tissue engineering. Biotechnol J 1(9):910–917
104. Fedorovich NE, De Wijn JR, Verbout AJ, Alblas J, Dhert WJ (2008) Three-dimensional fiber deposition of cell-laden, viable, patterned constructs for bone tissue printing. Tissue Eng Part A 14(1):127–133. doi:10.1089/ten.a.2007.0158

105. Shim JH, Lee JS, Kim JY, Cho DW (2012) Bioprinting of a mechanically enhanced threedimensional dual cell-laden construct for osteochondral tissue engineering using a multi-head tissue/organ building system. J Micromech Microeng 22(8):085014
106. Fedorovich NE, Swennen I, Girones J, Moroni L, van Blitterswijk CA, Schacht E, Alblas J, Dhert WJ (2009) Evaluation of photocrosslinked Lutrol hydrogel for tissue printing applications. Biomacromolecules 10(7):1689–1696. doi:10.1021/bm801463q
107. Shim JH, Kim JY, Park M, Park J, Cho DW (2011) Development of a hybrid scaffold with synthetic biomaterials and hydrogel using solid freeform fabrication technology. Biofabrication 3(3):034102. doi:10.1088/1758-5082/3/3/034102

Chapter 10
Biomimetic Approaches for the Engineering of Osteochondral Tissues

Le-Ping Yan

Abstract Osteochondral defects induced by trauma or pathology remain a big challenge in orthopedics. These defects comprise injuries in both the articular cartilage and the subchondral bone. Due to the limited regeneration ability of articular cartilage, this kind of defects is normally irreparable and would gradually deteriorate toward osteoarthritis which requires surgical intervention. The current clinical treatments for osteochondral defects, such as total joint replacement and osteochondral autograft transplantation, are not ideal long-term solutions for this problem. Tissue engineering and regenerative medicine provides a possible and prospective curative strategy for this clinical hinder. Nevertheless, In order to successfully solve this problem, there are still many basic and critical questions must be answered. These include the optimization on scaffold's composition and structure, selection of suitable cell source, modulation of primary/stem cell fate or engineered osteochondral tissue, incorporation of bioactive factors, and spatial regeneration of the chondral layer, subchondral layers and the interface. Biomimetic strategy has been employed to explore these fields and exciting progresses have been achieved. This chapter summarizes these interesting advances of the pre-clinical studies on engineering of osteochondral tissue in the last 5 years. Special attentions were given to the development of biomimetic layered scaffolds with dependences on composition, cell source, or bioactive factors. In addition, cell fate regulation, external stimuli and interface regeneration were discussed. New insights and promising directions for the future study in osteochondral regeneration were also provided.

L.-P. Yan (✉)
Tissue Engineering Research Group, Department of Anatomy, Royal College of Surgeons in Ireland, 123 St. Stephens Green, Dublin 2, Dublin, Ireland
e-mail: leping.yan@gmail.com; lepingyan@rcsi.ie

L.-P. Yan
Advanced Materials and Bioengineering Research (AMBER) Centre, RCSI&TCD, Dublin, Ireland

© Springer International Publishing AG 2017
J.M. Oliveira and R.L. Reis (eds.), *Regenerative Strategies for the Treatment of Knee Joint Disabilities*, Studies in Mechanobiology, Tissue Engineering and Biomaterials 21, DOI 10.1007/978-3-319-44785-8_10

10.1 Introduction

Osteochondral defect (OCD) is a common disease in the joint, including the hip, the knee and ankle [1–3]. It can be induced by the gradual deterioration of cartilage defects which ultimately expand into the subchondral bone. Traumas including osteochondral (OC) fracture constitute another main reason for OCD formation. Pathological diseases in the subchondral bone, such as osteochondritis dissecans and osteonecrosis, also lead to the development of OCD [4]. OCD brings pain and impaired physical mobility to the patients. Furthermore, it usually progressively develops into osteoarthritis (OA). OA is the main reason of pain and permanent work incapacity. It affects 9.6 % of male and 18 % of female who over 60 years old [5]. In 2000, around 151 million individuals globally suffered from OA [6]. In Euro, OA accounts for 20 % of the chronic pain to the population [7]. As the increasing of ageing population and life expectancy, the number and percentage of OA patients will keep growing. It has been estimated that OA will become the fourth leading cause of disability by 2020 [5]. Additionally, it will result in 130 million sufferers and around 40 million disabled people only in Asia by 2030 [8].

The management of OCD is still an unsolved and challenging problem in orthopedics. The difficulty lies in the fact that articular cartilage cannot spontaneously regenerate. Another change is that both chondral layer and subchondral layer must be repaired simultaneously, as further study revealed that the regeneration of chondral lay would fail if the subchondral bone was not repaired [4, 9]. The current clinical treatments for early-stage of OCD consist of debridement, microfracture, OC autograft transplantation (OAT), and autologous chondrocyte implantation (ACI) [3]. Debridement is used to treat OCD of small size and only alleviates the pain from the patients. Microfracture only allows for the formation of fibrocartilage in the OCD. OAT has limited donor and induces donor site morbidity. ACI is advantageous in generating hyaline cartilage, while it is a complex procedure which requires two operations, a long recovery time period and the harvest of periosteal flap to seal the chondrocytes in the defect. Total joint replacement, which replaces the whole joint by a prosthesis, is applied in the last stage of OCD.

The limitations of these treatments prompt the development of tissue engineering and regeneration medicine (TERM) for OC regeneration [10]. Matrix-associated autologous chondrocytes implantation (MACI) has been applied for the treatment of OCDs in clinics [11–13]. Even though long-term rehabilitation needs to be validated, short-term satisfied results were achieved. Researchers are striving to find better solution for OCD treatment. In the last few years, many new trends emerged and important progresses were achieved. Non-layered scaffolds with novel structure or incorporated with new bioactive factors for OCD repair were developed [14, 15]. In order to rebuild the stratified structure of OC tissue, biomimetic strategy has been used intensively in engineering of OC tissue [16]. Layered scaffolds presented component/structure, cell sources or bioactive factors dependences were designed [17–24]. The modulation of cell fate and stimuli of the engineered constructs received great emphases [25–28]. Interface engineering is a new challenge in regeneration of OC tissue [29–31]. This chapter aims to overview these interesting findings and discusses some basic questions on OC regeneration. It also gives new insights and future strategies in OC tissue engineering.

10.2 Non-layered Scaffolds for Engineering OC Tissue

Due to their high reproducibility and simplicity for fabrication and implantation, single layer scaffolds are widely used for OC regeneration. These scaffolds can be of one phase (polymeric or inorganic) or multi-phases [32–34]. They can be easily processed into porous structure or hydrogel format, and acted as bioactive factors carriers [14, 15, 35].

MACI is the first application of single layer scaffolds in clinics for OC regeneration. During the MACI procedure, collagen (Col) or hyaluronic acid scaffolds are used for loading autologous chondrocytes in the OCD. Following this method, other biomaterials are also explored for OC regeneration. In order to secure the success of OC regeneration, comprehensive study is necessary to investigate the interactions between biomaterials and host tissue, or biomaterials and cells.

10.2.1 Influences of Material's Intrinsic Property and Scaffold's Structure

The single phase scaffolds contains only one component and constitute an advantageous model for study the influence of biomaterial's intrinsic property on cellular behavior or host response. In the case of polymeric based scaffolds, the molecular weight, molecular weight distribution, purity, hydrolysis profiles, chemical groups and composite/monomer ratios of the scaffolds play important roles in cell attachment, proliferation, differentiation, immune response and concomitant OC regeneration.

In one study from Abarrategi et al., freeze-dried chitosan scaffolds of different molecular weight (7.9, 10.3, 11.49, 349.3, 507.4 and 510.6 kDa) and deacetylation degree (83 and 91 %) were developed and implanted in rabbit medial femoral condyle OCD for 3 months [32]. Results indicated that the scaffolds with intact mineral content (17.9 %), lower deacetylation degree (83 %) and molecular weight of 11.49 kDa presented the best regeneration outcomes, with a well structured subchondral bone and noticeable cartilaginous tissue regeneration. Igarashi et al. investigated the influence of ultra-purified alginate hydrogels on the in vitro proliferation of rabbit bone marrow stromal cells (BMSCs), as well as the regeneration capacity of the cell laden hydrogels on rabbit patellar groove OCD [33]. Commercial alginate was used as control. The 7 days proliferation study showed the ultra-pure alginate hydrogels induced higher living cell number compared to commercial alginate hydrogels. The 4 and 12 week implantation results showed that the ultra-pure alginate presented improved histological and mechanical properties compared with the control group. In another study performed by Lim et al., it was found that the pig BMSCs seeded rehydrated oligo[poly(ethylene glycol) fumarate] (OPF) hydrogels led to around 99 % defect filling and 84 % hyaline-like cartilage formation after implantation of the constructs in micropigs lateral and medial condyles for 4 months [36]. However, there was no regeneration on the subchondral bone. This indicated that osteoconductive/osteoinductive factors may be needed for this kind of hydrogels to regenerate the subchondral bone.

On the other hand, single phase inorganic scaffolds were also applied for OC regeneration. Regarding inorganic based scaffolds, the degradation ratio, the type and amount of released ions also have high influences on the mineral deposition and integration between scaffolds and host tissue. In one study, Bernstain et al. seeded sheep autologous chondrocytes in the microporous tricalcium phosphate (TCP) scaffolds and cultured the constructs for 4 weeks in vitro [34]. Following, the constructs were implanted in the sheep medial femoral condyle OCD for 12, 26 and 52 weeks. Results showed that there was around 81 % degradation for the scaffold and the cancellous bone was almost completely restored. Col II positive hyaline cartilage was detected at both 26 and 52 weeks. After 52 weeks, neocartialge integrated with the surrounding native cartilage.

By fully evaluation of the influences of the component and structure of biomaterials on cellular and tissue response in a standardized manner, we may build a library on the suitable formulations of biomaterials for OC regeneration.

Porosity of the scaffolds is considered to be important for OC regeneration. Eman et al. prepared poly(ethylene oxide terephthalate)/poly(butylenes terephthalate) (PEOT/PBT: 55/45 by weight) scaffolds by compression-moulding/salt-leaching (CM) and 3D fibre deposition (3DF) approaches, respectively [37]. The CM and 3DF scaffolds were different porosity (CM 75.6 % and 3DF 70.2 %), pore size (CM 182 μm and 3DF μm) and interconnectivity (CM 20 and 3DF 98). The scaffolds were seeded with rabbit allogenic chondrocytes and cultured for 3 weeks in vitro. Afterwards, the constructs were implanted in rabbit medial femoral condyles OCD for 3 weeks and 3 months. Empty defects were used as control. At week 3, there were no differences in cartilage repair among the three groups. After 3 months, cartilage was significantly improved in the 3DF group compared to the CM group. This study gave hints that the structure did have big influences on the final outcomes. More systematic comparison studies are in demand to study the impact of each aspect in OC regeneration.

10.2.2 Bioactive Factors

Bioactive factors, e.g. drugs or growth factors (GF), have showed their potential in direct the cells toward osteochondral differentiation [38]. The commonly used GF for chondrogenesis are transforming growth factor β (TGF-β), typically TGF-β1 and TGF-β3. Novel bioactive factors were also explored. They were incorporated into the scaffolds to improve the likelihood of OC regeneration.

Lee et al. impregnate PD98059, which is one of the extracellular signal-regulated kinase, into PLGA scaffolds and cultured human BMSCs (hBMSCs) in the scaffolds in vitro. Rabbit BMSCs were seeded in the scaffolds and then the constructs were implanted in OCD in rabbit patellar groove for 10 weeks [14]. TGF-β2 was used as control in both in vitro and in vivo studies. In vitro, PD98059 incorporated scaffolds showed superior capacity in suppressing hypertrophy of the cells than TGF-β2 group. After 10 weeks implantation, both groups showed defect regeneration. But Col X was not observed from regenerated cartilage in PD98059 impregnated scaffold, whereas it was detected in the chondrocytes around the TGF-β2 incorporated scaffold. Better subchondral plate reconstitution was also observed in the PD98059 group.

In another interesting study from Chen et al., radially oriented Col scaffold was developed by controlling the freezing of the scaffolds [15]. Stromal-cell derived factor 1 (SDF-1) was loaded into fibrin gel and injected into the cylindrical cavity of the scaffold. Rabbit BMSCs were cultured onto the scaffolds and then the constructs were implanted in rabbit patellar groove OCD for 6 and 12 weeks. Radially oriented scaffolds without SDF-1 and freeze-dried random Col scaffolds with or without SDF-1 were also implanted. The radially oriented scaffold presented better mechanical properties than the random scaffold before cell seeding. The radially oriented scaffold further facilitated the migration of the BMSCs in vitro compared to the random scaffold. The in vivo results demonstrated that both the radially oriented scaffold and SDF-1 effectively promoted the cartilage repair. The immunohistochemical staining of ossification indicator Col I and cartilage degradation marker MMP13 showed that radially oriented scaffold combined with SDF-1 could inhibit the expression of these two markers and therefore promote the formation of normal cartilage. In order fully understand the effect of these new bioactive factors, long-term evaluation and comparison with the common used GFs are worthy to perform.

10.3 Biomimetic Approaches for Engineering of OC Tissue

Biomimetic approaches have been used for many aspects related with OC tissue engineering, including scaffolds design, selection of cells and regulation of cell differentiation or engineered tissue function [17–19, 39, 40]. These approaches were also utilized for elucidation of the basic science on OC regeneration, such as the role of calcified cartilage in OC regeneration or building biomimetic models for OC interface regeneration [30, 41].

10.3.1 Layered Scaffolds for Engineering of OC Tissue

OC is a highly organized and stratified tissue comprising one chondral layer on the top and one subchondral layer at the bottom [3]. The components, structure, cells of these two integrated layers are completely different. The chondral layer comprises Col II and glycosaminoglycan (GAG), while the subchondral layer mainly contains cancerous bone. These two layers are connected by a calcified cartilage zone [17]. On the other hand, the chondral layer exhibits hydrogel-like format and contains only chondrocytes. The subchondral layer is of porous structure and is composed of varied cells, such as osteoblasts and bone marrow cells. Therefore, developing layered scaffolds which mimic the features of specific layers in OC tissue may provide a desirable microenvironment for cell homing and subsequently prompt the regeneration of the OCD. As a consequence, layered scaffolds (bilayered, multi-layered or gradient scaffolds) have been studied intensively in the last decade [3, 16]. Bilayered scaffolds are the predominant structure of layered scaffolds used for OC tissue engineering. Dependent on their characteristics, these bilayered scaffolds can be of components, cell sources or bioactive factors dependence for each layer.

10.3.1.1 Bilayered Scaffolds with Component Dependence

Varied strategies have been proposed for preparation of bilayered scaffolds with component dependence for each layer. Similar to MACI, Col and hyaluronic acid are commonly chosen for chondral layer fabrication [19, 25, 42, 43]. Besides, other polysaccharides, proteins, synthetic polymers and their composites have proven beneficial to support the growth and differentiation of chondrocytes and stem cells [17, 18, 22, 23, 26]. On the other hand, osteoconductive and osteoinductive inorganic components, such as hydroxyapatite (HA), TCP, biphasic calcium phosphate (BCP) and bioactive glass are able to enhance the binding capacity of the scaffolds to the bone defect [44, 45]. Hence, incorporation of these bioactive inorganic components in the subchondral layer of the scaffold is advantageous, as they can promote the fast integration between scaffolds and host tissue and thus provide mechanical support in the defect area in the early stage after implantation. Moreover, these bioactive inorganic components are also able to stimulate the in vitro differentiation of cells toward osteogenic differentiation [18, 46, 47].

In order to evaluate the influence of each layer of the bilayered scaffold on cell behavior, simplified models were built by culturing cells in separated layers or seeding only one cell source in the bilayered scaffolds. Oliveira et al. prepared bilayered HA and chitosan scaffolds and studied the goat BMSCs (GBMSCs) differentiations on the separated layers. The sintered porous HA layer supported the osteogenic differentiation of GBMSCs by increasing the alkaline phosphatase (ALP) content during the 2 weeks osteogenesis, and the freeze-dried chitosan layer showed enhanced amount of GAG by culturing the cells from day 7 to day 21. Zhou et al. developed integrated multilayered Col-HA scaffolds and studied the influence of the top Col layer and the bottom Col/HA layer on the differentiation of hBMSCs [48]. BMSCs were seeded onto the separated Col and Col/HA layer and subsequently the constructs underwent both chondrogeneic (28 days) and osteogenic differentiations (14 days). The results showed that Col layer was more efficient in inducing the chondrogenesis of BMSCs in terms of superior Sox-9, Col 2a1 and aggrecan genes expressions compared with the Col/HA layer. Regarding the osteogenic differentiation, the Col/HA layer demonstrated higher ALP content and higher amount of gene expression for Runx-2, ALP, Col 1a1, osteopontin and osteocalcin compared with the ones of the Col layer.

Intended to improve the scaffold mechanical properties, Yan et al. recently developed robust porous bilayered scaffolds including silk scaffolds in the chondral layer and silk/nano CaP scaffolds in the subchondral layer. Rabbit BMSCs were cultured in the whole scaffolds and underwent osteogenic differentiation for 2 weeks. The cells in the subchondral layer produced significantly higher ALP content than the cells in the chondral silk layer. The bilayered scaffolds were implanted in rabbit knee condyle OCD for 4 weeks. Large amount of new bone formation was observed in the subchondral layer of the scaffolds, and neocartilage regeneration was presented in the chondral layer [18].

These in vitro and in vivo studies showed that the combination of bioactive inorganic components in the subchondral layer of the bilayered scaffolds is promising strategy for OC regeneration.

10.3.1.2 Bilayered Scaffolds with Cell Source Dependence

It is believed that culturing proper cells in the scaffolds could promote the formation of extracellular matrix (ECM) similar to the one in the host tissue and thus speed up the in vivo regeneration progress after implantation of the constructs. The chondral layer and subchondral layer contain chondrocytes and bone cells, respectively. It makes sense to incorporate these cells into the specific layers of the scaffolds to engineer stratified tissue.

Only chondrocytes for chondral layer

In some studies, only chondrocytes were seeded in the chondral layer of the bilayered scaffolds. In these cases, the subchondral layer acted as anchors for fixing the whole scaffolds in the defects. This strategy is a step forward of the MACI procedure which only implants the scaffolds in the chondral layer and requires a periosteal membrane to seal the scaffolds in the wound.

Im et al. developed bilayered scaffolds composed of hyaluronant/Col in the chondral layer and HA/TCP in the subchondral layer [19]. Pig autologous chondrocytes were loaded into the chondral layer and the constructs were cultured under chondrogenic condition for 2 weeks before implantation into minipig medial and lateral condyle OCDs for 5 months. The defects were also treated by scaffold alone, autologous OC transplantation, chondrocytes implantation or left empty. International Cartilage Repair Society Macroscopic Score (ICRS) was used to evaluate the regeneration of defects. After 5 months, the ICRS value of the cell seeded scaffold group was similar to the ones of the OC transplantation and ACI groups. Seol et al. prepared BCP and alginate bilayered scaffolds by immersion of the BCP in alginate/rabbit chondrocytes/TGF-β suspension which was cross-linked afterwards [49]. The biphasic constructs were implanted into rabbit knee OCDs. The alginate/rabbit chondrocytes/TGF-β constructs were implanted in the cartilage defects. OCD without treatment was used as control. After 12 weeks, the defect control and the hydrogel group showed fibrous tissues in the defects. The biphasic group presented newly regenerated cartilage tissues which were morphologically similar to the native cartilage tissue. Both regenerated cartilage and subchondral bone integrated to the surrounding native tissues.

But in the large animal study from Sosio et al., different outcomes were showed. They prepared bilayered scaffolds including Col scaffold as chondral layer and Col coated HA scaffold as subchondral layer [21]. Pig chondrocytes were mixed with fibrinogen and seeded onto the chondral layer. Fibrin glue was formed by the addition of thrombin. Following, the bilayered constructs were cultured in vitro for 3 weeks and then implanted in the pig (around 70 kg) trochlea OCD for 12 weeks. Scaffolds without cells/fibrin and empty defect were used as control. After 3 months, the cell-free scaffolds showed superior ICRS macroscopic scores to the cell loaded group. The cell-free group also showed higher ICRS II histological scores for cell morphology and surface/superficial assessment with respect to the other groups. The scaffolds seeded with chondrocytes showed high cellularity but low GAG production. The possible reason was that the fibrin gel may act as an obstacle for the migration of host cells into the scaffolds.

Chondrocytes for chondral layer and other cell source for subchondral layer

Interaction between different cell sources may affect the function or differentiation of each cell type and subsequently plays some roles in the organization or regeneration of the tissues. Thus, co-culture of two kinds of cells in the layered scaffolds may present some synergistic influence. In a study from Cui et al., pig autologous chondrocytes and osteoblasts were seeded onto the poly(lactic-co-glycolic acid) (PLGA) and TCP scaffolds, respectively [20]. The two scaffolds were sutured to form bilayered construct before implantation into pig medial condyle OCD. PLGA scaffolds seeded with chondrocytes were implanted only in the chondral defect. Empty defects were used as control. After 6 months, the biphasic scaffolds showed the highest ICRS macroscopic scores and the ICRS histological scores among the three groups. This group also exhibited the highest compressive properties and GAG content than the other groups.

As subchondral bone also contains BMSCs and these cells are pluripotent cells which can be differentiated into cartilage and bone tissue, meriting their application for OC regeneration. They have been used for seeding in the subchondral layer of the porous scaffolds. Giannoni et al. engineered bilayered poly(ε-caprolactone) (PCL, chondral layer)) and PCL/HA (subchondral layer) scaffolds [22]. Bovine BMSCs and articular chondrocytes were seeded onto the subchondral and chondral layers, respectively. Afterwards, the constructs were implanted subcutaneously in the back of nude mice for 9 weeks. The results showed that the bilayered scaffolds presented good integrity as confirmed by tension fracture tests. In vivo data revealed that thick mature bone surrounding the ceramic granules in the subchondral layer and cartilaginous alcianophilic matrix presented in the chondral layer. Vascularization was mainly observed in the bony layer which possessed higher blood vessel density compared to the chondral layer.

Hydrogels based bilayered constructs were also developed. Agarose hydrogels encapsulated with porcine chondrocytes in the chondral layer and porcine BMSCs in the subchondral layer were prepared by Sheehy et al. [50]. The constructs were cultured under chondrogenesis condition for 3 weeks and then implanted subcutaneously in the back of nude mice for 4 weeks. The results demonstrated that the co-culture system enhanced the chondrogenesis in the chondral layer, and appeared to suppress hypertrophy of the chondrocytes and the mineralization in the bony layer. Furthermore, the in vivo data showed that endochondral ossification was only observed in the bony layer, leading to the development of an OC tissue. This study indicates that the structure of the biomaterials plays different roles for varied tissue regeneration.

Deng et al. studied the regeneration of large OCD defect in a rabbit model [51]. The rabbit chondrocytes and osteogenically differentiated BMSCs were injected into the chondral layer and bony layer of the distal femur shape bilayered scaffolds composed of gelatin/chondroitin sulphate/sodium hyaluronate (chondral layer) and gelatin/ceramic bovine bone (bony layer). These bilayered scaffolds with cells or without cells were kept in osteogenic medium for 48 h before implanting into large rabbit OCD ($15 \times 10 \times 5$ mm) in the patella of the right distal femur for 6, 12 and 24 weeks. Empty defects were used as control. At week 6 and 12, hyaline-like cartilage formed in the cell seeded constructs which were stained positively for Col II. This group also presented higher level of Col II expression by RT-PCR compared with other groups. Most of the component in the bony layer was replaced by new bone. At week 24, the bony layer was completely absorbed and a tidemark was observed in some areas. Scaffolds without cells and empty defect groups showed the formation of large amount of fibrous tissue and only a little of new bone.

Stem cells for chondral layer

Chondrocytes tends to lost their phenotype and differentiation capacity during the in vitro culture. Efforts are paid to seek other better cell sources for OC regeneration. Stem cells, such as BMSCs, adipose derived stromal cells (ADSCs), can be osteogenically and chondrogenically differentiated. Therefore, these cells alone can be used for generate the OC tissue.

Galperin et al. used sphere-templating technique to fabricate bilayered degradable poly(hydroxyethyl methacrylate) (PHEMA, chondral layer) and PHEMA/HA (bony layer) bilayered hydrogel based scaffolds [52]. hBMSCs and chondrogenically differentiated hBMSCs were seeded onto the bony layer and chondral layer, respectively. The chondral layer was decorated with hyaluronan and of 200 μm pore size, while the bony layer was of 38 μm pore size with surface coated with HA particles. The constructs were co-culture for 4 weeks in basal medium. In the absence of growth factors (GFs), the integrated scaffolds supported simultaneous matrix deposition and adequate cell growth of two distinct cell lineage in each layer. The bony layer provided a suitable environment for hBMSCs differentiation toward osteoblast and the cells in chondral layer retained the chondrocyte phenotype.

Apart from the BMSCs, other stem cells were also employed for OC regeneration. Rodrigues et al. generated agarose (chondral layer) and starch-PCL (SPCL, bony layer) bilayered scaffolds [53]. Human amniotic fluid-derived stem cells (AFSCs) were either encapsulated in the agarose hydrogel or seeded onto the SPCL scaffolds. Chondrogenesis and osteogenesis differentiations of the AFSCs were performed for the hydrogel and scaffold, respectively. After differentiations occurring in the two constructs, the separated hydrogel and scaffold were combined by agarose hydrogel to form the bilayered constructs which were subsequently cultured in a co-culture OC medium for 2 weeks. The results showed that pre-differentiated AFSCs seeded onto the SPCL scaffolds did not need OC medium to maintain the phenotype and they secreted abundant mineralized ECM for up to 2 weeks. However, the pre-chondrogenically differentiated cells still required OC medium to maintain their phenotype, but no need for insulin-like growth factor-1 (IGF-1).

In vivo studies were also performed to evaluate the performance of the BMSCs which were only encapsulated/seeded in the chondral layer of the bilayered constructs. Bal et al. encapsulated chondrogenically differentiated rabbit BMSCs into poly(ethylene glycol) (PEG) hydrogels and then combined the hydrogels with different porous scaffolds (tantalum scaffold, allograft bone or bioactive glass scaffold) to form the bilayered scaffolds [54]. The varied bilayered constructs were implanted in rabbit knee OCD for 6 and 12 weeks. The results showed that bioactive glass and porous tantalum were superior to the bone allograft regarding the integration to host tissue, regeneration of hyaline-like cartilage and the secretion of Col II.

Liu et al. firstly fabricated poly(L-lactic acid)-co-poly(ε-caprolactone) (PLLA-CL)/ Col I nanofiber yarn mesh [55]. And then immerse the yarn into Col I/hyaluronic acid solution and freeze-dried the mixture to prepare the scaffold. Afterwards, the yarn/Col/hyaluronic aicd scaffold (chondral layer, Yarn-CH) was combined with TCP (bony layer) to form bilayered scaffolds. Rabbit BMSCs were only seeded onto the chondral layer and underwent chondrogenesis differentiation in vitro for 3 weeks before implantation in OCD in rabbit patellar groove of the distal femur. Undifferentiated BMSCs in the yarn-CH/TCP bilayered scaffolds, differentiated

BMSCs seeded bilayered scaffolds without the yarn, undifferentiated BMSCs seeded bilayered scaffolds without the yarn and OC autograft were also implanted. Empty defects were used as control. The results showed that the yarn can greatly improve the mechanical strength of the chondral layer. Combining with BMSCs, the yarn-CH/TCP scaffolds were successfully used to repair the OCD in rabbit. The differentiated BMSCs with yarn-CH/TCP group presented the highest ICRS macroscopic and histologic scores.

Although further studies are necessary to maintain the chondrogenic phenotype of the differentiated stem cells, these studies showed that stem cells are promising alternative for chondrocytes in the regeneration of chondral layer in OCD.

10.3.1.3 Bilayered Scaffolds with Bioactive Factor Dependence

Bioactive factors incorporated scaffolds are of great interest as they can spatially modulate the fate of seeded cells toward OC differentiation and thus facilitate the regeneration journey. Majority of the studies focused on inducing the chondrogenesis of stem cells by bioactive factors. TGF-β family is the prominent GFs used for OCD regeneration, while other bioactive factors were also explored.

In one study from Ho et al., the influence of fibrin on the chondrogenesis of hBMSCs was studied by using the PCL/fibrin (chondral layer) and PCL/TCP/fibrin (bony layer) bilayered scaffolds [23]. The hBMSCs loaded fibrin or fibrin/alginate gel was added into the PCL scaffolds and cultured under chondrogenic condition supplemented with TGF-β1 for 28 days. hBMSCs loaded fibrin gel was added into the PCL/TCP scaffolds and cultured in osteoenic condition for 28 days. Afterwards, the two constructs were combined to form the bilayered scaffolds by using fibrin gel. The biphasic constructs were co-cultured in an OC co-culture medium. Results showed that fibrin promoted the chondrogenic differentiation of BMSCs, while fibrin/alginate declined the expression of Col II and aggrecan gene expressions. Mineralized tissue formed in the bone phase of the biphasic constructs, with boundary in the interface of the construct.

In another study from Guo et al., the osteogenically differentiated rabbit BMSCs (differentiated for 0, 3, 6 and 12 days) were mixed with OPF solution to form the bony layer, following a mixture of BMSCs, OPF solution and TGF-β3 was injected on top of the osteogenic layer [56]. The formed OC constructs were cultured under chondrogenic condition. In the chondrogenic domain, the growth factor greatly induced chondrogenic differentiation of BMSCs. Moreover, the cells with various osteogenic pre-culture time periods in the osteogenic layer together with TGF-β3, enhanced the chondrogenic gene marks expressions of the MSCs in the chondral layer. In the osteogenic layer, the cells maintained ALP activity during co-culture, while the mineralization was delayed with the presence of TGF-β3.

Dual GFs delivery was attempted for enhancing the chondral layer regeneration in OCD. Kim et al. prepared bilayered OPF hydrogels which contained IGF-1 and/or TGF-β3 in the chondral OPF layer and OPF alone as subchondral layer [57].

The growth factors were loaded into the hydrogel with the aid of gelatin microparticles (GMPs). Four groups of hydrogels were studies in a rabbit medial femoral condyle OCD model: blank control (no GFs), GMP-loaded IGF-1 alone, GMP-loaded IGF-1 and gel-loaded TGF-β3, GMP-loaded IGF-1 and GMP-loaded TGF-β3 in OPF hydrogels. After 12 weeks, all groups showed improvement in cartilage morphology compared to the control. Single delivery of IGF-1 showed higher scores in subchondral bone morphology, as well as in chondrocytes and GAG amount in adjacent cartilage tissue, when compared to a dual delivery of IGF-1 and TGF-β3.

Injection of hormone in the defect size together with scaffold implantation was studied by Zhang et al. [24]. Col and silk/HA bilayered scaffolds were firstly developed. Scaffolds were implanted into the rabbit patellar groove OCD for 16 weeks. During the implantation, intra-articular injection of the defects with parathyroid hormone-related protein (PTHrP) was performed (PBS as control) at the 4–6, 7–9, and 10–12 weeks with time windows every 7 days. After 16 weeks, better outcomes were showed in the defects treated with PTHrP at the 4–6 weeks time window, including reconstitution of cartilage and subchondral bone with minimal terminal differentiation (matrix degradation, ossification and hypertrophy), as well as enhanced chondrogenesis markers compared with treatment at other time windows.

Bilayered construct incorporated with varied bioactive factors in the chondral layer and the subchondral layer is advantageous for simultaneous regeneration of the chondral layer and the subchondral layer. In a recent study from Castro et al., TGF-β1 incorporated poly(dioxanone) (PDO) and BMP-2 incorporated PLGA nanospheres were prepared by electrospraying [58]. Following, the TGF-β1 incorporated PDO nanospheres were mixed with PEG hydrogels (chondral layer), and the BMP-2 incorporated PLGA nanospheres were combined with PCL/PEG/nano-HA scaffold (bony layer). The two layers were integrated by placing the chondral layer onto the bony layer and cured by UV. hBMSCs were seeded onto the separated chondral and bony layer, and chondrogenic and osteogenic differentiations were performed, respectively. Results showed that enhanced adhesion, proliferation and differentiation in bone and cartilage layers were observed.

Gene modification of the cells was explored for OC regeneration by Chen et al. [59]. Plasmid TGF-β1 gene and plasmid BMP-2 gene were loaded in the chitosan/gelatin scaffold (chondral layer) and chitosan/gelatin/HA scaffold (bony layer), respectively. The two scaffolds were combined by fibrin gel to form bilayered scaffolds. Rabbit BMSCs were seeded onto the separated chondral layer and bony layer firstly and then combined to form bilayered OC construct. Additionally, the bilayered scaffolds with or without genes, mono layer scaffolds with BMP-2 gene or TGF-β1 gene were used for implantation in rabbit patellar groove OCD for 4, 8 and 12 weeks. Non-treated defects acted as control. The in vitro results showed that BMSCs proliferated well in each layer. Highly expressed TGF-β1 protein and BMP-2 protein were observed from the chondral layer and bony layer, respectively. The bilayered scaffolds induced the differentiation of BMSCs from the bony layer and chondral layer into osteoblasts and chondrocytes in vitro, respectively. In vivo, the monolayer scaffolds with BMP-2 gene presented complete trabecular bone ingrowth within subchondral region and good integration with native bone tissue, but with abundant Col I in the cartilage layer. The monolayer with TGF-β1 gene showed similar cartilage surface with native cartilage, while the regeneration of subchondral bone was insufficient. The bilayered gene incorporated scaffolds showed successful reconstitution of cartilage and subchondral bone.

The incorporation of bioactive factors is a powerful tool to facilitate the regeneration procedure in OCD. Before transferring to clinical studies, several important aspects must be clarified, including the influences of incorporation manner/dose on the regeneration outcome, the in vitro and vivo release profiles.

10.3.1.4 Multilayered and Gradient Scaffolds

In order to closely mimic the fine stratified structure and the gradient component distribution in the OC tissue, multilayered and gradient scaffolds were developed. Ding et al. fabricated trilayered scaffolds which composed of a porous silk/HA bony layer, longitudinally oriented silk layer and an intermediate dense silk/HA layer [60]. For cell attachment study, chondrogenic and osteogenic-induced ADSCs were seeded onto the chondral and bony layer, respectively. For differentiation studies, chondrogenic or osteogenic-induced ADSCs were seeded onto the separated chondral and bony layer, and subsequently cultured in chondrogenic and osteogenic medium, respectively. The results indicated that the scaffolds possessed good biocompatibility by supporting the growth, proliferation and infiltration of ADSCs. Histological and immunohistochemical data, as well as the real-time polymerase chain reaction (RT-PCR) showed that the cells can be differentiated into chondrocytes and osteoblasts in the chondral and bony layers in vitro, respectively. The intermediate layer played a role in preventing the cells migrated between the two layers. In another study from Levingstone et al., multilayer scaffolds were developed which comprising Col I/Col II/hyaluronic acid as the chondral layer, Col I/Col II/HA as the intermediate layer, and Col I/HA as the bony layer [42]. MC3T3-E1 mouse pre-osteoblasts were seeded in the scaffolds. The scaffolds presented seamlessly integrated layer structures, high level porosity, homogeneous structure and high interconnectivity. Cellular distribution was homogeneous throughout the construct.

Scaffolds with gradient GFs distribution is of great interest since they may help to regenerate the highly organized OC tissue [41, 61]. Mohan et al. prepared PLGA microspheres loaded with TGF-β1 (chondral layer) and PLGA or PLGA/nano-HA microspheres loaded with BMP-2 (bony layer) [62]. Gradient scaffolds were prepared by filling a cylindrical mold with opposing gradients of two different types of microspheres. The microspheres were sintered to form the layered scaffolds. The scaffolds were implanted in rabbit medial condyle OCD for 6 and 12 weeks. Blank PLGA, PLGA scaffold with gradient GFs loading, PLGA and PLGA/HA scaffolds without GFs were also implanted. The gross morphology, histology data and MRI data showed that the best regeneration outcomes of cartilage and sub-chondral bone were presented in the PLGA and PLGA/HA scaffolds of gradient GFs. Similar GAG content and cartilage thickness to native cartilage were observed in this group. Additionally, this group displayed higher bone filling and superior edge integration with host bone tissue than the other group. In another interesting study, Jeon et al. used two programmable syringe pumps to prepare gradient hydrogels [63]. One syringe contained methacrylated alginate hydrogel precursor solution and heparin-alginate microparticles, the other one only contain the hydrogel precursor solution. By controlling the flow rate of each syringe, hydrogels containing gradient BMP-2 and TGF-β1 were developed. Additionally, hydrogels with gradient covalently coupled RGD-containing peptide were also prepared. Furthermore, stiffness gradient hydrogels can also be fabricated in this system. These hydrogels with gradient growth factors, RGD-containing peptide and stiffness were also used for hMSCs encapsulation. The results showed that encapsulated hMSCs number and/or osteochondral differentiation can be modulated. This technique is versatile platform for development of multi-dependent hydrogels and possesses high potential for different applications.

In the future, development of biomimetic and functional Layered scaffolds is still the predominant direction for OCD regeneration. Bioactive materials, such as ECM and its analogue, could be of great interesting since they can provide the biological cues for cell growth, differentiation, and maintaining their phenotypes. The basic science on the structure-function relationship of the scaffolds must be elucidated before their biological application. Emphasis should focus on the influences of scaffold components and structures on the cells' behaviors. Bioactive factor incorporated scaffolds is desirable for OC regeneration. But before the successful application of the bioactive factor loaded scaffolds, the incorporated dose and release profile of the drugs should be fully studied in order to match the clinical requirement.

10.3.2 *Modulation of Cell Fate and Engineered OC Tissue*

Harnessing the cell fate is critical for OC regeneration. Studies indicated that the way how the cells were cultured and the time for cell differentiation had big influence on their functions [26, 39]. Many approaches have been developed for tuning the behaviors of primary cells and stem cells in OC tissue engineering [25, 64]. Additionally, the external stimuli on the engineered OC tissue also affect the OC regeneration outcomes [40, 65, 66].

10.3.2.1 Modulation of Chondrocytes

Chondrocytes will lose their chondrogenic phenotype and dedifferentiate during in vitro expansion. The in vitro culture time period before implantation is important for the successful application of these cells. Miot et al. explored the influence of in vitro maturation of engineered cartilage on the in vivo OC regeneration. Goat autologous chondrocytes were cultured in hyaluronic acid scaffolds for varied time periods (2 days, 2 and 6 weeks) [39]. Then the constructs were implanted on top of a HA/hyaluronic acid sponge subchondral filler in goat trochlea groove OCD for 8 weeks and 8 months. Empty defects and scaffolds without cells acted as controls. It was found that increasing pre-culture time resulted in progressive maturation of the grafts in vitro. After 8 weeks in vivo, there was no obvious repair by any treatments. After 8 months, O'Driscoll histology scores showed poor cartilage architecture in untreated and cell-free scaffolds groups. The cellular grafts showed the best scores for grafts with 2 weeks pre-culture time. This study indicated it may not necessary to culture the chondrocytes for long time before implantation.

In an in vitro study, it was found that shorter time than 2 weeks is better to create cartilage ECM. Scotti et al. loaded chondrocytes into Col construct and pre-cultured the construct in chondrogenic medium for 0, 3 or 14 days before combining it with devitalized bone by fibrin gel [25]. The biphasic constructs were co-cultured in chondrogenic medium for 5 weeks in vitro. It was found that pre-culture of the chondral layer for 3 days prior to the generation of the bilayered construct resulted in more efficient cartilaginous matrix formation than that of the 0 pre-culture day, and superior biological bonding to the bony scaffold than the 14 days pre-culture group. The bony part scaffold induced the cells to secrete osteoblast-related gene-bone sialoprotein. To final validate the interesting finding from this work, future in vivo study should be conducted.

Freshly isolated chondrocytes were compared with pre-cultured chondrocytes for OC regeneration in vivo by Chiang et al. [67]. Chondrocytes underwent pre-culture in alginate beads for 3 weeks or were freshly isolated were seeded in the poly(D-lactic acid) layer (PDLA, chondral layer) of the bilayered PDLA and PDLA/TCP scaffolds. The bilayered constructs or bilayered scaffold alone were implanted in the pig femoral condyle OCD for 6 months. Empty defect acted as control. The results showed that the cell seeded group was repaired with hyaline cartilage. Scaffold alone group presented fibrocartilage. No regeneration was observed in the non-treated group. The two experimental groups showed comparable results with control group in subchondral bone regeneration. There were no significant differences between the two experimental groups in the ICRS scores in terms of surface, matrix, cell distribution, cell viability, subchondral bone and mineralization.

From these studies, it might reach a preliminary conclusion that the chondrocytes may not necessary to culture in vitro for long time before the implantation. Long in vitro culture time could lead to their dedifferentiation during the 2D expansion or 3D culture in a non-cartilaginous environment.

But what would happen if the chondrocytes are cultured in cartilaginous ECM scaffolds? Jin et al. studied the influence of the in vitro maturity of engineered cartilage on OCD regeneration [68]. Lyophilized ECM scaffolds derived from in vitro culture of rabbit chondrocytes were prepared. Then, rabbit chondrocytes were seeded onto the scaffolds and cultured for (A) 2 days, (B) 2 weeks and (C) 4 weeks in vitro before implantation in patella groove OCD in rabbit for 1 and 3 months. Un-treated defects acted as control group (D). After 1 month, group B and C showed hyaline-like cartilage regeneration. Fibrocartilage tissues were presented in group A and D. After 3 months, group C displayed striking features of hyaline cartilage containing mature matrix and columnar arrangement of chondrocytes and zonal distribution of Col II. This group also showed well regenerated subchondral bone. This study indicated that the more matured engineered cartilage was, the better repaired the OCD was, when using the cartilaginous ECM scaffolds.

10.3.2.2 Modulation of Stem Cells

Comparing with chondrocytes, stem cells are advantageous for their high proliferation rate and differentiation capacity, less possibility to be affected by the age of the patients. The most used adult stem cells are BMSCs. But there are many other adult stem cells can be chosen, such as ADSCs, AFSCs, synovial stem cells, umbilical cord mesenchymal stromal cells (UCMSCs) [3, 28, 41, 69]. Optimization the differentiation conditions for these stem cells toward OC regeneration is in great demand.

Grayson et al. investigated undifferentiated and pre-differentiated hBMSCs (osteogenic or chondrogenic differentiation for 1 week) for engineering OC tissue in vitro [26]. The undifferentiated and chondrogenically differentiated cells were encapsulated in agarose hydrogels. Meanwhile, undifferentiated and osteogenically differentiated cells were seeded onto trabecular bone. The trabecular bone was overlaid with the agarose to create the biphasic construct. Afterwards, the biphasic constructs were cultured in chondrogenic or cocktail medium, under static condition or in a bioreactor. The results indicated that the pre-differentiated BMSCs were only beneficial for bone formation. The perfusion condition and cocktail medium inhibited chondrogenesis of BMSCs. Perfusion improved the integration of the bone-cartilage interface. Lam et al. also performed a comprehensive study to understand the influence of pre-differentiation of BMSCs on their behavior on generation of OC tissue in vitro [42]. The BMSCs were chondrogenically (7 and 14 days) or osteogenically differentiated (6 days) before encapsulated into the OPF hydrogels to form bilayered constructs. The OPF/chondrogenic-induced BMSCs construct acted as chondral layer and OPF/osteogenic-induced BMSCs construct was used as subchondral layer. Four groups of bilayered constructs were generated: (1) Both layers comprised undifferentiated BMSCs; (2) undifferentiated BMSCs in the chondral layer and osteogenic BMSCs in the subchondral layer; (3) BMSCs with chondrogenic differentiation for 7 days in the chondral layer and osteogenic BMSCs in the subchondral layer; (4) BMSCs with chondrogenic differentiation for 14 days in the chondral layer and osteogenic BMSCs in the subchondral layer. After 28 days culturing in serum-free chondrogenic medium without GFs, the chondral layer of group (3) exhibited enhanced chondrogenic phenotype compared with other groups, in terms of Col II and aggrecan gene expressions, and GAG to hydroxyproline ratios. Osteogenic BMSCs co-cultured with chondrogenic BMSCs showed higher cellularity over time. The osteogenic BMSCs produced minerals in the subchondral layer. Additionally, the pre-differentiated BMSCs maintained their chondrogenic and osteogenic phenotypes during the co-culture without GFs.

In vivo studies have been conducted to validate the in vitro differentiation condition for BMSCs. Zscharnack et al. studied the differences between freshly isolated and pre-chondrogenically differentiated sheep BMSCs for chronic OCD regeneration [64]. Sheep medial femoral condyles OCDs were created 6 weeks before the implantation. The cells were expanded for 4 weeks and then loaded into Col gel. The constructs underwent 2 weeks chondrogenesis differentiation before implantation in the OCDs. Col gels without cells and with undifferentiated BMSCs were also implanted. Empty defects were used as control. After 6 months, the pre-differentiated MSCs group showed substantially better histologic scores with morphologic characteristics of hyaline cartilage and the presence of Col II. Chang et al. perform similar study by encapsulated pig autologous BMSCs in Col gel for pig medial condyle OCD regeneration [70]. The cells were loaded into the gel and cultured in vitro with or without TGF-β1for 2 weeks. Afterwards, the constructs were implanted for 6 months. Col gel without cells and empty defects were used as controls. The results demonstrated that Col gels with undifferentiated cells or chondrogenically induced cells presented similar repair outcomes. But the Col gel with undifferentiated cells showed better Pineda score grading compared with other groups. These differences could come from the different cells and different animal models used. Further comparison studies are necessary to reach a universal conclusion using standardized model.

ADSCs are a promising cell source for OC tissue engineering. They can be isolated easily in high quantity. The influence of differentiation on ADSCs was also studied by Jurgens et al. in a goat medial condyle/trochlear groove OCD model [28]. The freshly isolated adipose-derived stromal vascular fraction cells (SVF) and cultured ADSCs (passage 3) were seeded onto the Col scaffolds. The constructs were implanted for 1 and 4 months. Scaffolds without cells were used as control. After 1 month, constructs with cells showed more regeneration than acellular group. After 4 months, acellular group displayed enhanced regeneration, but inferior to the cell-seeded constructs. Scaffolds with cells showed extensive Col II and hyaline cartilage, as well as superior elastic moduli. These two groups presented GAG content closed to the value of native cartilage, and induced higher level of subchongral bone regeneration. SVF group showed better results than the ADSCs group.

Induced pluripotent stem cells (iPSCs) almost present unlimited differentiate potential and can circumvent the ethical argument of embryonic stem cells [71]. Therefore, they are highly attractive cell source for tissue engineering. Recently, iPSCs have been applied for cartilage and OC regeneration [72, 73]. Ko et al. performed the first study using iPSCs for OC regeneration [72]. The chondrogenesis performance of iPSCs was compared with the one of the hBMSCs in vitro. Following, the human iPSCs or iPSCs pellets were encapsulated in alginate gels and implanted in rat patellar groove OCD for 12 weeks. Empty defects and defects with only alginate gels acted as controls. The 21 days in vitro study showed that the iPSCs displayed superior chondrogenesis compared with induced BMSCs, in terms of higher amount of GAG expression, lower level of Col X and Col I. The in vivo results indicated that the cell encapsulated gel group showed better cartilage regeneration than the control groups.

Stem cells have showed enormous potential for OC regeneration. In spite of the promising results achieved, there are still some questions remained, such as how to "frozen" the chondrogenic phenotype of the induced stem cells, and what is the optimal condition for isolation, differentiation, or genetic modification of the stem cells.

10.3.2.3 Modulation of In Vivo Engineered Tissue

A lot of works have been performed to stimulate cells or cell seeded constructs in vitro before implantation. But little is known about the influences of the stimuli on the implanted constructs. A few attempts were performed to explore the influence of mechanical stimulus on OC regeneration in vivo.

Hannink et al. prepared PCL/polyurethane scaffolds of 3 or 4 mm in height and then implanted the scaffolds in rabbit trochlear groove OCD for 8 and 14 weeks [65]. The defect was 3 mm in depth. The 4 mm scaffolds were used to study the mechanical stimulus on OC regeneration. After 8 weeks, both the 3 and 4 mm scaffolds were flushed with the native cartilage. Center region had less matrix compared with the edge region and no differences between the two groups. At week 14, more cartilaginous tissue presented in the 4 mm scaffolds compared with the 3 mm scaffolds group. In the 4 mm scaffolds group, progression of the cartilaginous tissue grew from the surface toward the center of the scaffolds was observed over time. The 3 mm scaffolds group showed no difference in the central zone compared to its condition at week 8. Besides the internal mechanical stimulus, external stimulus was also performed. Chang et al. implanted PLGA scaffolds alone in the rabbit medial condyle OCD for 4 and 12 weeks [40]. Continuous passive motion (CPM) treatment was conducted and compared with the immobilization (Imm) and intermittent active motion (IAM) treatments in the PLGA scaffolds and empty defect groups. At week 12, the PLGA-CPM group showed the best results with normal cartilage surface, without contracture in the joint and no inflammation. Imm and IAM groups showed degenerated joints, abrasion cartilage surfaces and synovitis. The CPM group also showed enhanced bone volume compared with the IAM, Imm and empty defect groups.

The results from these studies gave hints that the external stimuli after implantation could also play important role on OC tissue formation. Further studies should be performed and explore the effects of other means.

10.3.3 Interface Regeneration

Chondral layer and subchondral layer are connected by a calcified cartilage zone also called OC interface which present unique properties in the components, cell morphology and Col orientation [29, 30, 41]. OC interface plays a role as a barrier to stop the synovial fluid from reaching the subchondral bone region and also prevents the invasion of the bone cells and vessels to the chondral layer. The regeneration of OC interface is critical for OC regeneration.

Researchers have used the bioactive inorganic component to induce the hypertrophy of the chondrocytes. Jiang et al. fabricated bilayered scaffolds composed of agarose as chondral layer and PLGA/bioactive glass microsphere scaffold as the subchondral layer. Chondrocytes seeded on the PLGA/bioactive glass scaffold showed superior ALP expression than chondrocytes seeded on the control PLGA scaffold. In another study, Cheng et al. prepared rabbit BMSCs loaded Col hydrogel microspheres and subsequently performed chondrogenesis and osteogenesis differentiation in vitro for 21 days, respectively [31]. Afterwards, the two layers were connected by Col gel containing undifferentiated BMSCs. The formed constructs were cultured in chondrogenic, osteogenic or normal medium from 14 to 21 days, respectively. The results showed that intact and continuous calcified cartilage zone was generated when co-cultured in chondrogenic medium. Chondrocytes in the interface region of this group presented hypertrophic phenotype (such as expression of Col II and X), produced calcium mineral and vertically oriented fibers in the ECM. Regarding the osteogenic medium, the upper layer chondrogenic tissue became calcified. In the normal medium group, undifferentiated BMSCs were still found in the interface, and the pre-differentiated cells were able to maintain their chondrogenic and osteogenic phynotypes.

Attempt was made to re-build the insulation function of the OC layer. Da et al. developed a tri-layered scaffolds composed of bovine cartilage ECM in the chondral layer, compact PLGA/β-TCP in the interface region and a rapid processed and Col wrapped PLGA/PLGA/β-TCP in the subchondral layer [74]. Rabbit BMSCs were firstly osteogenically or chondrogenically differentiated for 21 days and then seeded onto the bony layer or chondral layer, respectively. The constructs were then implanted in rabbit OCD for 3 and 6 months. Scaffolds without compact layer and non-treated defects were acted as controls. Mechanical analysis showed that the anti-tensile and anti-shear properties of the compact layer containing scaffolds were significantly higher than the group without compact layer. In vivo data indicated that better macroscopic scores, enhanced GAG and Col contents, superior micro-computed topography images and histological staining of the regenerated tissue were presented in the compact layer containing biphasic scaffolds compared to the controls.

Since the OC interface is very thin, novel analysis technique and standard method is required for the evaluation of the engineered OC interface. Following, the development of seamless and continuous OC interface is the main goal. The formation of a gradient interface and improvement of the interface strength are still big challenges.

10.4 Conclusions

TERM aims to engineer tissue with structural, functional, biomechanical and biochemical similarity to the native tissue. This chapter focuses on the application of TERM strategy for OC tissue regeneration. Emphasis is on current pre-clinical studies using biomimetic approaches. Even though a successful technique has not yet been developed for generation of OC tissue identical to the host tissue, progresses were achieved in many aspects of OC tissue engineering. Biomimetic layered scaffolds with spatially controlled distribution of component, structure, cell sources or bioactive factors have been developed and satisfied results were observed. In spite of the gold standard on cell source is not defined yet, the understanding of cell differentiation and cell–cell interaction is improved. Additionally, endeavors have been exerted to increase the comprehension of the interface using the biomimetic model. In the future, efforts should be paid on further optimization the scaffold composition to mimic the ECM in the OC tissue. The mechanical property of the scaffolds should be compatible to the biomechanical environment in OC tissue. Besides, the degradation profile of the scaffold must match the growth rate of the de novo tissue. For cell modulation, it still requires extensive study on how to maintain stable chondrogenesis and osteogenesis of the cells in the chondral layer and subchondral layer. Regarding in vivo study, novel biomimetic models are worthy to build in order to create a truly defect environment. The external stimuli on the implanted site in the experimental animals may be helpful to generate a healthy OC tissue. Despite there are still many unsolved problems, TERM is still the most promising strategy and considered as next generation technique for OC regeneration.

References

1. Mano JF, Reis RL (2007) Osteochondral defects: present situation and tissue engineering approaches. J Tissue Eng Regen Med 1(4):261–273. doi:10.1002/term.37
2. Zengerink M, Struijs PA, Tol JL, van Dijk CN (2010) Treatment of osteochondral lesions of the talus: a systematic review. Knee Surg Sports Traumatol Arthrosc 18(2):238–246. doi:10.1007/s00167-009-0942-6
3. Yan LP, Oliveira JM, Oliveira AL, Reis RL (2015) Current concepts and challenges in osteochondral tissue engineering and regenerative medicine. ACS Biomater Sci Eng 1(4):183–200. doi:10.1021/ab500038y
4. Gomoll AH, Madry H, Knutsen G, van Dijk N, Seil R, Brittberg M, Kon E (2010) The subchondral bone in articular cartilage repair: current problems in the surgical management. Knee Surg Sports Traumatol Arthrosc 18(4):434–447. doi:10.1007/s00167-010-1072-x
5. Woolf AD, Pfleger B (2003) Burden of major musculoskeletal conditions. Bull World Health Organ 81:646–656. http://dx.doi.org/10.1590/S0042-96862003000900007
6. Zhuo Q, Yang W, Chen J, Wang Y (2012) Metabolic syndrome meets osteoarthritis. Nat Rev Rheumatol 8(12):729–737. doi:10.1038/nrrheum.2012.135
7. Cedraschi C, Delézay S, Marty M, Berenbaum F, Bouhassira D, Henrotin Y, Laroche F, Perrot S (2013) "Let's talk about oa pain": a qualitative analysis of the perceptions of people

suffering from OA. Towards the development of a specific pain OA-related questionnaire, the Osteoarthritis Symptom Inventory Scale (OASIS). PLoS ONE 8(11):e79988. doi:10.1371/journal.pone.0079988

8. Lim K, Lau CS (2011) Perception is everything: OA is exciting. Int J Rheum Dis 14(2):111–112. doi:10.1111/j.1756-185X.2011.01614.x

9. Minas T, Gomoll AH, Rosenberger R, Royce RO, Bryant T (2009) Increased failure rate of autologous chondrocyte implantation after previous treatment with marrow stimulation techniques. Am J Sports Med 37(5):902–908. doi:10.1177/0363546508330137

10. Langer R, Vacanti JP (1993) Tissue engineering. Science 260(5110):920–926. doi:10.1126/science.8493529

11. Giannini S, Buda R, Vannini F, Di Caprio F, Grigolo B (2008) Arthroscopic autologous chondrocyte implantation in osteochondral lesions of the talus: Surgical technique and results. Am J Sports Med 36(5):873–880. doi:10.1177/0363546507312644

12. Bedi A, Foo LF, Williams Iii RJ, Potter HG (2010) The maturation of synthetic scaffolds for osteochondral donor sites of the knee: an MRI and T2-mapping analysis. Cartilage 1(1):20–28. doi:10.1177/1947603509355970

13. Giza E, Sullivan M, Ocel D, Lundeen G, Mitchell ME, Veris L, Walton J (2010) Matrix-induced autologous chondrocyte implantation of talus articular defects. Foot Ankle Int 31(9):747–753. doi:10.3113/fai.2010.0747

14. Lee JM, Kim JD, Oh EJ, Oh SH, Lee JH, Im GI (2014) PD98059-impregnated functional PLGA scaffold for direct tissue engineering promotes chondrogenesis and prevents hypertrophy from mesenchymal stem cells. Tissue Eng Part A 20(5–6):982–991. doi:10.1089/ten.tea.2013.0290

15. Chen P, Tao J, Zhu S, Cai Y, Mao Q, Yu D, Dai J, Ouyang H (2015) Radially oriented collagen scaffold with SDF-1 promotes osteochondral repair by facilitating cell homing. Biomaterials 39:114–123. doi:10.1016/j.biomaterials.2014.10.049

16. Nukavarapu SP, Dorcemus DL (2013) Osteochondral tissue engineering: current strategies and challenges. Biotechnol Adv 31(5):706–721. doi:10.1016/j.biotechadv.2012.11.004

17. Oliveira JM, Rodrigues MT, Silva SS, Malafaya PB, Gomes ME, Viegas CA, Dias IR, Azevedo JT, Mano JF, Reis RL (2006) Novel hydroxyapatite/chitosan bilayered scaffold for osteochondral tissue-engineering applications: scaffold design and its performance when seeded with goat bone marrow stromal cells. Biomaterials 27(36):6123–6137. doi:10.1016/j.biomaterials.2006.07.034

18. Yan LP, Silva-Correia J, Oliveira MB, Vilela C, Pereira H, Sousa RA, Mano JF, Oliveira AL, Oliveira JM, Reis RL (2015) Bilayered silk/silk-nanoCaP scaffolds for osteochondral tissue engineering: in vitro and in vivo assessment of biological performance. Acta Biomater 12:227–241. doi:10.1016/j.actbio.2014.10.021

19. Im GI, Ahn JH, Kim SY, Choi BS, Lee SW (2010) A hyaluronate-atelocollagen/beta-tricalcium phosphate-hydroxyapatite biphasic scaffold for the repair of osteochondral defects: a porcine study. Tissue Eng Part A 16(4):1189–1200. doi:10.1089/ten.TEA.2009.0540

20. Cui W, Wang Q, Chen G, Zhou S, Chang Q, Zuo Q, Ren K, Fan W (2011) Repair of articular cartilage defects with tissue-engineered osteochondral composites in pigs. J Biosci Bioeng 111(4):493–500. doi:10.1016/j.jbiosc.2010.11.023

21. Sosio C, Di Giancamillo A, Deponti D, Gervaso F, Scalera F, Melato M, Campagnol M, Boschetti F, Nonis A, Domeneghini C, Sannino A, Peretti GM (2015) Osteochondral repair by a novel interconnecting collagen-hydroxyapatite substitute: a large-animal study. Tissue Eng Part A 21(3–4):704–715. doi:10.1089/ten.TEA.2014.0129

22. Giannoni P, Lazzarini E, Ceseracciu L, Barone AC, Quarto R, Scaglione S (2012) Design and characterization of a tissue-engineered bilayer scaffold for osteochondral tissue repair. J Tissue Eng Regen Med 9(10):1182–1192. doi:10.1002/term.1651

23. Ho STB, Cool SM, Hui JH, Hutmacher DW (2010) The influence of fibrin based hydrogels on the chondrogenic differentiation of human bone marrow stromal cells. Biomaterials 31(1):38–47. doi:10.1016/j.biomaterials.2009.09.021

24. Zhang W, Chen J, Tao J, Hu C, Chen L, Zhao H, Xu G, Heng BC, Ouyang HW (2013) The promotion of osteochondral repair by combined intra-articular injection of parathyroid hormone-related protein and implantation of a bi-layer collagen-silk scaffold. Biomaterials 34(25):6046–6057. doi:10.1016/j.biomaterials.2013.04.055
25. Scotti C, Wirz D, Wolf F, Schaefer DJ, Burgin V, Daniels AU, Valderrabano V, Candrian C, Jakob M, Martin I, Barbero A (2010) Engineering human cell-based, functionally integrated osteochondral grafts by biological bonding of engineered cartilage tissues to bony scaffolds. Biomaterials 31(8):2252–2259. doi:10.1016/j.biomaterials.2009.11.110
26. Grayson WL, Bhumiratana S, Grace Chao PH, Hung CT, Vunjak-Novakovic G (2010) Spatial regulation of human mesenchymal stem cell differentiation in engineered osteochondral constructs: effects of pre-differentiation, soluble factors and medium perfusion. Osteoarthr Cartil 18(5):714–723. doi:10.1016/j.joca.2010.01.008
27. Lam J, Lu S, Meretoja VV, Tabata Y, Mikos AG, Kasper FK (2014) Generation of osteochondral tissue constructs with chondrogenically and osteogenically predifferentiated mesenchymal stem cells encapsulated in bilayered hydrogels. Acta Biomater 10(3):1112–1123. doi:10.1016/j.actbio.2013.11.020
28. Jurgens WJ, Kroeze RJ, Zandieh-Doulabi B, van Dijk A, Renders GA, Smit TH, van Milligen FJ, Ritt MJ, Helder MN (2013) One-step surgical procedure for the treatment of osteochondral defects with adipose-derived stem cells in a caprine knee defect: a pilot study. Biores Open Access 2(4):315–325. doi:10.1089/biores.2013.0024
29. Khanarian NT, Jiang J, Wan LQ, Mow VC, Lu HH (2012) A hydrogel-mineral composite scaffold for osteochondral interface tissue engineering. Tissue Eng Part A 18(5–6):533–545. doi:10.1089/ten.TEA.2011.0279
30. Khanarian NT, Haney NM, Burga RA, Lu HH (2012) A functional agarose-hydroxyapatite scaffold for osteochondral interface regeneration. Biomaterials 33(21):5247–5258. doi:10.1016/j.biomaterials.2012.03.076
31. Cheng HW, Luk KDK, Cheung KMC, Chan BP (2011) In vitro generation of an osteochondral interface from mesenchymal stem cell-collagen microspheres. Biomaterials 32(6):1526–1535. doi:10.1016/j.biomaterials.2010.10.021
32. Abarrategi A, Lopiz-Morales Y, Ramos V, Civantos A, Lopez-Duran L, Marco F, Lopez-Lacomba JL (2010) Chitosan scaffolds for osteochondral tissue regeneration. J Biomed Mater Res A 95(4):1132–1141. doi:10.1002/jbm.a.32912
33. Igarashi T, Iwasaki N, Kasahara Y, Minami A (2010) A cellular implantation system using an injectable ultra-purified alginate gel for repair of osteochondral defects in a rabbit model. J Biomed Mater Res, Part A 94(3):844–855. doi:10.1002/jbm.a.32762
34. Bernstein A, Niemeyer P, Salzmann G, Suedkamp NP, Hube R, Klehm J, Menzel M, von Eisenhart-Rothe R, Bohner M, Goerz L, Mayr HO (2013) Microporous calcium phosphate ceramics as tissue engineering scaffolds for the repair of osteochondral defects: histological results. Acta Biomater 9(7):7490–7505. doi:10.1016/j.actbio.2013.03.021
35. Guo X, Park H, Young S, Kretlow JD, van den Beucken JJ, Baggett LS, Tabata Y, Kasper FK, Mikos AG, Jansen JA (2010) Repair of osteochondral defects with biodegradable hydrogel composites encapsulating marrow mesenchymal stem cells in a rabbit model. Acta Biomater 6 (1):39–47. doi:10.1016/j.actbio.2009.07.041
36. Lim CT, Ren X, Afizah MH, Tarigan-Panjaitan S, Yang Z, Wu Y, Chian KS, Mikos AG, Hui JHP (2013) Repair of osteochondral defects with rehydrated freeze-dried oligo poly (ethylene glycol) fumarate hydrogels seeded with bone marrow mesenchymal stem cells in a porcine model. Tissue Eng Part A 19(15–16):1852–1861. doi:10.1089/ten.tea.2012.0621
37. Emans PJ, Jansen EJP, van Iersel D, Welting TJM, Woodfield TBF, Bulstra SK, Riesle J, van Rhijn LW, Kuijer R (2013) Tissue-engineered constructs: the effect of scaffold architecture in osteochondral repair. J Tissue Eng Regen Med 7(9):751–756. doi:10.1002/term.1477
38. Makris EA, Gomoll AH, Malizos KN, Hu JC, Athanasiou KA (2015) Repair and tissue engineering techniques for articular cartilage. Nat Rev Rheumatol 11(1):21–34. doi:10.1038/nrrheum.2014.157

39. Miot S, Brehm W, Dickinson S, Sims T, Wixmerten A, Longinotti C, Hollander AP, Mainil-Varlet P, Martin I (2012) Influence of in vitro maturation of engineered cartilage on the outcome of osteochondral repair in a goat model. Eur Cells Mater 23:222–236

40. Chang NJ, Lin CC, Li CF, Wang DA, Issariyaku N, Yeh ML (2012) The combined effects of continuous passive motion treatment and acellular PLGA implants on osteochondral regeneration in the rabbit. Biomaterials 33(11):3153–3163. doi:10.1016/j.biomaterials.2011.12.054

41. Dormer NH, Singh M, Wang L, Berkland CJ, Detamore MS (2010) Osteochondral interface tissue engineering using macroscopic gradients of bioactive signals. Ann Biomed Eng 38(6):2167–2182. doi:10.1007/s10439-010-0028-0

42. Levingstone TJ, Matsiko A, Dickson GR, O'Brien FJ, Gleeson JP (2014) A biomimetic multi-layered collagen-based scaffold for osteochondral repair. Acta Biomater 10(5):1996–2004. doi:10.1016/j.actbio.2014.01.005

43. Jiang Y, Chen L, Zhang S, Tong T, Zhang W, Liu W, Xu G, Tuan RS, Heng BC, Crawford R, Xiao Y, Ouyang HW (2013) Incorporation of bioactive polyvinylpyrrolidone–iodine within bilayered collagen scaffolds enhances the differentiation and subchondral osteogenesis of mesenchymal stem cells. Acta Biomater 9(9):8089–8098. doi:10.1016/j.actbio.2013.05.014

44. Dorozhkin SV, Epple M (2002) Biological and medical significance of calcium phosphates. Angew Chem Int Ed 41(17):3130–3146. doi:10.1002/1521-3773(20020902)41:17<3130:aid-anie3130>3.0.co;2-1

45. LeGeros RZ (2008) Calcium phosphate-based osteoinductive materials. Chem Rev 108(11):4742–4753. doi:10.1021/cr800427g

46. Zhang Y, Wu C, Friis T, Xiao Y (2010) The osteogenic properties of CaP/silk composite scaffolds. Biomaterials 31(10):2848–2856. doi:10.1016/j.biomaterials.2009.12.049

47. Jiang J, Tang A, Ateshian GA, Guo XE, Hung CT, Lu HH (2010) Bioactive stratified polymer ceramic-hydrogel scaffold for integrative osteochondral repair. Ann Biomed Eng 38(6):2183–2196. doi:10.1007/s10439-010-0038-y

48. Zhou J, Xu C, Wu G, Cao X, Zhang L, Zhai Z, Zheng Z, Chen X, Wang Y (2011) In vitro generation of osteochondral differentiation of human marrow mesenchymal stem cells in novel collagen-hydroxyapatite layered scaffolds. Acta Biomater 7(11):3999–4006. doi:10.1016/j.actbio.2011.06.040

49. Seol YJ, Park JY, Jeong W, Kim TH, Kim SY, Cho DW (2015) Development of hybrid scaffolds using ceramic and hydrogel for articular cartilage tissue regeneration. J Biomed Mater Res, Part A 103(4):1404–1413. doi:10.1002/jbm.a.35276

50. Sheehy EJ, Vinardell T, Buckley CT, Kelly DJ (2013) Engineering osteochondral constructs through spatial regulation of endochondral ossification. Acta Biomater 9(3):5484–5492. doi:10.1016/j.actbio.2012.11.008

51. Deng T, Lv J, Pang J, Liu B, Ke J (2014) Construction of tissue-engineered osteochondral composites and repair of large joint defects in rabbit. J Tissue Eng Regen Med 8(7):546–556. doi:10.1002/term.1556

52. Galperin A, Oldinski RA, Florczyk SJ, Bryers JD, Zhang M, Ratner BD (2013) Integrated Bi-Layered Scaffold for Osteochondral Tissue Engineering. Advanced Healthcare Materials 2(6):872–883. doi:10.1002/adhm.201200345

53. Rodrigues MT, Lee SJ, Gomes ME, Reis RL, Atala A, Yoo JJ (2012) Bilayered constructs aimed at osteochondral strategies: the influence of medium supplements in the osteogenic and chondrogenic differentiation of amniotic fluid-derived stem cells. Acta Biomater 8(7):2795–2806. doi:10.1016/j.actbio.2012.04.013

54. Bal BS, Rahaman MN, Jayabalan P, Kuroki K, Cockrell MK, Yao JQ, Cook JL (2010) In vivo outcomes of tissue-engineered osteochondral grafts. J Biomed Mater Res B Appl Biomater 93(1):164–174. doi:10.1002/jbm.b.31571

55. Liu S, Wu J, Liu X, Chen D, Bowlin GL, Cao L, Lu J, Li F, Mo X, Fan C (2015) Osteochondral regeneration using an oriented nanofiber yarn-collagen type I/hyaluronate hybrid/TCP biphasic scaffold. J Biomed Mater Res, Part A 103(2):581–592. doi:10.1002/jbm.a.35206

56. Guo X, Liao J, Park H, Saraf A, Raphael RM, Tabata Y, Kasper FK, Mikos AG (2010) Effects of TGF-beta3 and preculture period of osteogenic cells on the chondrogenic differentiation of rabbit marrow mesenchymal stem cells encapsulated in a bilayered hydrogel composite. Acta Biomater 6(8):2920–2931. doi:10.1016/j.actbio.2010.02.046

57. Kim K, Lam J, Lu S, Spicer PP, Lueckgen A, Tabata Y, Wong ME, Jansen JA, Mikos AG, Kasper FK (2013) Osteochondral tissue regeneration using a bilayered composite hydrogel with modulating dual growth factor release kinetics in a rabbit model. J Control Release 168 (2):166–178. doi:10.1016/j.jconrel.2013.03.013

58. Castro NJ, O'Brien CM, Zhang LG (2014) Biomimetic biphasic 3-D nanocomposite scaffold for osteochondral regeneration. AIChE J 60(2):432–442. doi:10.1002/aic.14296

59. Chen J, Chen H, Li P, Diao H, Zhu S, Dong L, Wang R, Guo T, Zhao J, Zhang J (2011) Simultaneous regeneration of articular cartilage and subchondral bone in vivo using MSCs induced by a spatially controlled gene delivery system in bilayered integrated scaffolds. Biomaterials 32(21):4793–4805. doi:10.1016/j.biomaterials.2011.03.041

60. Ding X, Zhu M, Xu B, Zhang J, Zhao Y, Ji S, Wang L, Wang L, Li X, Kong D, Ma X, Yang Q (2014) Integrated trilayered silk fibroin scaffold for osteochondral differentiation of adipose-derived stem cells. ACS Appl Mater Interfaces 6(19):16696–16705. doi:10.1021/am5036708

61. Dormer NH, Qiu Y, Lydick AM, Allen ND, Mohan N, Berkland CJ, Detamore MS (2012) Osteogenic differentiation of human bone marrow stromal cells in hydroxyapatite-loaded microsphere-based scaffolds. Tissue Eng Part A 18(7–8):757–767. doi:10.1089/ten.tea.2011. 0176

62. Mohan N, Dormer NH, Caldwell KL, Key VH, Berkland CJ, Detamore MS (2011) Continuous gradients of material composition and growth factors for effective regeneration of the osteochondral interface. Tissue Eng Part A 17(21–22):2845–2855. doi:10.1089/ten.tea. 2011.0135

63. Jeon O, Alt DS, Linderman SW, Alsberg E (2013) Biochemical and physical signal gradients in hydrogels to control stem cell behavior. Adv Mater 25(44):6366–6372. doi:10.1002/adma. 201302364

64. Zscharnack M, Hepp P, Richter R, Aigner T, Schulz R, Somerson J, Josten C, Bader A, Marquass B (2010) Repair of chronic osteochondral defects using predifferentiated mesenchymal stem cells in an ovine model. Am J Sports Med 38(9):1857–1869. doi:10. 1177/0363546510365296

65. Hannink G, de Mulder ELW, van Tienen TG, Buma P (2012) Effect of load on the repair of osteochondral defects using a porous polymer scaffold. J Biomed Mater Res Part B Appl Biomater 100B(8):2082–2089. doi:10.1002/jbm.b.32773

66. Chang NJ, Lin CC, Li CF, Su K, Yeh ML (2013) The effect of osteochondral regeneration using polymer constructs and continuous passive motion therapy in the lower weight-bearing zone of femoral trocheal groove in rabbits. Ann Biomed Eng 41(2):385–397. doi:10.1007/ s10439-012-0656-7

67. Chiang H, Liao CJ, Wang YH, Huang HY, Chen CN, Hsieh CH, Huang YY, Jiang CC (2010) Comparison of articular cartilage repair by autologous chondrocytes with and without in vitro cultivation. Tissue Eng Part C Methods 16(2):291–300. doi:10.1089/ten.tec.2009.0298

68. Jin CZ, Cho JH, Choi BH, Wang LM, Kim MS, Park SR, Yoon JH, Oh HJ, Min BH (2011) The maturity of tissue-engineered cartilage in vitro affects the repairability for osteochondral defect. Tissue Eng Part A 17(23–24):3057–3065. doi:10.1089/ten.TEA.2010.0605

69. Mohal JS, Tailor HD, Khan WS (2012) Sources of adult mesenchymal stem cells and their applicability for musculoskeletal applications. Curr Stem Cell Res Ther 7(2):103–109. doi:10. 2174/157488812799219027

70. Chang CH, Kuo TF, Lin FH, Wang JH, Hsu YM, Huang HT, Loo ST, Fang HW, Liu HC, Wang WC (2011) Tissue engineering-based cartilage repair with mesenchymal stem cells in a porcine model. J Orthop Res 29(12):1874–1880. doi:10.1002/jor.21461

71. Yamanaka S (2012) Induced pluripotent stem cells: past, present, and future. Cell stem cell 10 (6):678–684. doi:10.1016/j.stem.2012.05.005
72. Ko JY, Kim KI, Park S, Im GI (2014) In vitro chondrogenesis and in vivo repair of osteochondral defect with human induced pluripotent stem cells. Biomaterials 35(11):3571–3581. doi:10.1016/j.biomaterials.2014.01.009
73. Diekman BO, Christoforou N, Willard VP, Sun H, Sanchez-Adams J, Leong KW, Guilak F (2012) Cartilage tissue engineering using differentiated and purified induced pluripotent stem cells. Proc Natl Acad Sci 109(47):19172–19177. doi:10.1073/pnas.1210422109
74. Da H, Jia SJ, Meng GL, Cheng JH, Zhou W, Xiong Z, Mu YJ, Liu J (2013) The impact of compact layer in biphasic scaffold on osteochondral tissue engineering. PLoS One 8(1): e54838. doi:10.1371/journal.pone.0054838

Chapter 11
Osteochondral Tissue Engineering and Regenerative Strategies

Raphaël F. Canadas, Alexandra P. Marques, Rui Luís Reis
and Joaquim Miguel Oliveira

Abstract The orthopedic field has been facing challenging difficulties when it comes to regeneration of large and/or complex defects as we come across in osteochondral (OC) cases of lesions grade 4. Autologous OC mosaicplasty has proven to be a valid therapeutic option but donor site morbidity and the lack of long-term functionality remain sources of concern. OC tissue engineering has shown an increasing development to provide suitable strategies for the regeneration of damaged cartilage and underlying subchondral bone tissue. The use of two scaffolds with optimized properties for bone and cartilage architectures combined at the time of implantation as a multilayered structure was one of the first approaches for OC large defects regeneration. Last decade strategies using a bony-like scaffold supporting a cell layer for cartilage phase were introduced. Beyond the approaches already mentioned, three other strategies were reported for OCD regeneration. One methodology was the use of two different layers with a compact interface to create an integrated bilayered scaffold before cell seeding. A second strategy was the use of a single continuous structure but with different features in each layer. The last one was the combination of hydrogel phases creating this way the possibility to have injectable systems. These promising strategies for the regeneration of complex OCDs comprise the use of different biomaterials, growth factors, and cells alone or in combination, but the ideal solution is still to be found. The interface's mechanical properties have to be optimized. A different problem is related with the cell culture method within the 3D bilayered structures with heterogeneous properties. With the increasing demand of these stratified 3D structures new cell culture systems are required. Moreover these structures present the potential to be used as in vitro models, which is a need also because of the pressure resulting from the 3R's

R.F. Canadas · A.P. Marques · R.L. Reis · J.M. Oliveira (✉)
3B's Research Group—Biomaterials, Biodegradables and Biomimetics,
Headquarters of the European Institute of Excellence on TE
and Regenerative Medicine, University of Minho, AvePark,
4806-909 Taipas, Guimarães, Portugal
e-mail: miguel.oliveira@dep.uminho.pt

R.F. Canadas · A.P. Marques · R.L. Reis · J.M. Oliveira
ICVS/3B's—PT Government Associate Laboratory, Braga, Guimarães, Portugal

© Springer International Publishing AG 2017
J.M. Oliveira and R.L. Reis (eds.), *Regenerative Strategies for the Treatment of Knee Joint Disabilities*, Studies in Mechanobiology, Tissue Engineering and Biomaterials 21, DOI 10.1007/978-3-319-44785-8_11

principle implementation that is now occurring. Regarding this, adapted bioreactors are being developed, but more efforts are required to target this scientific demand.

11.1 Introduction

Articular cartilage lines the end of our joint surfaces and is composed by chondrocytes inside a matrix made of collagen fibers, associated to glycosaminoglycan chains and elastin fibers [1]. In healthy joints, this unique and durable tissue allows bones to move against each other with minimal friction. Although articular cartilage comprises just one type of cells, chondrocytes become less active with age and injury [2]. Furthermore, the avascular nature of cartilage together with the declining function of chondrocytes with age contributes to the inability of full-thickness defects to heal spontaneously.

Articular cartilage damage arises as a consequence of both acute and repetitive trauma resulting in pain, effusion and/or mechanical symptoms, affecting directly individuals' life style [3]. Cartilage lesions in joints are characterized with different degrees. Superficial lesions, as fissures or cracks, are classified as grade 1. A grade 2 abnormality is defined when cartilage is affected up to 50 % of its thickness while grade 3 lesions are characterized by defects in which more than 50 % of the cartilage thickness, down to the subchondral bone but without bone penetration, is damaged. When areas of cartilage are worn away or torn away, exposing underlying subchondral bone, an OC defect (OCD), graded 4, is created [4]. If untreated, these lesions can progress to more-serious degenerative joint conditions, such as osteoarthritis [5].

Current clinical treatments for OCDs involve surgical approaches, such as microfracture, autologous and allogeneic cartilage tissue grafts, and autologous chondrocytes implantation (ACI) [6]. Among those, ACI using collagen membrane (AMIC) [6, 7] and matrix-assisted chondrocyte implantation, which are the first clinically approved TE approaches for OCD treatment, have been the most successful in achieving long-lasting cartilage repair. In fact these treatments are well established and effective to reduce patients' pain and to slow down the disease progression, however are not able to regenerate hyaline cartilage and completely restore patient's mobility. Therefore, the demand for new therapeutic options to regenerate OCDs is significant. Since the most promising results were obtained through regenerative medicine strategies, such as cell therapy or TE applying ACI and MACI, respectively, the development of a therapy for OC lesions treatment is nowadays focused on these fields.

TE concept is based on the combination of cells with materials. Structures are made of biodegradable and biomimetic materials to create networks with architectural and biochemical features similar to native extracellular matrix (ECM) [8]. In opposition to the chondrocytes used in ACI and MACI approaches, the exploitation of stem cells in combination with those 3D ECM-like structures presents a huge potential. These stem cells can be extracted from specific cells niches and are promising for cell therapy and TE applications because of their high proliferative capacity and differentiation ability [9], as well as immunomodulatory role [10]. Moreover, improved in vitro 3D culture systems have been developed to maximize constructs features prior to implantation. The desired properties for a construct are being optimized to improve the ECM deposition and the mechanical performance.

This book chapter describes general regenerative medicine strategies mainly focusing in TE approaches for the treatment of OC lesions. The state-of-the-art of TE applied to complex OCD regeneration and the commonly used materials and their characteristics to create the support for host tissue invasion or transplanted cells growth is described and discussed. The strategies to recreate the bone and cartilage architectures as part of OC substitutes are disclosed and debated. Moreover, the sources of cells and their revealed efficacy for OC tissue regeneration will be analyzed. Finally the in vitro methodologies used to attain OC artificial constructs is scrutinized in terms of advantages and limitations.

11.2 Biomaterials-Based Strategies for OCD Regeneration

11.2.1 Strategies for OC Scaffolds Development

Biomaterials applied for bone and cartilage TE currently reported in literature present different physic-chemical characteristics for both parts. Most of the cartilage-like constructs are made of PGA meshes [11, 12], PCL prototyped structures [13], collagen [14, 15], hyaluronan [16, 17], chitosan [18] and gelatin porous sponges [19] and some approaches also applied external stimulus as, for example, a plasmid to induce TGF-β1 transcription [20]. For bony scaffolds, ceramics, polymeric blends and composites are being used, such as blends of PLGA and PEG [21], TCP-reinforced PCL [22], bioactive glass [23], hydroxyapatite (HAp) with chitosan [24] or gelatin [20]. Additionally, as in cartilage, also in the bone part external factors are being applied to reinforce the OC differentiation as, for example, a plasmid encoding BMP-2 gene [20].

Several strategies are being considered for restoring the biological and mechanical OC functionalities. Specific biomaterial-based strategies are being proposed, including (I) different scaffolds for the bone and cartilage sides, (II) a scaffold for the bone component, but a scaffold-free approach for the cartilage side, (III) a bilayered scaffold with integrated interface, which can be a gradient or a compact layer, and, finally, (IV) injectable biomaterials as hydrogels (Table 11.1).

Several techniques and combinations of techniques have been applied with a direct correlation to create the scaffold with the desired characteristics for OC regeneration and the specificities of this complex interface. Different architectures have been reported; nanofibers based structures, which can be created by electrospinning [25], sponges produced by freeze drying [24], or even agglomerated

Table 11.1 OC constructs design strategies for TE

Architecture	Advantages	Limitations	References
Two independent layers	Independent cell culture and differentiation	Poor integration between layers	[15, 28, 29]
Bone scaffold supporting cell free monolayer for cartilage	High control over cartilage layer phenotype	Poor integration and absence of transition zone	[30–32]
Bilayer structure with compact interface	Good integration between layers Control of cell migration Impaired vascularization in the cartilage part	Poor communication between subchondral bone and cartilage	[33–37]
Integrated bilayer structure	Good integration between layers Transition zone mimicking calcified cartilage part	Difficult cell culture regarding co-differentiation for osteo- and chondrogenesis	[14, 16, 18, 20, 27, 38–45]
Hydrogel based bilayered structures	Perfect fit and filling of deffect High hydrated environment	Poor diffusion of nutrients in several hydrogel systems Poor cell anchorage in high hydrated environments	[46–50]

particles and microparticles [26] as 3D scaffolds. Porous scaffolds can also be created by salt-leaching technique [27], while well-organized matrices can be produced by rapid prototyping technologies [13].

11.2.2 Bone- and Cartilage-Like Tissues

Bone was one of the first tissues focused by TE, so it is also one of the topics with most developed state-of-the-art. Because of the chronologic development of the studies for bone TE, and of the markedly different tissue properties in OC interface, researchers started by independently addressing different scaffolds for bone and cartilage and then their integration as a single structure [28, 51, 52]. This integration has been achieved by sutures [29] and press-fit [30, 53], taking advantage of the natural body weight pressure, since the mechanical load can help achieving good contact in the interface sites and between neotissue and host tissue [54, 55]. Also, the cell culture period influences the bonding at the interface between cartilage and bone parts [54, 55].

Schaefer et al. [28] investigated the use of cartilage constructs using PGA meshes cultured with bovine calf articular chondrocytes. Bone constructs were created with a blend of PLGA and PEG cultured with bovine calf periosteal cells.

After independent culture (1 or 4 weeks), the cartilage- and bone-like constructs were sutured together. The resulting structure was cultured for additional 4 weeks which allowed attaining OC-like substitutes in vitro [28]. An in vivo trial would be interesting to test this approach.

In a different approach, Shao et al. [53] combined bone- and cartilage-like constructs by press-fit implantation. The scaffolds comprised PCL for the cartilage component and TCP-reinforced PCL for the bone component. After implantation in a load-bearing lapine model, the PCL/PCL-TCP scaffolds seeded with mesenchymal stem cells (MSCs) showed better results for OCD regeneration than the acellular control group. New bone formation was observed between 12 and 24 weeks, leading to the integration to host tissue. After 24 weeks of implantation, subchondral bone filled the scaffold and glycosaminoglycans and collagen type II deposition were observed in the cartilage region. However, new cartilage tissue lacked zonal organization [53].

The main disadvantage of this strategy is the poor interface between the two layers [56]. Moreover this strategy implies more extensive work in laboratory for the in vitro culture of separated bone- and cartilage-like parts before the combination of both parts, which raises further issues in the translation to clinics when scale up is required.

11.2.3 Bone-Like Tissue Plus Chondrocytes Sheets or Layers

Based on the outcomes of ACI and MACI approaches researchers have considered the creation of a 3D bone-like construct able to attach to the host tissue that is then toped with chondrocytes or a cartilaginous tissue layer. These are then expected to integrate the 3D bone-like structure reinforcing the stability of the cartilage-bone interface.

Using this strategy, most of the studies have taken advantage of ceramics as scaffold for bone regeneration [30, 57]. These were then combined with a cell sheet of chondrocytes or stem cells, as for example synovial stem cells [58], to stimulate the regeneration of the cartilage part [57, 59].

Acellular porous calcium polyphosphate (CPP) scaffolds were used as a substrate to grow articular cartilage on top. After implantation, the structures successfully supported loading up to 36 weeks, allowed bone ingrowth in the CPP substrate and were fixed by host native cartilage. However, some implants presented cartilaginous tissue delaminated between the 12 and 16 weeks period of implantation because of a low cartilage/CPP interfacial shear strength in comparison with the native OC interface [30]. Furthermore the use of ceramic scaffolds for the bone part raises mechanical limitation because of their hardness and lack of flexibility in comparison to some polymers. Consequently, this mismatch of mechanical properties may lead to delamination between the bone scaffold and the new cartilage layer, which can be overcome by using polymer/ceramic composites.

Several biodegradable polymeric and composite materials were tested to overcome these mechanical problems, onto which neo-cartilage was produced in vitro. Porcine chondrocytes were seeded onto PLLA, PDLLA and collagen-HAp (Col-HA) scaffolds at high density in a closed and static bioreactor. PDLLA breakdown occurred in the first 11 days, leading to constructs of irregular shape and the highest amount of cell death. Low cell ingrowth and material breakdown was also evident in PLLA. Col-HA was the constructs formulation showing the superior results for OCD. These structures presented the highest cell viability and ingrowth and were also maintained with lowest degradation rate during the 15 days of cell culture, presenting the necessary integrity for further in vivo implantation after maturation. Furthermore Col-HA constructs also displayed collagen fibrils in neocartilage, contributing for the integration between cartilage- and subchondral bone-like tissue at the interface [32].

A different and important feature to reinforce the interface and to avoid the delamination problem is the calcified cartilage zone which was reported to be important for the interfacial shear properties [56]. In a study that anchored the cartilage tissue to CPP scaffold, the interface properties were enhanced as a result of the efficient integration of hyaline-like cartilage and the CPP phase by the calcified cartilage layer [60]. Although calcified cartilage could be formed by this strategy, the absence of a scaffold supporting the cell growth and ECM deposition of the cartilage layer can lead to a zonal organization failure of the neocartilage.

11.2.4 Bilayered Continuous Scaffolds-Based Strategies

There is also a category of bilayered scaffolds composed of two integrated layers for the cartilage and bone regions, or separated by a compact middle layer in the interface, to avoid mixing the two phenotypes. Like for the previously described approaches, ceramics and composites have been mainly used for the bone-like and natural or synthetic polymers for cartilage-like parts.

This strategy has been widely applied lately resulting in the development of several different structures for OC repair. A structure consisting of four layers, a porous CPP layer as bone component, a dense TCP layer to prevent blood vessel penetration, a porous CPP layer to fix bone and cartilage and a porous gelatin layer for the cartilage region [34].

To overcome the mechanical problems of ceramic-based scaffolds, a different study reported a structure composed of a composite and a blend of natural and synthetic polymers. The PLLA/HA or Bioglass® were used for the bone part and PLLA/starch blends for the cartilage part. This work considered that starch provides capability of water uptake and HA/Bioglass® to enhance bioactivity and thus HA formation on the bone side. The interface between cartilage and subchondral bone was integrated by a melt-based process (Fig. 11.1) [33].

Fig. 11.1 Bilayer structure presenting a compact interface layer obtained by melting. Reprinted with permission [33]. Copyright 2007, Elsevier

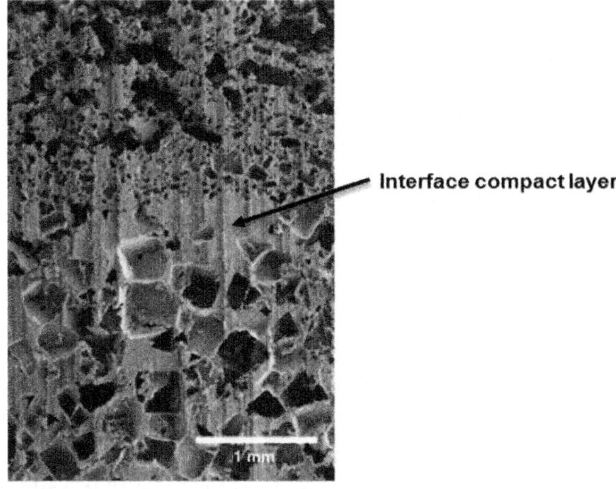

The compact interface presents the advantage of avoiding innervation and vascularization into the cartilage region. However this interface creates the disadvantage of interrupting the communication between both layers, impairing the paracrine effect from the MSCs of the subchondral bone.

An approach to solve some limitations associated to the poor invasion of host tissue into the bone-like part, combined a PGA woven for the cartilage part, a collagen I and HA coated porous PLLA/PCL foam for the bone part and a PLLA/PCL layer as the cartilage–bone interface. Vertical channels from the bottom layer to the upper border were created to allow the invasion of stem cells and blood from bone marrow after implanting following the mosaicplasty principle. The two phases of the scaffolds composed of PLLA and HA were assembled prior to cell seeding and implantation. These composite scaffolds were stabilized by using two bonded cylinders of PLLA and a thin PGA film was deposited between those two layers to prevent cell migration. These bilayers were co-cultured with osteoblasts and chondrocytes [35].

Despite this alternative thin but compact interface, the lack of communication between both new formed cartilage-like and bone-like tissue, which is important for good cell signaling, is also an issue. The poor connection present in most of the cases can compromise the natural integration of cartilage and subchondral bone through the calcified cartilage tidemark, which will not be created naturally.

A different strategy based on bilayered structures is based on the idea of having a homogeneous single integrated structure used for bone and cartilage regeneration, without a compact interface. The host tissue play the role of invading and defining the architecture between bone- and cartilage-like layers. Scaffolds incorporating or coated with various GFs, creating gradients have been proposed to stimuli simultaneous bone and cartilage regeneration [41]. Moreover, seeding with different cell lineages or stem cells stimulated with different GFs will promote the stratification inside the structure [61].

Bilayered scaffolds with integrated interface are usually formed by a composite and polymeric phase in a continuous structure for bone and cartilage layers, respectively. In contrast to a bony-like scaffold plus a cell-sheet for cartilage where neocartilage is generated from seeded chondrocytes on a subchondral support to form cartilaginous layer, bilayered scaffolds with integrated interface are designed to repair OCDs by using tailored bilayered structure, which mimic the structure of articular cartilage and subchondral bone tissue. The interface between engineered cartilage and subchondral bone parts is developed via fabrication methods including sintering [18], freeze-drying [14], salt leaching [27], emulsion [39], microspheres agglomeration [41], or even a CAD/CAM based process as the TheriForm™ [38, 62].

Oliveira et al. developed a HA/chitosan (HA/CS) bilayered scaffold by combining sintering and freeze drying techniques. The interface of the HA/CS bilayered scaffolds was achieved by partially impregnating the porous ceramic layer with the polymer one. Two distinct porous layers were obtained. Moreover, in vitro cell culture studies using MSCs demonstrated that both HA and CS layers provided an adequate 3D support for attachment, proliferation and differentiation of MSCs (Fig. 11.1) into osteoblasts and chondrocytes, respectively [18] (Fig. 11.2).

The first strategies proposing bilayered structures for OC regeneration are now being converted in gradient multiple layered structures to achieve a continuous layered structure. Levingstone et al. produced a layered construct by an "iterative layering" freeze-drying technique. The construct mimics the inherent gradient structure of healthy OC tissue: a bone layer composed of type I collagen and HAp, an intermediate layer composed of type I collagen, type II collagen and HAp, and a cartilaginous region composed of type I and type II collagen and hyaluronic acid (HA). This scaffold is currently being commercialized through the SurgaColl Technologies company, named ChondroColl™.

Recently, a multi-layered structure was developed by assembling a gelatin layer with layers containing different amounts of gelatin and HAp nanocrystals in which

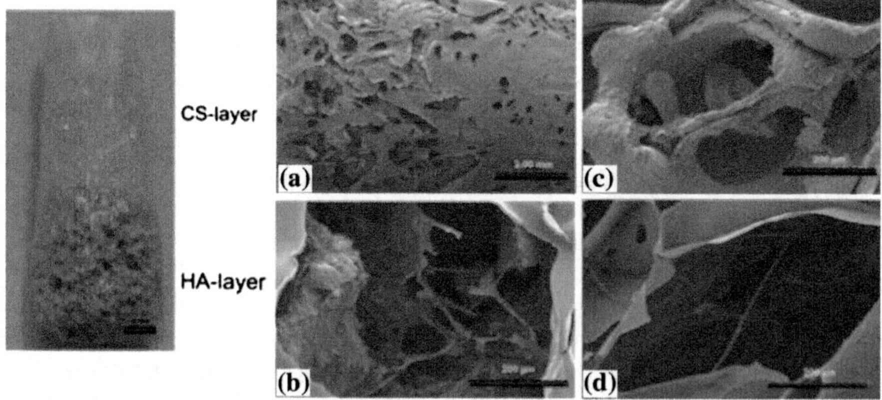

Fig. 11.2 Bilayered scaffold composed by a CS-based cartilage-like layer and a HA(sintered)/ CS-based bone-like layer. Reprinted with permission [18]. Copyright 2006, Elsevier

a gelatin solution was used to stick the layers together. These scaffolds exhibit a high and interconnected porosity, and show mechanical properties that vary with the composition along the scaffolds. The in vitro co-culture of hMSCs results demonstrated that osteogenic and chondrogenic-differentiated hMSC influenced each other's behavior [45].

The bilayered structures seem to be promising for the regeneration of a tissue that presents two very distinct phases in terms of biology, but also in terms of architecture and physic-chemical properties. The presence of a hard interface can be an advantage to avoid the vascularization and innervation of cartilage tissue, however this hard interface avoids the crosstalk with the MSCs from subchondral bone. MSCs present a key paracrine effect over tissue regeneration, which seems to be more important than the anti-vascularization and –innervation role of the compact layer. Thus the continuous bilayered structure, presenting a gradient of structural and chemical characteristics from subchondral bone up to the top cartilage layer is a promising strategy for OCD regeneration.

11.2.5 Injectable Approaches

Commonly, biomaterials are used in OCD to create a 3D solid porous support for cell growth. Hydrogels were one of the last routes being explored in the field and are now produced using not only chemical crosslinking but also ionic [63] or photo-crosslinking [48, 64]. The use of these different types of crosslinking agents or starters fostered the development of injectable strategies for OCD regeneration because of the lower cytotoxic risk. The main advantage of this strategy relies in the ability of the hydrogel to occupy the defect shape. Moreover as part of a minimally invasive procedure, this is a more friendly method for a clinical approach.

Due to major concerns regarding the use of non-cytotoxic crosslinking agents and mild conditions, cells have been also easily mixed within the material before injection and polymerization, avoiding in vitro cell culture. By injecting a gel with encapsulated cells, an one-step procedure can be achieved introducing both components in a homogenized way [63]. The crosslinking can occur for example by ionic reaction with the host blood [65].

A robust integration of two different hydrogels without mixing both layers, designed for bone and cartilage has been difficult to obtain. Therefore, there are some studies combining hydrogels with hard scaffolds, such as composites or ceramics. Chondrocytes-containing agarose hydrogel was created for cartilage and osteoblasts-containing microspheres of PLGA and 45S5 Bioglass® for bone were fabricated in a cylindrical mould. An interface formed by chondrocytes embedded within a hybrid phase of gel and microspheres was achieved [46].

Recently, the proof of concept of a bilayer hydrogel integrating two different layers of low acyl gellan gum (LAGG) for cartilage part and LAGG with HAp particles for bone part, that can be ionically crosslinged in vivo was presented (Fig. 11.3) [63].

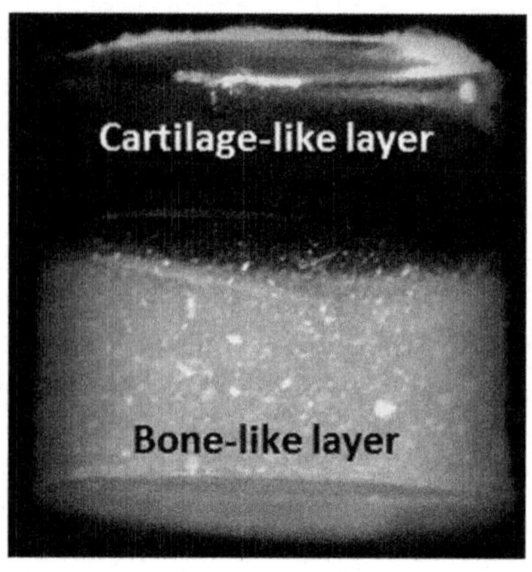

Fig. 11.3 Bilayered hydrogel composed of LAGG in the cartilage layer and LAGG incorporating HAp in the bone layer. Reprinted with permission [63]. Copyright 2014, Trans Tech Publications

The advantage of this last approach is mostly concerned with the defect shape, which can be completely fit by the injection of the hydrogels. The bone part can be easily adjusted in volume by eye, without requiring to previously knowing the exact volume of the defect, and then the cartilage part filled with the second layer of the hydrogel. Moreover this also allows encapsulating different cell types in the bone and cartilage layers at the time of implantation. The difficult host tissue invasion into the hydrogel, which can act as a barrier for tissue ingrowth and ECM deposition, can be seen as a main disadvantage of this approach.

11.3 Cells, Growth Factors and Gene Therapy for OCD Regeneration

Different cell sources have been used to obtain primary cells, as osteoblasts or chondrocytes, and stem cells for OCD regeneration. Osteoblasts and chondrocytes present the advantage of express the phenotype of the OC tissue, however low number of cells is obtained after isolation, requiring expansion in vitro, which is also very limited in the case of primary cells and usually leads to cell dedifferentiation [66].

Recently a new promising cell niche for articular cartilage regeneration was investigated by Pelttari et al. [67]. Adult human neuroectoderm-derived nasal chondrocytes, constitutively distinguished from mesoderm-derived articular chondrocytes by lack of expression of specific HOX genes, including HOXC4 and HOXD8, were shown, in contrast to articular chondrocytes, to be continuously reverted from differentiated to dedifferentiated states, conserving the ability to form cartilage tissue in vitro and in vivo. Moreover, those nasal chondrocytes are also reprogrammed to stably express HOX genes, typical of articular chondrocytes.

Stem cells can be isolated from several niches, as bone marrow, umbilical cord or abdominal fat. Recently, for OC TE, niches as fat pad, from Hoffa's body close to the knee [68, 69], or synovial fluid [70, 71] have been tested. The limitation of low proliferation of primary cells and cell lineage dedifferentiation can be overcome using stem cells. Bone marrow, umbilical cord and abdominal fat are the common sources of stem cells, however new niches are being explored for OC application as mentioned above.

Scientists are now gaining interest in infrapatellar fat pad (Hoffa's body) to obtain MSCs for OC application, since the niche is present in the knee and the proximity to the lesion place plays an important role for the regeneration performed by cells [72]. Hoffa's body, which is a fat pad of adipose-derived stem cells (ASCs), has to be removed during an arthroscopy to facilitate the visualization of the knee and surgery handling, and also to avoid tissue inflammation as was explained before. This way, this tissue can also be considered as a promising source of ASCs with great potential to differentiate into chondrocytes and osteoblasts.

While primary cells represent native tissue phenotypes, stem cells have been either triggered and differentiated in vitro towards the lineages of interest or transplanted in an undifferentiated state. In this case a major concern related to in situ differentiation has supported the concomitant use of GFs [66, 73].

Designed structures for OCD regeneration are carrying GFs as TGF-β1 for the chondrogenesis or BMPs for osteogenesis [74, 75]. Furthermore, structures presenting gradients of GFs are being explored. Microspheres are used to deliver the GF after implantation of the structures in the OCD [74]. For example, gelatine microparticles were used to carry IGF-1 and TGF-β3 for cartilage phase. The results suggest that the dual delivery of TGF-β3 and IGF-1, does not synergistically enhance the quality of engineered tissue [50].

GFs are also being used to promote selective cell differentiation and achieve the desired cell phenotype. However, with the more recently emerged gene therapy, this effect can be performed in a more constitutive and long term way [76]. Using a genetic modification strategy, chondrocytes overexpressing IGF-1 were cultured on biodegradable PGA scaffolds in dynamic flow rotating bioreactor up to 28 days. The resulting cartilaginous constructs implanted into OCD in rabbit knee joints lead to a spatially defined overexpression of IGF-1 enhancing articular cartilage repair and reducing osteoarthritic changes in the cartilage adjacent to the defect [77], which includes single parameters of cellularity, staining intensity and cluster formation [78]. Cellular morphology and architecture were significantly improved for defects receiving IGF-I constructs compared with those receiving lacZ constructs.

SOX trio is a genetic sequence of SOX-5,-6 and -9 very interesting for chondrogenesis and cartilage repair. Plasmid DNA (pDNA) containing the SOX trio genes was incorporated into a PLGA scaffold to slowly release, transfect ASCs and trigger chondrogenic differentiation. The in vivo study showed enhanced cartilage regeneration in ASCs/SOX trio pDNA-incorporated PLGA scaffolds [79]. In a different approach, SOX trio genes were also used for chondrogenesis, but in combination with Runt-related transcription factor 2 (RUNX2) for osteogenesis. A branched poly(ethylenimine) (bPEI)-HA delivery vector was loaded in a bilayered hydrogel mimicking native OC tissue. The spatially loaded combination of

RUNX2 and SOX trio DNA particularly in the bone part, significantly improved healing in relation to controls, hydrogels and factor alone [80].

11.4   3D In Vitro Cell Culture Methods

With the development of TE in the last decade, the need for 3D in vitro culture methods largely increased. The limited success obtained from the strategies tested until now to overcome disorders as osteoarthristis, for example, made the need to overcome their limitations using more realistic cell culture methods experiments. 3D cell culture, in vitro, present a huge potential to replicate in a better way several physiologic conditions that were not possible to correctly mimic in 2D or static cell culture. Furthermore, as a consequence of this huge introduction of new structures and methodologies to produce 3D scaffolds with different architectures, adaptable 3D in vitro culture systems have been developed as a need. There is a paradigm shifting occurring related with cell culture, which is changing from 2D to 3D, from static to dynamic conditions and from time-point analysis to real-time monitoring. The cell culture method is being adapted and bioreactors are emerging for TE field, which will be also useful to make the field of diseases in vitro modelling more realistic in the future.

A bioreactor can be described as a dynamic device or system for culturing cells or tissues under controlled conditions, either biochemical or mechanically. Several systems have been created; shake flasks ("mixed flasks") [81] evolved to rotary vessels ("rotating vessels") [82] and then to perfused chamber ("perfused cartridges") [83]. Lately the incorporation of mechanical stimuli in the bioreactors has been followed in order to mimic the physical stress that occurs naturally in cellular environment [84, 85]. Generically in 3D cell culture, including the 3D structures for OC tissue engineering, cell sedimentation during the phase of cell adhesion is usually a problem, so the bioreactor should avoid this sedimentation [86]. Furthermore the cellular waste has to exit the interior of the structure being replaced by the fresh culture medium. Furthermore, bioreactors for OC TE have to be adapted for gradient or multi-layered structures and dual-environment cell culture conditions.

For OC TE and in vitro modelling, the mono-chamber bioreactors are evolving to dual-chamber bioreactors and compressive stimuli are also being included to create some dynamics mimicking the in vivo conditions. The first's bioreactors used to mature chondrogenic and osteogenic constructs were not adapted for bilayered scaffolds, thus cells were cultured in separate environments using independent scaffolds for each part [85, 87, 88]. If the structure is multilayered, the culture chamber has to be adapted to offer the optimal culture environment to each layer giving rise to a different and specific engineered tissue [86].

In order to address the above mentioned need and despite the proposed above solutions, it is necessary to develop a reactor and a method to obtain engineered OC tissue grafts with 3D bi- or multilayered architecture that mimics the native tissue. Kuiper et al. designed a dual-chamber bioreactor for OC plugs and demonstrated that the dual-chamber perfusion bioreactor positively influenced the co-culture of primary human chondrocytes and osteoblasts in the biphasic scaffold, in terms of cell viability, cell proliferation, ECM production and gene expression, however longer-term experiments have to be performed to evaluate the mechanical integrity of the cultured tissues [89].

For in vitro modelling of the osteoarthritic condition, a bioreactor presenting dual-chambers was designed for a high-throughput approach. This bioreactor system was fitted into a microfluidic device (Fig. 11.4). Each dual-chamber and insert of the 24 culture positions was fabricated using a stereolithography apparatus. The OC construct is supplied by two different culture mediums. The medium conduits are critical to create tissue-specific microenvironments in which chondral and osseous tissues develop and mature [90].

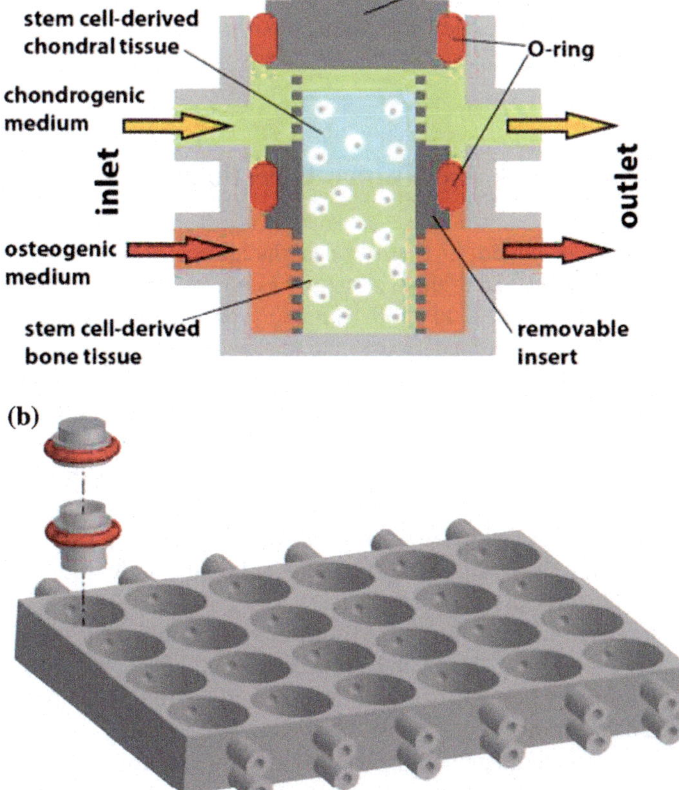

Fig. 11.4 Schematic of the bioreactor for OC in vitro model. **a** An individual bioreactor composed of the removable insert (*dark gray*) within a chamber (*light gray*) of the microfluidic plate (**b**) and fixed in place with two O-rings. Reprinted with permission [90]. Copyright 2013, American Chemical Society

In a different perspective, but also important in terms of physical stimulus, Nam et al. used a compressive system to test in vitro the expression of BMPs by osteoblasts and articular chondrocytes in 3D OC constructs culture under static and dynamic biomechanical stimulation (cyclic compressive strain). Biomechanical stimulation led to enhanced tissue morphogenesis possibly through this BMP regulation [91].

Combining the features of those systems reported above, a rotational dual-chamber bioreactor was patented describing the combination of a multi-chamber with physical compression and added also to the design the concept of rotational movements to improve cell distribution in 3D structures. This system can be used for OC TE flowing two different culture mediums and creating compression on top of cartilage-like layer [93].

Although some bioreactors are being developed for OC TE, the majority of them are being applied for constructs production. However bioreactors have the potential to be used as platforms for in vitro modelling of diseases as for example OA. These systems can contribute to recreate a dynamic and 3D environment which is more appropriate to mimic the natural OC tissue than a static and 2D cell culture of chondrocytes or osteoblasts.

11.5 Final Remarks and Future Directions

Despite the variety of materials, scaffold designs and cells that have been investigated for OC applications, an optimal strategy has not yet emerged. Therefore, more research efforts are needed to find suitable combinations of materials and methodologies that can be transferred to clinical practice.

Scaffolds for individual bone and cartilage tissue regeneration combined at the time of implantation represent one of the strategies for OC repair. These are based on the combination of different materials and cells specified for individual cartilage and subchondral bone tissues, or different biological factors capable to induce the selective differentiation of stem cells into chondrocytes and osteoblasts. The major limitation of this strategy is related with the lack of an interface, which contribute to the failure of the two layers under the stress created by the body weight, resulting in constructs delamination.

Another approach involves a scaffold for the bone component, but none for the cartilage component, have shown that bilayered scaffold-free cartilage constructs exhibit in vitro formation of cartilaginous-like tissue by chondrocytes seeded without the aid of biomaterial support. However, low interfacial shear strength at the interface between cartilage and the underlying bone scaffold is still a potentially vulnerable aspect of such systems. The formation of a mineralized layer in engineered cartilage has been suggested to solve this problem, considering that calcified cartilage is important for the integration of soft tissue (nonmineralized hyaline-like cartilage) and hard tissue (mineralized subchondral bone), and it can distribute the mechanical load across the interface. Although calcified cartilage could be formed by this strategy, the generation of zonal organization in new articular cartilage might be inhibited by the lack of a cartilage-like scaffold for cell accommodation and tissue framework development.

Recently biphasic and multiphasic scaffold for OCD repair have been developed by TE. The biphasic scaffolds may present a compact interface. This compact interface can present the advantage of avoiding cartilage vascularization and innervation; however results in a non-naturally formed tidemark interface on the calcified cartilage transition zone. Moreover the communication between new cartilage and subchondral bone is important for the architecture, integration and maturation of the new formed cartilage. Following the progression of the bilayered structures, many researchers are now showing deeper interest in bilayered gradient scaffolds with continuous interfaces. With a graded scaffold, the interface would more closely resemble the native environment, allowing the interconnection of cartilage and subchondral bone, as occurs in vivo. Although there are some reported gradient biphasic scaffolds, it is extremely complicated to make a continuously gradient structure that allow smooth bone–cartilage interface. The combination of a continuous bilayered structure with a chemical ability to avoid vascularization and innervation in the cartilage layer can be a promising and key way to improve the current results using this strategy.

With the increasing understanding of the mechanical strengths, general structure, and the biology of bone and cartilage, the reconstruction of these two individual areas has led to improved OC constructs. Beyond the development of the scaffolds architecture, the use of GFs to form gradients and also the introduction of gene therapy for OC TE will potentially improve the quality of the engineered tissues.

Still, OC tissue repair needs the integration and interconnection between both tissues, which requires an advanced knowledge of how bone and cartilage interact. Understanding these two components separately allowed for the current state of the art. However, the true challenge in OC repair lies in the comprehension of the OC interface and its combined yet separate mechanical strengths, structure, and biology. This said OC TE tools and results improve with the understanding of OC interface architecture and phenotype. To further improve the mechanical strength, future studies were suggested to focus not only on using biochemical factors, but also mechanical stimuli.

The huge development of TE techniques and the now-how related with OC tissue can now be applied to the area of disease modeling. The in vivo (animal) models are nowadays the main models for drug effect screening before the clinical trials. However, there are efforts to decrease the use of animal models. The three Rs' (3Rs) principle, first described by Russell and Burch [92], is now being under higher attention to be applied for a better ethical use of animals in testing. This will create the need of new in vitro models to emerge in the drug development market to replace gradually the animal models to a certain extent. However, since the failure in predicting the efficacy or toxicity of a new drug carries huge costs for the industry, more reliable and realistic in vitro models are needed when compared to the existing ones, which are commonly 2D and static systems. This is a new born research field in OC related diseases.

More recently, iPS cells, which are generated by reprogramming somatic cells through the exogenous expression of transcription factors [93], improved the potential of autologous cell replacement therapies for regenerative medicine [94]. Nevertheless, and because of iPS technology is recent and is not completely controlled in terms of cell phenotype and in vivo functionality, no OCD regeneration studies are reported the use of this cells.

The huge variability of the OA tissues between individuals and the several joint tissues interplay are proven to be the most important challenges to be overcome. Thus, the development of a 3D OC model, with induced OA to provide the required reproducibility is expected to contribute to advances in the OA knowledge. In fact, we envision that the investigation of the disease in a controlled and reproducible way, will allow to identify new biomarkers for early OA diagnosis and to open up new possibilities for discovering new drugs for new therapies, with different efficiencies in the progressive stages of disease.

Acknowledgments Thanks are due to the Portuguese Foundation for Science and Technology and POPH/FSE program for the fellowship grant of Raphaël Canadas (SFRH/BD/92565/2013). The FCT distinction attributed to J.M. Oliveira under the Investigator FCT program (IF/00423/2012) is also greatly acknowledged.

References

1. Sophia Fox AJ, Bedi A, Rodeo SA (2009) The basic science of articular cartilage: structure, composition, and function. Sports Health 1(6):461–468. doi:10.1177/1941738109350438
2. Martin JA, Buckwalter JA (2001) Roles of articular cartilage aging and chondrocyte senescence in the pathogenesis of osteoarthritis. Iowa Orthop J 21:1–7
3. Bhosale AM, Richardson JB (2008) Articular cartilage: structure, injuries and review of management. Br Med Bull 87(1):77–95. doi:10.1093/bmb/ldn025
4. Kleemann RU, Krocker D, Cedraro A, Tuischer J, Duda GN (2005) Altered cartilage mechanics and histology in knee osteoarthritis: relation to clinical assessment (ICRS Grade). Osteoarthr Cartil 13(11):958–963. doi:10.1016/j.joca.2005.06.008
5. Thomas CM, Fuller CJ, Whittles CE, Sharif M (2007) Chondrocyte death by apoptosis is associated with cartilage matrix degradation. Osteoarthr Cartil 15(1):27–34. doi:10.1016/j.joca.2006.06.012
6. Peterson L, Minas T, Brittberg M, Lindahl A (2003) Treatment of osteochondritis dissecans of the knee with autologous chondrocyte transplantation, vol 85. Results at two to ten years. vol suppl 2
7. Steinwachs MR, Guggi T, Kreuz PC (2008) Marrow stimulation techniques. Injury 39(1, Supplement):26–31
8. Langer R, Vacanti J (1993) Tissue engineering. Science 260(5110):920–926. doi:10.1126/science.8493529
9. Caplan AI (2007) Adult mesenchymal stem cells for tissue engineering versus regenerative medicine. J Cell Physiol 213(2):341–347. doi:10.1002/jcp.21200
10. Zhao Q, Ren H, Han Z (2016) Mesenchymal stem cells: immunomodulatory capability and clinical potential in immune diseases. J Cell Immunother. doi:10.1016/j.jocit.2014.12.001
11. Mahmoudifar N, Doran PM (2013) Osteogenic differentiation and osteochondral tissue engineering using human adipose-derived stem cells. Biotechnol Prog 29(1):176–185. doi:10.1002/btpr.1663

12. Li WJ, Cooper JA, Mauck RL, Tuan RS (2006) Fabrication and characterization of six electrospun poly(alpha-hydroxy ester)-based fibrous scaffolds for tissue engineering applications. Acta Biomater 2(4):377–385. doi:10.1016/j.actbio.2006.02.005

13. Shor L, Guceri S, Wen XJ, Gandhi M, Sun W (2007) Fabrication of three-dimensional polycaprolactone/hydroxyapatite tissue scaffolds and osteoblast-scaffold interactions in vitro. Biomaterials 28(35):5291–5297. doi:10.1016/j.biomaterials.2007.08.018

14. Levingstone TJ, Matsiko A, Dickson GR, O'Brien FJ, Gleeson JP (2014) A biomimetic multi-layered collagen-based scaffold for osteochondral repair. Acta Biomater 10(5):1996–2004. doi:10.1016/j.actbio.2014.01.005

15. Zhou JA, Xu CX, Wu G, Cao XD, Zhang LM, Zhai ZC, Zheng ZW, Chen XF, Wang YJ (2011) In vitro generation of osteochondral differentiation of human marrow mesenchymal stem cells in novel collagen-hydroxyapatite layered scaffolds. Acta Biomater 7(11):3999–4006. doi:10.1016/j.actbio.2011.06.040

16. Galperin A, Oldinski RA, Florczyk SJ, Bryers JD, Zhang MQ, Ratner BD (2013) Integrated bi-layered scaffold for osteochondral tissue engineering. Adv Healthc Mater 2(6):872–883. doi:10.1002/adhm.201200345

17. Antunes JC, Oliveira JM, Reis RL, Soria JM, Gómez-Ribelles JL, Mano JF (2010) Novel poly (L-lactic acid)/hyaluronic acid macroporous hybrid scaffolds: characterization and assessment of cytotoxicity. J Biomed Mater Res, Part A 94A(3):856–869. doi:10.1002/jbm.a.32753

18. Oliveira JM, Rodrigues MT, Silva SS, Malafaya PB, Gomes ME, Viegas CA, Dias IR, Azevedo JT, Mano JF, Reis RL (2006) Novel hydroxyapatite/chitosan bilayered scaffold for osteochondral tissue-engineering applications: Scaffold design and its performance when seeded with goat bone marrow stromal cells. Biomaterials 27:6123–6137

19. Lien SM, Chien CH, Huang TJ (2009) A novel osteochondral scaffold of ceramic-gelatin assembly for articular cartilage repair. Mater Sci Eng C Biomim Supramol Syst 29(1):315–321. doi:10.1016/j.msec.2008.07.017

20. Chen J, Chen H, Li P, Diao H, Zhu S, Dong L, Wang R, Guo T, Zhao J, Zhang J (2011) Simultaneous regeneration of articular cartilage and subchondral bone in vivo using MSCs induced by a spatially controlled gene delivery system in bilayered integrated scaffolds. Biomaterials 32(21):4793–4805. doi:10.1016/j.biomaterials.2011.03.041

21. Sidney LE, Heathman TRJ, Britchford ER, Abed A, Rahman CV, Buttery LDK (2015) Investigation of localized delivery of diclofenac sodium from Poly(D,L-lactic acid-co-glycolic acid)/poly(ethylene glycol) scaffolds using an in vitro osteoblast inflammation model. Tissue Eng Part A 21(1–2):362–373. doi:10.1089/ten.tea.2014.0100

22. Schumann D, Ekaputra AK, Lam CXF, Hutmacher DW (2007) Biomaterials/scaffolds. Design of bioactive, multiphasic PCL/collagen type I and type II-PCL-TCP/collagen composite scaffolds for functional tissue engineering of osteochondral repair tissue by using electrospinning and FDM techniques. Methods Mol Med 140:101–124

23. Wu Y, Zhu SA, Wu CT, Lu P, Hu CC, Xiong S, Chang J, Heng BC, Xiao Y, Ouyang HW (2014) A bi-lineage conducive scaffold for osteochondral defect regeneration. Adv Funct Mater 24(28):4473–4483

24. Oliveira JM, Rodrigues MT, Silva SS, Malafaya PB, Gomes ME, Viegas CA, Dias IR, Azevedo JT, Mano JF, Reis RL (2006) Novel hydroxyapatite/chitosan bilayered scaffold for osteochondral tissue-engineering applications: Scaffold design and its performance when seeded with goat bone marrow stromal cells. Biomaterials 27(36):6123–6137. doi:10.1016/j. biomaterials.2006.07.034

25. Mellor LF, Mohiti-Asli M, Williams J, Kannan A, Dent MR, Guilak F, Loboa EG (2015) Extracellular calcium modulates chondrogenic and osteogenic differentiation of human adipose-derived stem cells: a novel approach for osteochondral tissue engineering using a single stem cell source. Tissue Eng Part A 21(17–18):2323–2333. doi:10.1089/ten.tea.2014. 0572

26. Malafaya PB, Pedro AJ, Peterbauer A, Gabriel C, Redl H, Reis RL (2006) Chitosan particles agglomerated scaffolds for cartilage and osteochondral tissue engineering approaches with

adipose tissue derived stem cells. J Mater Sci Mater Med 17(7):675. doi:10.1007/s10856-006-9231-9

27. Yan LP, Silva-Correia J, Oliveira MB, Vilela C, Pereira H, Sousa RA, Mano JF, Oliveira AL, Oliveira JM, Reis RL (2015) Bilayered silk/silk-nanoCaP scaffolds for osteochondral tissue engineering: In vitro and in vivo assessment of biological performance. Acta Biomater 12:227–241. doi:10.1016/j.actbio.2014.10.021

28. Schaefer D, Martin I, Shastri P, Padera RF, Langer R, Freed LE, Vunjak-Novakovic G (2000) In vitro generation of osteochondral composites. Biomaterials 21(24):2599–2606

29. Cui WD, Wang Q, Chen G, Zhou SX, Chang Q, Zuo Q, Ren KW, Fan WM (2011) Repair of articular cartilage defects with tissue-engineered osteochondral composites in pigs. J Biosci Bioeng 111(4):493–500. doi:10.1016/j.jbiosc.2010.11.023

30. Kandel RA, Grynpas M, Pilliar R, Lee J, Wang J, Waldman S, Zalzal P, Hurtig M (2006) Repair of osteochondral defects with biphasic cartilage-calcium polyphosphate constructs in a Sheep model. Biomaterials 27(22):4120–4131. doi:10.1016/j.biomaterials.2006.03.005

31. Shimomura K, Moriguchi Y, Ando W, Nansai R, Fujie H, Hart DA, Gobbi A, Kita K, Horibe S, Shino K, Yoshikawa H, Nakamura N (2014) Osteochondral repair using a scaffold-free tissue-engineered construct derived from synovial mesenchymal stem cells and a hydroxyapatite-based artificial bone. Tissue Eng Part A 20(17–18):2291–2304

32. Wang X, Grogan SP, Rieser F, Winkelmann V, Maquet V, Berge ML, Mainil-Varlet P (2004) Tissue engineering of biphasic cartilage constructs using various biodegradable scaffolds: an in vitro study. Biomaterials 25(17):3681–3688. doi:10.1016/j.biomaterials.2003.10.102

33. Ghosh S, Viana JC, Reis RL, Mano JF (2008) Bi-layered constructs based on poly(L-lactic acid) and starch for tissue engineering of osteochondral defects. Mater Sci Eng C Biomim Supramol Syst 28(1):80–86. doi:10.1016/j.msec.2006.12.012

34. Lien S-M, Chien C-H, Huang T-J (2009) A novel osteochondral scaffold of ceramic–gelatin assembly for articular cartilage repair. Mater Sci Eng, C 29(1):315–321

35. Aydin HM (2011) A three-layered osteochondral plug: structural, mechanical, and in vitro biocompatibility analysis. Adv Eng Mater 13(12):B511–B517. doi:10.1002/adem.201180005

36. Ding X, Zhu M, Xu B, Zhang J, Zhao Y, Ji S, Wang L, Wang L, Li X, Kong D, Ma X, Yang Q (2014) Integrated trilayered silk fibroin scaffold for osteochondral differentiation of adipose-derived stem cells. ACS Appl Mater Interfaces 6(19):16696–16705. doi:10.1021/am5036708

37. Da H, Jia S-J, Meng G-L, Cheng J-H, Zhou W, Xiong Z, Mu Y-J, Liu J (2013) The impact of compact layer in biphasic scaffold on osteochondral tissue engineering. PLoS ONE 8(1): e54838. doi:10.1371/journal.pone.0054838

38. Sherwood JK, Riley SL, Palazzolo R, Brown SC, Monkhouse DC, Coates M, Griffith LG, Landeen LK, Ratcliffe A (2002) A three-dimensional osteochondral composite scaffold for articular cartilage repair. Biomaterials 23(24):4739–4751. doi:10.1016/S0142-9612(02)00223-5

39. Jiang J, Tang A, Ateshian G, Guo XE, Hung C, Lu H (2010) Bioactive stratified polymer ceramic-hydrogel scaffold for integrative osteochondral repair. Ann Biomed Eng 38(6):2183–2196. doi:10.1007/s10439-010-0038-y

40. Deng T, Lv J, Pang J, Liu B, Ke J (2014) Construction of tissue-engineered osteochondral composites and repair of large joint defects in rabbit. J Tissue Eng Regen Med 8(7):546–556. doi:10.1002/term.1556

41. Mohan N, Dormer NH, Caldwell KL, Key VH, Berkland CJ, Detamore MS (2011) Continuous gradients of material composition and growth factors for effective regeneration of the osteochondral interface. Tissue Eng Part A 17(21–22):2845–2855. doi:10.1089/ten.tea.2011.0135

42. Liu XD, Liu S, Liu SH, Cui WG (2014) Evaluation of oriented electrospun fibers for periosteal flap regeneration in biomimetic triphasic osteochondral implant. J Biomed Mater Res, Part B 102(7):1407–1414. doi:10.1002/jbm.b.33119

43. H-w Cheng, Luk KDK, Cheung KMC, Chan BP (2011) In vitro generation of an osteochondral interface from mesenchymal stem cell–collagen microspheres. Biomaterials 32(6):1526–1535

44. Erisken C, Kalyon DM, Wang H (2010) Viscoelastic and biomechanical properties of osteochondral tissue constructs generated from graded polycaprolactone and beta-tricalcium phosphate composites. J Biomech Eng 132(9):091013. doi:10.1115/1.4001884
45. Amadori S, Torricelli P, Panzavolta S, Parrilli A, Fini M, Bigi A (2015) Multi-layered scaffolds for osteochondral tissue engineering: in vitro response of co-cultured human mesenchymal stem cells. Macromol Biosci 15(11):1535–1545. doi:10.1002/mabi.201500165
46. Yunos DM, Ahmad Z, Boccaccini AR (2010) Fabrication and characterization of electrospun poly-DL-lactide (PDLLA) fibrous coatings on 45S5 Bioglass (R) substrates for bone tissue engineering applications. J Chem Technol Biotechnol 85(6):768–774. doi:10.1002/jctb.2283
47. Ju Young P, Jong-Cheol C, Jin-Hyung S, Jung-Seob L, Hyoungjun P, Sung Won K, Junsang D, Dong-Woo C (2014) A comparative study on collagen type I and hyaluronic acid dependent cell behavior for osteochondral tissue bioprinting. Biofabrication 6(3):035004
48. Mazaki T, Shiozaki Y, Yamane K, Yoshida A, Nakamura M, Yoshida Y, Zhou D, Kitajima T, Tanaka M, Ito Y, Ozaki T, Matsukawa A (2014) A novel, visible light-induced, rapidly cross-linkable gelatin scaffold for osteochondral tissue engineering. Sci Rep 4:4457. doi:10.1038/srep04457. http://www.nature.com/articles/srep04457#supplementary-information
49. Lam J, Lu S, Meretoja VV, Tabata Y, Mikos AG, Kasper FK (2014) Generation of osteochondral tissue constructs with chondrogenically and osteogenically predifferentiated mesenchymal stem cells encapsulated in bilayered hydrogels. Acta Biomater 10(3):1112–1123
50. Kim K, Lam J, Lu S, Spicer PP, Lueckgen A, Tabata Y, Wong ME, Jansen JA, Mikos AG, Kasper FK (2013) Osteochondral tissue regeneration using a bilayered composite hydrogel with modulating dual growth factor release kinetics in a rabbit model. J Control Release 168 (2):166–178. doi:10.1016/j.jconrel.2013.03.013
51. Schek R, Taboas J, Segvich S, Hollister S, Krebsbach P (2004) Engineered osteochondral grafts using biphasic composite solid free-form fabricated scaffolds. Tissue Eng 10(9–10):1376–1385. doi:10.1089/ten.2004.10.1376
52. Abrahamsson CK, Yang F, Park H, Brunger JM, Valonen PK, Langer R, Welter JF, Caplan AI, Guilak F, Freed LE (2010) Chondrogenesis and mineralization during in vitro culture of human mesenchymal stem cells on three-dimensional woven scaffolds. Tissue Eng Part A 16(12):3709–3718. doi:10.1089/ten.tea.2010.0190
53. Shao X, Goh JCH, Hutmacher DW, Lee EH, Zigang G (2006) Repair of large articular osteochondral defects using hybrid scaffolds and bone marrow-derived mesenchymal stem cells in a rabbit model. Tissue Eng 12(6):1539–1551. doi:10.1089/ten.2006.12.1539
54. Scotti C, Wirz D, Wolf F, Schaefer DJ, Bürgin V, Daniels AU, Valderrabano V, Candrian C, Jakob M, Martin I, Barbero A (2010) Engineering human cell-based, functionally integrated osteochondral grafts by biological bonding of engineered cartilage tissues to bony scaffolds. Biomaterials 31(8):2252–2259. doi:10.1016/j.biomaterials.2009.11.110
55. Schaefer D, Martin I, Jundt G, Seidel J, Heberer M, Grodzinsky A, Bergin I, Vunjak-Novakovic G, Freed L (2002) Tissue-engineered composites for the repair of large osteochondral defects. Arthritis Rheum 46(9):2524–2534
56. Nooeaid P, Salih V, Beier JP, Boccaccini AR (2012) Osteochondral tissue engineering: scaffolds, stem cells and applications. J Cell Mol Med 16(10):2247–2270. doi:10.1111/j.1582-4934.2012.01571.x
57. Niyama K, Ide N, Onoue K, Okabe T, Wakitani S, Takagi M (2011) Construction of osteochondral-like tissue graft combining beta-tricalcium phosphate block and scaffold-free centrifuged chondrocyte cell sheet. J Orthop Sci 16(5):613–621. doi:10.1007/s00776-011-0120-9
58. Ito S, Sato M, Yamato M, Mitani G, Kutsuna T, Nagai T, Ukai T, Kobayashi M, Kokubo M, Okano T, Mochida J (2012) Repair of articular cartilage defect with layered chondrocyte sheets and cultured synovial cells. Biomaterials 33(21):5278–5286. doi:10.1016/j.biomaterials.2012.03.073
59. Shimizu R, Kamei N, Adachi N, Hamanishi M, Kamei G, Mahmoud EE, Nakano T, Iwata T, Yamato M, Okano T, Ochi M (2014) Repair mechanism of osteochondral defect promoted by

bioengineered chondrocyte sheet. Tissue Eng Part A 21(5–6):1131–1141. doi:10.1089/ten.tea. 2014.0310
60. Allan KS, Pilliar RM, Wang J, Grynpas MD, Kandel RA (2007) Formation of biphasic constructs containing cartilage with a calcified zone interface. Tissue Eng 13(1):167–177. doi:10.1089/ten.2006.0081
61. Sheehy EJ, Vinardell T, Buckley CT, Kelly DJ (2013) Engineering osteochondral constructs through spatial regulation of endochondral ossification. Acta Biomater 9(3):5484–5492
62. Sachs E, Cima M, Cornie J (1990) Three-dimensional printing: rapid tooling and prototypes directly from a CAD model. CIRP Ann Manuf Technol 39(1):201–204
63. Pereira D, Canadas R, Silva-Correia J, Marques A, Reis R, Oliveira J (2014) Gellan gum-based hydrogel bilayered scaffolds for osteochondral tissue engineering. Key Eng Mater 587:255–260
64. Xiao W, He J, Nichol J, Wang L, Hutson C, Wang B (2011) Synthesis and characterization of photocrosslinkable gelatin and silk fibroin interpenetrating polymer network hydrogels. Acta Biomater 7:2384–2393
65. Silva-Correia J, Zavan B, Vindigni V, Silva T, Oliveira J, Abatangelo G, Reis R (2013) Biocompatibility evaluation of ionic- and photo-crosslinked methacrylated gellan gum hydrogels: in vitro and in vivo study. Adv Healthc Mater 2:568–575
66. Dahlin RL, Ni M, Meretoja VV, Kasper FK, Mikos AG (2014) TGF-β3-induced chondrogenesis in co-cultures of chondrocytes and mesenchymal stem cells on biodegradable scaffolds. Biomaterials 35(1):123–132. doi:10.1016/j.biomaterials.2013.09.086
67. Pelttari K, Pippenger B, Mumme M, Feliciano S, Scotti C, Mainil-Varlet P, Procino A, von Rechenberg B, Schwamborn T, Jakob M, Cillo C, Barbero A, Martin I (2014) Adult human neural crest–derived cells for articular cartilage repair. Sci Transl Med 6(251):251ra119
68. English A, Jones EA, Corscadden D, Henshaw K, Chapman T, Emery P, McGonagle D (2007) A comparative assessment of cartilage and joint fat pad as a potential source of cells for autologous therapy development in knee osteoarthritis. Rheumatology 46(11):1676–1683
69. Khan WS, Adesida AB, Tew SR, Longo UG, Hardingham TE (2012) Fat pad-derived mesenchymal stem cells as a potential source for cell-based adipose tissue repair strategies. Cell Prolif 45(2):111–120. doi:10.1111/j.1365-2184.2011.00804.x
70. Marsano A, Millward-Sadler SJ, Salter DM, Adesida A, Hardingham T, Tognana E, Kon E, Chiari-Grisar C, Nehrer S, Jakob M, Martin I (2007) Differential cartilaginous tissue formation by human synovial membrane, fat pad, meniscus cells and articular chondrocytes. Osteoarthr Cartil 15(1):48–58. doi:10.1016/j.joca.2006.06.009
71. Kim YS, Lee HJ, Yeo JE, Kim YI, Choi YJ, Koh YG (2015) Isolation and characterization of human mesenchymal stem cells derived from synovial fluid in patients with osteochondral lesion of the talus. Am J Sports Med 43(2):399–406. doi:10.1177/0363546514559822
72. Maumus M, Manferdini C, Toupet K, Peyrafitte J-A, Ferreira R, Facchini A, Gabusi E, Bourin P, Jorgensen C, Lisignoli G, Noël D (2013) Adipose mesenchymal stem cells protect chondrocytes from degeneration associated with osteoarthritis. Stem Cell Res 11(2):834–844
73. Meng F, He A, Zhang Z, Zhang Z, Lin Z, Yang Z, Long Y, Wu G, Kang Y, Liao W (2014) Chondrogenic differentiation of ATDC5 and hMSCs could be induced by a novel scaffold-tricalcium phosphate-collagen-hyaluronan without any exogenous growth factors in vitro. J Biomed Mater Res, Part A 102(8):2725–2735. doi:10.1002/jbm.a.34948
74. Wang XQ, Wenk E, Zhang XH, Meinel L, Vunjak-Novakovic G, Kaplan DL (2009) Growth factor gradients via microsphere delivery in biopolymer scaffolds for osteochondral tissue engineering. J Control Release 134(2):81–90. doi:10.1016/j.jconrel.2008.10.021
75. Luo ZW, Jiang L, Xu Y, Li HB, Xu W, Wu SC, Wang YL, Tang ZY, Lv YG, Yang L (2015) Mechano growth factor (MGF) and transforming growth factor (TGF)-beta 3 functionalized silk scaffolds enhance articular hyaline cartilage regeneration in rabbit model. Biomaterials 52:463–475
76. Evans CH, Huard J (2015) Gene therapy approaches to regenerating the musculoskeletal system. Nat Rev Rheumatol 11(4):234–242. doi:10.1038/nrrheum.2015.28

77. Madry H, Kaul G, Zurakowski D, Vunjak-Novakovic G, Cucchiarini M (2013) cartilage constructs engineered from chondrocytes overexpressing IGF-I improve the repair of osteochondral defects in a rabbit model. Eur Cells Mater 25:229–247

78. Im G-I, Kim H-J, Lee JH (2011) Chondrogenesis of adipose stem cells in a porous PLGA scaffold impregnated with plasmid DNA containing SOX trio (SOX-5,-6 and -9) genes. Biomaterials 32(19):4385–4392

79. Im G-I, Kim H-J, Lee JH (2011) Chondrogenesis of adipose stem cells in a porous PLGA scaffold impregnated with plasmid DNA containing SOX trio (SOX-5,-6 and -9) genes. Biomaterials 32(19):4385–4392. doi:10.1016/j.biomaterials.2011.02.054

80. Needham CJ, Shah SR, Dahlin RL, Kinard LA, Lam J, Watson BM, Lu S, Kasper FK, Mikos AG (2014) Osteochondral tissue regeneration through polymeric delivery of DNA encoding for the SOX trio and RUNX2. Acta Biomater 10(10):4103–4112

81. Vunjak-Novakovic G, Freed LE, Biron RJ, Langer R (1996) Effects of mixing on the composition and morphology of tissue-engineered cartilage. AIChE J 42(3):850–860. doi:10.1002/aic.690420323

82. Freed LE, Hollander AP, Martin I, Barry JR, Langer R, Vunjak-Novakovic G (1998) Chondrogenesis in a cell-polymer-bioreactor system. Exp Cell Res 240(1):58–65. doi:10.1006/excr.1998.4010

83. Carrier RL, Rupnick M, Lange R, Schoen FJ, Freed LE, Vunjak-Novakovic G (2002) Perfusion improves tissue architecture of engineered cardiac muscle. Tissue Eng 8(2):175–188. doi:10.1089/107632702753724950

84. Altman GH, Lu HH, Horan RL, Calabro T, Ryder D, Kaplan DL, Stark P, Martin I, Richmond JC, Vunjak-Novakovic G (2002) Advanced bioreactor with controlled application of multi-dimensional strain for tissue engineering. J Biomech Eng 124(6):742–749. doi:10.1115/1.1519280

85. Datta N, Pham QP, Sharma U, Sikavitsas VI, Jansen JA, Mikos AG (2006) In vitro generated extracellular matrix and fluid shear stress synergistically enhance 3D osteoblastic differentiation. Proc Natl Acad Sci USA 103(8):2488–2493. doi:10.1073/pnas.0505661103

86. Chang C-H, Lin C-C, Chou C-H, Lin F-H, Liu H-C (2005) Novel bioreactors for osteochondral tissue engineering. Biomed Eng Appl Basis Commun 17(01):38–43. doi:10.4015/s101623720500007x

87. Marolt D, Augst A, Freed LE, Vepari C, Fajardo R, Patel N, Gray M, Farley M, Kaplan D, Vunjak-Novakovic G (2006) Bone and cartilage tissue constructs grown using human bone marrow stromal cells, silk scaffolds and rotating bioreactors. Biomaterials 27(36):6138–6149. doi:10.1016/j.biomaterials

88. Mahmoudifar N, Doran PM (2005) Tissue engineering of human cartilage and osteochondral composites using recirculation bioreactors. Biomaterials 26(34):7012–7024. doi:10.1016/j.biomaterials.2005.04.062

89. Kuiper NJ, Wang QG, Cartmell SH (2014) A perfusion co-culture bioreactor for osteochondral tissue engineered plugs. J Biomater Tissue Eng 4(2):162–171. doi:10.1166/jbt.2014.1145

90. Lin H, Lozito TP, Alexander PG, Gottardi R, Tuan RS (2014) Stem Cell-Based Microphysiological Osteochondral System to Model Tissue Response to Interleukin-1β. Mol Pharm 11(7):2203–2212

91. Nam J, Perera P, Rath B, Agarwal S (2013) Dynamic regulation of bone morphogenetic proteins in engineered osteochondral constructs by biomechanical stimulation. Tissue Eng Part A 19(5–6):783–792. doi:10.1089/ten.tea.2012.0103

92. Russell W, Burch R (1959) The principles of humane experimental technique. Universities Federation for Animal Welfare

93. Takahashi K, Yamanaka S (2006) Induction of pluripotent stem cells from mouse embryonic and adult fibroblast cultures by defined factors. Cell 126(4):663–676. doi:10.1016/j.cell.2006.07.024

94. Kiskinis E, Eggan K (2010) Progress toward the clinical application of patient-specific pluripotent stem cells. J Clin Investig 120(1):51–59. doi:10.1172/jci40553

Part III
Meniscus Injuries: Replacement, Repair and Regeneration

Chapter 12
Basics of the Meniscus

Ibrahim Fatih Cengiz, Joana Silva-Correia, Helder Pereira,
João Espregueira-Mendes, Joaquim Miguel Oliveira
and Rui Luís Reis

Abstract The meniscus is a fibro-cartilaginous tissue located between the femoral
condyles and the tibial plateau in the knee. The presence of the meniscus tissue is
vital for the proper function of the knee. Meniscal injuries are very frequent cases in
the orthopaedics, and they have limited self-healing capacity. The importance of the
basic science of meniscus has been acknowledged for the meniscus development of
regenerative strategies, and the knowledge is increasing over time. Herein, the
biology, anatomy, and biochemistry of meniscus tissue are overviewed.

12.1 Introduction

The meniscus is a vital component of the knee. It is a fibro-cartilaginous tissue
present between the femoral condyles and the tibial plateau. Each knee has two
menisci, the lateral and medial meniscus. The meniscus has a C-like shape with a
wedge-like cross-section. It is composed of different types of meniscal cells [1, 2]
and the extracellular matrix (ECM) collagens fibres that are mainly the type I,
proteoglycans, glycoproteins, noncollagenous and water [3, 4]. It is a complex

I.F. Cengiz · J. Silva-Correia · H. Pereira · J.M. Oliveira (✉) · R.L. Reis
3B's Research Group—Biomaterials, Biodegradables and Biomimetics, Headquarters of the
European Institute of Excellence on Tissue Engineering and Regenerative Medicine,
University of Minho, Avepark—Parque de Ciência E Tecnologia, Zona Industrial Da Gandra,
4805-017 Barco GMR, Portugal
e-mail: miguel.oliveira@dep.uminho.pt

I.F. Cengiz · J. Silva-Correia · H. Pereira · J. Espregueira-Mendes · J.M. Oliveira · R.L. Reis
ICVS/3B's—PT Government Associate Laboratory, Braga/Guimarães, Portugal

H. Pereira · J. Espregueira-Mendes
Clínica Espregueira-Mendes F.C. Porto Stadium—FIFA Medical Centre of Excellence, Porto,
Portugal

H. Pereira
Orthopedic Department, Centro Hospitalar Póvoa de Varzim, Vila do Conde, Portugal

© Springer International Publishing AG 2017
J.M. Oliveira and R.L. Reis (eds.), *Regenerative Strategies for the Treatment
of Knee Joint Disabilities*, Studies in Mechanobiology, Tissue Engineering
and Biomaterials 21, DOI 10.1007/978-3-319-44785-8_12

tissue with a particular cell distribution [5] and specific blood supply that is important for the self-healing potential [4].

The meniscus serves certain purposes in the knee biomechanics so that the knee can function normally [6, 7]. Meniscus injuries are the most common knee injury [8] Injuries of meniscus cause pain, catching and locking in the knee. It was common to remove meniscus partially or totally in the past [4, 9]. The removal of the meniscus in the knee brings important consequences such as flattened femoral condyle, and narrowed joint space [10]. This can promote early degenerative changes in the knee [10–13]. Meniscal injuries are very frequent cases in the orthopaedics. In the clinics, treatments for the meniscus injuries depend on the patient condition and the injury [4, 14]. Due to the limited vascularity the complete self-healing of the meniscus is difficult [15, 16].

12.2 The Embryology and Development of the Meniscus

The typical crescent shape of the meniscus is achieved by the 8th–10th gestational week [17, 18]. Gray and Gardner [18] reported their observation on meniscus samples from different gestational weeks. By the gestational week 8, few collagen strands can be observed in the meniscus [18]. The condensation of the intermediate layer of mesenchymal tissue leads attachments to form towards the encircling joint capsule. Cellularity and vascularity get higher throughout the meniscus during the development. By the 9th week, numerous cells, and thin strands of collagen were observed [18]. The cells located on the surface have thin and flat shape, and the cells from more deep layers appear fusiform. Blood vessels progressively get formed, and got spread almost within the entire meniscus at birth [19]. Later with a progressive decrease in the cellularity, the content of collagen increases. The collagen fibres are aligned depending on the motion of the joint and weight bearing pattern within the knee [20]. They are mainly organized in a circumferential manner. Gray and Gardner [18] reported their observation on meniscus samples from different gestational weeks. In week 10, it was observed that a Wrisberg ligament extends from the posterior portion of the lateral meniscus to the medial femoral condyle [18]. In week 12, a transverse ligament connected to the anterior parts of menisci and, a ligamentous band connects the lateral meniscus to the fibula. The meniscus is similar in week 9 and 12, only the collagen amount is higher in week 12 [18]. By week 13, the cellularity of the meniscus is higher, and the collagenous fibres were more noticeable, particularly at the horns. Another Wrisberg ligament was seen in week 14. In week 21, the collagenous fibres bundles were more ordered. In week 23, the bundles were generally parallel to each other. The appearance of the menisci at 34 and 35 weeks was similar to that of previous weeks [18].

12.3 The Gross Anatomy of the Meniscus

Lateral meniscus and medial meniscus are two smooth, shiny-surfaced whitish coloured fibro-cartilaginous tissues found between the tibial plateau and femoral condyles in each knee joint, the largest synovial joint of the human body. The menisci are crescent-shaped where the medial meniscus is less circular (Fig. 12.1), and they have a wedge-shaped cross-section with a concave top surface in accordance with the convex femoral condyles (Fig. 12.2).

The horns of the meniscus are the anchorage points to the tibial plateau. The stability of meniscus is ensured by several ligaments. Transverse ligament connects the lateral and medial meniscus from their anterior parts. The medial meniscus is attached to the tibia by its horns and merged with the knee joint capsule from its outer periphery [21]. The outer periphery of the anterior horn of the lateral meniscus enters into the tibial intercondylar eminence, and the posterior horn of the lateral meniscus continues and connects to the medial femoral condyle by the menisco-femoral ligaments [22, 23], i.e. ligaments of Wrisberg, and of Humphrey which respectively run behind and in front of posterior cruciate ligaments. Coronary ligaments are found around the menisci and enhance the attachment to the tibial plateau. However, it was reported that not all individuals have all of these ligaments [24, 25]. The medial meniscus is attached to the medial collateral ligament and thus has less ability to move. The lateral meniscus has more mobility, and it is not attached to the lateral collateral ligament. For this reason, lateral meniscus has relatively less tendency to be injured than the medial meniscus [21]. It should be

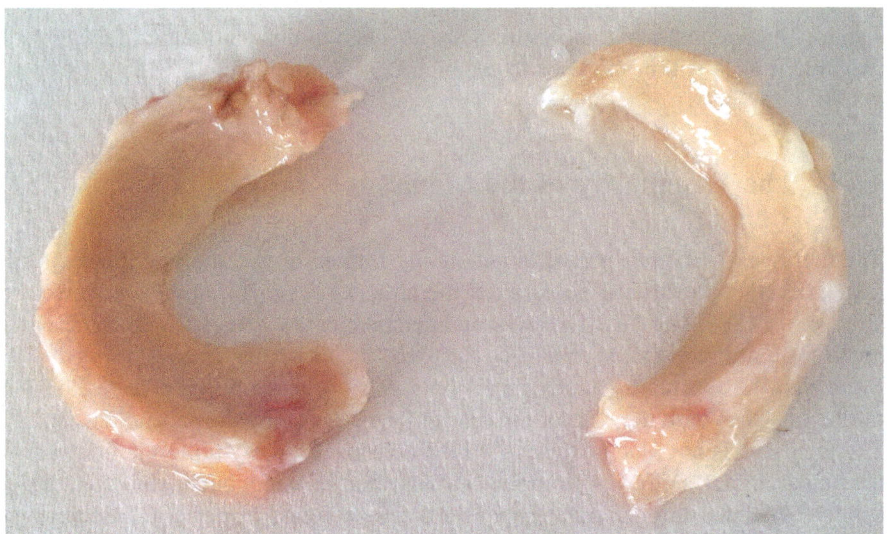

Fig. 12.1 Photographs of lateral (*left*), and medial (*right*) meniscus harvested from the right knee of a 77-years old female human donor

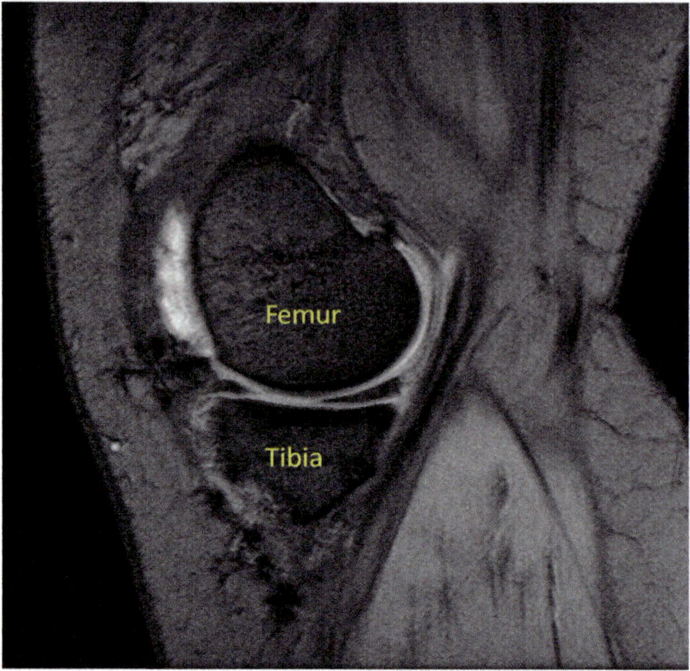

Fig. 12.2 A sagittal view from the T2-weighted MRI of the left knee of an 18-years old female subject showing the *wedge-shaped* cross-section in *black* with a *top* and *bottom* surfaces in accordance with the femoral condyle and the tibial plateau

also highlighted that while the menisci are attached to tibia and femur, they are still dynamic tissues within the knee to maintain a safe articulation [26, 27].

12.4 The Vascularity of the Meniscus

The meniscus tissue has partial blood supply limited to the outer periphery. More than a decade ago vascular anatomy of the human knee meniscus was reviewed by Gray et al. [28]. The blood vessels and lymphatics are present within the entire meniscus until the age of one. At age two, an avascular area is formed [19]. Moreover, the vascularity and lymph supply gets limited within the outer 25–33 % of the meniscus with the start of the role in load-bearing by the second decennium. The meniscus is not much exposed to biomechanical forces during the first year of the human infant. Diffusion from the synovial fluid is not enough, and direct blood supply is needed throughout the meniscus for an infant to perform standing and walking activities. With the start of the bipedal walking, the stress from the weight of the body and the muscle forces is thought to be the underlying reason of avascularization of the inner regions [19, 28, 29]. On the other hand, the horns that have high vascularity and neural innervation, remain vascularized; this might be because they are not under weight-bearing forces [28].

The branches of the popliteal artery, i.e. medial and lateral inferior and middle geniculate arteries supply the meniscus [30, 31]. The vessels are mainly arranged in a circumferential manner with radial branches oriented to the central region of the meniscus. The perimeniscal capillary plexus extends inside the meniscus across the synovium around the meniscosynovial junction, and supplies up to one-quarter of the periphery of the meniscus. The area next to the popliteus tendon is avascular [31]. Vascular synovial tissue covers the anterior and posterior horns, and the horns receive rich blood supply [30, 31]. Endoligamentous blood vessels from the horns extend with a short-range into the bulk of the meniscus, and a direct nourishment is provided [32].

Lymphatics accompany the blood vessels throughout the meniscus. The avascular regions are nourished through the synovial fluid by diffusion or by mechanical pumping during the motion of the joint [19, 20]. And after the age fifty, the vascularity is present only within the outer 10–33 % of the meniscus [19, 28]. Therefore, while injuries in the vascular region can have the capacity to heal, the injuries where the blood supply cannot reach have limited healing capacity [28].

12.5 The Innervation of the Meniscus

The meniscus also has a role in deep sensitivity by being able to send and receive proprioceptive signals [33]. In meniscus, the nerve fibres are mostly associated with vascularity [34]. Accordingly, the nerve fibres and sensory receptors are found mainly vascular regions and get denser at the horn regions [31, 35, 36]. Like blood vessels, the neural innervation is not seen in the inner third of the meniscus [36]. The meniscus is innervated by the recurrent peroneal branch of common peroneal nerve [20, 37, 38]. The circumferential nerve fibres in the vascular region are relatively thicker while the fibres that enter radially into the meniscus are thinner. Free nerve endings, i.e. nociceptors, and three mechanoreceptors, i.e. Pacinian corpuscles, Ruffini corpuscles, and Golgi tendon organ are present in the peripheral two-thirds of the meniscus and horns [28, 34]. Pacinian corpuscles give information about acceleration and deceleration of the knee. Ruffini corpuscles provide information on the static position of the knee, the change in intra-articular pressure, and changes in the parameters of the knee motion, i.e. direction, amplitude and velocity. Golgi tendon organ acts as a protective reflex inhibitor of the motor activity of the muscles related to the knee. Nociceptors create impulses that are interpreted as pain in the brain [28].

Mine et al. [34] explained the pain sensation in the course of a fresh meniscal tear. If the tear is in the avascular region, the pain is caused by the deterioration of the micromilovia around the tear stimulating nociceptors in the synovia and the joint capsule, while in the case of a tear in the vascular region, the pain additionally derives from the nerves within the tissue [34]. Substance P-immunofluorescent nerves are found in the knee synovial membrane and meniscus [35]. Since they have a role in pain transmission, relief of pain after meniscectomy is a commonly observed as a denervation effect [35].

12.6 The Cells of Meniscus

In what concerns cellularity, different types of cells have been observed in the meniscus, i.e. chondrocyte-like, fibroblast-like and intermediate cells [39]. However, there is no consensus regarding the classification of meniscus cells and several names such as fibrocytes, fibroblasts, meniscus cells, fibrochondrocytes, and chondrocytes are being used [40]. The outer zone cells present an oval, fusiform shape, resembling to fibroblasts in appearance and behaviour. For this reason, they may be described as fibroblast-like cells [2]. These cells display long cell extensions and are enclosed within a matrix largely composed by type I collagen, with a small percentage of glycoproteins and types III and V collagen [41]. In the inner portion of meniscus, cells have a round morphology and thus are commonly referred as fibrochondrocytes or chondrocyte-like cells [2]. These cells are embedded in a ECM consisting mainly of type II collagen, interlinked with a smaller but significant quantity of type I collagen and a higher concentration of GAGs. The outer zone cells display higher migration ability when compared to inner cells, and also seem to exhibit lower adhesion strengths [42]. A third type of cell population, with a flattened and fusiform morphology and no cell extensions, has also been identified at the superficial region of the meniscus. It has been suggested that these cells might be specific progenitor cells [43].

The phenotype and distribution of cells within the different segments and zones of meniscus has recently been investigated by Pereira et al. [44] in 44 patients. A gradual decrease of cell density from the vascular (outer) to the avascular (inner) zones was observed by histomorphometry analysis, in all the segments (i.e., anterior, middle, and posterior) of both lateral and medial menisci. Complementarily, it was found a lower cell density in the anterior sections of lateral and medial menisci when compared to the middle or posterior sections. In that work, cells with rounded and fusiform-like morphology were isolated and characterized by flow cytometry. The phenotypic analysis showed that the surface markers CD44, CD73, CD90, and CD105 were positive in more than 97 % of cells. CD31 and CD34 were being expressed in 2.3 ± 0.8 and 3.2 ± 1.0 % of cells, whereas the CD45 marker for hematopoietic stem cells was present in 0.2 ± 0.1 % of cells.

Meniscus cells isolated from vascular, avascular and mixed zone present cell plasticity and thus can be induced towards chondrogenic, adipogenic and osteogenic lineages. Outer cells also can be induced to osteogenesis lineage [45].

12.7 The Ultrastructure of the Meniscus

Meniscus composition studies have demonstrated that it possesses high water content (72 %), being the remaining 28 % portion composed by an organic component, namely ECM and asymmetrically distributed cells [15]. The ECM that surrounds meniscus cells is largely composed by several types of collagen (75 % of total organic matter). The other constituents consist in glycosaminoglycans (GAGs, 17 %) and small percentages of adhesion glycoproteins (<1 %) and elastin (<1 %) [15]. Meniscus composition differ with age, in injury or under pathological conditions [46].

Several types of collagen are present in various amounts in each meniscus segment. Type I collagen is the main type of collagen present in the vascular region (80 % dry weight), but other variants are present in small proportions (<1 %), namely type II, III, IV, VI and XVIII. In the avascular zone, collagen constitutes up to 70 % dry weight, being 60 % type II collagen and 40 % type I collagen [47].

The role of the other fibrillar component of meniscus, i.e. elastin, is not completely understood, but in adult meniscus, the combination of mature and immature fibers has been observed in very low proportions (<0.6 %) [3, 48].

Proteoglycans comprise a core protein decorated with GAGs, being their main function to enable water absorption [15]. Chondoitin-6-sulfate (60 %), dermatan sulfate (20–30 %), chondroitin-4-sulfate (10–20 %) and keratin sulfate (15 %) are the main types of GAGs present in normal human meniscus [49]. Aggrecan is the most important large proteoglycan found in meniscus tissue, while biglycan and decorin are the main small proteoglycans [50]. Also these molecules present an irregular distribution within the tissue, with a higher proportion of proteoglycans in the inner two-thirds as compared to the outer one-third.

The major adhesion glycoproteins present in human meniscus are fibronectin, thrombospondin and type VI collagen and these components serve as anchor sites between ECM and cells [51].

12.8 The Functions of the Meniscus

The meniscus has an important role on preserving knee joint stability and load transfer. Each knee has lateral and medial menisci that are attached between the tibial and femoral surfaces, thus covering two-thirds of the tibia plateau. Kurosawa et al. showed that upon total excision of meniscus, the total contact area is decreased by a third to a half in the fully extended knee [52]. Another study demonstrated the possible deleterious effects of meniscectomy in articular cartilage, subchondral bone, proximal tibia's trabecular bone and cortex [53].

The medial meniscus present reduced mobility as compared to the lateral one, due to its unique anatomy (including stronger attachment to medial collateral ligament) [54]. In a stable knee, where central pivot ligaments are functioning, the medial meniscus has reduced involvement on anterior tibial displacement constraint. The anterior cruciate ligament impairs anterior knee motion prior to significant contact of femoral condyle with posterior horn of medial meniscus and tibial plateau [54]. Significant differences have been recognized between both femorotibial compartments on the knee joint. Lateral tibial plateau tend to have a more convex form, differing to a more concave shape on the medial compartment [54, 55]. Due to this fact, the loss of the lateral meniscus results in a significant decrease on femorotibial congruence. Additionally, most of the load transfer on the lateral compartment is supported by lateral meniscus (70 % as compared to 50 % in the medial), whereas in the medial compartment, force transmission is disseminated between articular cartilage and respective meniscus [56, 57].

The biomechanical performance of menisci to loads acting on tibiofemoral joints, results from their macro-geometry, fine architecture and insertional ligaments [55]. The collagen bundles present in the superficial layer mimic those of articular hyaline cartilage, i.e. they are randomly orientated, providing lower friction between menisci, femur and tibia during joint motion [58]. Under the superficial layer, the bulk of meniscal tissue consists in two distinct zones of collagen fibers: the inner one-third bundles that present a radial pattern, and the outer two-thirds that are circumferentially orientated. Therefore, it has been suggested that the inner third function mainly in compression, whereas the outer two-thirds may function mostly in tension. Furthermore, the bulk of meniscal tissue also presents some radially-orientated collagen fibers that act as "tie fibers", resisting longitudinal splitting of the circumferentially-orientated collagen bundles [59].

The viscoelastic behavior of the meniscus can be correlated with its ECM composition [15]. It can be described as a rubber-like pattern at high loading frequencies, whereas at lower frequencies viscous dissipation occurs [15, 44]. This behavior is not so dependent on collagen content, but it is mainly related with GAGs and water content, i.e. higher with increasing GAGs and lower with increasing water content [60]. Segmental and zonal differences in relation to GAGs content, size and cellular density has been observed in animal meniscal tissue [61]. In a similar way to the asymmetric distribution of blood vessels and cells, the mechanical properties also differ within the different menisci [60]. Recently, Pereira et al. [44] analyzed the mechanical properties in the different segments of fresh human menisci (lateral and medial), at physiological conditions (37 °C and pH 7.4). Significant differences were observed between the medial (higher E0 and tan d) and lateral menisci. Also, when analysis was performed in combination for both menisci (medial and lateral), significant differences were observed between posterior, middle and anterior segments (posterior segments are stiffer than the middle ones, and these are significantly stiffer as compared to the anterior).

12.9 Final Remarks

The meniscus is crucial for the normal function of the knee. It is a tissue with high complexness and restricted capability for self-healing. The current treatments of meniscus lesions are not entirely satisfactory, and there is a great clinical need for appropriate tissue engineered implants. The knowledge on the biology, anatomy, and biochemistry of meniscus is expanding and is of great importance for developing improved meniscus regeneration strategies.

Acknowledgments The authors thank the financial support of the MultiScaleHuman project (Contract Number: MRTN-CT-2011-289897) in the Marie Curie Actions—Initial Training Networks. I. F. Cengiz thanks the Portuguese Foundation for Science and Technology (FCT) for the Ph.D. scholarship (SFRH/BD/99555/2014). J. M. Oliveira also thanks the FCT for the funds provided to under the program Investigador FCT 2012 (IF/00423/2012).

References

1. Sanchez-Adams J, Athanasiou KA (2009) The knee meniscus: a complex tissue of diverse cells. Cell Mol Bioeng 2(3):332–340
2. Verdonk PC, Forsyth R, Wang J, Almqvist KF, Verdonk R, Veys EM, Verbruggen G (2005) Characterisation of human knee meniscus cell phenotype. Osteoarthr Cartil 13(7):548–560
3. Mcdevitt CA, Webber RJ (1990) The ultrastructure and biochemistry of meniscal cartilage. Clin Orthop Relat Res 252:8–18
4. Tudor F, McDermott ID, Myers P (2014) Meniscal repair: a review of current practice. Orthop Trauma 28(2):88–96
5. Cengiz IF, Pereira H, Pêgo JM, Sousa N, Espregueira-Mendes J, Oliveira JM, Reis RL (2015) Segmental and regional quantification of 3D cellular density of human meniscus from osteoarthritic knee. J Tissue Eng Regen Med. doi:10.1002/term.2082
6. Greis PE, Bardana DD, Holmstrom MC, Burks RT (2002) Meniscal injury: I. Basic science and evaluation. J Am Acad Orthop Surg 10(3):168–176
7. Brindle T, Nyland J, Johnson DL (2001) The meniscus: review of basic principles with application to surgery and rehabilitation. J Athl Train 36(2):160
8. Clayton RAE, Court-Brown CM (2008) The epidemiology of musculoskeletal tendinous and ligamentous injuries. Injury 39(12):1338–1344
9. Fetzer GB, Spindler KP, Amendola A, Andrish JT, Bergfeld JA, Dunn WR, Flanigan DC, Jones M, Kaeding CC, Marx RG (2009) Potential market for new meniscus repair strategies: evaluation of the MOON cohort. J Knee Surg 22(3):180
10. Fairbank TJ (1948) Knee joint changes after meniscectomy. J Bone Joint Surg Br 30(4):664–670
11. Allen PR, Denham RA, Swan AV (1984) Late degenerative changes after meniscectomy. Factors affecting the knee after operation. J Bone Joint Surg Br 66(5):666–671
12. Jackson JP (1968) Degenerative changes in the knee after meniscectomy. Br Med J 2 (5604):525
13. McDermott ID, Amis AA (2006) The consequences of meniscectomy. J Bone Joint Surg Br 88 (12):1549–1556
14. Mordecai SC, Al-Hadithy N, Ware HE, Gupte CM (2014) Treatment of meniscal tears: an evidence based approach. World J Orthop 5(3):233

15. Makris EA, Hadidi P, Athanasiou KA (2011) The knee meniscus: structure–function, pathophysiology, current repair techniques, and prospects for regeneration. Biomaterials 32 (30):7411–7431
16. Scotti C, Hirschmann MT, Antinolfi P, Martin I, Peretti GM (2013) Meniscus repair and regeneration: review on current methods and research potential. Eur Cells Mater 26:150–170
17. Allen AA, Caldwell GL Jr, Fu FH (1995) Anatomy and biomechanics of the meniscus. Oper Tech Orthop 5(1):2–9
18. Gray DJ, Gardner E (1950) Prenatal development of the human knee and superior tibiofibular joints. Am J Anat 86(2):235–287
19. Petersen W, Tillmann B (1995) Age-related blood and lymph supply of the knee menisci: a cadaver study. Acta Orthop 66(4):308–312
20. Fox AJS, Wanivenhaus F, Burge AJ, Warren RF, Rodeo SA (2014) The human meniscus: a review of anatomy, function, injury, and advances in treatment. Clin Anat 28(2):269–287
21. Rath E, Richmond JC (2000) The menisci: basic science and advances in treatment. Br J Sports Med 34(4):252–257
22. Heller L, Langman J (1964) The menisco-femoral ligaments of the human knee. J Bone Joint Surg Br 46(2):307–313
23. Yamamoto M, Hirohata K (1991) Anatomical study on the menisco-femoral ligaments of the knee. Kobe J Med Sci 37(4–5):209
24. Gupte CM, Smith A, McDermott ID, Bull AMJ, Thomas RD, Amis AA (2002) Meniscofemoral ligaments revisited. J Bone Joint Surg Br 84-B:846–851
25. Kohn D, Moreno B (1995) Meniscus insertion anatomy as a basis for meniscus replacement: a morphological cadaveric study. Arthrosc J Arthrosc Relat Surg 11(1):96–103
26. Thompson WO, Thaete FL, Fu FH, Dye SF (1991) Tibial meniscal dynamics using three-dimensional reconstruction of magnetic resonance images. Am J Sport Med 19(3):210–216
27. Vedi V, Spouse E, Williams A, Tennant SJ, Hunt DM, Gedroyc WMW (1999) Meniscal movement an in-vivo study using dynamic MRI. J Bone Joint Surg Br 81(1):37–41
28. Gray JC (1999) Neural and vascular anatomy of the menisci of the human knee. J Orthop Sport Phys Ther 29(1):23–30
29. Renström P, Johnson RJ (1990) Anatomy and biomechanics of the menisci. Clin Sports Med 9 (3):523–538
30. Arnoczky SP, Warren RF (1982) Microvasculature of the human meniscus. Am J Sport Med 10(2):90–95
31. Day B, Mackenzie WG, Shim SS, Leung G (1985) The vascular and nerve supply of the human meniscus. Arthrosc J Arthrosc Relat Surg 1(1):58–62
32. Danzig L, Resnick D, Gonsalves M, Akeson W (1983) Blood supply to the normal and abnormal menisci of the human knee. Clin Orthop Relat Res 172:271–276
33. Assimakopoulos AP, Katonis PG, Agapitos MV, Exarchou EI (1992) The innervation of the human meniscus. Clin Orthop Relat Res 275:232–236
34. Mine T, Kimura M, Sakka A, Kawai S (2000) Innervation of nociceptors in the menisci of the knee joint: an immunohistochemical study. Arch Orthop Trauma Surg 120(3–4):201–204
35. Grönblad M, Korkala O, Liesl P, Karaharju E (1985) Innervation of synovial membrane and meniscus. Acta Orthop 56(6):484–486
36. Zimny ML, Albright DJ, Dabezies E (1988) Mechanoreceptors in the human medial meniscus. Acta Anat (Basel) 133(1):35–40
37. Gardner E (1948) The innervation of the knee joint. Anat Rec (Hoboken, NJ: 2007) 101 (1):109–130
38. Kennedy JC, Alexander IJ, Hayes KC (1982) Nerve supply of the human knee and its functional importance. Am J Sport Med 10(6):329–335
39. Ghadially FN, Thomas I, Yong N, Lalonde JM (1978) Ultrastructure of rabbit semilunar cartilages. J Anat 125(Pt 3):499–517

40. Nakata K, Shino K, Hamada M, Mae T, Miyama T, Shinjo H, Horibe S, Tada K, Ochi T, Yoshikawa H (2001) Human meniscus cell: characterization of the primary culture and use for tissue engineering. Clin Orthop Relat Res 391(Suppl):S208–S218

41. Melrose J, Smith S, Cake M, Read R, Whitelock J (2005) Comparative spatial and temporal localisation of perlecan, aggrecan and type I, II and IV collagen in the ovine meniscus: an ageing study. Histochem Cell Biol 124(3):225–235

42. Gunja NJ, Dujari D, Chen A, Luengo A, Fong JV, Hung CT (2012) Migration responses of outer and inner meniscus cells to applied direct current electric fields. J Orthop Res 30(1):103–111

43. Van der Bracht H, Verdonk R, Verbruggen G, Elewaut D, Verdonk P (2007) Cell based meniscus tissue engineering. In: Ashammakhi N, Reis RL, Chiellini E (eds) Topics in tissue engineering, vol 3. Biomaterials and tissue engineering group, Oulu, Finland. Available at http://www.oulu.fi/spareparts/ebook_topics_in_t_e_vol3/

44. Pereira H, Caridade SG, Frias AM, Silva-Correia J, Pereira DR, Cengiz IF, Mano JF, Oliveira JM, Espregueira-Mendes J, Reis RL (2014) Biomechanical and cellular segmental characterization of human meniscus: building the basis for Tissue Engineering therapies. Osteoarthr Cartil 22(9):1271–1281

45. Mauck RL, Martinez-Diaz GJ, Yuan X, Tuan RS (2007) Regional multilineage differentiation potential of meniscal fibrochondrocytes: implications for meniscus repair. Anat Rec (Hoboken, NJ: 2007) 290(1):48–58

46. Sweigart MA, Athanasiou KA (2001) Toward tissue engineering of the knee meniscus. Tissue Eng 7(2):111–129

47. Cheung HS (1987) Distribution of type I, II, III and V in the pepsin solubilized collagens in bovine menisci. Connect Tissue Res 16(4):343–356

48. Ghosh P, Taylor TK (1987) The knee joint meniscus. A fibrocartilage of some distinction. Clin Orthop Relat Res 224:52–63

49. Herwig J, Egner E, Buddecke E (1984) Chemical changes of human knee joint menisci in various stages of degeneration. Ann Rheum Dis 43(4):635–640

50. Scott PG, Nakano T, Dodd CM (1997) Isolation and characterization of small proteoglycans from different zones of the porcine knee meniscus. Biochim Biophys Acta (BBA) Gen Subj 1336(2):254–262

51. Miller RR, McDevitt CA (1991) Thrombospondin in ligament, meniscus and intervertebral disc. Biochim Biophys Acta (BBA) Gen Subj 1115(1):85–88

52. Kurosawa H, Fukubayashi T, Nakajima H (1980) Load-bearing mode of the knee joint: physical behavior of the knee joint with or without menisci. Clin Orthop Relat Res 149:283–290

53. Fukubayashi T, Kurosawa H (1980) The contact area and pressure distribution pattern of the knee. A study of normal and osteoarthrotic knee joints. Acta Orthop Scand 51(6):871–879

54. Levy IM, Torzilli PA, Warren RF (1982) The effect of medial meniscectomy on anterior-posterior motion of the knee. J Bone Joint Surg Am 64(6):883–888

55. McDermott ID, Masouros SD, Amis AA (2008) Biomechanics of the menisci of the knee. Curr Orthop 22(3):193–201

56. Walker PS, Hajek JV (1972) The load-bearing area in the knee joint. J Biomech 5(6):581–589

57. Bourne RB, Finlay JB, Papadopoulos P, Andreae P (1984) The effect of medial meniscectomy on strain distribution in the proximal part of the tibia. J Bone Joint Surg Am 66(9):1431–1437

58. Beaupre A, Choukroun R, Guidouin R, Garneau R, Gerardin H, Cardou A (1986) Knee menisci. Correlation between microstructure and biomechanics. Clin Orthop Relat Res 208:72–75

59. Bullough PG, Munuera L, Murphy J, Weinstein AM (1970) The strength of the menisci of the knee as it relates to their fine structure. J Bone Joint Surg Br 52-B(3):564–570

60. Bursac P, Arnoczky S, York A (2009) Dynamic compressive behavior of human meniscus correlates with its extra-cellular matrix composition. Biorheology 46(3):227–237

61. Killian ML, Lepinski NM, Haut RC, Haut Donahue TL (2010) Regional and zonal histo-morphological characteristics of the lapine menisci. Anat Rec (Hoboken, NJ: 2007) 293 (12):1991–2000

Chapter 13
Biomaterials in Meniscus Tissue Engineering

João B. Costa, Joaquim Miguel Oliveira and Rui Luís Reis

Abstract Meniscus is a complex tissue that plays important roles on the knee performance and homeostasis. The meniscus tissue is very susceptible to injury and despite the great advances in meniscus regeneration area, none of the current strategies for the treatment of meniscus lesions are completely effective. To overcome such great challenge, tissue engineering-based strategies have been attempted. One of the main targets in this research area is to find out a biomaterial or formulations that are able to mimic as much as possible the meniscus native extracellular matrix. In the last few years the characteristics and behaviors of different biomaterials were explored and several processing routes attempted to obtain an adequate architecture for proper cells adhesion, ingrowths, proliferation and differentiation. Herein, a panoply of biomaterials that have been used in meniscus tissue engineering strategies are overviewed.

13.1 Introduction

Meniscus plays an important role on the knee performance and it has been described as "functionless remnants of leg muscle origin" [1]. Due to forces that meniscus can be subjected and his location, he is very susceptible to injury. The mean incidence of meniscal injury in the United States is 66/100,000 [2, 3], which implies over 1 million surgical procedures each year in the United States [4]. Several strategies to repair and replace meniscus have been proposed, but only few of them have been shown to be effective [5]. Currently, the irreparable meniscal

J.B. Costa (✉) · J.M. Oliveira · R.L. Reis
3B's Research Group—Biomaterials, Biodegradables and Biomimetics,
Headquarters of the European Institute of Excellence on Tissue Engineering
and Regenerative Medicine, University of Minho, AvePark,
Zona Industrial Da Gandra, 4805-017 Barco GMR, Portugal
e-mail: jotapcosta14@hotmail.com

J.B. Costa · J.M. Oliveira · R.L. Reis
ICVS/3B's—PT Government Associate Laboratory, Braga, Guimarães, Portugal

© Springer International Publishing AG 2017

249

J.M. Oliveira and R.L. Reis (eds.), *Regenerative Strategies for the Treatment of Knee Joint Disabilities*, Studies in Mechanobiology, Tissue Engineering and Biomaterials 21, DOI 10.1007/978-3-319-44785-8_13

lesions are usually treated by partial or (sub) total meniscectomy. The clinical follow-up results showed knee osteoarthritis in all patients 14 years after partial meniscectomy [6], especially in young and middle aged patients [1], which is dramatic in all possible ways.

This is a major orthopedic operation with concomitant risks and costs. Since the current methods are not able to resolve this issue, it is very important to discover new strategies for this common problem in our daily life. Therefore, there has been an increase in scientific and clinical interest to find a meniscal substitute aimed to minimize the risk for developing knee osteoarthritis but also to offer a solution for patients suffering from enduring symptoms post meniscectomy [7]. The panoply of biomaterials that have been tested as a meniscus substitute/scaffold can be divided in two different categories, i.e. the non-resorbable polymers and resorbable polymers as depicted in Fig. 13.1. A non resorbable polymer is used as a permanent implant with biomechanical properties similar of the native tissue. The resorbable polymers, that can be natural or synthetic, are used in strategies which is formed new meniscal tissue while occurs a slow degradation of the scaffold.

As non-resorbable polymers, can be identified biomaterials such as polyethylene terephthalate (PET), poly(vinyl alcohol) (PVA) and polycarbonate-urethane (PCU). Elsner et al., for example, developed a PCU synthetic meniscal implant that does not require surgical attachment but still provides the biomechanical function necessary for joint preservation [8]. Regarding to the resorbable polymers, in particular the synthetic, the Polyurethanes (PU) have been widely used in the meniscus regeneration area. A scaffold composed of PU along with Polycaprolactone (PCL) that is another resorbable synthetic polymer, has been tested in clinical environment. A couple of studies have demonstrated that this synthetic scaffold commercially named as Actifit, showed good clinical results [9, 10]. In the last few years, the scientific community is concentrating their efforts on the discovery of biomaterials from natural sources. Natural polymers as collagen, silk, cellulose and gelatin, have shown interesting characteristics to be used as a meniscus substitute biomaterial. Rodkey et al., using a collagen based scaffold, showed that it is safe to implant the collagen meniscal implant leading to a protection of the articular cartilage from damage and a better knee function compared with partial meniscectomy

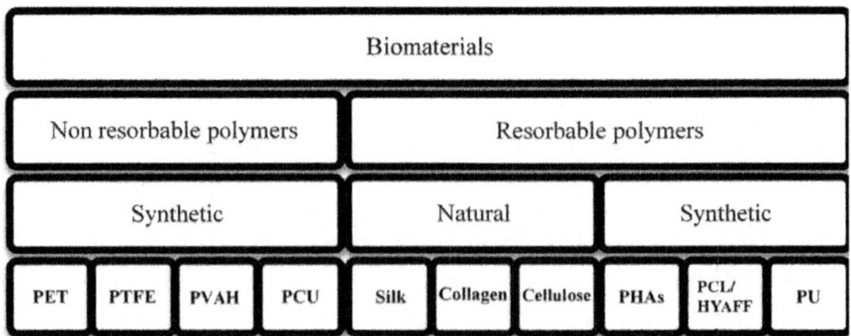

Fig. 13.1 Biomaterials in meniscus tissue engineering

[11]. Herein, it will be made an overview about the different biomaterials and formulations that have been used in meniscus regeneration.

13.2 Non-resorbable Polymers

13.2.1 Synthetic

13.2.1.1 Polyethylene Terephthalate (PET)

A permanent prosthesis was evaluated by Sommerlath et al. [12]. This meniscus implant was composed by PET and coated with polyurethane. Dacron (PET) is non-resorbable polymer constituted by fibers with an outstanding crease and abrasion resistance [13]. Despite the potential characteristics of this PET implant, the studies reveal disappointing results. Sommerlath et al. showed that Dacron implant led to altered joint mechanics, osteophyte formation and synovitis caused by the non-normal biomechanics properties, the improper sizing and the impossibility to match the stress-relaxation of the native meniscus tissue [12].

13.2.1.2 Polytetrafluorethylene Terephthalate (PTFE)

PTFE, also known as Teflon, is a commercial polymer formed by the polymerization of tetrafluorethylene. It has a high molecular weight, is hydrophobic and has a very low coefficient of friction compared with any other solids [13].

Comparing PET and PTFE, it can be said that PTFE has similar compliance to the normal meniscus, showing compressive results more close to native meniscus [14]. However, some studies made by Messner et al. [15] in rabbit models proved that Teflon prosthesis has some issues. The PTFE implant lost its shape after implantation and it is prone to wear, resulting in debris formation and synovitis. A coated Teflon prosthesis was also tested. Polyurethane was the polymer chosen to coat the implant. The coated PTFE prosthesis gave the best overall results showing, on macroscopic examination, that the cartilage appeared similar to meniscus native tissue. Although the tendency to cartilage softening, osteophyte formation, and synovitis indicates that the procedure was not able to prevent all these problems [15, 16]. These disappointing results lead to a decrease of interest from the scientific community in this meniscus substitutes.

13.2.1.3 Poly(Vinyl Alcohol) Hydrogel (PVA-H)

PVA-H has been shown to have cartilage like viscoelasticity and excellent biocompatibility [17]. Hydrogels are biomaterials that consist of a water-swollen network of crosslinked polymer chains [18]. They can be made from chains of from synthetic polymers such as poly(vinyl alcohol) (PVA). Due to PVA-H hydrogels biocompatibility, ease of fabrication and viscoelastic properties several attempts at tissue engineering have previously been examined, including cartilage tissue engineering [19]. Kabayashi et al. [20] combined the PVA with the characteristics of a hydrogel and tested a poly(vinyl alcohol) hydrogel with a water content of 90 %, a degree of polymerization of 17,500 and cross linked as an implant in a rabbit model. In that study, with a timeframe of 2 year, it has proven that PVA-H implant was able to reduce the articular cartilage damaged compared to meniscectomy [20]. Kelly et al. [21] took the next step and assessed the PVA-H implant (Fig. 13.2) in a sheep. The results showed that, compared with meniscectomy, the

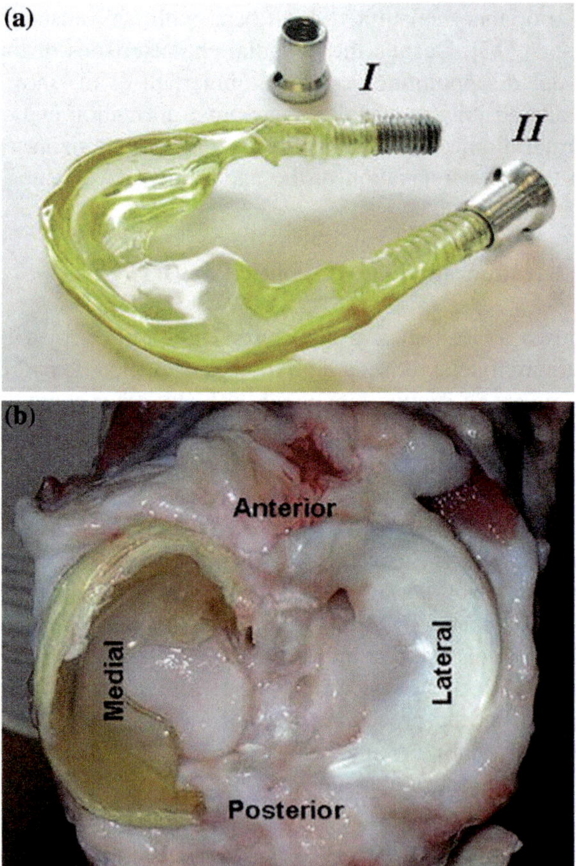

Fig. 13.2 Polycarbonate-urethane meniscal implant, with the stainless steel fixation bolt in the unfastened (*I*) and fastened state (*II*). Reprinted with permission from [25]. Copyright 2010, Springer

PVA-H meniscal replacement resulted in a significantly decrease cartilage degeneration. Compared with meniscal allograft transplantation animals, after 4 months, the PVA-H implant had significantly increased cartilage degeneration in the peripheral zone of the tibial plateau [21].

The significantly cartilage degeneration and implant failure after 1 year as compared to allograft transplantation corroborate that, improvements in PVA-H properties, surface characteristics and a more accurate size matching must be done in the future in order to improve the clinical outcomes.

In order to overtake these challenges, a reinforced PVA-H implant was developed. A fibrous reinforcement of poly(vinyl alcohol) hydrogels is capable of increasing the hydrogel tensile modulus within the range of 0.23 and 260 MPa depending on fiber volume fraction [22]. However the interface of the reinforced PVA-H implant has been identified as a potential issue in meniscal replacements. Holloway et al. [23] developed a novel PVA grafting technique in order to create a covalent linkage and improve the interfacial properties. This approach led to an increase of the interfacial shear strength and more efficient stress transfer. These studies show that this biomaterial has some potential to be used as a meniscal implant, however further pre-clinical tests must be accessed.

13.2.1.4 Polycarbonate-Urethane (PCU)

PCU is a tough polymer whose compliance can be customized by the mixture of hard and soft segments [24]. Due to his relatively low elastic modulus (10–100 MPa) it can potentially improve conformity and load distribution by permitting local material deformation [25]. Hereupon, polycarbonate-urethane possesses some features which make it a potential biomaterial to be used as a meniscal implant. PCU is durable, bio-stable and has excellent mechanical properties [26, 27].

Zur et al. [25] accessed the ability of a non-degradable, anatomically shaped artificial meniscal implant in a sheep model. This experience occurred in a time frame of 6 months to test the hypothesis that PCU implant could provide a chondroprotective effect on cartilage. The PCU implants remained well secured throughout the experimental period and showed no signs of wear or changes in structural or material properties providing a chondroprotective protection to the cartilage. That study provided preliminary evidence for the ability of an artificial PCU meniscal implant to delay or prevent osteoarthritic changes in knee joint following complete medial meniscectomy [25].

The viscoelastic properties of an improved PCU implant were accessed by measuring and characterization of the strain-rate response, after simulated use, by subjecting the implant to realistic joint loads [28]. This meniscus implant made of a compliant polycarbonate-urethane matrix reinforced with high modulus ultrahigh molecular weight polyethylene fibers can redistribute joint loads in a similar pattern to natural meniscus, without risking the integrity of the implant materials due to the optimal pressure distribution and similar shape of natural meniscus [8]. Regarding to the viscoelastic properties, the PCU implant was very flexible and able to deform relatively easily under low compressive loads (E = 120–200 kPa). Although, when the compressive load increase, the implant became stiffer (E = 3.8–5.2 MPa) to resist deformation. The meniscus implant behaved as a non-linear viscoelastic material. The meniscus implant appears well-matched to the viscoelastic properties of the natural meniscus, and importantly, these properties were found to remain stable and minimally affected by potentially degradative and loading conditions associated with long-term use [28].

The main synthetic non-resorbable polymers and outcome that have been explored as meniscus implants is summarized in Table 13.1.

Table 13.1 Summary of main studies concerning non-resorbable synthetic biomaterials

Biomaterial	Follow up	Model	Results	References
Pre-clinical trial				
PET (Dacron)	3 months	Rabbit	Dacron implant led to altered joint mechanics, osteophyte formation and synovitis caused by the non-normal biomechanics properties, the improper sizing and the impossibility to match the stress-relaxation of the native meniscus tissue	[12]
PTFE (Teflon)	3 months	Rabbit	The cartilage appeared similar to meniscus native tissue, however PTFE implant lost its shape after implantation and it is prone to wear, resulting in debris formation and synovitis PTFE implant lost its shape after implantation and it is prone to wear, resulting in debris formation and synovitis	[15, 16]
PVA-H	12 months	Sheep	Compared with meniscectomy, the PVA-H meniscal replacement resulted in a significantly decrease cartilage degeneration. Compared with meniscal allograft transplantation animals, after 4 months, the implant had significantly increased cartilage degeneration in the peripheral zone of the tibial plateau	[21]
PCU	6 months	Sheep	PCU implants remained well secured throughout the experimental period and showed no signs of wear or changes in structural or material properties providing a chondroprotective protection to the cartilage	[25]

13.3 Resorbable Polymers

13.3.1 Synthetic

13.3.1.1 Poly(α-Hydroxy Acids) (PHAs)

Poly(lactic acid) (PLA), poly(glycolic acid) (PGA), and their copolymer poly [(lactic acid)-*co*-(glycolic acid)] (PLGA), are part of a group of polymers called poly(α-hydroxy acids), that due to their well-controlled architecture, inter pore connectivity, and mechanical properties, are widely used in tissue engineering applications [29].

Testa Pezzin et al. [30] developed bioreabsorbable polymer scaffold made of poly (L-lactic acid) (PLLA) and poly(p dioxanone) (PPD) to be used as a temporary meniscal prosthesis to stimulate the formation of an in situ meniscal replication. That study was made in a rabbit model and revealed that PLLA has great potential to be used as a meniscal prosthesis, especially because, the new meniscus promoted a significant cartilage protection, allowed tissue ingrowth and induces the fibrocartilage [30].

Poly(lactic acid) has a long degradation time. Keeping this in mind, Esposito et al. [31] used the copolymer PLDLA [poly(L-co-D,L-lactic acid)] that has similar mechanical properties to poly(lactic acid) but with a short degradation time due to the greater degradation of poly(D,L lactic acid). Poly(caprolactone-triol) was added, yielding a more hydrophilic polymer that results in an improvement of the inter-action with cells and tissue. The results showed that the polymer biodegradability capability and the fact that allow formation fibrocartilaginous tissue, makes the PLDLA/PCL-T scaffolds very promising in the meniscus regeneration area [31].

Poly(glycolic acid) is a propitious synthetic material under investigation as a synthetic scaffold for meniscus tissue engineering. PGA is a biomaterial that is biocompatible and has the ability to load isolated fibrochondrocytes from menisci [32, 33]. A study by Kang et al. [26], using a meniscus-shaped PGA scaffold reinforced with PLGA (75:25) and seeded with fibrochondrocytes before implan-tation in rabbits, demonstrated that the regenerated neomenisci were similar to the native meniscus as well as zonal production of collagen I and II as seen in native menisci. However the tibial articular cartilage was not prevented and the novel meniscus tissue showed differences in collagen content and aggregate modulus in comparison with native meniscus [34].

Fox et al. [35] developed a fibroblast-like synoviocytes (FLS) strategy that consists in the seeding of these cells on PGA/PLLA scaffolds under the influence of growth factors. FLS were seeded onto synthetic scaffolds in a rotating bioreactor under the influence of three growth factor regimens: none, basic fibroblast growth factor (b-FGF) alone, and b-FGF plus transforming growth factor (TGF-b1) and insulin-like growth factor (IGF-1). The data suggests that this novel strategy may constitutively signal for production of type I collagen, and can be induced to signal for collagen II and aggrecan. This may prove favorable for in vitro fibrocartilage tissue engineering under appropriate conditions [35].

Another approach, developed by Freymann et al. [36], was the production of a three-dimensional (3-D) bioresorbable polymer graft made of PGA and hyaluronic acid. The scaffold material was shown to be biocompatible and retained its initial shape stability over the cultivation period of 3 weeks. The focus of the in vitro study is limited to a first proof of biocompatibility of the meniscal cells with the scaffold material, along with a first insight into the differentiation of cells and potential meniscus matrix formation [36].

The poly[(lactic acid)-co-(glycolic acid)] scaffold comes up when PLA was added to PGA scaffolds in order to improve its mechanical properties. Gu et al. used a PLGA scaffold loaded with autologous myoblasts and cultured in a chondrogenic medium for 14 days and tested it in a canine model. Comparing with cell-free PLGA scaffolds, the scaffolds loaded with myoblasts and pre-cultured in chondrogenic medium prior to the insertion into meniscal defects resulted in significantly better meniscal defect filling and meniscal regeneration. However, did not occur the full restoration of the surface area and the tissue quality of the normal meniscus [37].

Recently, Kwak et al. [38] test the hypothesis that platelet-rich plasma (PRP) pretreatment on a poly-lactic-*co*-glycolic acid (PLGA) mesh scaffold enhances the healing capacity of the meniscus with human chondrocyte-seeded scaffolds in vivo, even when the seeded number of cells was reduced from 10 million to one million. The results showed PRP exerts a positive stimulatory effect on attachment of human chondrocytes onto the PLGA mesh scaffold. This is a clinically relevant finding because the number of donor cells can be reduced and donor site morbidity may be minimized. The objective of this study was not to restore the native tissue architecture but to encourage healing of the injured meniscal tissue. Once again PLGA scaffold helps because is easily degradable in in vivo conditions within several weeks and can work as a cell-delivery device [38].

13.3.1.2 Polycaprolactone/Hyaluronic Acid (PCL/HYAFF)

Polycaprolactone (PCL), a semi-crystalline linear resorbable aliphatic polyester, is subjected to biodegradation because of the susceptibility of its aliphatic ester linkage to hydrolysis. PCL is regarded as a soft and hard-tissue compatible material including resorbable suture, drug delivery system, and recently bone graft substitutes [39].

Extensive in vitro and in vivo biocompatibility and efficacy studies have been performed, resulting in US Food and Drug Administration approval of a number of medical and drug delivery devices [40, 41].

Hyaluronic acid (HYAFF) is a biodegradable and biocompatible polymer that is a suitable substrate to grow a variety of cell types [42].

Chiari et al. developed a PCL/HYFF scaffold (Fig. 13.3) capable to be used as a partial or total meniscus substitute and was reinforced with circumferential poly-lactic acid (PLA) fibers and with a polyethylene terephthalate (PET) net respectively. The scaffold was introduced for meniscus replacement in a sheep model. The

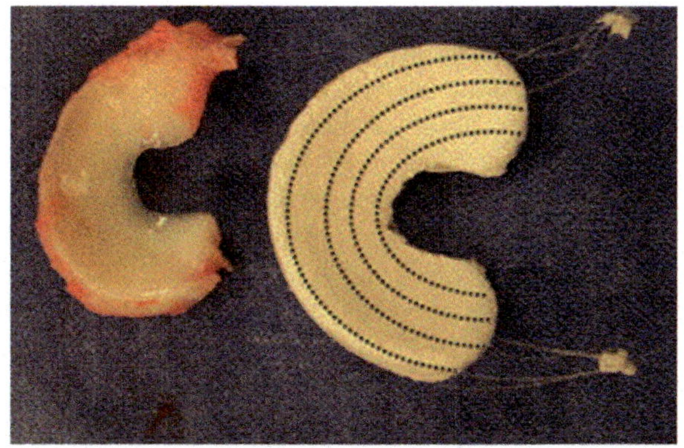

Fig. 13.3 The total meniscus implant is augmented with circumferential PLA fibers. Reprinted with permission from [43]. Copyright 2006, Elsevier

properties of the biomaterial in terms of tissue meniscus regeneration are promising: the implants remained in position, retained their shape, and showed adequate mechanical properties. However, compression of the implant led to extrusion, which mainly occurred in the posterior aspect. Complete ingrowth to the capsule and the formation of tissue between the implant and the original meniscus in the partial replacements, seen on gross inspection as well as histology, provided evidence of implant integration after 6 weeks [43].

In subsequent studies the implant show some issues. After 6 months, the PCL/HYAFF scaffold reveals insufficient biomechanics properties and fixation related problems [44]. Kon et al. [44] also demonstrated that, after 12 months, occurred an implant dislocation, a sight extrusion and wrinkling of the scaffold in the posterior region. Despite being obtained some promising results, damage to articular cartilage was not prevented and the biomechanical properties were not accessed in the subsequent studies.

13.3.1.3   Polyurethanes (PU)

Polyurethane is a polymer that is formed by reacting a diisocyanate or a polymeric isocyanate (hard segment) with a polyol (soft segment) in the presence of suitable catalysts and additives. The wide range of possibilities in the components that can be used to produce PU, gives the possibility to produce a broad spectrum of materials to meet the needs of specific applications.

Polyurethane scaffolds shown its biodegradability, biocompatibility and elasticity in a couple of studies. The studies that used poly(urethane)-poly(L-lactide) (PU-PLLA) and PU scaffolds (Estane 5701F) attempted to understand the potential of these implants in meniscus reconstruction. The authors used a canine model and realized that occurred a good tissue ingrowth due to the porosity of the PU scaffold. In conclusion, the repair of meniscal lesions in the avascular part of the meniscus is possible when a porous PU implant is used to guide repair tissue and blood vessels from the vascular peripheral part of the meniscus into the defect [45, 46].

Tienen et al. [47] accessed if an improved PU scaffold (Estane 5701F) was able to protects the cartilage from degeneration and develop into a neomeniscus. The implant was completely filled with fibrovascular tissue after 3 months and, after 6 months was observed cartilage-like tissue in the central areas of the implant. However, the PU scaffold was not able to mimic the mechanical properties of the native meniscus tissue and prevent the articular cartilage damage [47]. Afterwards, due to some carcinogenic problems with the Estane implant, Heijkants et al. developed a polyurethane scaffold with polycaprolactone (PCL) as soft segment and 1,4-butanediisocyanate and 1,4-butanediol as uniform hard segments, resulting in better compressive characteristics [48]. Despite the lack of prevention on the articular cartilage damage, this new PU scaffold showed a faster tissue ingrowth 2 weeks after implantation [49].

A subsequent study, with a bigger timeframe (24 months) demonstrated the previous issues [50]. Meanwhile, the polyurethane implant, as it can be seen in Fig. 13.4, became commercialized and named Actifit (Orteq Ltd, London, United Kingdom). Therefore, Brophy et al. tried a different approach, and used the PU scaffold for partial meniscal replacement, instead of total meniscal replacement. First, a study was made in a cadaver model and reveals that polyurethane scaffold improved knee-contact mechanics. Further, an in vivo study demonstrated that there were no differences in the prevention of articular cartilage damage between scaffold and partial meniscectomy [51, 52].

Actifit astoundingly managed to reach clinical trials. A set of clinical trials using Actifit was made over the past 5 years. The study of Verdonk et al. revealed successful tissue ingrowth and biocompatibility and, for the first time, occurred a consistent regeneration of tissue when using an acellular polyurethane scaffold to treat irreparable partial meniscus tissue lesions [53]. Further, a clinical case using 18 patients, proved again, that Actifit increases and promotes the meniscal regeneration by normal chondrocytes and fibrochondrocytes and reduces the risk of progression to knee osteoarthritis [10]. Bouyarmane et al. [54] carried out a study,

Fig. 13.4 Actifit: appearance of the implant to the naked eye

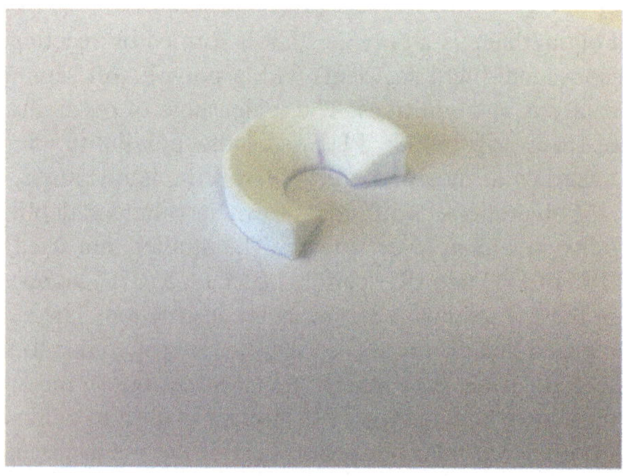

involving 54 patients, and accessed the clinical outcomes at 24 months follow-up. Clinical results of this study demonstrate clinically and statistically significant improvements of pain and function scores and showed that Actifit scaffold is safe and effective in treating lateral meniscus defects [54]. A longer clinical trial with a timeframe of 48 months and involving 18 patients was made by Schuttler et al. [55] Until this clinical case, no data concerning a longer follow-up than 2 years for Actifit scaffold are available, and, once again, the study revealed that polyurethane meniscal scaffold in patients with chronic segmental medial meniscus deficiency is not only a safe procedure but leads to good clinical results at a 4-year follow-up [9, 55]. Recently, Gelber et al. [56] tried to evaluate the influence of different degrees of articular chondral injuries on the imaging aspect of Actifit scaffold. As expected, patients without chondral injuries showed a better MRI aspect of the polyurethane scaffold in terms of size and morphology and the implant provided also significant pain relief and functional improvement regardless the presence of advanced cartilage injuries after a 2 follow-up [56].

The main synthetic resorbable polymers and outcomes that have been explored as meniscus implants is summarized in Table 13.2.

Table 13.2 Summary of main studies concerning resorbable synthetic biomaterials

Biomaterial	Follow up	Model	Results	References
Clinical trial				
PU (Actifit)	24 months	Human	Significant improvements of pain and function scores; scaffold is safe and effective in treating lateral meniscus defects	[54]
PU (Actifit)	24 months	Human	Safe implant procedure and leads to good clinical results	[9]
PU (Actifit)	24 months	Human	Inducing and promoting meniscal regeneration by normal chondrocytes and fibrochondrocytes and decreases the risk of progression to knee osteoarthritis	[10]
Pre-clinical trial				
PCL/HYAFF	12 months	Sheep	Osteoarthritis was less in cell-seeded group than in meniscectomy group, however occurred an implant dislocation, a sight extrusion and wrinkling of the scaffold in the posterior region	[44]
PGA	36 weeks	Rabbit	Reported proteoglycan types I and II collagen in neomenisci, however there was some differences in collagen content and aggregate modulus in	[26]

(continued)

Table 13.2 (continued)

Biomaterial	Follow up	Model	Results	References
			comparison with native meniscus	
PLDLA/PCL-T	24 weeks	Rabbit	Without apparent rejection, infection, or chronic inflammatory response and good integration to native tissue	[31]
PLGA	12 weeks	Canine	Fibrocartilage formation with hyaline-like regions with collagen I, II and aggrecan production.	[37]
In vitro				
PGA-PLLA	6 weeks	Equine	This novel strategy may prove favorable for in vitro fibrocartilage tissue engineering under appropriate conditions No integration of cell-scaffold construct to meniscal tissue. No measurable collagen or GAGs.	[35]
PGA-Hyaluronan	3 weeks	Human	Increase in matrix protein expression compared with control. Decrease in collagen X expression for all groups. Suggested redifferentiation of meniscus cells by scaffold.	[36]

13.3.2 Natural

13.3.2.1 Silk

Silk is a natural resorbable material that has been applied in meniscal tissue engineering. Silks, as one group of fibrous proteins, are produced by a wide range of insects and spiders and can consist of helical, β-sheet (the chain axis is parallel to the fiber axis) or cross-β-sheet (the chain axis is perpendicular to the fiber axis) secondary structures, depending on the organism [57]. Its versatility allied to its biocompatibility, good biomechanical properties and controlled durability make silk a suitable natural biomaterial for the production of scaffolds.

Mandal et al. [58] used silk fibroin from Bombyx mori silkworm cocoons to develop a multilayered, multiporous scaffold with the aim to mimic native meniscal architecture and morphology. In that study, human primary chondrocytes were seeded into the inside of the silk scaffold and human primary fibroblasts were seeded into the outside. The study showed that the constructs increased cellularity and provided the support to form an extra cellular matrix similar to native tissue.

The compressive modulus and tensile modulus increased with time, however, they remained inferior to those of the native meniscus [58]. The same authors tried to use bone marrow stem cells as an alternative approach. The silk scaffold had behavior similar to the previous study [59]. The poor mechanical properties of the silk scaffolds are the biggest obstacle and, with that in mind, Yan et al. [60] showed that it is possible to improve mechanical and structural properties (Fig. 13.5). The results revealed that mechanical properties of the silk fibroin scaffolds increased with increasing silk fibroin concentration. The scaffolds presented favorable stability as their structure integrity, morphology and mechanical properties were maintained after in vitro degradation for 30 days. However, with the increase in the silk fibroin concentration, the scaffold porosity and interconnectivity decreasing which can be a problem in cell encapsulation process [60].

Gruchenberg et al. [61] performed an in vivo trial using a sheep model. The silk scaffold caused no inflammatory reaction in the joint 6 months postoperatively, and there were no significant differences in cartilage degeneration between the scaffold and sham groups. The compressive properties of the scaffold approached those of meniscal tissue and there is preliminary evidence of chondroprotective properties. However, the scaffolds were not always stably fixed in the defect, leading to gapping between implant and host tissue or to total loss of the implant [61].

13.3.2.2 Gelatin

Grogan et al. [62] present a three-dimensional methacrylated gelatin (GelMA) scaffolds patterned via projection stereolithography to emulate the circumferential alignment of cells in native meniscus tissue. That study showed that micropatterned GelMA scaffolds are non-toxic, provide organized cellular alignment, and promote meniscus-like tissue [62]. A couple of studies, added chitosan to gelatin to produce macroporous scaffolds. Han et al. [63] proved that gelatin/chitosan scaffolds are promising candidates for stem-cell-based tissue engineering. Sarem et al. [64], used the same biomaterials, to produce a scaffold that were prepared using genipin as a biocompatible crosslinker. The results showed that all composite scaffolds showed

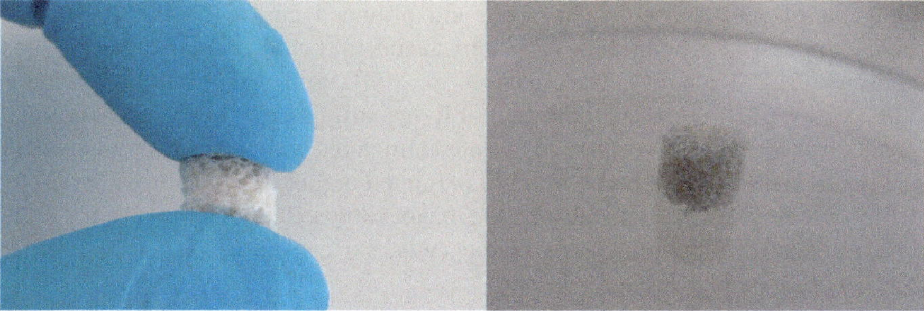

Fig. 13.5 Silk meniscus scaffold

favorable interaction with mixed population of meniscus derived cells, leading to a proliferation behavior that were directly correlating with the increase fraction of gelatin in the composite [64, 65].

An in vivo trial was made by Ishida et al. in order to accesses if platelet-rich plasma (PRP) combined with gelatin hydrogel (GH) scaffolds could enhance meniscal regeneration. Histological scoring of the defect sites at 12 weeks revealed significantly better meniscal repair in animals that received PRP with GH, which confirms the hypothesis proposed by the authors [66]. Narita et al. [67], also made an in vivo trial using a rabbit model. The aim of this study was to investigate the in vivo effects of gelatin hydrogels (GHs) incorporating fibroblast growth factor 2 (FGF-2) on meniscus repair. The results showed that GHs incorporating FGF-2 can significantly stimulate the cell proliferation and inhibited the death of meniscal cells until 4 weeks, thereby increasing meniscal cell density and enhancing meniscal repair [67].

13.3.2.3 Cellulose

Bacterial cellulose (BC) is an organic polysaccharide synthesized extracellularly as nanosized fibrils by the bacterium Gluconacetobacter xylinus bacterium. BC consists in a linear chain of several hundred to over ten thousand linked glucose units [68]. The in vivo biocompatibility of bacterial cellulose is good [69] and already showed that has potential as a scaffold for tissue engineering of cartilage [70]. Bodin et al. [71] accessed the potential of BC gel be used as a scaffold in meniscus tissue engineering. As advantages, BC scaffold is almost inexpensive, can be produced in a meniscus shape and promotes the cell migration. The results showed that compression modulus of the BC gel (1.8 kPa) was five times better than that of the collagen meniscal implant (0.23 kPa). However, it was inferior to the native pig meniscus (21 kPa). Further tests should be done to investigate the characteristics of the material [71].

A cellulose perforated by micro-channels has been developed by Martinez et al. [72] as a potential future scaffold material for meniscus implants. The BC scaffold was seeded with mouse fibroblasts and compared dynamic compression to that of a static culture. The results showed that the microchannel structure facilitated the alignment of cells and collagen fibers and provided guided tissue growth. This guiding process was important to obtain an ultrastructure mimicking that of the meniscus [72]. Recently, Markstedt et al. [73] used nanofibrillated cellulose (NFC) with alginate for 3D bioprinting of living soft tissue with cells. In this study, the authors were able to produce 3D printed structures with meniscus shape and the results showed that NFC-based scaffold exhibited a good cell viability after 7 days of 3D culture [73]. As conclusion, the nanocellulose-based bioink is a suitable hydrogel for 3D bioprinting with living cells and can be subsequently used for meniscus tissue engineering.

13.3.2.4 Collagen

Resorbable collagen meniscus scaffolds are made from processed bovine Achilles tendon tissue from which type I collagen fibers are extracted and later crosslinked with glutaraldehyde to form a matrix-like scaffold material. The resulting product is a flexible disk that can be trimmed and shaped to fit a meniscus defect [74]. The Collagen Meniscus Implant (CMI, Ivy Sports Medicine GmbH, Gräfelfing, Germany, Fig. 13.6) is the first biologic scaffold used for the treatment of partial meniscal deficiencies [75].

Several clinical trials were made in the last few years. Like all scaffolds, CMI has different goals in meniscus tissue engineering. According to the literature [76], the resorption of the collagen scaffold results in a decrease of the final meniscus volume in terms of implant-new tissue complex. Steadman and Rodkey, showed that occurs a progressive invasion of the CMI with meniscus cells after 6 and 12 months [77]. Another study from Rodkey et al., revealed similar results to the previous study occurring a continuous and uniform fibrocartilaginous matrix formation [11]. This newly formed tissue can counterbalance the resorption of the meniscus implant ensuring restoration of the normal volume of meniscus tissue. In terms of prevention further degenerative changes associated with the loss of meniscus tissue, Zaffagnini et al. [78] results noted only a few patients with degenerative changes 120 months after collagen scaffold implantation. Monllau et al., had the same results of the previous study, only a few patients had degenerative changes after CMI implantation. That study showed that meniscus implantation provided significant pain relief and functional improvement after a minimum 10-year follow-up period [79]. In comparison with partial meniscectomy, some studies showed that CMI provided an improvement in the clinical outcomes. Rodkey et al. [11] showed that only patients with chronic injuries were found to have regained a significantly greater proportion of their lost activity compared to patients that were submitted to a partial medial meniscectomy alone. In this study, the collagen scaffold revealed a biomechanical stability for more than 5 years [11]. On the other hand, Zaffagnini et al. [80] showed that collagen meniscus implantation resulted in improved pain, activity level, and radiologic outcomes after a minimum 10-year follow-up period when compared with

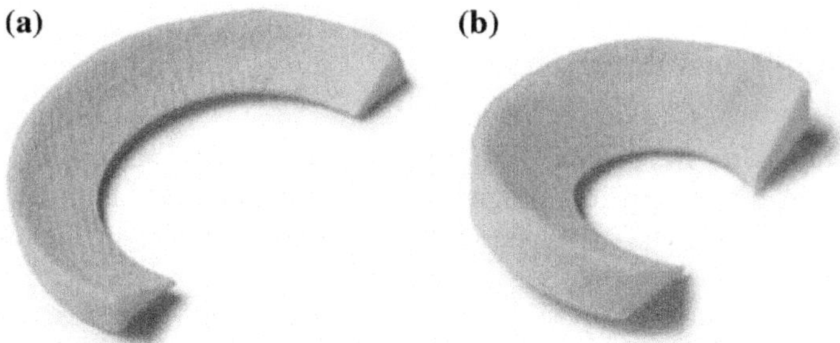

Fig. 13.6 CMI for the lateral meniscus (**a**) and CMI for the medial meniscus (**b**)

Table 13.3 Summary of main studies concerning resorbable natural biomaterials

Biomaterial	Follow up	Model	Results	References
Clinical trial				
Collagen (CMI)	10 years	Human	Significant pain relief and functional improvement followed by no development or progression of degenerative knee joint disease in most of the cases	[79]
Collagen (CMI)	10 years	Human	Pain, activity level, and radiological outcomes are significantly improved	[80]
Pre-clinical trial				
Silk	6 months	Sheep	The compressive properties of the scaffold approached those of meniscal tissue. However, the scaffolds were not always stably fixed in the defect	[61]
Gelatin	12 weeks	Rabbit	Gelatin hydrogel scaffold can significantly stimulate the cell proliferation and inhibited the death of meniscal cells until 4 weeks, thereby increasing meniscal cell density and enhancing meniscal repair	[67]
In vitro				
Cellulose	28 days		Scaffold microchannels facilitated the alignment of cells and collagen fibers providing guided tissue growth	[72]

partial medial meniscectomy alone. This study, unlike the previous one, had improvements in all patients with meniscus injuries (chronic and acute), not only in chronic injuries [80].

As summarized in Table 13.3, from natural based polymers, collagen is the unique natural biomaterial that has reached clinical trials stage. Collagen meniscal implants offer good clinical and structural outcomes providing higher clinical outcomes compared with partial meniscectomy alone.

13.4 Final Considerations

The research area of meniscus tissue engineering has achieved tremendous advancements, but there is still progress to be made to move with most developed biomaterials into clinical trial phase. Different biomaterials in meniscus tissue engineering were addressed as a potential meniscus substitute/scaffold with promising results in pre-clinical stages. Many authors developed scaffolds using different biomaterials that possess many attractive characteristics and behaviors. Regarding synthetic non-resorbable materials, the most important features is that the implant is biocompatible, intrinsically stable, safe, and mimic the biomechanical properties of the native meniscus. Concerning synthetic resorbable materials, the outcomes are more complex. In this type of materials it is important to access the tissue ingrowth and the degradation profile of the material. It is noteworthy that most of the literature has focused on biodegradable tissue engineered strategies, leaving a little aside the search for a permanent implant meniscus. As mentioned above, the only biomaterials that reached clinical trials as meniscal substitute/scaffold were two resorbable materials (Polyurethane and Collagen). However, any scaffold has fully mimic the physical and chemical properties of native meniscus and scaffold fixation still presents a big challenge in meniscus tissue engineering area. Future research trends should focus on optimizing the processing routes to better tune the mechanical and chemical properties of meniscus substitute/scaffold and produce patient-specific implants, and in particular using natural-based biomaterials, in order to provide a better treatment for meniscus lesions.

Acknowledgments Authors acknowledge the financial support from FCT/MCTES (Fundação para a Ciência e a Tecnologia/Ministério da Ciência, Tecnologia, e Ensino Superior) and the Fundo Social Europeu através do Programa Operacional do Capital Humano (FSE/POCH), PD/59/2013. J.M. Oliveira also thanks to the Portuguese Foundation for Science and Technology (FCT) for the funds provided to under the program Investigador FCT 2012 (IF/00423/2012)

References

1. Scotti C, Hirschmann MT, Antinolfi P, Martin I, Peretti GM (2013) Meniscus repair and regeneration: review on current methods and research potential. Eur Cells Mater 26:150–170
2. Baker BE, Peckham AC, Pupparo F, Sanborn JC (1985) Review of meniscal injury and associated sports. Am J Sports Med 13(1):1–4
3. Hede A, Jensen DB, Blyme P, Sonne-Holm S (1990) Epidemiology of meniscal lesions in the knee. 1,215 open operations in Copenhagen 1982–84. Acta Orthop Scand 61(5):435–437
4. Khetia EA, McKeon BP (2007) Meniscal allografts: biomechanics and techniques. Sports Med Arthrosc Rev 15(3):114–120. doi:10.1097/JSA.0b013e3180dca217
5. Longo UG, Loppini M, Forriol F, Romeo G, Maffulli N, Denaro V (2012) Advances in meniscal tissue engineering. Stem Cells Int 2012:11. doi:10.1155/2012/420346
6. Guo W, Liu S, Zhu Y, Yu C, Lu S, Yuan M, Gao Y, Huang J, Yuan Z, Peng J, Wang A, Wang Y, Chen J, Zhang L, Sui X, Xu W, Guo Q (2015) Advances and prospects in tissue-engineered meniscal scaffolds for meniscus regeneration. Stem Cells Int 2015:13. doi:10.1155/2015/517520
7. Rongen JJ, van Tienen TG, van Bochove B, Grijpma DW, Buma P (2014) Biomaterials in search of a meniscus substitute. Biomaterials 35(11):3527–3540. doi:10.1016/j.biomaterials. 2014.01.017
8. Elsner JJ, Portnoy S, Zur G, Guilak F, Shterling A, Linder-Ganz E (2010) Design of a free-floating polycarbonate-urethane meniscal implant using finite element modeling and experimental validation. J Biomech Eng 132(9):095001. doi:10.1115/1.4001892
9. Schuttler KF, Pottgen S, Getgood A, Rominger MB, Fuchs-Winkelmann S, Roessler PP, Ziring E, Efe T (2015) Improvement in outcomes after implantation of a novel polyurethane meniscal scaffold for the treatment of medial meniscus deficiency. Knee Surg Sports Traumatol Arthrosc 23(7):1929–1935. doi:10.1007/s00167-014-2977-6

10. Baynat C, Andro C, Vincent JP, Schiele P, Buisson P, Dubrana F, Gunepin FX (2014) Actifit synthetic meniscal substitute: experience with 18 patients in Brest, France. Orthop Traumatol Surg Res 100(8 Suppl):S385–S389. doi:10.1016/j.otsr.2014.09.007

11. Rodkey WG, DeHaven KE, Montgomery WH 3rd, Baker CL Jr, Beck CL Jr, Hormel SE, Steadman JR, Cole BJ, Briggs KK (2008) Comparison of the collagen meniscus implant with partial meniscectomy. A prospective randomized trial. J Bone Joint Surg 90(7):1413–1426. doi:10.2106/jbjs.g.00656

12. Sommerlath K, Gillquist J (1992) The effect of a meniscal prosthesis on knee biomechanics and cartilage. An experimental study in rabbits. Am J Sports Med 20(1):73–81

13. Odian G (2004) Principles of polymerization. Wiley, London

14. Sommerlath K, Gallino M, Gillquist J (1992) Biomechanical characteristics of different artificial substitutes for rabbit medial meniscus and effect of prosthesis size on knee cartilage. Clin Biomech (Bristol Avon) 7(2):97–103. doi:10.1016/0268-0033(92)90022-v

15. Messner K, Lohmander LS, Gillquist J (1993) Cartilage mechanics and morphology, synovitis and proteoglycan fragments in rabbit joint fluid after prosthetic meniscal substitution. Biomaterials 14(3):163–168. doi:10.1016/0142-9612(93)90018-W

16. Messner K, Gillquist J (1993) Prosthetic replacement of the rabbit medial meniscus. J Biomed Mater Res 27(9):1165–1173. doi:10.1002/jbm.820270907

17. Noguchi T, Yamamuro T, Oka M, Kumar P, Kotoura Y, Hyon S, Ikada Y (1991) Poly(vinyl alcohol) hydrogel as an artificial articular cartilage: evaluation of biocompatibility. J Appl Biomater 2(2):101–107. doi:10.1002/jab.770020205

18. Anseth KS, Bowman CN, Brannon-Peppas L (1996) Mechanical properties of hydrogels and their experimental determination. Biomaterials 17(17):1647–1657

19. Yamaoka H, Asato H, Ogasawara T, Nishizawa S, Takahashi T, Nakatsuka T, Koshima I, Nakamura K, Kawaguchi H, Chung UI, Takato T, Hoshi K (2006) Cartilage tissue engineering using human auricular chondrocytes embedded in different hydrogel materials. J Biomed Mater Res Part A 78(1):1–11. doi:10.1002/jbm.a.30655

20. Kobayashi M, Chang YS, Oka M (2005) A two year in vivo study of polyvinyl alcohol-hydrogel (PVA-H) artificial meniscus. Biomaterials 26(16):3243–3248. doi:10.1016/j.biomaterials.2004.08.028

21. Kelly BT, Robertson W, Potter HG, Deng XH, Turner AS, Lyman S, Warren RF, Rodeo SA (2007) Hydrogel meniscal replacement in the sheep knee: preliminary evaluation of chondroprotective effects. Am J Sports Med 35(1):43–52. doi:10.1177/0363546506292848

22. Holloway JL, Lowman AM, Palmese GR (2010) Mechanical evaluation of poly(vinyl alcohol)-based fibrous composites as biomaterials for meniscal tissue replacement. Acta Biomater 6(12):4716–4724. doi:10.1016/j.actbio.2010.06.025

23. Holloway JL, Lowman AM, VanLandingham MR, Palmese GR (2014) Interfacial optimization of fiber-reinforced hydrogel composites for soft fibrous tissue applications. Acta Biomater 10(8):3581–3589. doi:10.1016/j.actbio.2014.05.004

24. Scholes SC, Unsworth A, Jones E (2007) Polyurethane unicondylar knee prostheses: simulator wear tests and lubrication studies. Phys Med Biol 52(1):197–212. doi:10.1088/0031-9155/52/1/013

25. Zur G, Linder-Ganz E, Elsner JJ, Shani J, Brenner O, Agar G, Hershman EB, Arnoczky SP, Guilak F, Shterling A (2011) Chondroprotective effects of a polycarbonate-urethane meniscal implant: histopathological results in a sheep model. Knee Surg Sports Traumatol Arthrosc 19(2):255–263. doi:10.1007/s00167-010-1210-5

26. Khan I, Smith N, Jones E, Finch DS, Cameron RE (2005) Analysis and evaluation of a biomedical polycarbonate urethane tested in an in vitro study and an ovine arthroplasty model. Part II: in vivo investigation. Biomaterials 26(6):633–643. doi:10.1016/j.biomaterials.2004.02.064

27. Bigsby RJA, Auger DD, Jin ZM, Dowson D, Hardaker CS, Fisher JA (1998) Comparative tribological study of the wear of composite cushion cups in a physiological hip joint simulator. J Biomech 31(4):363–369. doi:10.1016/S0021-9290(98)00034-7

28. Shemesh M, Asher R, Zylberberg E, Guilak F, Linder-Ganz E, Elsner JJ (2014) Viscoelastic properties of a synthetic meniscus implant. J Mech Behav Biomed Mater 29:42–55. doi:10.1016/j.jmbbm.2013.08.021

29. Ma PX, Choi JW (2001) Biodegradable polymer scaffolds with well-defined interconnected spherical pore network. Tissue Eng 7(1):23–33. doi:10.1089/107632701300003269

30. Testa Pezzin AP, Cardoso TP, do Carmo Alberto Rincon M, de Carvalho Zavaglia CA, de Rezende Duek EA (2003) Bioreabsorbable polymer scaffold as temporary meniscal prosthesis. Artif Organs 27(5):428–431

31. Esposito AR, Moda M, Cattani SM, de Santana GM, Barbieri JA, Munhoz MM, Cardoso TP, Barbo ML, Russo T, D'Amora U, Gloria A, Ambrosio L, Duek EA (2013) PLDLA/PCL-T scaffold for meniscus tissue engineering. BioResearch 2(2):138–147. doi:10.1089/biores.2012.0293

32. Ibarra C, Jannetta C, Vacanti CA, Cao Y, Kim TH, Upton J, Vacanti JP (1997) Tissue engineered meniscus: a potential new alternative to allogeneic meniscus transplantation. Transpl Proc 29(1–2):986–988

33. Aufderheide AC, Athanasiou KA (2005) Comparison of scaffolds and culture conditions for tissue engineering of the knee meniscus. Tissue Eng 11(7–8):1095–1104. doi:10.1089/ten.2005.11.1095

34. Kang SW, Son SM, Lee JS, Lee ES, Lee KY, Park SG, Park JH, Kim BS (2006) Regeneration of whole meniscus using meniscal cells and polymer scaffolds in a rabbit total meniscectomy model. J Biomed Mater Res Part A 78(3):659–671. doi:10.1002/jbm.a.30904

35. Fox DB, Warnock JJ, Stoker AM, Luther JK, Cockrell M (2010) Effects of growth factors on equine synovial fibroblasts seeded on synthetic scaffolds for avascular meniscal tissue engineering. Res Vet Sci 88(2):326–332. doi:10.1016/j.rvsc.2009.07.015

36. Freymann U, Endres M, Neumann K, Scholman HJ, Morawietz L, Kaps C (2012) Expanded human meniscus-derived cells in 3-D polymer-hyaluronan scaffolds for meniscus repair. Acta Biomater 8(2):677–685. doi:10.1016/j.actbio.2011.10.007

37. Gu Y, Zhu W, Hao Y, Lu L, Chen Y, Wang Y (2012) Repair of meniscal defect using an induced myoblast-loaded polyglycolic acid mesh in a canine model. Exp Ther Med 3(2):293–298. doi:10.3892/etm.2011.403

38. Kwak HS, Nam J, J-h L, Kim HJ, Yoo JJ (2014) Meniscal repair in vivo using human chondrocyte-seeded PLGA mesh scaffold pretreated with platelet-rich plasma. J Tissue Eng Regener Med. doi:10.1002/term.1938

39. Kweon H, Yoo MK, Park IK, Kim TH, Lee HC, Lee HS, Oh JS, Akaike T, Cho CS (2003) A novel degradable polycaprolactone networks for tissue engineering. Biomaterials 24(5):801–808

40. Bezwada RS, Jamiolkowski DD, Lee IY, Agarwal V, Persivale J, Trenka-Benthin S, Erneta M, Suryadevara J, Yang A, Liu S (1995) Monocryl suture, a new ultra-pliable absorbable monofilament suture. Biomaterials 16(15):1141–1148

41. Darney PD, Monroe SE, Klaisle CM, Alvarado A (1989) Clinical evaluation of the Capronor contraceptive implant: preliminary report. Am J Obstet Gynecol 160(5 Pt 2):1292–1295

42. Milella E, Brescia E, Massaro C, Ramires PA, Miglietta MR, Fiori V, Aversa P (2002) Physico-chemical properties and degradability of non-woven hyaluronan benzylic esters as tissue engineering scaffolds. Biomaterials 23(4):1053–1063

43. Chiari C, Koller U, Dorotka R, Eder C, Plasenzotti R, Lang S, Ambrosio L, Tognana E, Kon E, Salter D, Nehrer S (2006) A tissue engineering approach to meniscus regeneration in a sheep model. Osteoarthritis Cartilage 14(10):1056–1065. doi:10.1016/j.joca.2006.04.007

44. Kon E, Filardo G, Tschon M, Fini M, Giavaresi G, Marchesini Reggiani L, Chiari C, Nehrer S, Martin I, Salter DM, Ambrosio L, Marcacci M (2012) Tissue engineering for total meniscal substitution: animal study in sheep model–results at 12 months. Tissue Eng Part A 18(15–16):1573–1582. doi:10.1089/ten.TEA.2011.0572

45. Elema H, Groot JH, Nijenhuis AJ, Pennings AJ, Veth RPH, Klompmaker J, Jansen HWB (1990) Use of porous biodegradable polymer implants in meniscus reconstruction. 2)

Biological evaluation of porous biodegradable polymer implants in menisci. Colloid Polym Sci 268(12):1082–1088. doi:10.1007/bf01410673

46. Klompmaker J, Jansen HW, Veth RP, de Groot JH, Nijenhuis AJ, Pennings AJ (1991) Porous polymer implant for repair of meniscal lesions: a preliminary study in dogs. Biomaterials 12(9):810–816

47. Tienen TG, Heijkants RG, de Groot JH, Pennings AJ, Schouten AJ, Veth RP, Buma P (2006) Replacement of the knee meniscus by a porous polymer implant: a study in dogs. Am J Sports Med 34(1):64–71. doi:10.1177/0363546505280905

48. Heijkants RG, van Calck RV, De Groot JH, Pennings AJ, Schouten AJ, van Tienen TG, Ramrattan N, Buma P, Veth RP (2004) Design, synthesis and properties of a degradable polyurethane scaffold for meniscus regeneration. J Mater Sci Mater Med 15(4):423–427

49. Tienen TG, Heijkants RG, de Groot JH, Schouten AJ, Pennings AJ, Veth RP, Buma P (2006) Meniscal replacement in dogs. Tissue regeneration in two different materials with similar properties. J Biomed Mater Res B Appl Biomater 76(2):389–396. doi:10.1002/jbm.b.30406

50. Hannink G, van Tienen TG, Schouten AJ, Buma P (2011) Changes in articular cartilage after meniscectomy and meniscus replacement using a biodegradable porous polymer implant. Knee Surg Sports Traumatol Arthrosc 19(3):441–451. doi:10.1007/s00167-010-1244-8

51. Maher SA, Rodeo SA, Doty SB, Brophy R, Potter H, Foo L-F, Rosenblatt L, Deng X-H, Turner AS, Wright TM, Warren RF (2010) Evaluation of a porous polyurethane scaffold in a partial meniscal defect ovine model. Arthroscopy 26(11):1510–1519. doi:10.1016/j.arthro.2010.02.033

52. Brophy RH, Cottrell J, Rodeo SA, Wright TM, Warren RF, Maher SA (2010) Implantation of a synthetic meniscal scaffold improves joint contact mechanics in a partial meniscectomy cadaver model. J Biomed Mater Res Part A 92(3):1154–1161. doi:10.1002/jbm.a.32384

53. Verdonk R, Verdonk P, Huysse W, Forsyth R, Heinrichs EL (2011) Tissue ingrowth after implantation of a novel, biodegradable polyurethane scaffold for treatment of partial meniscal lesions. Am J Sports Med 39(4):774–782. doi:10.1177/0363546511398040

54. Bouyarmane H, Beaufils P, Pujol N, Bellemans J, Roberts S, Spalding T, Zaffagnini S, Marcacci M, Verdonk P, Womack M, Verdonk R (2014) Polyurethane scaffold in lateral meniscus segmental defects: clinical outcomes at 24 months follow-up. Orthop Traumatol Surg Res 100(1):153–157. doi:10.1016/j.otsr.2013.10.011

55. Schuttler KF, Haberhauer F, Gesslein M, Heyse TJ, Figiel J, Lorbach O, Efe T, Roessler PP (2015) Midterm follow-up after implantation of a polyurethane meniscal scaffold for segmental medial meniscus loss: maintenance of good clinical and MRI outcome. Knee Surg Sports Traumatol Arthrosc. doi:10.1007/s00167-015-3759-5

56. Gelber PE, Petrica AM, Isart A, Mari-Molina R, Monllau JC (2015) The magnetic resonance aspect of a polyurethane meniscal scaffold is worse in advanced cartilage defects without deterioration of clinical outcomes after a minimum two-year follow-up. Knee 22(5):389–394. doi:10.1016/j.knee.2015.01.008

57. Valluzzi R, Winkler S, Wilson D, Kaplan DL (2002) Silk: molecular organization and control of assembly. Philos Trans R Soc Lond B Biol Sci 357(1418):165–167. doi:10.1098/rstb.2001.1032

58. Mandal BB, Park SH, Gil ES, Kaplan DL (2011) Multilayered silk scaffolds for meniscus tissue engineering. Biomaterials 32(2):639–651. doi:10.1016/j.biomaterials.2010.08.115

59. Mandal BB, Park S-H, Gil ES, Kaplan DL (2011) Stem cell-based meniscus tissue engineering. Tissue Eng Part A 17(21–22):2749–2761. doi:10.1089/ten.tea.2011.0031

60. Yan LP, Oliveira JM, Oliveira AL, Caridade SG, Mano JF, Reis RL (2012) Macro/microporous silk fibroin scaffolds with potential for articular cartilage and meniscus tissue engineering applications. Acta Biomater 8(1):289–301. doi:10.1016/j.actbio.2011.09.037

61. Gruchenberg K, Ignatius A, Friemert B, von Lubken F, Skaer N, Gellynck K, Kessler O, Durselen L (2015) In vivo performance of a novel silk fibroin scaffold for partial meniscal replacement in a sheep model. Knee Surg Sports Traumatol Arthrosc 23(8):2218–2229. doi:10.1007/s00167-014-3009-2

62. Grogan SP, Chung PH, Soman P, Chen P, Lotz MK, Chen S, D'Lima DD (2013) Digital micromirror device projection printing system for meniscus tissue engineering. Acta Biomater 9(7):7218–7226. doi:10.1016/j.actbio.2013.03.020

63. Thein-Han WW, Saikhun J, Pholpramoo C, Misra RD, Kitiyanant Y (2009) Chitosan-gelatin scaffolds for tissue engineering: physico-chemical properties and biological response of buffalo embryonic stem cells and transfectant of GFP-buffalo embryonic stem cells. Acta Biomater 5(9):3453–3466. doi:10.1016/j.actbio.2009.05.012

64. Sarem M, Moztarzadeh F, Mozafari M (2013) How can genipin assist gelatin/carbohydrate chitosan scaffolds to act as replacements of load-bearing soft tissues? Carbohydr Polym 93(2):635–643. doi:10.1016/j.carbpol.2012.11.099

65. Sarem M, Moztarzadeh F, Mozafari M, Shastri VP (2013) Optimization strategies on the structural modeling of gelatin/chitosan scaffolds to mimic human meniscus tissue. Mater Sci Eng C Mater Biol Appl 33(8):4777–4785. doi:10.1016/j.msec.2013.07.036

66. Ishida K, Kuroda R, Miwa M, Tabata Y, Hokugo A, Kawamoto T, Sasaki K, Doita M, Kurosaka M (2007) The regenerative effects of platelet-rich plasma on meniscal cells in vitro and its in vivo application with biodegradable gelatin hydrogel. Tissue Eng 13(5):1103–1112. doi:10.1089/ten.2006.0193

67. Narita A, Takahara M, Sato D, Ogino T, Fukushima S, Kimura Y, Tabata Y (2012) Biodegradable gelatin hydrogels incorporating fibroblast growth factor 2 promote healing of horizontal tears in rabbit meniscus. Arthroscopy 28(2):255–263. doi:10.1016/j.arthro.2011.08.294

68. Brown RM, Willison JH, Richardson CL (1976) Cellulose biosynthesis in Acetobacter xylinum: visualization of the site of synthesis and direct measurement of the in vivo process. Proc Natl Acad Sci USA 73(12):4565–4569

69. Helenius G, Backdahl H, Bodin A, Nannmark U, Gatenholm P, Risberg B (2006) In vivo biocompatibility of bacterial cellulose. J Biomed Mater Res Part A 76(2):431–438. doi:10.1002/jbm.a.30570

70. Svensson A, Nicklasson E, Harrah T, Panilaitis B, Kaplan DL, Brittberg M, Gatenholm P (2005) Bacterial cellulose as a potential scaffold for tissue engineering of cartilage. Biomaterials 26(4):419–431. doi:10.1016/j.biomaterials.2004.02.049

71. Bodin A, Concaro S, Brittberg M, Gatenholm P (2007) Bacterial cellulose as a potential meniscus implant. J Tissue Eng Regen Med 1(5):406–408. doi:10.1002/term.51

72. Martinez H, Brackmann C, Enejder A, Gatenholm P (2012) Mechanical stimulation of fibroblasts in micro-channeled bacterial cellulose scaffolds enhances production of oriented collagen fibers. J Biomed Mater Res Part A 100(4):948–957. doi:10.1002/jbm.a.34035

73. Markstedt K, Mantas A, Tournier I, Martínez Ávila H, Hägg D, Gatenholm P (2015) 3D bioprinting human chondrocytes with nanocellulose-alginate bioink for cartilage tissue engineering applications. Biomacromolecules 16(5):1489–1496. doi:10.1021/acs.biomac.5b00188

74. Warth RJ, Rodkey WG (2015) Resorbable collagen scaffolds for the treatment of meniscus defects: a systematic review. Arthroscopy 31(5):927–941. doi:10.1016/j.arthro.2014.11.019

75. Rodkey WG, Stone K, Steadman J (1992) Prosthetic meniscal replacement. Biology and biomechanics of the traumatized synovial joint: the knee as a model. American Academy of Orthopaedic Surgeons, Rosemont, pp 221–231

76. Hansen R, Bryk E, Vigorita V (2013) Collagen scaffold meniscus implant integration in a canine model: a histological analysis. J Orthop Res 31(12):1914–1919. doi:10.1002/jor.22456

77. Steadman JR, Rodkey WG (2005) Tissue-engineered collagen meniscus implants: 5- to 6-year feasibility study results. Arthroscopy 21(5):515–525. doi:10.1016/j.arthro.2005.01.006

78. Zaffagnini S, Marcheggiani Muccioli GM, Bulgheroni P, Bulgheroni E, Grassi A, Bonanzinga T, Kon E, Filardo G, Busacca M, Marcacci M (2012) Arthroscopic collagen meniscus implantation for partial lateral meniscal defects: a 2-year minimum follow-up study. Am J Sports Med 40(10):2281–2288. doi:10.1177/0363546512456835

79. Monllau JC, Gelber PE, Abat F, Pelfort X, Abad R, Hinarejos P, Tey M (2011) Outcome after partial medial meniscus substitution with the collagen meniscal implant at a minimum of 10 years' follow-up. Arthroscopy 27(7):933–943. doi:10.1016/j.arthro.2011.02.018
80. Zaffagnini S, Marcheggiani Muccioli GM, Lopomo N, Bruni D, Giordano G, Ravazzolo G, Molinari M, Marcacci M (2011) Prospective long-term outcomes of the medial collagen meniscus implant versus partial medial meniscectomy: a minimum 10-year follow-up study. Am J Sports Med 39(5):977–985. doi:10.1177/0363546510391179

Chapter 14
Advanced Regenerative Strategies for Human Knee Meniscus

Ibrahim Fatih Cengiz, Joana Silva-Correia, Helder Pereira, João Espregueira-Mendes, Joaquim Miguel Oliveira and Rui Luís Reis

Abstract The meniscus tissue has important roles in the function and biomechanics of the knee. Despite the great advances in the treatment of meniscus lesions, the clinical need is still not fulfilled. To overcome the challenges of regeneration, tissue engineering-based strategies have been attempted with limited success. The process of meniscus tissue regeneration is very complex and has many parameters that are evident only to a certain degree. Today, the regenerative strategies have been advancing beyond the traditional tissue engineering concept by growing the utilization of the expertises of complementary areas that include, but not limited to, bioreactor engineering, bioprinting coupled to reverse engineering, biology, nanotechnology and gene therapy approaches. Herein, the recent reported advanced strategies involving bioreactors, self-assembling process, and somatic gene therapy for meniscus regeneration are overviewed.

I.F. Cengiz (✉) · J. Silva-Correia · H. Pereira · J.M. Oliveira · R.L. Reis
3B's Research Group—Biomaterials, Biodegradables and Biomimetics,
Headquarters of the European Institute of Excellence on Tissue Engineering
and Regenerative Medicine, University of Minho, Avepark—Parque de Ciência e
Tecnologia, Zona Industrial da Gandra, 4805-017 Barco GMR, Portugal
e-mail: fatih.cengiz@dep.uminho.pt

I.F. Cengiz · J. Silva-Correia · H. Pereira · J. Espregueira-Mendes · J.M. Oliveira · R.L. Reis
ICVS/3B's—PT Government Associate Laboratory, Braga/Guimarães, Portugal

H. Pereira · J. Espregueira-Mendes
Clínica Espregueira-Mendes F.C. Porto Stadium—FIFA Medical Centre of Excellence,
Porto, Portugal

H. Pereira
Orthopedic Department, Centro Hospitalar Póvoa de Varzim, Vila do Conde, Portugal

© Springer International Publishing AG 2017
J.M. Oliveira and R.L. Reis (eds.), *Regenerative Strategies for the Treatment of Knee Joint Disabilities*, Studies in Mechanobiology, Tissue Engineering and Biomaterials 21, DOI 10.1007/978-3-319-44785-8_14

14.1   Introduction

The knee meniscus has critical roles in the knee function [1, 2]. Until some time ago, meniscectomy had been performed as a standard treatment for meniscus lesions [3, 4]. Recognizing the consequences of meniscectomy [5, 6] the meniscus started to be preserved as much as it is possible. Clinical management of the lesions depends on many parameters about the lesion itself (e.g. extension of lesion, area of lesion) and the condition of the patient [4, 7]. Blood supply is critical, and it defines the healing potential within the meniscus [4]. The vascularity is limited to the outer 25–33 % of the meniscus [8], and avascular regions have low cellular density (Fig. 14.1). Therefore, the complete healing of meniscus to remain as a clinical challenge [9, 10].

To overcome the challenge in the clinics, several tissue engineering and regenerative medicine strategies that are aiming to repair or regenerate the meniscus tissue have been developed [10, 11]. These include the use of cellular scaffolds [12, 13], acellular scaffolds [14, 15], growth factors [16, 17], and bioreactors [18, 19], alone or in combination. There have been many progresses, but each strategy has its own shortcomings [11] and the ideal tissue engineered implant is not yet developed [20, 21]. A pioneering strategy has been proposed by Oliveira et al. [22] aiming to emulate the segmental vascularization of the tissue engineered meniscus. The combinatorial use of scaffolds and gellan gum hydrogels showed promising results in an *in ovo* model, as possibly the maintenance of cells' phenotype, and control of blood vessels infiltration into the implants. This is a major step-forward since excessive vascularization [15] is one of the causes of implant failure.

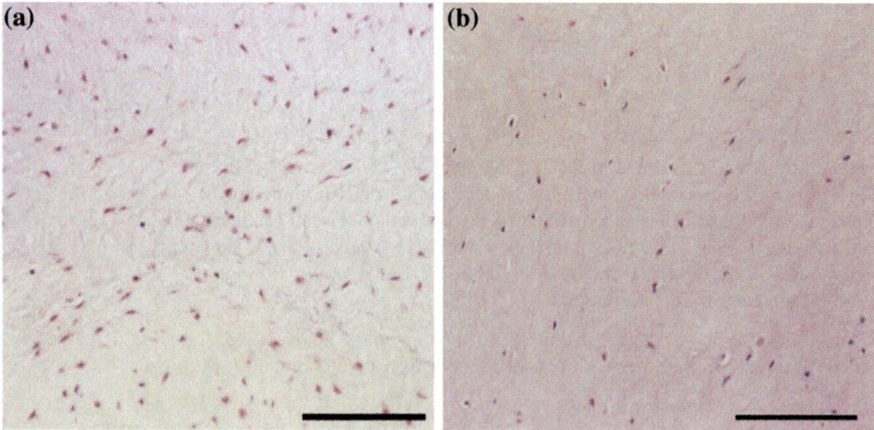

Fig. 14.1 Microscopy images of hematoxylin and eosin stained histological sections from vascular region (**a**) and avascular region (**b**) of human meniscus showing that the cellularity is much lower in the avascular region than in the vascular region. The *scale bars* indicate 100 μm

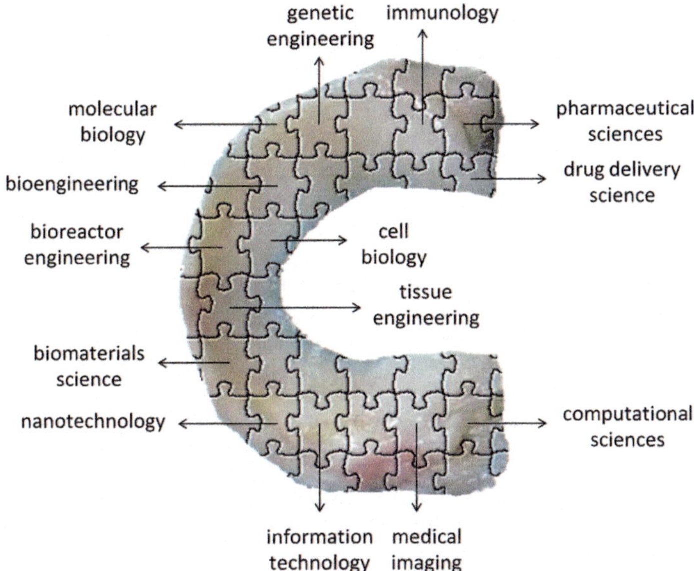

Fig. 14.2 It is a need to have different areas working together and contributing regenerative strategies to advance because of them is a unique piece of the whole. Today, the advanced strategies for meniscus regeneration use the expertise of more than one area, and in the future strategies many areas will be directly collaborating. These areas include, but not limited to, cell biology, molecular biology, biomaterials science, bioreactor engineering, tissue engineering, nanotechnology, bioengineering, genetic engineering, drug delivery science, immunology, pharmaceutical sciences, medical imaging, information technology, and computational science

Advanced regenerative strategies refer to the strategies that utilize the expertise of many areas with the aim of the tissue regeneration (Fig. 14.2). These areas include, but not limited to, cell biology, molecular biology, biomaterials science, bioreactor engineering, tissue engineering, nanotechnology, bioengineering, genetic engineering, drug delivery science, immunology, pharmaceutical sciences, medical imaging, information technology, and computational sciences.

A conceptual illustration of complex patient-specific all-in-one multi-science strategy for meniscus regeneration is depicted in Fig. 14.3. Briefly, it involves the use of autologous cells that are transduced with gene therapy either in a self-assembling process or in a scaffold-based tissue engineering approach, where in both cases a bioreactor can be used.

Regenerative strategies have been advancing beyond the traditional tissue engineering concept to develop better solutions where the native vascular architecture and tissue organization, namely the extracellular matrix and cellular distribution are mimicked, and the appropriate shape, size, and the biomechanical properties. In this chapter, strategies involving bioreactors, self-assembling process, and somatic gene therapy for meniscus regeneration are overviewed.

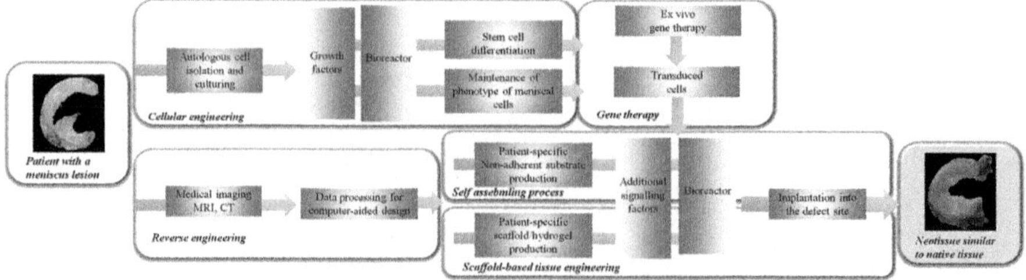

Fig. 14.3 A conceptual illustration of complex patient-specific all-in-one multi-science strategy for meniscus regeneration. Autologous cells are isolated from a piece of tissue obtained from the patient via biopsy. Based on the source, cells can be either stem cells or meniscal cells. In the case of stem cells, the cells will be differentiated into meniscal cells, whereas cells' phenotypes are maintained throughout the process with the use of growth factors and bioreactor in an adequate way. Cells are transduced with a well-established gene therapy protocol. The data obtained from medical imaging, for instance magnetic resonance imaging (MRI) or computed tomography (CT) is processed further either for non-adherent substrate the design or for scaffold/hydrogel design respectively for self-assembling process strategy or for scaffold based tissue engineering strategy. Generation of patient-specific anatomically correct neotissue is achieved by using one of the strategies in the presence of additional signals and bioreactor, and then it is implanted into the defect site of the patient

14.2 Bioreactors for Meniscus Tissue Regeneration

Meniscus tissue is under mechanical stimulation during its normal function in the body. Thus, the inclusion of mechanical stimulation appears to be necessary for proper neotissue formation. Bioreactors are specific equipments that can be employed in tissue regeneration strategies to mimic the native conditions for proper cell culturing and maintenance of cells phenotype [23–27]. Bioreactors can be designed to regulate the conditions in the culture media including mechanical stress conditions, supply of oxygen, pH, temperature, nutrients, and osmolality. Several kinds of bioreactors exist, such as spinner flasks [28], flow perfusion bioreactors [29], rotating wall vessels [30], and mechanical stimulation bioreactor [31]. Besides providing appropriate mechanical stimulation, bioreactors can allow uniform cell seeding and provide an improved mass transfer between the cells and the media.

Operating parameters of a bioreactor should be optimized for achieving the appropriate tissue regeneration conditions. With this aim, Neves et al. [29] showed that magnetic resonance imaging and spectroscopy can used as a tool for the optimization of a perfusion bioreactor that was used for meniscal cartilage constructs. Fox et al. [32] showed the potential of a rotating bioreactor for the cultivation of fibroblast-like synoviocyte-laden synthetic scaffolds with incorporation of growth factors avascular meniscus regeneration. In a study with rabbit knee meniscus fibrochondrocytes [33], it was reported that the use of rotating wall bioreactor did not provide a statistically significant increase in cell growth and matrix production. Similarly, in another study with perfused rotating wall vessel bioreactor [30], glycosaminoglycan (GAG) and collagen synthesis were not found to be significantly increased. However, its distribution within the construct was affected, and a bi-zonal structure could be obtained. It was reported that such a bioreactor system could potentially find application for meniscus generation, that is similar to the native tissue to some extent [30].

Gunja and Athanasiou [34] showed the advantageous effect of hydrostatic pressure on both biochemical and biomechanical features of cell-scaffold constructs for meniscus tissue engineering. In another study [35], it was proposed the culturing conditions combining TGF-β1 exposure and hydrostatic pressure cells maintenance. The authors reported the synergistic effects on the improvement of biomechanical and compositional features of the engineered meniscal constructs. The impacts of continuous perfusion and cyclic compression stimulation on collagen meniscus implants with human bone marrow stromal cells were investigated by Petri et al. [31]. That work has shown the enhancement of cells differentiation upon mechanical stimulation. In addition, cell proliferation was also stimulated by continuous perfusion [31].

Similarly, stirring-induced mixing in spinner flasks increases cell proliferation rate, the synthesis of GAG and collagen in generated meniscal cartilage constructs with sheep meniscus fibrochondrocytes [28]. Probably this is because the mixing approach enhances the transport of nutrients, growth factors, and metabolic wastes. The intensity of stirring does not have the same effect on different matrix components. To emulate the native meniscus, and to allow a better integration in the knee repair site, the culturing of meniscus tissue-engineered constructs shouldn't be longer than 7 days because the level of GAG increases only little after long periods of time, whereas the collagen level almost doubles [28]. Ballyns et al. [18] investigated the effects of media mixing either on the mechanical properties and extracellular matrix (ECM) synthesis of anatomically-shaped engineered meniscus constructs in a bioreactor. The main effect of mixing was reported to be the redistribution of ECM and improved homogeneity in constructs with large volume. On the other hand, mixing also caused DNA and ECM loss, and led to faster degradation of the scaffolds. With an optimal intensity of mixing, appropriate constructs can be developed [18].

Puetzer et al. [19] reported that effect of dynamic compression on the regulation of matrix synthesis by meniscal fibrochondrocytes is dependent on the load duration. Furthermore, GAG and collagen synthesis are affected differently. Load duration of 2 weeks was advantageous as 4 weeks for the formation of matrix, accumulation of collagen, and for the mechanical properties. Nevertheless, accumulation of GAG was negatively affected or not affected at all from the load duration. Under prolonged static culture conditions subsequent to the 2 weeks of loading, the maintenance and improvement of the functional properties can occur [19].

In a study [36], efficiency of centrifugal cell immobilization and dynamic oscillating motion were compared as cell seeding techniques using a low number of cells. The cell-scaffold constructs were used as implants to obtain cellular bonding between to two pieces of devitalized meniscus. Dynamic oscillating motion provided higher cell load and a more uniform distribution of cells. In dynamic oscillating technique, the cell-scaffold suspension is in a seeding tube, which is located on a platform that makes cyclic horizontal motion, oscillation. It was reported that the best seeding conditions were obtained by means of using 35–45 oscillations per minute [36]. Using a bioreactor which provides dynamic unconfined compression, McNulty et al. [37] showed that dynamic loading promotes the integrative repair in porcine meniscus explants in the presence of interleukin-1 by means of blocking the catabolic effects of interleukin-1.

In the study of Ballyns and Bonassar [38] investigated the effects of dynamic unconfined compression with a particular loading regime for three times a week. After 2 weeks of culturing, a significant increase in matrix accumulation was reported with a compression modulus 60–80 % of the native tissue. Prolonged dynamic compression caused loss of ECM to the culture, and a decrease in mechanical properties and GAG's levels [38]. In another study, Martinez et al. [39] used a dynamic compression bioreactor to cultivate fibroblast-laden micro-channeled cellulose scaffolds, and reported an increased collagen production as well as the alignment of collagen fibers and cells. Liu et al. [40] combined perfusion and on-off cyclic compression loading, and reported that this bioreactor system can be used for improving the functional properties of the cell-laden meniscal scaffold.

Chondrogenic stimuli and cyclic tensile loading were together reported to be useful in chondrogenesis of bone marrow stromal cells and fibrocartilage matrix development [41]. Further optimization of the parameters related to both tensile loading and chondrogenic stimuli would be beneficial for fibrocartilage tissue generation [41]. In another study [42], it was shown that static and dynamic compression differently regulate the RNA levels of certain matrix proteins in pig meniscus [42].

The pro-inflammatory gene expression in meniscus fibrochondrocytes can be regulated by biomechanical signals. It was reported that dynamic tensile strain suppressed the pro-inflammatory gene transcription in rat meniscus fibrochondrocytes [43].

The studies focused on meniscus are relatively more recent than that for other musculoskeletal tissues. For this reason, the use of bioreactors for meniscus regeneration is still at its infancy. With its own complex structure, meniscus possesses specific biomechanical characteristics [44]. The optimal stimulation regime for meniscus is to be determined by the contribution of many future studies.

14.3 Self-Assembling Process for Meniscus Tissue Regeneration

Tremendous amount of research has been performed in the field of tissue engineering and regenerative medicine, and scaffolding approaches are being used in most of these studies. The concept of advanced regeneration strategies has been remodeling, and nowadays it involves the self-assembling process. Self-assembling process is one of the scaffold-free methods that also include traditional methods (e.g. pellet culture, aggregate culture); and self-organization methods (e.g. bio-printing, and cell-sheet engineering) [45]. Athanasiou et al. [45] defined self-assembly as "a scaffoldless technology that produces tissues that demonstrate spontaneous organization without external forces". It is mediated by the minimization of free energy via cell–cell interactions [45], and the cells coalesce into a cohesive structure thus serving as each other's scaffold [46].

In a concise way, four sequential phases can be named for tissue formation: (1) seeding cells in a non-adherent substrate in high-density, (2) binding of cell adhesion receptors resulting minimization of free energy, (3) migration of cells and production of extracellular matrix, and (4) formation of distinct regional matrix and maturation of the tissue [45]. Hu and Athanasiou [46] stated that the self-assembling process, owing to being a scaffold-free method, can eliminate the possible complications associated with the use of scaffolds in tissue engineering, including the following: (1) biomaterial-induced inflammatory response, (2) toxicity of the degrading biomaterial, (3) stress-shielding, (4) loss of cell phenotype, (5) inhibition of cell migration, and (6) cell-to-cell communication, hindering the remodeling of extracellular matrix [46]. Accordingly, self-assembling process can provide highly desirable circumstances such as higher cell viability, biomimetic micro-environment and higher biocompatibility [45].

In a monolayer primary culture study, it was shown that rabbit meniscus cells can grow and form rapidly and uniformly cellular aggregates similar to the cellular nodules with three developmental stages [47]. These are: (1) formation of the cellular nodules between day 1 and day 3, (2) nodular growth, highest at day 5, and (3) nodular regression as of day 8. It was concluded that the fibrochondrocytes maintained its capacity to form chondro-like structures in vitro [47]. Hoben et al. [48, 49] showed the feasibility of fibrocartilage tissue formation with an extracellular matrix to some extent similar to native meniscus, despite the construct contraction, using the self-assembling process of a high-density co-culture of fibrochondrocytes and chondrocytes. In the self-assembling strategy of Aufderheide and Athanasiou [50], a ring-shaped well was used for a tensile force to be provided upon the self-assembly and the contraction of the construct, and also for the guidance of fiber orientation within the construct. Another feature of the mold had a slope at the bottom to emulate the normal meniscus cross-section. In this study, co-cultures of bovine meniscal fibrochondrocytes and articular chondrocytes were combined in different ratios. It was reported that different cell population ratio would affect the biomechanical, biochemical and morphology of the self-assembled meniscal construct [50].

MacBarb et al. [51] and Huey and Athanasiou [52] successfully incorporated catabolic enzyme chondroitinase-ABC and TGF-β1 with bovine meniscal and articular cartilage cells by means of using the self-assembling process. Maturation growth of meniscal neotissue was achieved by which the biomechanical properties were enhanced to some extent, and the concentration of collagen and GAG's were increased [52]. In another study of Huey and Athanasiou [53], the biochemical and biomechanical properties of the self-assembled anatomically-shaped meniscus constructs were aimed to be further enhanced through mechanical stimulation in addition to the aforementioned biochemical stimulation, i.e. chondroitinase-ABC and TGF-β1. A significant increase in biomechanical properties and collagen per net weight were reported upon combining biochemical and mechanical stimulation, i.e. in order to mimic the native meniscus loading conditions, both tensile and compression loading [53]. That study would be a guide for further studies that aim to develop successfully advanced strategies for meniscus regeneration.

14.4 Somatic Gene Therapy for Meniscus Tissue Regeneration

Orthopedics is one of the areas that somatic gene therapy has been attempted [54–59]. The concept of human gene therapy [57, 60, 61] is based on the transfer of exogenous genes or its complementary deoxyribonucleic acid (cDNA) into target somatic cells directly or by means of using viral or non-viral methods. These gene transfer methods could be applied in vivo or ex vivo. In an in vivo strategy, the genes are transferred directly into the target cells or tissue of the individual; whereas in ex vivo strategies the genes are first transferred into the isolated cells in culture, then the cells are modified and re-administered to the individual.

The interesting concept of gene therapy is the transfer of the genes encoding the relevant growth factors to the cells of the tissue to be treated. Like other tissues, meniscus is made of two main components: the cells, and the extracellular matrix that is synthesized by the cells. The functioning of the cells and its phenotype is greatly influenced by the biological and physical signals to which they are exposed. Growth factors are polypeptides that transmit biological signals by binding to the specific receptors that are on the surface or inside the cells [62, 63]. These molecules can affect the activity of the target cells in a specific way [64, 65]. The effects can be diverse, such as inhibition or stimulation of differentiation, proliferation, adhesion, migration, and gene expression of the cells, also affecting secretion and activation of other growth factors. The down-regulation of expression of lubricin that is a lubricating protein was reported in injured menisci as compared to the intact menisci [66]. Different signaling molecules are known to up- or down-regulate the expression of this protein [67]. In another study, it was reported that the cytokines interleukin-1 and tumor necrosis factor-α can inhibit the intrinsic repair response in porcine meniscus [68]. Accordingly, if the level of these pro-inflammatory cytokines is decreased, the repair response may be assisted in vivo [68]. For this reason, it can be concluded that once the effects of the signaling molecules are well defined, new strategies can be envisioned to up- or down-regulate their synthesis within the meniscus, alone or in combination with mechanical stimulation. In this way, both degeneration and regeneration process of meniscus tissue could be controlled.

With the use of gene therapy, the growth factors can be synthesized by the cells locally thus contributing to the healing/regeneration process of meniscus. The study of Goto et al. [69] is a leading work on this strategy for meniscus tissue in which the lacZ marker gene that expresses the ß-galactosidase enzyme was used. In that study, the genetic transduction was achieved in two ways by means of using two different vector types, either an adenovirus or a retrovirus. In the case of an adenovirus, a suspension of adenovirus having the gene and the blood was prepared, and then a clot was placed into the meniscus defect. In another approach, the allogenic meniscal cells carrying the gene retrovirally transduced were embedded into collagen gels and then introduced to the meniscus defect site. It was shown in vitro and in vivo that the expression of genes can be observed for several weeks [69]. Based on this innovative approach, transfer of the genes encoding the relevant growth factors can be effectively used to attain meniscus tissue regeneration.

Delivery of lacZ gene was also studied by Madry et al. [70]. It was shown that recombinant adeno-associated viruses having the gene can be used in vitro and in vivo to achieve efficient direct gene transfer to meniscus cells [70]. In another study, recombinant adeno-associated viruses were used in a direct gene delivery strategy in human meniscus with a lesion [71]. A human FGF-2 was delivered to human meniscal fibrochondrocytes in vitro, and also to healthy meniscus and to the lesions that were created in the human meniscus in situ [71]. The treatment did not increase the synthesis of major extracellular matrix components. Yet, there was a decrease in the matrix/DNA contents. However, there was an enhancement in the cell proliferation and a superior expression of the α-smooth muscle actin. It was concluded that the proposed treatment may bring favorable responses in meniscus defects [71]. Hidaka et al. [72] investigated the delivery of hepatocyte growth factor gene (AdHGF) using an adenovirus vector for its potential of inducing blood vessel formation within engineered meniscus. In that study, transduced meniscus cells were seeded onto a poly-glycolic acid scaffold and subcutaneously implanted in mice. It was reported that 2.5-fold more blood vessels were formed at 8 weeks post-transplantation by the transduced meniscus cells as compared to controls [72]. Martinek et al. [73] applied a gene transfer method to treat meniscus allograft pre-transplantation in a rabbit model. Rabbit meniscus allografts were transduced by retroviral vectors ex vivo. LacZ gene expression was detected in deeper layers of the menisci and at the menisco-synovial junction for up to two months post-transplantation. Pre-treatment of meniscus allografts provided a faster healing and remodeling. The authors showed immunocompatibility can also be improved [59]. In another gene delivery strategy, Steinert et al. [74] modified the meniscus cells and mesenchymal stem cells using adenoviral vectors encoding transforming growth factor (TGF)-β1 cDNA. The genetic modified cells were seeded into collagen type I-glycosaminoglycan scaffolds and cultured in vitro to evaluate their potential for the healing of tears located in the avascular region of the meniscus. It was reported that delivery of TGF-β1 cDNA increased cellularity and synthesis of GAG. It also strengthens the staining for proteoglycans and collagen type II, and expression of meniscal genes was also enhanced [74].

Bonadio [75] described and reviewed a local plasmid gene transfer technology and its potential use in the tissue engineering field. It is also known as gene activated matrix (GAM). It was reported that this strategy has been proven that it allows infection of the invading repair cells in situ. In the studies with animals, plasmid genes can be delivered to several tissues including bone, tendon, and ligament. A feature of this strategy is that the action mechanism of plasmid gene transfer is similar to the sequence of normal wound healing events [75]. Since GAM has been employed for several musculoskeletal tissues, it may also be possible to use for meniscus regeneration.

Gene transfer for meniscus tissue can be also achieved by using cells from other tissues as a gene delivery vector. In the study of Kasemkijwattane et al. [65] efficient gene transfer was possible by direct injection of the isolated myoblasts, which were transduced with adenovirus having the β-galactosidase expressing gene, at the meniscus injury. The underlying reasons to prefer myoblasts over meniscus cells for gene delivery were stated as the easiness of isolating the cells, higher efficiency of transduction, possibility of longer term gene expression, and possibility of isolating muscle derived stem cells that may be capable of differentiating into meniscus fibrochondrocytes [65]. Goto et al. [76] also showed that human and canine meniscus cells that were transduced with the TGF-β1 gene by retroviral vector can express the transgene and respond accordingly by significantly increasing the synthesis of the matrix components like collagens, non-collagenous proteins and proteoglycan [76].

It has been shown that an injectable ex vivo gene therapy strategy can be used to repair full-thickness avascular meniscal defects in a goat model [77]. Defects treated with the combination of calcium alginate gel and autologous bone marrow stromal cells with human insulin-like growth factor-1 (hIGF-1) gene transfection have also been reported. This strategy allowed repairing the defect with a white neotissue that is to some extent similar to the native meniscus tissue [77]. That work supported the idea that gene therapy can be an advanced and efficient strategy for meniscus regeneration. Despite the great advances in gene therapy, many issues thus remain to be solved, and in this way tissue engineering strategies could be combined with gene delivery strategies to achieve higher regenerative effects.

14.5 Final Remarks

Nowadays, the gold standard of meniscus regeneration still remains to be achieved. Most of the factors that affect the tissue regeneration are also unknown, and probably they will be discovered gradually with deeper studies in a near the future. One of the future trends in advanced tissue regeneration strategies will be focused on the development of patient-specific strategies. Patient-specific does not only mean the use of autologous cells (genetic engineered or not; and differentiated or undifferentiated), but also obtaining anatomically correct size and shape neo-tissue. With the use of medical imaging and computer-aided design, anatomically correct implants can be generated. The bioreactors have been in the area of interest to achieve appropriate cell culturing, and to preserve of cells phenotype. Even though, strong research efforts are targeting the development of appropriate scaffolds and hydrogels, and selection of best biomaterials, self-assembling process is an alternative to the traditional processing methodologies currently being used in tissue engineering.

Gene therapy has been moving forward from the conceptual idea to the clinics. Although there are limitations, challenges and pending issues, there are over two thousand gene therapy clinical trials for all indications globally (August 2014, http://www.wiley.com//legacy/wileychi/genmed/clinical/). Among those, cancer diseases have been addressed in more than half of the studies. However, there are only a few orthopedic gene therapy clinical trials. Nevertheless, once the challenges are overcome and the process is optimized, gene therapy may enhance the clinical management of tens of millions of orthopedic patients. Transfer of genes encoding the relevant growth factors into cells and optimization of culturing conditions by means of using bioreactors may bring great clinical achievements in meniscus regeneration.

The extremely complex and multi-parametric process of meniscus tissue regeneration is highly challenging with many unknowns. A deeper knowledge on the biology and biomechanics of native human meniscus including its anchoring to bone tissue is required. In addition, a broadened understanding of the intricate synergetic effects of the biological and physical signals would comprise one of the other challenges to be addressed in the scientific roadmap for "saving" the meniscus.

Acknowledgments The authors thank to the financial support of the MultiScaleHuman project (Contract number: MRTN-CT-2011-289897) in the Marie Curie Actions—Initial Training Networks. I. F. Cengiz thanks the Portuguese Foundation for Science and Technology (FCT) for the Ph.D. scholarship (SFRH/BD/99555/2014). J. M. Oliveira also thanks to the FCT for the funds provided to under the program Investigador FCT 2012 (IF/00423/2012).

References

1. Allen AA, Caldwell GL Jr, Fu FH (1995) Anatomy and biomechanics of the meniscus. Oper Tech Orthop 5(1):2–9
2. McDermott ID, Masouros SD, Amis AA (2008) Biomechanics of the menisci of the knee. Curr Orthop 22(3):193–201
3. Fetzer GB, Spindler KP, Amendola A, Andrish JT, Bergfeld JA, Dunn WR, Flanigan DC, Jones M, Kaeding CC, Marx RG (2009) Potential market for new meniscus repair strategies: evaluation of the MOON cohort. J Knee Surg 22(3):180
4. Tudor F, McDermott ID, Myers P (2014) Meniscal repair: a review of current practice. Orthop Trauma 28(2):88–96
5. Fairbank T (1948) Knee joint changes after meniscectomy. J Bone Joint Surg Br 30(4): 664–670
6. McDermott I, Amis A (2006) The consequences of meniscectomy. J Bone Joint Surg Br 88(12):1549–1556

7. Mordecai SC, Al-Hadithy N, Ware HE, Gupte CM (2014) Treatment of meniscal tears: an evidence based approach. World J Orthop 5(3):233

8. Arnoczky SP, Warren RF (1982) Microvasculature of the human meniscus. Am J Sports Med 10(2):90–95

9. Makris EA, Hadidi P, Athanasiou KA (2011) The knee meniscus: structure–function, pathophysiology, current repair techniques, and prospects for regeneration. Biomaterials 32(30):7411–7431

10. Scotti C, Hirschmann MT, Antinolfi P, Martin I, Peretti GM (2013) Meniscus repair and regeneration: review on current methods and research potential. Eur Cells Mater 26:150–170

11. Pereira H, Frias AM, Oliveira JM, Espregueira-Mendes J, Reis RL (2011) Tissue engineering and regenerative medicine strategies in meniscus lesions. Arthroscopy 27(12):1706–1719

12. Kang SW, Sun-Mi S, Jae-Sun L, Eung-Seok L, Kwon-Yong L, Sang-Guk P, Jung-Ho P, Byung-Soo K (2006) Regeneration of whole meniscus using meniscal cells and polymer scaffolds in a rabbit total meniscectomy model. J Biomed Mater Res Part A 77(4):659–671

13. Zellner J, Mueller M, Berner A, Dienstknecht T, Kujat R, Nerlich M, Hennemann B, Koller M, Prantl L, Angele M (2010) Role of mesenchymal stem cells in tissue engineering of meniscus. J Biomed Mater Res Part A 94(4):1150–1161

14. Stone KR, Rodkey WG, Webber R, McKinney L, Steadman JR (1992) Meniscal regeneration with copolymeric collagen scaffolds in vitro and in vivo studies evaluated clinically, histologically, and biochemically. Am J Sports Med 20(2):104–111

15. Verdonk R, Verdonk P, Huysse W, Forsyth R, Heinrichs E-L (2011) Tissue ingrowth after implantation of a novel, biodegradable polyurethane scaffold for treatment of partial meniscal lesions. Am J Sports Med 39(4):774–782

16. Gu Y, Wang Y, Dai H, Lu L, Cheng Y, Zhu W (2012) Chondrogenic differentiation of canine myoblasts induced by cartilage-derived morphogenetic protein-2 and transforming growth factor-β1 in vitro. Mol Med Rep 5(3):767–772

17. Ishida K, Kuroda R, Miwa M, Tabata Y, Hokugo A, Kawamoto T, Sasaki K, Doita M, Kurosaka M (2007) The regenerative effects of platelet-rich plasma on meniscal cells in vitro and its in vivo application with biodegradable gelatin hydrogel. Tissue Eng 13(5):1103–1112

18. Ballyns JJ, Wright TM, Bonassar LJ (2010) Effect of media mixing on ECM assembly and mechanical properties of anatomically-shaped tissue engineered meniscus. Biomaterials 31(26):6756–6763

19. Puetzer JL, Ballyns JJ, Bonassar LJ (2012) The effect of the duration of mechanical stimulation and post-stimulation culture on the structure and properties of dynamically compressed tissue-engineered menisci. Tissue Eng Part A 18(13–14):1365–1375

20. Liu C, Toma IC, Mastrogiacomo M, Krettek C, von Lewinski G, Jagodzinski M (2013) Meniscus reconstruction: today's achievements and premises for the future. Arch Orthop Trauma Surg 133(1):95–109

21. Pereira H, Silva-Correia J, Oliveira J, Reis R, Espregueira-Mendes J (2013) Future trends in the treatment of meniscus lesions: from repair to regeneration. In: Verdonk R, Espregueira-Mendes J, Monllau JC (eds) Meniscal transplantation. Springer, Berlin, pp 103–112

22. Oliveira J, Pereira H, Yan L, Silva-Correia J, Oliveira A, Espregueira-Mendes J, Reis R (2013) Scaffold that enables segmental vascularization for the engineering of complex tissues and methods of making the same, PT Patent 106174, Priority date: 161/2013, 26-08-2013

23. Zhong J-J (2010) Recent advances in bioreactor engineering. Korean J Chem Eng 27(4): 1035–1041

24. Wang D, Liu W, Han B, Xu R (2005) The bioreactor: a powerful tool for large-scale culture of animal cells. Curr Pharm Biotechnol 6(5):397–403

25. Hansmann J, Groeber F, Kahlig A, Kleinhans C, Walles H (2013) Bioreactors in tissue engineering—principles, applications and commercial constraints. Biotechnol J 8(3):298–307

26. Pörtner R, Nagel-Heyer S, Goepfert C, Adamietz P, Meenen NM (2005) Bioreactor design for tissue engineering. J Biosci Bioeng 100(3):235–245

27. Martin Y, Vermette P (2005) Bioreactors for tissue mass culture: design, characterization, and recent advances. Biomaterials 26(35):7481–7503

28. Neves AA, Medcalf N, Brindle KM (2005) Influence of stirring-induced mixing on cell proliferation and extracellular matrix deposition in meniscal cartilage constructs based on polyethylene terephthalate scaffolds. Biomaterials 26(23):4828–4836
29. Neves AA, Medcalf N, Brindle K (2003) Functional assessment of tissue-engineered meniscal cartilage by magnetic resonance imaging and spectroscopy. Tissue Eng 9(1):51–62
30. Marsano A, Wendt D, Quinn T, Sims T, Farhadi J, Jakob M, Heberer M, Martin I (2006) Bi-zonal cartilaginous tissues engineered in a rotary cell culture system. Biorheology 43(3):553–560
31. Petri M, Ufer K, Toma I, Becher C, Liodakis E, Brand S, Haas P, Liu C, Richter B, Haasper C (2012) Effects of perfusion and cyclic compression on in vitro tissue engineered meniscus implants. Knee Surg Sports Traumatol Arthrosc 20(2):223–231
32. Fox DB, Warnock JJ, Stoker AM, Luther JK, Cockrell M (2010) Effects of growth factors on equine synovial fibroblasts seeded on synthetic scaffolds for avascular meniscal tissue engineering. Res Vet Sci 88(2):326–332
33. Aufderheide AC, Athanasiou KA (2005) Comparison of scaffolds and culture conditions for tissue engineering of the knee meniscus. Tissue Eng 11(7–8):1095–1104
34. Gunja NJ, Athanasiou KA (2010) Effects of hydrostatic pressure on leporine meniscus cell-seeded PLLA scaffolds. J Biomed Mater Res Part A 92(3):896–905
35. Gunja NJ, Uthamanthil RK, Athanasiou KA (2009) Effects of TGF-β1 and hydrostatic pressure on meniscus cell-seeded scaffolds. Biomaterials 30(4):565–573
36. Weinand C, Xu JW, Peretti GM, Bonassar LJ, Gill TJ (2009) Conditions affecting cell seeding onto three-dimensional scaffolds for cellular-based biodegradable implants. J Biomed Mater Res B Appl Biomater 91(1):80–87
37. McNulty AL, Estes BT, Wilusz RE, Weinberg JB, Guilak F (2010) Dynamic loading enhances integrative meniscal repair in the presence of interleukin-1. Osteoarthr Cartil 18(6):830–838
38. Ballyns JJ, Bonassar LJ (2011) Dynamic compressive loading of image-guided tissue engineered meniscal constructs. J Biomech 44(3):509–516
39. Martínez H, Brackmann C, Enejder A, Gatenholm P (2012) Mechanical stimulation of fibroblasts in micro-channeled bacterial cellulose scaffolds enhances production of oriented collagen fibers. J Biomed Mater Res Part A 100(4):948–957
40. Liu C, Abedian R, Meister R, Haasper C, Hurschler C, Krettek C, von Lewinski G, Jagodzinski M (2012) Influence of perfusion and compression on the proliferation and differentiation of bone mesenchymal stromal cells seeded on polyurethane scaffolds. Biomaterials 33(4):1052–1064
41. Connelly JT, Vanderploeg EJ, Mouw JK, Wilson CG, Levenston ME (2010) Tensile loading modulates bone marrow stromal cell differentiation and the development of engineered fibrocartilage constructs. Tissue Eng Part A 16(6):1913–1923
42. Upton ML, Chen J, Guilak F, Setton LA (2003) Differential effects of static and dynamic compression on meniscal cell gene expression. J Orthop Res 21(6):963–969
43. Ferretti M, Madhavan S, Deschner J, Rath-Deschner B, Wypasek E, Agarwal S (2006) Dynamic biophysical strain modulates proinflammatory gene induction in meniscal fibrochondrocytes. Am J Physiol Cell Physiol 290(6):C1610–C1615
44. Pereira H, Caridade SG, Frias AM, Silva-Correia J, Pereira DR, Cengiz IF, Mano JF, Oliveira JM, Espregueira-Mendes J, Reis RL (2014) Biomechanical and cellular segmental characterization of human meniscus: building the basis for tissue engineering therapies. Osteoarthr Cartil 22(9):1271–1281
45. Athanasiou KA, Eswaramoorthy R, Hadidi P, Hu JC (2013) Self-organization and the self-assembling process in tissue engineering. Annu Rev Biomed Eng 15:115–136
46. Hu JC, Athanasiou KA (2006) A self-assembling process in articular cartilage tissue engineering. Tissue Eng 12(4):969–979
47. Araujo V, Figueiredo C, Joazeiro P, Mora O, Toledo O (2002) In vitro rapid organization of rabbit meniscus fibrochondrocytes into chondro-like tissue structures. J Submicrosc Cytol Pathol 34(3):335–343

48. Hoben GM, Athanasiou KA (2008) Creating a spectrum of fibrocartilages through different cell sources and biochemical stimuli. Biotechnol Bioeng 100(3):587–598
49. Hoben GM, Hu JC, James RA, Athanasiou KA (2007) Self-assembly of fibrochondrocytes and chondrocytes for tissue engineering of the knee meniscus. Tissue Eng 13(5):939–946
50. Aufderheide AC, Athanasiou KA (2007) Assessment of a bovine co-culture, scaffold-free method for growing meniscus-shaped constructs. Tissue Eng 13(9):2195–2205
51. MacBarb RF, Makris EA, Hu JC, Athanasiou KA (2013) A chondroitinase-ABC and TGF-β1 treatment regimen for enhancing the mechanical properties of tissue-engineered fibrocartilage. Acta Biomater 9(1):4626–4634
52. Huey DJ, Athanasiou KA (2011) Tension-compression loading with chemical stimulation results in additive increases to functional properties of anatomic meniscal constructs. PLoS ONE 6(11):e27857
53. Huey DJ, Athanasiou KA (2011) Maturational growth of self-assembled, functional menisci as a result of TGF-β1 and enzymatic chondroitinase-ABC stimulation. Biomaterials 32(8): 2052–2058
54. Adachi N, Pelinkovic D, Lee CW, Fu FH, Huard J (2001) Gene therapy and the future of cartilage repair. Oper Tech Orthop 11(2):138–144
55. Chen Y (2001) Orthopedic applications of gene therapy. J Orthop Sci 6(2):199–207
56. Evans C (1995) Current concepts review: possible orthopaedic applications of gene therapy. J Bone Joint Surg Am 77(7):1103–1114
57. Evans C, Ghivizzani S, Robbins P (2012) Orthopedic gene therapy—lost in translation? J Cell Physiol 227(2):416–420
58. Evans C, Ghivizzani S, Smith P, Shuler F, Mi Z, Robbins P (2000) Using gene therapy to protect and restore cartilage. Clin Orthop Rel Res (379 Suppl):S214
59. Giannoudis PV, Tzioupis CC, Tsirids E (2006) Gene therapy in orthopaedics. Injury 37(1): S30–S40
60. Anderson WF (1998) Human gene therapy. Nature 392(6679):25
61. Kaufmann KB, Büning H, Galy A, Schambach A, Grez M (2013) Gene therapy on the move. EMBO Mol Med 5(11):1642–1661
62. Babensee JE, McIntire LV, Mikos AG (2000) Growth factor delivery for tissue engineering. Pharm Res 17(5):497–504
63. Nimni M (1997) Polypeptide growth factors: targeted delivery systems. Biomaterials 18(18):1201–1225
64. Bhargava MM, Attia ET, Murrell GA, Dolan MM, Warren RF, Hannafin JA (1999) The effect of cytokines on the proliferation and migration of bovine meniscal cells. Am J Sports Med 27(5):636–643
65. Kasemkijwattana C, Menetrey J, Goto H, Niyibizi C, Fu FH, Huard J (2000) The use of growth factors, gene therapy and tissue engineering to improve meniscal healing. Mater Sci Eng C 13(1):19–28
66. Musumeci G, Loreto C, Carnazza ML, Cardile V, Leonardi R (2012) Acute injury affects lubricin expression in knee menisci. An immunohistochemical study. Ann Anatomy Anat Anz 95(2):151–158
67. Jones A, Flannery C (2007) Bioregulation of lubricin expression by growth factors and cytokines. Eur Cells Mater 13:40
68. Hennerbichler A, Moutos FT, Hennerbichler D, Weinberg JB, Guilak F (2007) Interleukin-1 and tumor necrosis factor alpha inhibit repair of the porcine meniscus in vitro. Osteoarthr Cartil 15(9):1053–1060
69. Goto H, Shuler FD, Lamsam C, Moller HD, Niyibizi C, Fu FH, Robbins PD, Evans CH (1999) Transfer of LacZ marker gene to the meniscus. J Bone Joint Surg 81(7):918–925
70. Madry H, Cucchiarini M, Kaul G, Kohn D, Terwilliger EF, Trippel SB (2004) Menisci are efficiently transduced by recombinant adeno-associated virus vectors in vitro and in vivo. Am J Sports Med 32(8):1860–1865

71. Cucchiarini M, Schetting S, Terwilliger E, Kohn D, Madry H (2009) rAAV-mediated overexpression of FGF-2 promotes cell proliferation, survival, and α-SMA expression in human meniscal lesions. Gene Ther 16(11):1363–1372
72. Hidaka C, Ibarra C, Hannafin JA, Torzilli PA, Quitoriano M, Jen S-S, Warren RF, Crystal RG (2002) Formation of vascularized meniscal tissue by combining gene therapy with tissue engineering. Tissue Eng 8(1):93–105
73. Martinek V, Usas A, Pelinkovic D, Robbins P, Fu FH, Huard J (2002) Genetic engineering of meniscal allografts. Tissue Eng 8(1):107–117
74. Steinert AF, Palmer GD, Capito R, Hofstaetter JG, Pilapil C, Ghivizzani SC, Spector M, Evans CH (2007) Genetically enhanced engineering of meniscus tissue using ex vivo delivery of transforming growth factor-β1 complementary deoxyribonucleic acid. Tissue Eng 13(9):2227–2237
75. Bonadio J (2000) Tissue engineering via local gene delivery. J Mol Med 78(6):303–311
76. Goto H, Shuler F, Niyibizi C, Fu F, Robbins P, Evans C (2000) Gene therapy for meniscal injury: enhanced synthesis of proteoglycan and collagen by meniscal cells transduced with a TGFβ1 gene. Osteoarthr Cartil 8(4):266–271
77. Zhang H, Leng P, Zhang J (2009) Enhanced meniscal repair by overexpression of hIGF-1 in a full-thickness model. Clin Orthop Relat Res 467(12):3165–3174

Part IV
Knee Ligament Injuries

Chapter 15
Fundamentals on Injuries of Knee Ligaments in Footballers

Hélder Pereira, Sérgio Gomes, Luís Silva, António Cunha,
Joaquim Miguel Oliveira, Rui Luís Reis
and João Espregueira-Mendes

Abstract The ligament injuries around the knee, represent some of the most frequent lesions in football. Moreover, anterior cruciate ligament (ACL) tear is known to be amongst the most frequent prolonged absence from competition in this subgroup of patients. In the last decade we have been experiencing an important change on the treatment of these tears, focused on the technical development of surgical treatment. Nevertheless, return to play on some level is far from being completely guaranteed after ACL reconstruction and a number of factors might play a role on this major goal of treatment. Herein it will be discussed the current concepts of dealing with ACL ruptures in footballers. The second most relevant injury around the knee in this populations is medial collateral ligament sprain. In this case, most patients are dealt successfully with conservative treatment and only specific cases might require surgery. Posterior cruciate ligament (PCL) and posterolateral corner (PLC) injuries are less frequent then the former in footballers.

H. Pereira (✉) · J.M. Oliveira · R.L. Reis · J. Espregueira-Mendes
3B's Research Group—Biomaterials, Biodegradables and Biomimetics,
Headquarters of the European Institute of Excellence on Tissue Engineering
and Regenerative Medicine, University of Minho, AvePark, Zona Industrial da Gandra,
4805-017 Barco GMR, Portugal
e-mail: helderduartepereira@gmail.com

H. Pereira · J.M. Oliveira · J. Espregueira-Mendes
ICVS/3B's—PT Government Associate Laboratory, Braga, Guimarães, Portugal

H. Pereira
Orthopedic Department, Centro Hospitalar Póvoa de Varzim, Vila do Conde, Portugal

H. Pereira · J.M. Oliveira
Ripoll y de Prado Sports Clinic—FIFA Medical Centre of Excellence, Murcia-Madrid, Spain

S. Gomes · L. Silva · J.M. Oliveira · J. Espregueira-Mendes
Clínica do Dragão, Espregueira-Mendes Sports Centre—FIFA Medical Centre of Excellence,
Porto, Portugal

H. Pereira · A. Cunha · J.M. Oliveira
International Sports Traumatology Centre of Ave, Taipas Termal, Caldas Taipas, Portugal

© Springer International Publishing AG 2017   289
J.M. Oliveira and R.L. Reis (eds.), *Regenerative Strategies for the Treatment
of Knee Joint Disabilities*, Studies in Mechanobiology, Tissue Engineering
and Biomaterials 21, DOI 10.1007/978-3-319-44785-8_15

However, they have the potential to cause important influence on athlete's performance. Isolated PCL injuries, particularly up to grade 2 are suitable for conservative treatment and return to play. A brief comment on the "state of the art" on PLC lesions is also included.

15.1 Introduction

Sports, namely football is a powerful global health promoter [116]. It has proven to assist in the prevention of a wide spectrum of diseases and this way it can have a major role in preventing work absenteeism, improve quality of life and decrease the costs with health care provided either by governments or insurance companies [62]. Physical activity can help reduce healthcare costs and increase productivity which are key issues in emerging economies. Football is by far the most popular sport in the all world with more than 300 million participants [7].

However, there is some risk inherent to sports activity in any sports. The global injury rate of professional footballers is approximately 1000 times higher compared to that of typical high risk industrial occupations [59].

Given the impressive number of people who play football at amateur or professional level, prevention programs like FIFA 11+ have proven their value in reducing the injury risk (including knee ligaments) thus minimizing the risks of football with substantial health benefits [7].

The FIFA 11+ is a simple, and easy to implement, sports injury prevention program comprising a warm up of 10 conditioning exercises [7].

Moreover, at any competitive level, healthy teams with lower injury rate are more prone to competitive success and higher performance [56].

In its last report from 2015, the UEFA study group reported that 15 % of all injuries during one season corresponded to ligaments [34]. Muscle strain, ligament sprain (knee and ankle) and contusion are the most common injury types in elite football [33]. Moreover, there is a tendency for higher frequency of lesions during games as opposing to training periods [33].

Amongst the ones related to longer absence from competition are knee ligament lesions, particularly anterior cruciate ligament (ACL) tear [33]. The most frequent ligament injuries around the knee in football are ACL tears and medial collateral ligament (MCL) sprain. Severe injuries in football (defined as causing absence longer than 28 days) account for 16 % of all injuries. The most common subtypes of severe lesions were hamstring strains (12 %) followed by MCL injuries (9 %) [33, 98]. Considering the former, ACL and MCL tears will be particularly considered throughout this text.

Although less frequent, posterior cruciate ligament (PCL) and posterolateral corner injuries might have severe consequences for a football player's career and present some challenges for surgeons' and team physicians. Hence they will also be briefly described.

There is a tendency for increased incidence over time for traumatic injuries in football and a significant percentage (around 20 %) are linked to fouls [33, 34]. Both these aspects must be considered in the development of preventions strategies.

15.2 Anterior Cruciate Ligament

Anterior cruciate ligament (ACL) rupture is one of the most frequent orthopaedic sports-related injuries, with a yearly incidence of 35 per 100,000 people [45] and also one of the most frequent causes of orthopedic surgery [16, 53]. It is likely that more than 100.000 of these procedures are performed each year in the United States alone [53, 117].

Considering the former and the fact that sports participation and football practice keeps increasing [118], it is undoubted that it represents significant socio-economic impact [16]. This is also must be taken into account in the development of strategies for early diagnosis and prevention based in detection of risk factors [62]. Football is the most popular sports all over the world and has a major social impact either because of the amount of people enrolled in the phenomenon but also because it is capable to reach and influence people towards health promotion.

Currently, both athletes and general population have higher expectations concerning medical care. The present-day demand is complete repair of anatomy and function with full return to high demanding activities [48, 155].

Most ACL ruptures (around 75 %) occur with minimal or no contact at the time of injury [11]. This is also true during football play [165]. In these cases, most often a change in velocity and/or multidirectional force across the knee joint while weight bearing is implicated [11]. Rapid deceleration moments, including planting the injured leg to cut and rapidly change direction, landing from a jump, twisting, and pivoting have also been related to such injuries [145].

Contact-type mechanisms usually result from higher energy mechanisms including traumatic knee dislocations or high-energy on-field injuries [11]. Some of these might result on the so-called O'Donoghue's Triad which has been described as a mechanism resulting in ACL, medial collateral ligament, and medial meniscus injuries [143]. However, multiple ligaments, lateral meniscus, cartilage or bony fractures might also occur.

Ongoing research enrolls the use of video analysis, computer simulation, cadaveric studies, as well as epidemiologic in order to enhance our understanding of the mechanism of injury for ACL tears. However, despite all the recent insights, the mechanism of ACL injury remains incompletely understood.

Screening strategies aiming to detect pre-participation individuals and athletes with increased risk of ACL injury could enable to develop their coaching methods accordingly. This could help to lower the incidence of this pathology with a positive impact in sports performance and healthcare [16, 53, 105, 120].

Despite all advances in surgical management, there is still some controversy around conservative versus surgical treatment for an ACL rupture. It must be acknowledged that risk of further knee lesions and osteoarthritis remains regardless the treatment option, surgical or conservative [28, 149]. Conservative treatment is based in neuromuscular rehabilitation, including proprioceptive training and muscular strengthening [28].

However, once football is a high-demand activity with frequent pivoting movements involved, in the vast majority of cases, ACL ruptures in footballers are treated by adequate ACL reconstruction [45]. The return to high-level participation in strenuous sports is often not possible without surgical reconstruction which also enables to address frequent comorbidities (e.g. meniscus and/or cartilage lesions) [45].

However, it has not been proven that ACL reconstruction is effective in impairing consistently further development of osteoarthritis (OA). One study at more than ten years follow-up described that up to 50 % of patients with a mean age of 31 had developed symptomatic OA, and at this age group we lack effective treatments for this condition [96].

During the nineties, single-bundle ACL reconstruction intended to reconstruct the antero-medial (AM) bundle of the ACL aiming for an "isometric" graft placement and the reported results were globally satisfactory [43, 179].

The need to improve rotational stability led to the current concept of anatomical graft placement (lower and more distal comparing to over-the-top position on the femur in a flexed knee). Most likely this change in trends (from "isometric" to "anatomic" graft positioning) aroused from the recognition that some patients treated in the past had limited return to same level of pivoting activities after ACL surgery [109].

Recent developments arising from basic sciences dedicated to ACL research (anatomy, biology, physiopathology or biomechanics) have changed the current clinical approach [110].

New insights on graft selection, tunnel placement, graft fixation and rehabilitation protocols have been established [23]. The biology of graft incorporation and the so called "ligamentization" process continues under research, however progressively growing knowledge has helped in preventing complications such as excessive graft elongation, pullout or slippage due to technical errors or inadequate rehabilitation [102].

Technical advances have also enlarged our options concerning graft-fixation systems (screws, cortical fixations, sliding suspensory loops) (Fig. 15.1) [24].

Despite remaining controversial, the most frequent graft in footballers is still the patellar tendon but hamstrings and quadriceps autografts are also valid options to consider in specific situations [122]. It has not been shown, undoubtedly, that any graft consistently represents an advantage over the other in elite footballers [165].

The "double bundle concept" considers two functional bundles—anteromedial (AM) and posterolateral (PL)—according to the relative location of the tibial insertion site [45, 51].

Previous ACL reconstruction techniques were based on the presumption that all ACLs are of similar size, shape and have the same correlation to other anatomic structures, such as the distance to posterior cruciate ligament and menisci.

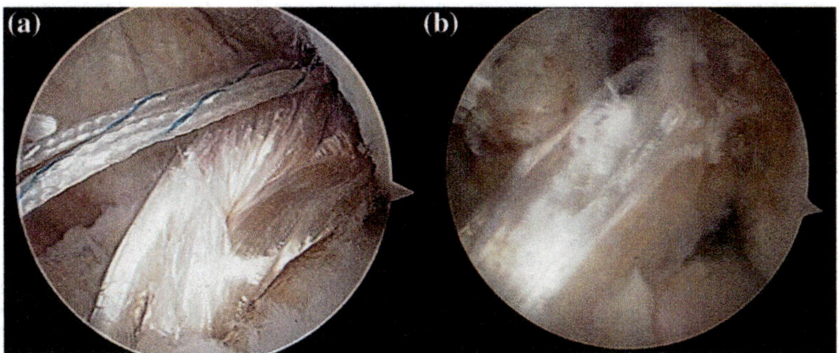

Fig. 15.1 **a** Arthroscopic view of quadruple hamstrings graft using suspensory fixation (this graft is somewhat rounder and bulkier). **b** Arthroscopic view of patellar tendon graft using interference screw fixation. This graft is more flat and somewhat closer to the native

However, it has been recently demonstrated that there is substantial variation in the size and shape of the ACL insertion site [45]. Transtibial arthroscopic ACL repair has been associated to tibial tunnels consistently positioned medial to the anatomic PL position and femoral tunnels positioned anterior to both the AM and PL anatomic insertions [45].

Several landmarks have been described in order to help the surgeon in placing the ACL femoral tunnels the more anatomically possible. The lateral intercondylar ridge had been identified Clancy as the most anterior border of the ACL insertion site on the femur [68]. More recently, the lateral bifurcate ridge which separates the AM and PL bundle insertion sites on the femur was also described [45]. These ridges are helpful for optimal placing of the femoral tunnels for ACL. These ridges seem to persist even in chronic ACL injuries, thereby being helpful even when the soft tissue remnants of the chronically disrupted ACL are no longer identified [45].

Some authors claim that the rotational control demands anatomical double-bundle ACL reconstruction [45, 174, 175], at least in some selected cases [109]. Other authors defend advantages of augmentation or "partial repair" [114, 115, 119, 130, 150].

Partial tears of the ACL (Fig. 15.2) account for 10–43 % of all ACL tears [44, 90, 113, 138] and in the pediatric population this prevalence seem to be even higher [128]. Although MRI is quite effective in distinguishing the normal from abnormal ACL, it is less reliable in the diagnosis of partial tears [88] even with 3-Tesla MRI devices [159, 160].

Preservation of the ACL remnants may aid the biological healing process that follows ACL repair or augmentation surgery. The remnants might preserve some partial mechanical function in anterior knee stability [115]. Moreover, histological studies revealed improved healing potential due to the vascular support provided by the epiligamentous tissue [115]. Similarly, the remnants have neural mechanoreceptors which might assist in restoring the proprioception [114, 115].

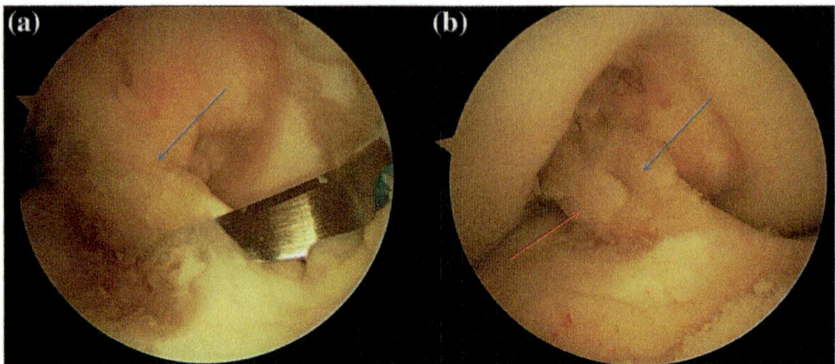

Fig. 15.2 Partial postero-lateral bundle reconstruction. *Blue arrow* demonstrates the remnant of antero-medial bundle (**a**, **b**); Hamstring graft was used to replace the damaged PL bundle (**b** *red arrow*)

What has become generally accepted is the need for Individualized ACL repair strategies depending on patient's characteristics, specific demands and surgeons' experience [4, 122].

Outcomes for ACL repair also depends on a careful and effective diagnosis and pre-operative planning. Improved diagnostic options are in required in order to assist in the choice for the best course of treatment for each patient.

Treatment algorithms for injuries of the anterior cruciate ligament (ACL) are constantly improving as the best available evidence from higher-level studies upraises. New techniques for reconstruction or augmentation keep being reported based on scientific, technical and biologic developments.

Further research is required to define the role of growth factors, platelet rich plasma (PRP), fibrin clot, and other regenerative medicine-based augmentation techniques [123, 124].

Doctors must follow these constant advances in technique to provide a more complete analysis of the injury, and keep participating in the ultimate goal of restoring each patient's unique anatomy by individualized ACL reconstruction.

15.2.1 Developments on ACL Assessment

Clinical examination one of the most important steps when evaluating the injured knee and requires training and skills from surgeons as well as team doctors [121]. Grading knee laxity is considered a key point to success when dealing with ACL injuries [6, 64]. The most frequently used clinical tests are the Lachman test (most sensitive) and the Pivot shift test (most specific) [129].

Nevertheless, manual clinical examination is difficult to quantify, is examiner dependent and, lacks intra-tester reliability [6, 81, 82, 111, 154].

Several methods to achieve objective instrumented laxity assessment have been attempted in order to quantify the Lachman test [121] such as the KT-1000/2000 (MEDmetric Corp., San Diego, CA, USA) or the Rolimeter (Aircast, Vista, CA) [77]. However, poor correlation with clinical outcome have been referred [77].

The pivot shift is considered more specific than the Lachman test [9] and might also be useful in the clinical diagnosis of partial tears [26]. However, in a recent study, the clinical grading of the pivot shift has been considered as subjective and inconsistent [64]. Poor correlations were found between the quantitative measurements and clinical pivot shift grade [64]. Many descriptions of the maneuver have been proposed and many devices have been developed in an attempt to objectively quantify the pivot shift test [111, 119, 121].

If a positive pivot shift test remains after ACL repair, this fact has been correlated with poor subjective and objective outcome as well as lower rates of return to sports and higher development of degenerative changes have been reported [72, 91].

Some biomechanical considerations should be taken into account: first, the pivot-shift test is a non-weight bearing examination and cannot replicate the rotatory knee laxity in dynamic weight-bearing conditions [115]; second, bony morphology also plays a role on the pivot shift phenomenon. Smaller lateral tibial plateaus might be related to higher grades on pivot shift testing [111]. An increased degree of posterior-inferior tibial slope is also related to higher pivot shift grade [13]. Additionally, the distal femoral geometry can influence dynamic rotatory laxity [63].

Characteristics of the ACL ligament itself are also implicated in this pivot shift phenomenon. The postero-lateral (PL) bundle of the ACL ligament was thought to be the primary responsible for rotational stability control, however, the antero-medial (AM) also plays a relevant role [75, 177]. The proportional contributions of each bundle are dependent on the degrees of knee flexion [177]. This is another point that favors the fact that, ideally we should aim for simultaneous anatomical and functional evaluation.

The suppression of a positive pivot shift test is considered a major goal of the ACL repair surgery.

Radiographic assessment of all patients is routinely used with bilateral standing X-rays (frontal, lateral, schuss and long axis) as well as skyline patella view. This radiological protocol is a valuable, low cost, accessible tool which helps gathering varied and fundamental information concerning bone morphology [3].

Although somewhat debatable, in our practice, moreover in footballers all patients are submitted to dynamic MRI evaluation with Porto Knee Testing Device (PKTD®) for ACL assessment as well as possible comorbidities (Fig. 15.3). CT scan are mainly used for pre-operative planning of revision cases [121–123].

Literature reports a general 78–100 % sensitivity and 68–100 % specificity of MRI for the diagnosis of isolated ACL tears [50, 134, 157, 168]. However, when considering only the most recent data an accuracy of approximately 95 % has been reported [8]. The distinction of proximal, partial, or chronic tears has been considered the most challenging and account for most of the interpretations inaccuracies [8]. Sensitivity is also significantly decreased in cases of multiple ligamentous injuries of the knee [8, 139].

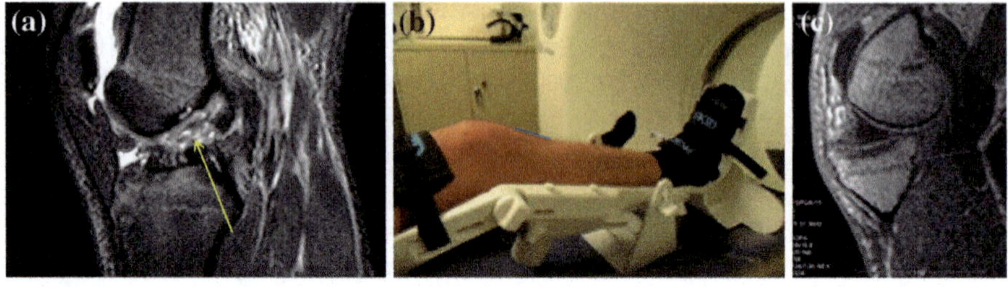

Fig. 15.3 MRI of an ACL tear (**a** *yellow line*); PKTD device for dynamic evaluation inside regular MRI equipment (**b**); the anterior tibial translation is visible (**c**)

3-Tesla devices have improved the distinction between AM and PL bundles, however have not significantly increased the MRI accuracy for global detection of ACL injuries [161]. About 70 % of ACL tears occur in the middle aspect of the ligament, 7–20 % occur near its femoral origin and only 3–10 % are identified at the tibial insertion [132, 133].

MRI protocols for the knee joint are designed to yield diagnostic images of the ACL as well as the menisci, bones, articular cartilage, and other ligamentous structures of the knee.

MRI has proved its value in the static anatomic study of the knee. Several attempts are continuously ongoing to enhance the capacity of imaging assessment. One of the most promising fields of research is the possibility for dynamic evaluation either in X-ray [69, 151], ultra-sound [55] or MRI [36, 79] in order to test simultaneously the anatomy and the functional features of ligaments.

The PKTD® (Fig. 15.3) has shown promising results and is part of our daily practice protocol once it is cost-effective in improving dynamic evaluation of the knee while providing object assessment of knee laxity [36]. It combines the assessment of "anatomy" and "function" during the same examination [119]. The PKTD enables study at different degrees of knee flexion, and different degrees of external/internal rotation inflected by the footplate. When required, it can also be used for PCL evaluation. PKTD might play a role in prevention strategies by assisting in the detection of risk factors and/or identifying those patients presenting higher rotational instability, who might require a surgical technique which provides higher rotational constraint (e.g., double-bundle or extra-articular plasty) [61].

Robotics [19, 180] and electronic devices [108] have also been proposed.

15.2.2 Risk Factors

Young, active, skeletally mature patients (under 18 years old) have higher failure rates after primary ACL reconstruction [131]. Moreover, even after revision surgery successful in restoring knee stability, only up to 50 % of such patients have returned to their prior level of activity or sport [131] after revision ACL repair. Young patients have been considered a particularly demanding population when dealing with ACL injuries [131]. As conclusion, young age seems to be a risk factor.

There is a lot of ongoing research dedicated to ACL ruptures in female. It has been recognized that the risk of a primary ACL injury is up to three times higher in women athletes when compared to men [153]. When performing primary ACL repair (particularly in youngsters and female), maximal preservation of bone stock and minimal surgical aggression should be kept in mind considering the possibility for ACL revision through life time [97]. In elite footballers, female seem to have two times the risk for ACL injury and a tendency so suffer such injuries in a younger age [165].

Other intrinsic risk factors include anatomy (e.g., morphology, notch width, lower extremity valgus), hormonal factors, and biomechanical factors (e.g., hamstring weakness, jump-land pattern) besides genetic, cognitive function, previous injury, and extrinsic risk factors [3, 147, 148].

Femoral notch characteristics and tibial plateau slope and/or depth have been proposed as morphologic risk factors possible to identify from standard X-ray evaluation [148, 164, 171]. Females are more likely than males to have a narrow A-shaped intercondylar notch which has been associated to gender-specific risk factor [153]. However, notch width index has not been considered a feasible method [164]. It has been recently concluded that Type A femoral notch appears to be a risk factor for ACL injury, whereas a reduced notch index has no significant correlation to ACL injury [2]. When compared with controls, men with ACL tears had deeper medial and lateral tibial plateaus, as well as an increased posterior slope of the lateral tibial plateau [10]. Women with shallower medial tibial plateau depth and men with steeper lateral tibial plateau slope (LTPS) are at higher risk of sustaining ACL injury. Overall, steeper LTPS might be considered a significant risk factor for sustaining ACL injury [76].

15.2.3 ACL Revision Surgery

Revision of ACL reconstruction is always a challenge in orthopaedic practice given the multiple possible scenarios and problems possibly involved in any specific case [122]. However, given the increasing number of primary ACL repairs the incidence of failures subsequently increases and any surgeon dedicated to knee reconstruction must be prepared to use the multiple possible options [15, 17, 20, 26, 31, 35, 52, 57, 71, 102, 112, 144, 156, 158, 169, 170]. Revision of ACL repair, irrespective of the cause, must always be considered a demanding procedure. Its results have been considered globally less satisfactory than those of primary repairs [29]. This has been the case on patient-reported outcome scores [172] but also when considering higher remaining laxity, higher graft failure rate, meniscal or cartilage degeneration [49, 105]. Despite several reports of favorable results and return to sports at same level after ACL revision [123], clinical failure rates as low as 25 % have also been reported [156]. Considering the former, we suggest that ACL revision should be performed on dedicated centers with experience in the field.

15.2.4 Return to Play

Return to sports (RTS) is a primary goal for ACL repair, mainly for young active individuals and athletes (regardless the level of play). There is scarce data in literature concerning ACL repair outcome in football players [136, 162, 165].

Despite there is a general perception that ACL reconstruction successfully enables most athletes to return to their previous sports at the same level, the reality sometimes is quite different.

In some series, as few as 45 % of patients actually returned to their pre-injury level of sports participation [45]. This is something to acknowledge, study and a background for continuous research and improvement. Besides, patient-specific and injury-specific factors, the surgical experience of the team plays a significant role in this rate of return to play [122].

In a recent report combining surgery with patient-specific rehabilitation protocol, 95 and 62 % of professional male soccer players were able to return to the same professional sport activity 1 and 4 years after surgery respectively [181].

At 4 years after ACL reconstruction, 71 % of patients were still playing competitive soccer, 62 % at the same pre-injury professional division, and 9 % in a lower division compared to the pre-injury status [181]. The former described issues not related to knee performance as cause for this lower level of play. Personal issues (not knee-related) were the main reason supporting the decision to abandon soccer career for most cases. Furthermore the age at final follow-up of those athletes whom retired were significantly higher compared to active players (30.4 $\pm$ 7.2 vs 25.5 $\pm$ 4.0; p = 0.0311). However, all the retired players were still involved in sport activities (e.g. Jogging, swimming and gym) [181]. At 6 months after surgery the clinical scores were found to be similar to pre-injury status, while return to play in official match was reported 186 $\pm$ 52 days after surgery [181].

This study assumes the importance of surgery, however highlights the role for patient-specific dedicated rehabilitation protocol.

Moreover, return to play should be dictated not by the calendar but by the achievement of individual objectives during the rehabilitation process. However, further research is also required on the final criteria of return to play after ACL injury.

15.3 Medial Collateral Ligament

Medial collateral ligament (MCL) injury is one of the most common knee injuries, especially in young athletic patients [126] including football [98]. The largest series of MCL injuries in professional football players suggested that the mean time loss from football for all MCL injury is around 23 days [98]. Moreover, the MCL injury rate decreased significantly during the 11-year study period which might indicate that appropriate training and prevention program as well as some changes in the rules of the game (more severe punishments for severe fouls) can help to reduce its incidence [98].

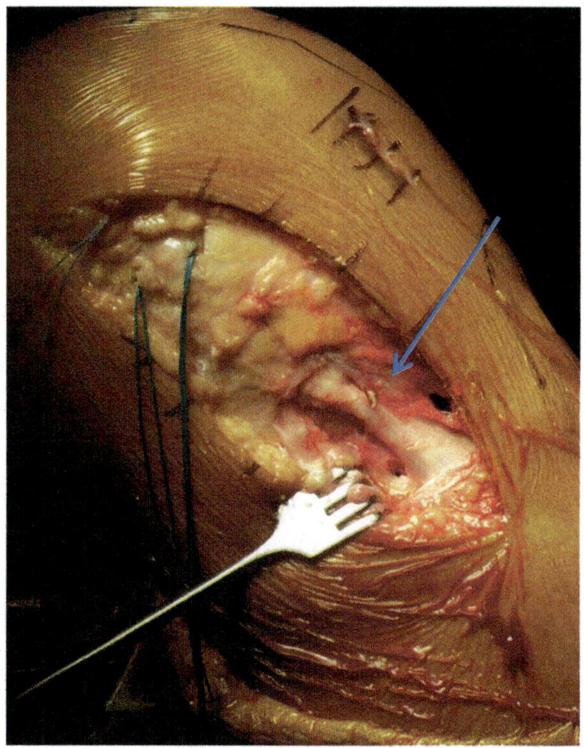

Fig. 15.4 Dissection of superficial medial collateral ligament (*blue arrow*)

Most MCL injuries can be managed conservatively with good results [126].

The MCL (Fig. 15.4) is a static stabilizer of the knee and is a strong ligament with attachments to the medial meniscus and capsule. It is divided into two portions: superficial MCL and deep MCL. The posterior oblique ligament (POL) is also an important medial stabilizer structure that is often neglected. The anatomy of the medial side of the knee is classically described in tree layers, with the superficial MCL and posterior oblique ligament (POL) being in the second layer and the deep MCL in the third [166]. The superficial MCL has a mean width of 1.5 cm and a mean length of 11 cm. It's runs from the medial femoral epicondyle and has two distinct distal attachments: one around 1 cm distal to the joint line and the second and strongest attachment occurs approximately 6 cm more distally [83, 87, 126].

The deep MCL is sometimes referred as a reinforcement of the medial capsule and thus is also entitled "middle capsular ligament" (classically is described in two portions—meniscofemural and meniscotibial). It extends from 1 cm distal to the origin of the superficial MCL, then runs distally to attach the medial meniscus and finally inserts in the tibia 3 to 4 mm distal to the joint line [87, 126].

The MCL is mainly described as a valgus restraining structure, howevert recent studies have reported that the superficial MCL also plays a role in rotational stability [167].

Biomechanically, the superficial MCL is the strongest, resisting up to 534 N; deep MCL holds 194 N and the POL 425 N [135].

15.3.1   Injury Mechanism and Evaluation

The two most frequent mechanisms leading to MCL injury are: direct valgus blow to the lateral side of the knee (usually associated to more severe lesions) or a noncontact rotational injury [98]. Noncontact rotational injury can be essentially a valgus force but most frequently it comprises a combinations of movements (flexion/valgus/external rotation) [60, 70, 111, 125].

This might occur during a sudden change of direction or when the foot gets trapped to the ground [98].

The femoral attachment is the most frequent injury site for the superficial MCL but interstitial failure was more common in posterior oblique ligament and deep MCL [135].

There is no clear classification of MCL injuries. However, Hughston's classification combining severity and laxity is probably still one of the most frequently used [66, 126]:

- Grade 1—is described as stable, involving only few fibers with local tenderness.
- Grade 2—higher amount of disrupted fibers causing more severe pain but still without instability.
- Grade 3—there is complete disruption of the ligament causing instability of Medial compartment.

Concerning the laxity assessment on unstable injuries (grade 3 lesions), they are further described as:

- Grade 1—3 to 5 mm of absolute medial separation;
- Grade 2—6 to 10 mm of medial opening;
- Grade 3—more than 10 mm of medial joint line opening.

Valgus stress tests should be performed in 0 and 30 degrees of flexion and compared to the opposite knee. Any laxity at 0 degrees of flexion should raise suspicion of severe lesion, possibly combined with injury of other structures such as the posterior cruciate ligament [42, 54]. Isolated laxity at 30 degrees usually suggests injury of the superficial portion of the MCL.

In chronic MCL injuries the patient often complains of medial knee pain and valgus instability [99, 100] which sometimes might limit high-demand activities.

For a complete assessment of MCL injuries imaging studies are often necessary. Radiographic evaluation should only be obtained if the requirements on the Ottawa knee rules are fulfilled, with exception of stress X-rays in skeletally immature individuals [152, 182].

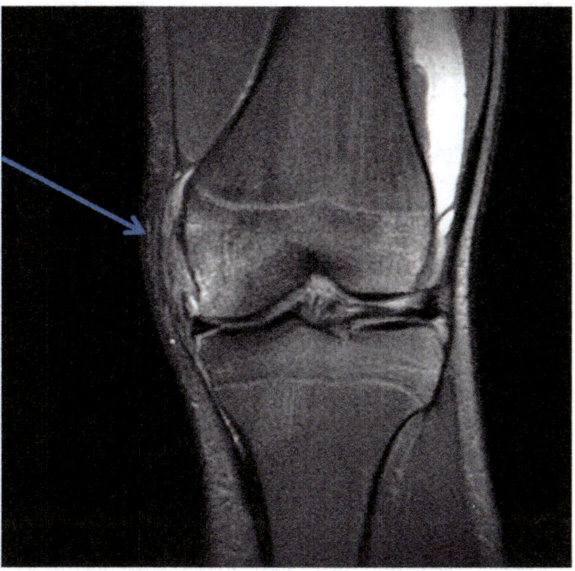

Fig. 15.5 MRI image of superficial medial collateral ligament injury

Ultrasound can be a useful tool in evaluating isolated medial collateral ligament injuries and also predicting the outcome on the basis of the location of the injury [89]. Moreover, ultrasound can be used to guide percutaneous treatments such as local application of growth factors.

MRI (Fig. 15.5) might not be mandatory to evaluate low grade MCL injuries but it has an important role in more severe lesions and in the exclusion of other associated injuries [126].

15.3.2 Treatment Strategy

Most isolated MCL injuries, including most grade 3 without associated injuries are treated conservatively.

The type of rehabilitation treatment indicated for a medial collateral ligament (MCL) injury depends on the severity of the injury. Our group performs platelet rich plasma PRP with ongoing promising results such as it has been referred in the literature [32, 178]. PRP seems to have a direct effect in diminishing pain and swelling and might accelerate the time required to return to play.

Once more, rehabilitation and physiotherapy is patient-tailored according to achievement of specific goals in different stages of rehabilitation: pain and inflammation management neuromuscular rehabilitation, proprioception and strengthening.

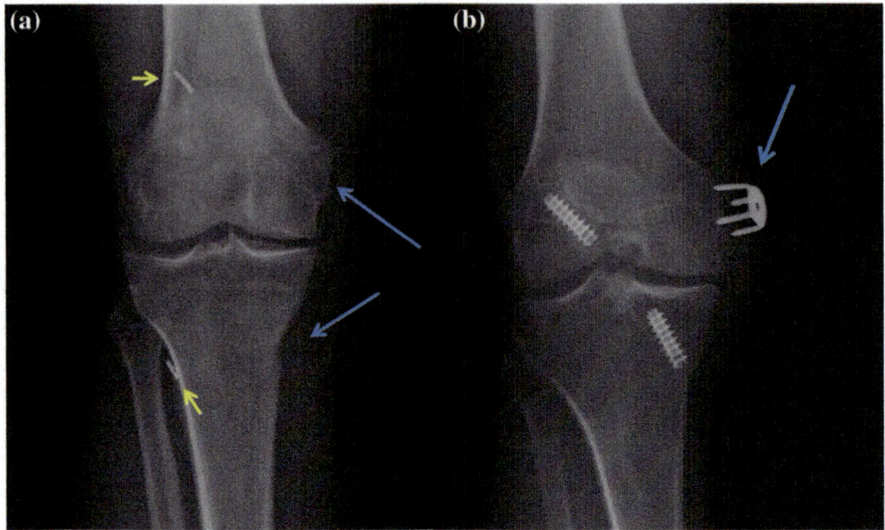

Fig. 15.6 Reconstruction of medial collateral ligament with hamstring grafts with suspensory cortical fixation (*yellow arrows*). The short tunnels (*blue arrows*) are also shown 5 years after surgery (**a**); 15 years after O'Donoghue's Triad surgery with ACL repair and MCL femoral re-attachment with a staple which became loose (*blue arrow*) (**b**)

When comparing isolated grade 3 injuries treated non-operatively versus surgically, several reports described better results in subjective scores and earlier return to play with conservative treatment [98, 111].

Surgical options are indicated in acute grade 3 injuries with multiligament injuries or other specific situations such as: intra-articular ligament entrapment; large bony avulsion; MRI finding of complete tibial side avulsion in athletes [87, 100, 126].

There are many different surgical techniques described (Fig. 15.6) in the literature including re-attachment, several types of reinforcement or some ligament's plasty requiring the use of allografts or autografts with good results [94, 95].

15.4 Posterior Cruciate Ligament

Posterior cruciate ligament (PCL) injuries and PCL-based multiple ligament knee injuries have low frequency, particularly in football. This fact is somewhat connected to some limitations in clinical studies and a subsequent delay in basic science and clinical research [38]. However it has been stated that the incidence of these lesions have not been correctly established over time, and in fact it might be to some extent underestimated [104].

The PCL is a large ligament extending from the lateral surface of the medial femoral condyle to the posterior aspect of the tibia. It is closely related to other structures such as the joint capsule, the ACL, the menisci, the ligaments of Humphrey and Wrisberg and the posterior neurovascular structures. The average length ranges from 35 to 38 mm and its width is 11 to 13 mm. The ligament is described as having two bundles with distinct footprints at both the femoral and tibial side. These are the anterolateral (AL) and the posteromedial (PM) bundles. The anterolateral bundle is larger and is the primary stabilizer of the knee when a posterior drawer test is applied [163].

15.4.1 Diagnosis, Mechanism of Injury and Classification

Two main mechanisms have been implicated in PCL rupture, however other possibilities must be considered. The most common is the 'dashboard injury' (40 %) [163]. Since the PCL is the primary restraint to posterior translation of the tibia relative to the femur, the PCL is the first ligament to be injured in these dashboard injuries. In this setting, the knee is in a flexed position and a posteriorly directed force is applied to the proximal tibia as the joint strikes the dashboard (often the case in car accidents). When the knee is in external rotation, the traumatic forces will be directed towards the posterolateral and lateral structures of the knee. In football or rugby, falls on the flexed knee with the foot in plantarflexion is the second most common mechanism of injury (25 %) [163]. If the foot is in dorsiflexion, the force is transmitted more through the patella and distal femur, which might protect the PCL. Hyperextension is typically reported to be a cause of PCL rupture and may result in disruption of the posterior capsule (12 %) but a forced valgus or varus movement can also cause PCL rupture [163].

Clinical diagnosis and grading is not an easy task even by experienced surgeons.

In a grade I injury, the tibial plateau remains anterior to the condyle, preserving an anterior step-off. A grade II injury is probable when the anterior border of the tibia sits flush with the femoral condyle (5–10 mm translation). Grades I and II usually refer to partial tears. When the anterior border of the tibial plateau rests posterior to the femoral condyle (more than 10 mm of translation), a complete tear (grade III) is present. In such cases it is mandatory to rule out an associated posterolateral structure lesion [163].

Clinical tests directed for assessment of PCL include posterior drawer test, posterior sag test and the quadriceps active test. Dial test at 30° and 90° is also required to assess comorbidities. The complete clinical assessment of the joint further includes the reverse pivot shift test, the external rotation of the tibia test, the Lachman test and collateral ligament examination.

It has been recognized the need to combine imaging with functional assessment for correct classification [93].

Stress radiography techniques have been used [46, 103] however this method does not provide direct assessment of PCL neither of associated injuries of

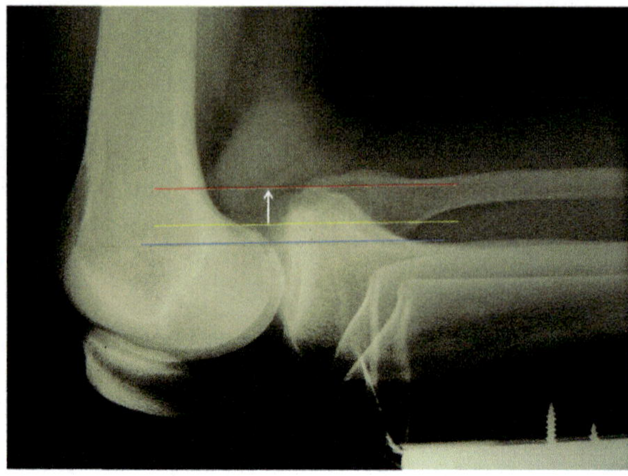

Fig. 15.7 Stress X-ray according to Barttlets's description (the patient uses its own weigh to stress the anterior tibia in an adequate device. Notice the amount of posterior tibial displacement (*red line*) taking as reference the posterior limit of the femoral condyle (*yellow line*). The *blue line* represents the line of the posterior tibial cortex

ligaments, meniscus or cartilage (Fig. 15.7). Fluoroscopic analysis under anesthesia might be helpful prior to final decision for treatment in selected cases (Fig. 15.8). Recent developments from Porto School have enabled dynamic and functional evaluation of PCL during regular MRI imaging protocol [36]. A device has been developed which permits MRI imaging while posterior stress is put in the tibia at different knee flexion angles and foot rotation (Fig. 15.9).

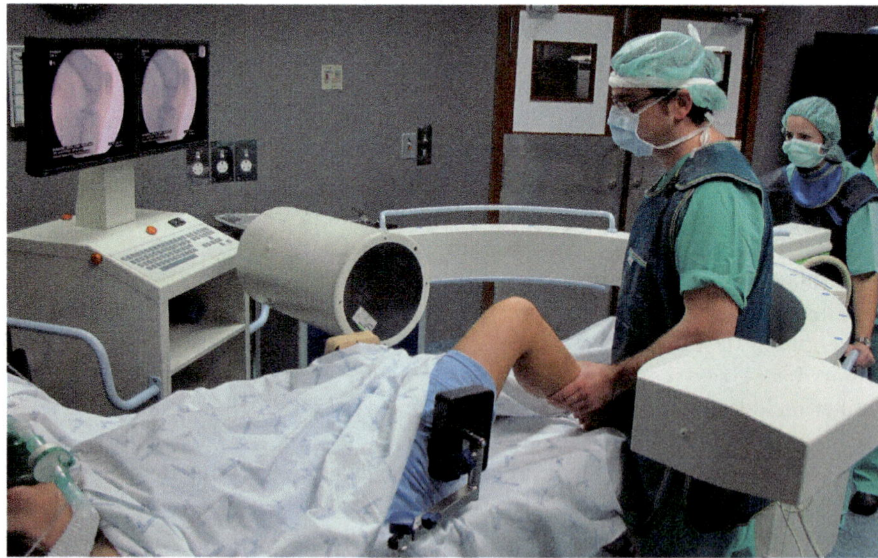

Fig. 15.8 Dynamic fluoroscopic evaluation on a patient with PCL injury. The images are recorded and can be presented to the patient prior to definitive decision concerning treatment

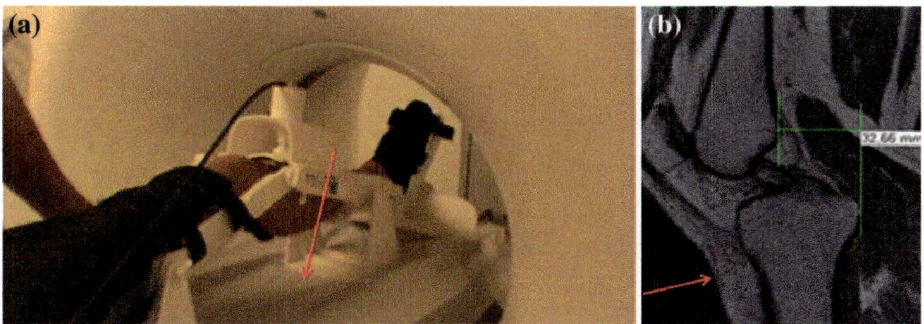

Fig. 15.9 **a** PKTD used for posterior tibial translation to assess PCL function. **b** Notice the amount of posterior tibial translation which can be measured. Red lines represent the vector of force (**a, b**)

15.4.2 Treatment Options

There is still controversy considering conservative versus surgical treatment [137, 142].

In a prospective study conservative treatment, Shelbourne reported the outcome of 271 athletes with acute, isolated, PCL injuries treated conservatively [142]. 76 % were able to return to sport or activity at a similar level [142]. Several authors have described the intrinsic healing potential of the PCL, return to competitive sport, lack of symptomatic instability and good outcomes at mid-term follow-up with non-operative treatment including footballers [163, 173]. Some have tried PRP, hyperbaric chamber to enhance healing however there is no clinical evidence in literature to support these techniques. Conservative treatment, based on a physiotherapy protocol, gives good results if the PCL rupture occurs as an isolated injury, with return to sport which can, in some cases, be in less than 2 months.

Technical advancements in allograft tissue, surgical instrumentation, and fixation methods contributed to improve the outcome in PCL reconstruction and PCL-based multiple ligament surgeries [38].

Despite improved surgical techniques, better postoperative rehabilitation methods have also contributed to this development.

Posterior cruciate ligament reconstruction improves patient-reported outcomes and return to sport although stability and knee kinematics may not be fully restored [163]. Athletic population presents an additional challenge considering their specific demands [101].

One important step for surgical decision is the choice of the graft. This will determine several aspects including the required length, fixation or method for passage of tendon grafts through tunnels (soft tissue alone, soft tissues with bone block in one or both extremities) [146]. Furthermore the technique to prepare a graft also depends if a single- or double-bundle technique will be used. Recent laboratory and cadaveric studies have suggested that double-bundle reconstructions of the posterior cruciate ligament could provide better joint kinematics than single-bundle reconstructions although this founds no support in clinical results [163].

It has been reported that arthroscopic single-bundle PCL reconstruction produces satisfactory return of function and decrease in symptoms [141]. All patients in this study had improved laxity of at least 1 grade. When compared with chronic reconstructions, acute reconstructions had statistically significant better activities of daily living scale (ADLS) and sports activities scale (SAS) scores [141].

A recent systematic review concluded that the superiority of single-bundle or double-bundle posterior cruciate ligament reconstruction remains uncertain [78]. Harner's group stated that "tailoring the PCL reconstruction technique to the individual injury pattern will likely yield a reconstruction that better replicates the natural biomechanics of the native knee, thereby resulting in better functional outcomes" [18].

Reconstruction of the posterior cruciate ligament (PCL) using the tibial inlay fixation (Fig. 15.10) either open or arthroscopic) has been reported as an alternative to the transtibial tunnel technique, being the preferred method for several authors [21, 127]. The arguments behind this choice include the attempt to avoid the killer turn" and graft laxity with cyclic loading.

The best graft source for posterior cruciate ligament (PCL) reconstruction is also debated and includes autografts, allografts, and synthetic ligaments [137]. In the knee with multiple injured ligaments a combination of autograft and some source of allograft (e.g. allograft Achilles tendon, allograft BPTB, anterior tibialis) are often required [1, 39]. The possibility for tissue-engineered grafting is under intense research but in the early clinical experience of last-generation implants [14, 41]. It has been reported a superior reduction of posterior tibial laxity after PCL reconstruction in female patients when compared with males [73]. Consequently it has been hypothesized that gender-related differences might exist although further evidence is limited by the lower incidence of these injuries comparing ACL tears [73].

No definitive clinical differences between allograft and autograft in PCL reconstruction have been reported [65]. Satisfactory outcome have been obtained with both graft sources [65].

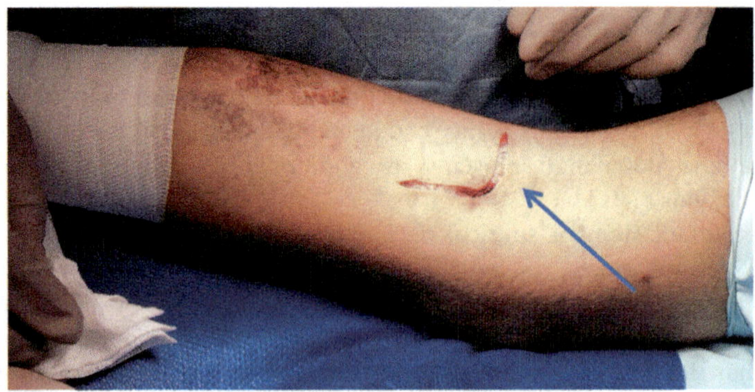

Fig. 15.10 Surgical posterior approach for inlay PCL reconstruction

One series achieved better results of isolated PCL reconstructions performed with a mixed graft (hamstrings autograft plus tibialis anterior allograft tendon) comparing to Achilles tendon allograft considering functional knee scores, posterior stability, and the graft macroscopic appearance. The use of allograft alone in that series presented a relatively higher rate of partial re-tear and less synovialization in the femoral aperture area [176]. This might be explained by superior biologic characteristics of fresh autografts when compared to allografts.

15.5 Posterolateral Corner Injuries

Injuries of the posterolateral corner (PLC) of the knee are rare, however might potentially lead to relevant chronic disability due to persistent instability and subsequent articular cartilage degeneration if not adequately treated. Recent anatomic and biomechanical studies have demonstrated its importance to knee stability [106, 140].

Isolated PLC injuries represent less than 2 % of all acute knee ligamentous injuries. More commonly, PLC injuries are found in combination with tibial plateau fractures (in up to 68 % of them), anterior cruciate ligament (ACL) tears and mainly posterior cruciate ligament (PCL) (from 43 to 80 %) tears [22, 27].

Prompt diagnosis might enable primary repair of the injured structures of the PLC rather than delayed reconstruction [47, 107]. PLC injuries left untreated increase the failure rate for both ACL and PCL repairs [12, 58, 85]. Considering the previous, high attentiveness is required to diagnose and appropriately treat injuries of the PLC in order to achieve the best possible outcome.

Recent data suggests that early operative treatment yields improved outcomes compared with non-operative or delayed surgery of the PLC in multi-ligament injured knees [47, 92, 107].

The assessment of results is very difficult to evaluate once there is a high percentage of multi-ligament knee injuries in the repair-technique groups. Moreover, if the cruciate ligaments were torn, they were ofte reconstructed in a second stage.

The immediate mobilization rehabilitation protocol may have also influence the oucome in the patients undergoing repair.

15.5.1 Summary of Anatomy and Biomechanics

To understand the posterolateral corner of the knee and provide successful repairs or reconstructions, a thorough knowledge of its anatomy and biomechanics is mandatory. The anatomy and biomechanics of the PLC are required to understand the clinical examination, imaging and treatment (conservative or surgical). While the biceps femoris, the iliotibial tract, the lateral head of the gastrocnemius and the popliteus complex provide the dynamic stability of the PLC, its static stability is mainly provided by the fibular collateral ligament (FCL) and the popliteofibular ligament (PFL). The detailed description of anatomy and biomechanics of PLC is considered out of the scope of this work and is properly detailed elsewhere [25].

15.5.2 Clinical Examination

In the presence of an acute isolated PLC injury, patients typically complain of tenderness, ecchymosis and swelling in the posterolateral aspect of the knee. Peroneal nerve injury has to be ruled out once an injury rate up to 30 % has been described. The nerve is more commonly affected in more sever PLC injuries in which the FCL and biceps femoris are also disrupted.

Clinically, instability is more truthfully assessed once the acute swelling and pain have subsided and the patient is more prone to cooperate. Posterolateral instability patterns are easier to identify in chronic injuries to the PLC. Ligamentous examination is always performed on both knees to provide a comparison between the injured and normal opposite knee.

Varus testing is performed at 0° and 30° of flexion. The distal femur is stabilized with one hand while a finger is placed along the lateral joint line. The other hand is used to apply gentle varus force at the lower leg. The amount of opening is quantified. Grade 1 opens 0–5 mm, grade 2 up to 6–10 mm, and grade 3 opens >10 mm. Until proven otherwise, varus opening at 0° is indicative of a severe posterolateral injury and an associated cruciate ligament tear. Isolated PLC injuries result in maximum varus opening at 30° of flexion, but low-degree PLC injuries can be observed with minimal or no varus deformity with significant rotational instability [30].

Different tests, on physical examination, have been described to help diagnose PLC injuries. The most commonly used test is the dial test or posterolateral rotation test. It assesses asymmetry in tibial external rotation and is usually performed prone. The test compares the degree of external rotation of both knees at 30° and 90° of knee flexion. A 10° difference at 30° of knee flexion is considered positive. When examination at 90° of flexion reveals a decreased amount (more than 50 %) of external rotation compared to 30°, the injury to the PLC is probably isolated. If the external rotation at 90° is further increased, a combined PCL/PLC injury is likely. The posterolateral drawer test is performed supine with the hip flexed at 45°, the knee at 80°, and the tibia externally rotated 15°. A posteriorly directed force is applied similar to the posterior drawer test. This test is considered positive when the lateral tibial plateau rotates posteriorly and externally relative to the medial tibial plateau. A knee with a PLC injury will fall into relative hyperextension and the tibia will externally rotate into relative varus. This is known as the external rotation recurvatum test (Hughston test) [30]. Finally, the reverse pivot shift test, is performed with the knee being taken from 90° into extension under valgus load during simultaneous external rotation of the foot. When the posteriorly subluxated lateral tibial plateau is abruptly reduced at 20°–30° as the IT band changes from being a flexor to became an extensor of the knee, the test is considered positive. This obviously requires an intact IT band, which may be disrupted in high-grade injuries. Nevertheless, the test can be also positive in up to 30 % of normal knees examined under anaesthesia.

15.5.3 Classification

The first important division is between acute and chronic injuries. Acute injuries are traditionally defined as those with less than 3 weeks following the onset [47].

However PLC injuries can be repaired for up to 4 to 6 weeks following the traumatic event. After this period, they are considered as chronic injuries.

The most commonly used classification system has been described by Hughston et al. [67]. This system particularly is useful in the assessment of FCL injuries. Grade 1 is defined as a varus opening from 0 to 5 mm with stress manoeuvres and grade 2 shows an opening from 6 to 10 mm. A higher opening is categorized as Grade 3 and reflects combined injury of cruciate ligament. The limitation of Hughston classification is that doesn't take into account the rotational instability. Considering that several PLC lesions present significant rotational instability with minimal varus laxity, a grading system that considers both characteristics should be preferred and was described by Fanelli and Larsen [40]:

Type A: Isolated rotational injury to the popliteofibular ligament (PFL) and popliteus complex. There is an increase in external rotation with no varus or a minimal varus component (this kind of injury can be missed in the Hughston's classification once no provides limited varus instability).

Type B: Rotational injury plus a mild varus instability with a firm end point to varus stress at 30° knee flexion. It represents a type A plus FCL weakening.

Type C: Significant rotational and varus component without a firm end point at both 0° and 30° of knee flexion with varus stress. This occurs as consequence of complete disruption of the PFL, popliteus complex, FCL, lateral capsule, and cruciate ligaments.

15.5.4 Indications for Treatment

The first step for deciding whether surgical treatments are indicated is to rule out other associated injuries, specifically referring to the cruciate ligaments. In these cases, surgical treatment is mandatory. It has been verified that grade 3 PLC injuries have minimal spontaneous healing and poor outcome from conservative treatment is expected [86].

Conservative treatment is indicated when dealing with isolated PLC grade 1 lesions (rare) [80]. Conservative treatment can be performed either by 4 weeks of cast immobilization with the knee in extension or by early mobilization [74, 80]. Acceptable functional results after conservative treatment have also been described for grade 2 injuries with minimal radiographic changes after 8 years follow-up, however with persistent instability [74]. Given this fact, currently this option is reserved only for low-demand patients.

PLC injuries are most commonly combined with PCL damage and are frequently part of a multiple ligament injured knee. These combined instability knees are best treated with surgical correction of all components of the instability.

In summary, patients with PLC injuries grading 3 or C, those isolated PLC injuries grade B in high demand patients, and Fanelli grade A (which might be classified without any lesion according to Hughston classification) and B with concomitant cruciate ligament injuries should undergo surgical treatment.

In cases with bony varus high tibial valgus osteotomy should be considered before knee ligament reconstruction, however this is most likely career ending for any active high-level football player [37]. A treatment algorithm is presented in. However, personalization of the treatment should be done with an eye on the general condition of the patient, patient expectations, age, level of activity and non-knee related injuries.

Considering the effects of PLC injuries on ACL reconstruction, the ACL graft force is considerably higher from 0° to 30° when the PLC is injured [12, 85]. This force is additionally increased with a coupled loading in varus and internal rotation.

Significant increases in force have also been demonstrated on PCL reconstructed graft when the PLC is injured for both varus moment and a coupled posterior drawer and external rotation torque [58].

This helps to understand that repair or reconstruction of the torn PLC structures is recommended at the time of an ACL or PCL reconstruction in order to reduce the risk for secondary graft failure.

Several techniques (Fig. 15.11) have been described for surgical repair of PLC however they are out of the scope of this text [5, 47, 84, 107].

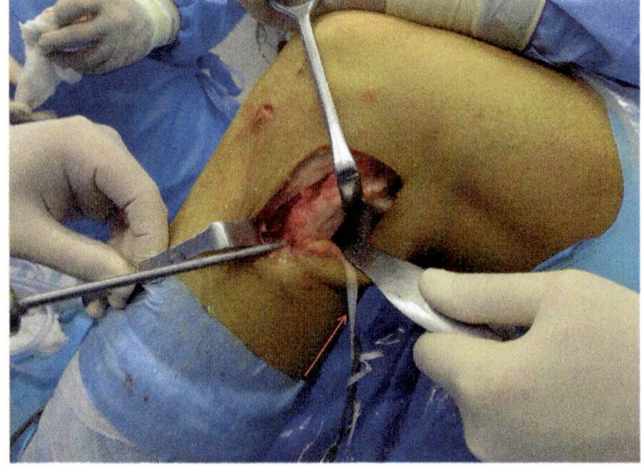

Fig. 15.11 Posterolateral corner reconstruction while making fibular tunnels as described by Arciero. The peroneal nerve is identified and protected through all the procedure by a nylon or cloth (*red arrow*)

15.6 Take-Home Message

The ligament injuries around the knee, represent some of the most frequent lesions in football and are linked to absent to sports participation and subsequent joint degeneration.

For high demand footballers, surgical repair of a torn ACL is usually the best option in order to return to same previous level of activity.

Opposing to this, most MCL tears can be adequately addressed through conservative treatment.

Despite representing a not so common injury in football, PCL tears in most cases can be treated non-operatively with return to sports.

However, the even rarer PLC injuries, usually occurring in combination with other ligament injuries around the knee, often requires operative treatment and represents a sever risk for returning to competition at high-level.

References

1. Adler GG (2013) All-inside posterior cruciate ligament reconstruction with a graftlink. Arthrosc Tech 2(2):e111–e115
2. Al-Saeed O, Brown M, Athyal R, Sheikh M (2013) Association of femoral intercondylar notch morphology, width index and the risk of anterior cruciate ligament injury. Knee Surg Sports Traumatol Arthrosc 21(3):678–682. doi:10.1007/s00167-012-2038-y
3. Andrade R, Vasta S, Sevivas N, Pereira R, Leal A, Papalia R, Pereira H, Espregueira-Mendes J (2016) Notch morphology is a risk factor for ACL injury: a systematic review and meta-analysis. J ISAKOS. doi:10.1136/jisakos-2015-000030
4. Araujo PH, Kfuri Junior M, Ohashi B, Hoshino Y, Zaffagnini S, Samuelsson K, Karlsson J, Fu F, Musahl V (2014) Individualized ACL reconstruction. Knee Surg Sports Traumatol Arthrosc 22(9):1966–1975. doi:10.1007/s00167-014-2928-2
5. Arciero RA (2005) Anatomic posterolateral corner knee reconstruction. Arthroscopy 21(9):1147. doi:10.1016/j.arthro.2005.06.008
6. Arilla FV, Yeung M, Bell K, Rahnemai-Azar AA, Rothrauff BB, Fu FH, Debski RE, Ayeni OR, Musahl V (2015) Experimental execution of the simulated pivot-shift test: a systematic review of techniques. Arthroscopy. doi:10.1016/j.arthro.2015.06.027
7. Barengo NC, Meneses-Echavez JF, Ramirez-Velez R, Cohen DD, Tovar G, Bautista JE (2014) The impact of the FIFA 11+ training program on injury prevention in football players: a systematic review. Int J Environ Res Public Health 11(11):11986–12000. doi:10.3390/ijerph111111986
8. Behairy NH, Dorgham MA, Khaled SA (2009) Accuracy of routine magnetic resonance imaging in meniscal and ligamentous injuries of the knee: comparison with arthroscopy. Int Orthop 33(4):961–967. doi:10.1007/s00264-008-0580-5
9. Benjaminse A, Gokeler A, van der Schans CP (2006) Clinical diagnosis of an anterior cruciate ligament rupture: a meta-analysis. J Orthop Sports Phys Ther 36(5):267–288. doi:10.2519/jospt.2006.2011
10. Bisson LJ, Gurske-DePerio J (2010) Axial and sagittal knee geometry as a risk factor for noncontact anterior cruciate ligament tear: a case–control study. Arthroscopy 26(7):901–906. doi:10.1016/j.arthro.2009.12.012

11. Boden BP, Dean GS, Feagin JA Jr, Garrett WE Jr (2000) Mechanisms of anterior cruciate ligament injury. Orthopedics 23(6):573–578
12. Bonanzinga T, Zaffagnini S, Grassi A, Marcheggiani Muccioli GM, Neri MP, Marcacci M (2014) management of combined anterior cruciate ligament-posterolateral corner tears: a systematic review. Am J Sports Med 42(6):1496–1503. doi:10.1177/0363546513507555
13. Brandon ML, Haynes PT, Bonamo JR, Flynn MI, Barrett GR, Sherman MF (2006) The association between posterior-inferior tibial slope and anterior cruciate ligament insufficiency. Arthroscopy 22(8):894–899. doi:10.1016/j.arthro.2006.04.098
14. Brunet P, Charrois O, Degeorges R, Boisrenoult P, Beaufils P (2005) Reconstruction of acute posterior cruciate ligament tears using a synthetic ligament. Rev Chir Orthop Reparatrice Appar Mot 91(1):34–43
15. Burks RT, Friederichs MG, Fink B, Luker MG, West HS, Greis PE (2003) Treatment of postoperative anterior cruciate ligament infections with graft removal and early reimplantation. Am J Sports Med 31(3):414–418
16. Chadwick CC, Rogowski J, Joyce ETJ (2008) The economics of anterior cruciate ligament reconstruction. In: Prodromos C, Brown C, Fu FH, Georgoulis AD, Gobbi A, Howell SM et al (eds) The anterior cruciate ligament: reconstruction and basic science. Saunders Elsevier, Philadelphia, pp 79–83
17. Carson EW, Anisko EM, Restrepo C, Panariello RA, O'Brien SJ, Warren RF (2004) Revision anterior cruciate ligament reconstruction: etiology of failures and clinical results. J Knee Surg 17(3):127–132
18. Chhabra A, Kline AJ, Harner CD (2006) Single-bundle versus double-bundle posterior cruciate ligament reconstruction: scientific rationale and surgical technique. Instr Course Lect 55:497–507
19. Citak M, Suero EM, Rozell JC, Bosscher MR, Kuestermeyer J, Pearle AD (2011) A mechanized and standardized pivot shifter: technical description and first evaluation. Knee Surg Sports Traumatol Arthrosc 19(5):707–711. doi:10.1007/s00167-010-1289-8
20. Colosimo AJ, Heidt RS Jr, Traub JA, Carlonas RL (2001) Revision anterior cruciate ligament reconstruction with a reharvested ipsilateral patellar tendon. Am J Sports Med 29(6):746–750
21. Cooper DE, Stewart D (2004) Posterior cruciate ligament reconstruction using single-bundle patella tendon graft with tibial inlay fixation: 2- to 10-year follow-up. Am J Sports Med 32(2):346–360
22. Covey DC (2001) Injuries of the posterolateral corner of the knee. J Bone Joint Surg 83-A(1):106–118
23. Dargel J, Gotter M, Mader K, Pennig D, Koebke J, Schmidt-Wiethoff R (2007) Biomechanics of the anterior cruciate ligament and implications for surgical reconstruction. Strategies Trauma Limb Reconstr 2(1):1–12. doi:10.1007/s11751-007-0016-6
24. Dargel J, Schmidt-Wiethoff R, Heck M, Bruggemann GP, Koebke J (2008) Comparison of initial fixation properties of sutured and nonsutured soft tissue anterior cruciate ligament grafts with femoral cross-pin fixation. Arthroscopy 24(1):96–105. doi:10.1016/j.arthro.2007.07.031
25. Davies H, Unwin A, Aichroth P (2004) The posterolateral corner of the knee. Anatomy, biomechanics and management of injuries. Injury 35(1):68–75
26. DeFranco MJ, Bach BR Jr (2009) A comprehensive review of partial anterior cruciate ligament tears. J Bone Joint Surg 91(1):198–208. doi:10.2106/JBJS.H.00819
27. DeLee JC, Riley MB, Rockwood CA Jr (1983) Acute posterolateral rotatory instability of the knee. Am J Sports Med 11(4):199–207
28. Delince P, Ghafil D (2012) Anterior cruciate ligament tears: conservative or surgical treatment? A critical review of the literature. Knee Surg Sports Traumatol Arthrosc 20(1):48–61. doi:10.1007/s00167-011-1614-x
29. Denti M, Lo Vetere D, Bait C, Schonhuber H, Melegati G, Volpi P (2008) Revision anterior cruciate ligament reconstruction: causes of failure, surgical technique, and clinical results. Am J Sports Med 36(10):1896–1902. doi:10.1177/0363546508318189

30. Devitt BM, Whelan DB (2015) Physical examination and imaging of the lateral collateral ligament and posterolateral corner of the knee. Sports Med Arthrosc Rev 23(1):10–16. doi:10.1097/JSA.0000000000000046
31. E.W. C, C.J. B (2003) Revision anterior cruciate ligament surgery. The adult knee. Lippincot, Williams & Wilkins, Philadelphia
32. Eirale C, Mauri E, Hamilton B (2013) Use of platelet rich plasma in an isolated complete medial collateral ligament lesion in a professional football (soccer) player: a case report. Asian J Sports Med 4(2):158–162
33. Ekstrand J, Hagglund M, Walden M (2011) Injury incidence and injury patterns in professional football: the UEFA injury study. Br J Sports Med 45(7):553–558. doi:10.1136/bjsm.2009.060582
34. UEFA Elite Club Injury Study Report 2014/15 (2015)
35. Espregueira-Mendes J (2005) Revision of failures after reconstruction of the anterior cruciate ligament. In: J.S. L, F. H, R. V (eds) EFORT—European Instructional Course Lectures, 7th ed. The British Editorial Society of Bone and Joint Surgery, Springer, London, pp 184–189
36. Espregueira-Mendes J, Pereira H, Sevivas N, Passos C, Vasconcelos JC, Monteiro A, Oliveira JM, Reis RL (2012) Assessment of rotatory laxity in anterior cruciate ligament-deficient knees using magnetic resonance imaging with Porto-knee testing device. Knee Surg Sports Traumatol Arthrosc 20(4):671–678. doi:10.1007/s00167-012-1914-9
37. Fanelli GC (2006) Surgical treatment of lateral posterolateral instability of the knee using biceps tendon procedures. Sports Med Arthrosc Rev 14(1):37–43
38. Fanelli GC, Beck JD, Edson CJ (2010) Current concepts review: the posterior cruciate ligament. J Knee Surg 23(2):61–72
39. Fanelli GC, Edson CJ (2002) Arthroscopically assisted combined anterior and posterior cruciate ligament reconstruction in the multiple ligament injured knee: 2- to 10-year follow-up. Arthroscopy 18(7):703–714
40. Fanelli GC, Larson RV (2002) Practical management of posterolateral instability of the knee. Arthroscopy 18(2 Suppl 1):1–8
41. Fare S, Torricelli P, Giavaresi G, Bertoldi S, Alessandrino A, Villa T, Fini M, Tanzi MC, Freddi G (2013) In vitro study on silk fibroin textile structure for Anterior Cruciate Ligament regeneration. Mater Sci Eng C Mater Biol Appl 33(7):3601–3608. doi:10.1016/j.msec.2013.04.027
42. Fetto JF, Marshall JL (1978) Medial collateral ligament injuries of the knee: a rationale for treatment. Clin Orthop Relat Res 132:206–218
43. Freedman KB, D'Amato MJ, Nedeff DD, Kaz A, Bach BR Jr (2003) Arthroscopic anterior cruciate ligament reconstruction: a metaanalysis comparing patellar tendon and hamstring tendon autografts. Am J Sports Med 31(1):2–11
44. Fruensgaard S, Johannsen HV (1989) Incomplete ruptures of the anterior cruciate ligament. J Bone Joint Surg 71(3):526–530
45. Fu FH, van Eck CF, Tashman S, Irrgang JJ, Moreland MS (2015) Anatomic anterior cruciate ligament reconstruction: a changing paradigm. Knee Surg Sports Traumatol Arthrosc 23(3):640–648. doi:10.1007/s00167-014-3209-9
46. Garavaglia G, Lubbeke A, Dubois-Ferriere V, Suva D, Fritschy D, Menetrey J (2007) Accuracy of stress radiography techniques in grading isolated and combined posterior knee injuries: a cadaveric study. Am J Sports Med 35(12):2051–2056. doi:10.1177/0363546507306466
47. Geeslin AG, Moulton SG, LaPrade RF (2015) A systematic review of the outcomes of posterolateral corner knee injuries, Part 1: surgical treatment of acute injuries. Am J Sports Med. doi:10.1177/0363546515592828
48. Georgoulis AD, Ristanis S, Chouliaras V, Moraiti C, Stergiou N (2007) Tibial rotation is not restored after ACL reconstruction with a hamstring graft. Clin Orthop Relat Res 454:89–94. doi:10.1097/BLO.0b013e31802b4a0a

49. Getelman MH, Friedman MJ (1999) Revision anterior cruciate ligament reconstruction surgery. J Am Acad Orthop Surg 7(3):189–198
50. Giaconi JC, Allen CR, Steinbach LS (2009) Anterior cruciate ligament graft reconstruction: clinical, technical, and imaging overview. Top Magn Reson Imaging 20(3):129–150. doi:10.1097/RMR.0b013e3181d657a7
51. Girgis FG, Marshall JL, Monajem A (1975) The cruciate ligaments of the knee joint. Anatomical, functional and experimental analysis. Clin Orthop Relat Res 106:216–231
52. Greis PE, Johnson DL, Fu FH (1993) Revision anterior cruciate ligament surgery: causes of graft failure and technical considerations of revision surgery. Clin Sports Med 12(4):839–852
53. Griffin LY, Agel J, Albohm MJ, Arendt EA, Dick RW, Garrett WE, Garrick JG, Hewett TE, Huston L, Ireland ML, Johnson RJ, Kibler WB, Lephart S, Lewis JL, Lindenfeld TN, Mandelbaum BR, Marchak P, Teitz CC, Wojtys EM (2000) Noncontact anterior cruciate ligament injuries: risk factors and prevention strategies. J Am Acad Orthop Surg 8(3): 141–150
54. Grood ES, Noyes FR, Butler DL, Suntay WJ (1981) Ligamentous and capsular restraints preventing straight medial and lateral laxity in intact human cadaver knees. J Bone Joint Surg 63(8):1257–1269
55. Grzelak P, Podgorski MT, Stefanczyk L, Domzalski M (2015) Ultrasonographic test for complete anterior cruciate ligament injury. Indian J Orthop 49(2):143–149. doi:10.4103/0019-5413.152432
56. Hagglund M, Walden M, Magnusson H, Kristenson K, Bengtsson H, Ekstrand J (2013) Injuries affect team performance negatively in professional football: an 11-year follow-up of the UEFA Champions League injury study. Br J Sports Med 47(12):738–742. doi:10.1136/bjsports-2013-092215
57. Harner CD, Giffin JR, Dunteman RC, Annunziata CC, Friedman MJ (2001) Evaluation and treatment of recurrent instability after anterior cruciate ligament reconstruction. Instr Course Lect 50:463–474
58. Harner CD, Vogrin TM, Hoher J, Ma CB, Woo SL (2000) Biomechanical analysis of a posterior cruciate ligament reconstruction. Deficiency of the posterolateral structures as a cause of graft failure. Am J Sports Med 28(1):32–39
59. Hawkins RD, Fuller CW (1999) A prospective epidemiological study of injuries in four English professional football clubs. Br J Sports Med 33(3):196–203
60. Hayes CW, Brigido MK, Jamadar DA, Propeck T (2000) Mechanism-based pattern approach to classification of complex injuries of the knee depicted at MR imaging. Radiographics: a review publication of the Radiological Society of North America, Inc 20 Spec No:S121-134. doi:10.1148/radiographics.20.suppl_1.g00oc21s121
61. Hemmerich A, van der Merwe W, Batterham M, Vaughan CL (2011) Knee rotational laxity in a randomized comparison of single- versus double-bundle anterior cruciate ligament reconstruction. Am J Sports Med 39(1):48–56. doi:10.1177/0363546510379333
62. Hootman JM, Dick R, Agel J (2007) Epidemiology of collegiate injuries for 15 sports: summary and recommendations for injury prevention initiatives. J Athl Train 42(2):311–319
63. Hoshino Y, Kuroda R, Nagamune K, Araki D, Kubo S, Yamaguchi M, Kurosaka M (2012) Optimal measurement of clinical rotational test for evaluating anterior cruciate ligament insufficiency. Knee Surg Sports Traumatol Arthrosc 20(7):1323–1330. doi:10.1007/s00167-011-1643-5
64. Hoshino Y, Musahl V, Irrgang JJ, Lopomo N, Zaffagnini S, Karlsson J, Kuroda R, Fu FH (2015) Quantitative evaluation of the pivot shift—relationship to clinical pivot shift grade: a prospective international multicenter study. Orthop J Sports Med. doi:10.1177/2325967115s00108
65. Hudgens JL, Gillette BP, Krych AJ, Stuart MJ, May JH, Levy BA (2013) Allograft versus autograft in posterior cruciate ligament reconstruction: an evidence-based systematic review. J Knee Surg 26(2):109–115. doi:10.1055/s-0032-1319778

66. Hughston JC (1994) The importance of the posterior oblique ligament in repairs of acute tears of the medial ligaments in knees with and without an associated rupture of the anterior cruciate ligament. Results of long-term follow-up. J Bone Joint Surg 76(9):1328–1344
67. Hughston JC, Andrews JR, Cross MJ, Moschi A (1976) Classification of knee ligament instabilities. Part II. The lateral compartment. J Bone Joint Surg 58(2):173–179
68. Hutchinson MR, Ash SA (2003) Resident's ridge: assessing the cortical thickness of the lateral wall and roof of the intercondylar notch. Arthroscopy 19(9):931–935
69. Isberg J, Faxen E, Brandsson S, Eriksson BI, Karrholm J, Karlsson J (2006) KT-1000 records smaller side-to-side differences than radiostereometric analysis before and after an ACL reconstruction. Knee Surg Sports Traumatol Arthrosc 14(6):529–535. doi:10.1007/s00167-006-0061-6
70. Jesse C, DeLee DD, Miller MD (2003) DeLee & Drez's orthopaedic sports medicine: principles and practice. Saunders, Philadelphia
71. Johnson DL, Swenson TM, Irrgang JJ, Fu FH, Harner CD (1996) Revision anterior cruciate ligament surgery: experience from Pittsburgh. Clin Orthop Relat Res 325:100–109
72. Jonsson H, Riklund-Ahlstrom K, Lind J (2004) Positive pivot shift after ACL reconstruction predicts later osteoarthrosis: 63 patients followed 5-9 years after surgery. Acta Orthop Scand 75(5):594–599. doi:10.1080/00016470410001484
73. Jung TM, Lubowicki A, Wienand A, Wagner M, Weiler A (2011) Knee stability after posterior cruciate ligament reconstruction in female versus male patients: a prospective matched-group analysis. Arthroscopy 27(3):399–403. doi:10.1016/j.arthro.2010.08.019
74. Kannus P (1989) Nonoperative treatment of grade II and III sprains of the lateral ligament compartment of the knee. Am J Sports Med 17(1):83–88
75. Kato Y, Maeyama A, Lertwanich P, Wang JH, Ingham SJ, Kramer S, Martins CQ, Smolinski P, Fu FH (2013) Biomechanical comparison of different graft positions for single-bundle anterior cruciate ligament reconstruction. Knee Surg Sports Traumatol Arthrosc 21(4):816–823. doi:10.1007/s00167-012-1951-4
76. Khan MS, Seon JK, Song EK (2011) Risk factors for anterior cruciate ligament injury: assessment of tibial plateau anatomic variables on conventional MRI using a new combined method. Int Orthop 35(8):1251–1256. doi:10.1007/s00264-011-1217-7
77. Kocher MS, Steadman JR, Briggs KK, Sterett WI, Hawkins RJ (2004) Relationships between objective assessment of ligament stability and subjective assessment of symptoms and function after anterior cruciate ligament reconstruction. Am J Sports Med 32(3):629–634
78. Kohen RB, Sekiya JK (2009) Single-bundle versus double-bundle posterior cruciate ligament reconstruction. Arthroscopy 25(12):1470–1477. doi:10.1016/j.arthro.2008.11.006
79. Kothari A, Haughom B, Subburaj K, Feeley B, Li X, Ma CB (2012) Evaluating rotational kinematics of the knee in ACL reconstructed patients using 3.0 Tesla magnetic resonance imaging. Knee 19(5):648–651. doi:10.1016/j.knee.2011.12.001
80. Krukhaug Y, Molster A, Rodt A, Strand T (1998) Lateral ligament injuries of the knee. Knee Surg Sports Traumatol Arthrosc 6(1):21–25. doi:10.1007/s001670050067
81. Kuroda R, Hoshino Y, Kubo S, Araki D, Oka S, Nagamune K, Kurosaka M (2011) Similarities and differences of diagnostic manual tests for anterior cruciate ligament insufficiency: a global survey and kinematics assessment. Am J Sports Med. doi:10.1177/0363546511423634
82. Lane CG, Warren R, Pearle AD (2008) The pivot shift. J Am Acad Orthop Surg 16(12):679–688
83. LaPrade RF, Engebretsen AH, Ly TV, Johansen S, Wentorf FA, Engebretsen L (2007) The anatomy of the medial part of the knee. J Bone Joint Surg 89(9):2000–2010. doi:10.2106/JBJS.F.01176
84. LaPrade RF, Engebretsen L, Marx RG (2015) Repair and reconstruction of medial- and lateral-sided knee injuries. Instr Course Lect 64:531–542
85. LaPrade RF, Resig S, Wentorf F, Lewis JL (1999) The effects of grade III posterolateral knee complex injuries on anterior cruciate ligament graft force. A biomechanical analysis. Am J Sports Med 27(4):469–475

86. LaPrade RF, Wentorf FA, Crum JA (2004) Assessment of healing of grade III posterolateral corner injuries: an in vivo model. J Orthop Res 22(5):970–975. doi:10.1016/j.orthres.2004.01.001

87. Laprade RF, Wijdicks CA (2012) The management of injuries to the medial side of the knee. J Orthop Sports Phys Ther 42(3):221–233. doi:10.2519/jospt.2012.3624

88. Lawrance JA, Ostlere SJ, Dodd CA (1996) MRI diagnosis of partial tears of the anterior cruciate ligament. Injury 27(3):153–155

89. Lee JI, Song IS, Jung YB, Kim YG, Wang CH, Yu H, Kim YS, Kim KS, Pope TL Jr (1996) Medial collateral ligament injuries of the knee: ultrasonographic findings. J Ultrasound Med 15(9):621–625

90. Lee K, Siegel MJ, Lau DM, Hildebolt CF, Matava MJ (1999) Anterior cruciate ligament tears: MR imaging-based diagnosis in a pediatric population. Radiology 213(3):697–704. doi:10.1148/radiology.213.3.r99dc26697

91. Leitze Z, Losee RE, Jokl P, Johnson TR, Feagin JA (2005) Implications of the pivot shift in the ACL-deficient knee. Clin Orthop Relat Res 436:229–236

92. Levy BA, Dajani KA, Whelan DB, Stannard JP, Fanelli GC, Stuart MJ, Boyd JL, MacDonald PA, Marx RG (2009) Decision making in the multiligament-injured knee: an evidence-based systematic review. Arthroscopy 25(4):430–438. doi:10.1016/j.arthro.2009.01.008

93. Levy BA, Stuart MJ (2012) Treatment of PCL, ACL, and lateral-side knee injuries: acute and chronic. J Knee Surg 25(4):295–305. doi:10.1055/s-0032-1324813

94. Lind M, Jakobsen BW, Lund B, Hansen MS, Abdallah O, Christiansen SE (2009) Anatomical reconstruction of the medial collateral ligament and posteromedial corner of the knee in patients with chronic medial collateral ligament instability. Am J Sports Med 37(6):1116–1122. doi:10.1177/0363546509332498

95. Liu X, Feng H, Zhang H, Hong L, Wang XS, Zhang J, Shen JW (2013) Surgical treatment of subacute and chronic valgus instability in multiligament-injured knees with superficial medial collateral ligament reconstruction using Achilles allografts: a quantitative analysis with a minimum 2-year follow-up. Am J Sports Med 41(5):1044–1050. doi:10.1177/0363546513479016

96. Lohmander LS, Ostenberg A, Englund M, Roos H (2004) High prevalence of knee osteoarthritis, pain, and functional limitations in female soccer players twelve years after anterior cruciate ligament injury. Arthritis Rheum 50(10):3145–3152. doi:10.1002/art.20589

97. Lubowitz JH, Schwartzberg R, Smith P (2013) Randomized controlled trial comparing all-inside anterior cruciate ligament reconstruction technique with anterior cruciate ligament reconstruction with a full tibial tunnel. Arthroscopy 29(7):1195–1200. doi:10.1016/j.arthro.2013.04.009

98. Lundblad M, Walden M, Magnusson H, Karlsson J, Ekstrand J (2013) The UEFA injury study: 11-year data concerning 346 MCL injuries and time to return to play. Br J Sports Med 47(12):759–762. doi:10.1136/bjsports-2013-092305

99. Pj M (2007) Current diagnosis & treatment in sports medicine. Lange medical book. Lange Medical Books/McGraw Hill Medical Pub, New York

100. Marchant MH Jr, Tibor LM, Sekiya JK, Hardaker WT Jr, Garrett WE Jr, Taylor DC (2011) Management of medial-sided knee injuries, part 1: medial collateral ligament. Am J Sports Med 39(5):1102–1113. doi:10.1177/0363546510385999

101. Margheritini F, Rihn J, Musahl V, Mariani PP, Harner C (2002) Posterior cruciate ligament injuries in the athlete: an anatomical, biomechanical and clinical review. Sports Med 32(6):393–408

102. Menetrey J, Duthon VB, Laumonier T, Fritschy D (2008) "Biological failure" of the anterior cruciate ligament graft. Knee Surg Sports Traumatol 16(3):224–231. doi:10.1007/s00167-007-0474-x

103. Menetrey J, Garavaglia G, Fritschy D (2007) Definition, classification, rationale and treatment for PCl deficient knee. Rev Chir Orthop Reparatrice Appar Mot 93(8 Suppl):5S70–5S73

104. Miller MD, Johnson DL, Harner CD, Fu FH (1993) Posterior cruciate ligament injuries. Orthop Rev 22(11):1201–1210
105. Mohtadi NG, Chan DS, Dainty KN, Whelan DB (2011) Patellar tendon versus hamstring tendon autograft for anterior cruciate ligament rupture in adults. Cochrane Database Syst Rev (9):CD005960. doi:10.1002/14651858.CD005960.pub2
106. Moorman CT 3rd, LaPrade RF (2005) Anatomy and biomechanics of the posterolateral corner of the knee. J Knee Surg 18(2):137–145
107. Moulton SG, Geeslin AG, LaPrade RF (2015) A systematic review of the outcomes of posterolateral corner knee injuries, Part 2: surgical treatment of chronic injuries. Am J Sports Med. doi:10.1177/0363546515593950
108. Muller B, Hofbauer M, Rahnemai-Azar AA, Wolf M, Araki D, Hoshino Y, Araujo P, Debski RE, Irrgang JJ, Fu FH, Musahl V (2015) Development of computer tablet software for clinical quantification of lateral knee compartment translation during the pivot shift test. Comput Methods Biomech Biomed Eng. doi:10.1080/10255842.2015.1006210
109. Muller B, Hofbauer M, Wongcharoenwatana J, Fu FH (2013) Indications and contraindications for double-bundle ACL reconstruction. Int Orthop 37(2):239–246. doi:10.1007/s00264-012-1683-6
110. Musahl V, Becker R, Fu FH, Karlsson J (2011) New trends in ACL research. Knee Surg Sports Traumatol Arthrosc 19(Suppl 1):S1–S3. doi:10.1007/s00167-011-1688-5
111. Musahl V, Voos J, O'Loughlin PF, Stueber V, Kendoff D, Pearle AD (2010) Mechanized pivot shift test achieves greater accuracy than manual pivot shift test. Knee Surg Sports Traumatol Arthrosc 18(9):1208–1213. doi:10.1007/s00167-009-1004-9
112. Thomas NP, Pandit HG (2008) Revision anterior cruciate ligament. In: Prodromos B, Fu FH, Georgoulis AD, Gobbi A, Howell SM et al (eds) The anterior cruciate ligament: reconstruction and basic science. Saunders Elsevier, Philadelphia, pp 443–457
113. Noyes FR, Mooar LA, Moorman CT 3rd, McGinniss GH (1989) Partial tears of the anterior cruciate ligament. Progression to complete ligament deficiency. J Bone Joint Surg 71(5): 825–833
114. Ochi M, Adachi N, Deie M, Kanaya A (2006) Anterior cruciate ligament augmentation procedure with a 1-incision technique: anteromedial bundle or posterolateral bundle reconstruction. Arthroscopy 22(4):463 e461–463 e465. doi:10.1016/j.arthro.2005.06.034
115. Ohashi B, Ward J, Araujo P, Kfuri M, Pereira H, Espregueira-Mendes J, Musahl V (2014) Partial ACL ruptures: knee laxity measurements and pivot shift. In: Doral MN, Karlsson J (eds) Sports injuries. Springer, Berlin, pp 1–16. doi:10.1007/978-3-642-36801-1_85-1
116. Organization WH (2010) Global recommendations on physical activity for health
117. Owings MF, Kozak LJ (1998) Ambulatory and inpatient procedures in the United States, 1996. Vital Health Stat 13(139):1–119
118. Palacios-Cena D, Fernandez-de-Las-Penas C, Hernandez-Barrera V, Jimenez-Garcia R, Alonso-Blanco C, Carrasco-Garrido P (2012) Sports participation increased in Spain: a population-based time trend study of 21 381 adults in the years 2000, 2005 and 2010. Br J Sports Med 46(16):1137–1139. doi:10.1136/bjsports-2012-091076
119. Pereira H, Fernandes M, Pereira R, Jones H, Vasconcelos JC, Oliveira JM, Reis RL, Musahl V, Espregueira-Mendes J (2014) ACL Injuries identifiable for pre-participation imagiological analysis: risk factors. In: Doral MN, Karlsson J (eds) Sports Injuries. Springer, Berlin, pp 1–15. doi:10.1007/978-3-642-36801-1_80-1
120. Pereira H, Fernandes M, Pereira R, Jones H, Vasconcelos JC, Oliveira JM, Reis RL, Musahl V, Espregueira-Mendes J (2015) Risk factors for ACL injuries identifiable for pre-participation imagiological analysis. In: Doral MN, Karlsson J (eds) Sports injuries: prevention, diagnosis, treatment and rehabilitation. Springer, Berlin
121. Pereira H, Sevivas N, Pereira R, Monteiro A, Oliveira JM, Reis RL, Espregueira-Mendes J (2012) New tools for diagnosis, assessment of surgical outcome and follow-up. In: Hernández JA, Monllau JC (eds) Lesiones ligamentosas de la rodilla. Marge Médica Books Barcelona, Spain, pp 185–198

122. Pereira H, Sevivas N, Pereira R, Monteiro A, Sampaio R, Oliveira J, Reis R, Espregueira-Mendes J (2014) Systematic approach from Porto school. In: Siebold R, Dejour D, Zaffagnini S (eds) Anterior cruciate ligament reconstruction. Springer, Berlin, pp 367–386. doi:10.1007/978-3-642-45349-6_34

123. Pereira H, Sevivas N, Varanda P, Monteiro A, Monllau JC, Espregueira-Mendes J (2013) Failed anterior cruciate ligament repair. In: Pre-Pub (ed) EFORT textbook—European surgical orthopaedics and traumatology. Springer, Berlin

124. Pereira HM, Correlo VM, Silva-Correia J, Oliveira JM, Reis Ceng RL, Espregueira-Mendes J (2013) Migration of "bioabsorbable" screws in ACL repair. How much do we know? A systematic review. Knee Surg Sports Traumatol Arthrosc 21(4):986–994. doi:10.1007/s00167-013-2414-2

125. Perryman JR, Hershman EB (2002) The acute management of soft tissue injuries of the knee. Orthop Clin North Am 33(3):575–585

126. Phisitkul P, James SL, Wolf BR, Amendola A (2006) MCL injuries of the knee: current concepts review. Iowa Orthop J 26:77–90

127. Piedade SR, Munhoz RR, Cavenaghi G, Miranda JB, Mischan MM (2006) Knee P.C.L. reconstruction: a tibial bed fixation ("INLAY") technique. Objective and subjective evaluation of a 30-cases series. Acta Ortop Bras 14(2):92–96

128. Prince JS, Laor T, Bean JA (2005) MRI of anterior cruciate ligament injuries and associated findings in the pediatric knee: changes with skeletal maturation. AJR Am J Roentgenol 185(3):756–762. doi:10.2214/ajr.185.3.01850756

129. Prins M (2006) The Lachman test is the most sensitive and the pivot shift the most specific test for the diagnosis of ACL rupture. Aust J Physiother 52(1):66. doi:10.1016/S0004-9514(06)70069-1

130. Pujol N, Colombet P, Cucurulo T, Graveleau N, Hulet C, Panisset JC, Potel JF, Servien E, Sonnery-Cottet B, Trojani C, Djian P (2012) Natural history of partial anterior cruciate ligament tears: a systematic literature review. Orthop Traumatol Surg Res 98(8 Suppl):S160–S164. doi:10.1016/j.otsr.2012.09.013

131. Reinhardt KR, Hammoud S, Bowers AL, Umunna BP, Cordasco FA (2012) Revision ACL reconstruction in skeletally mature athletes younger than 18 years. Clin Orthop Relat Res 470(3):835–842. doi:10.1007/s11999-011-1956-1

132. Remer EM, Fitzgerald SW, Friedman H, Rogers LF, Hendrix RW, Schafer MF (1992) Anterior cruciate ligament injury: MR imaging diagnosis and patterns of injury. Radiographics 12(5):901–915. doi:10.1148/radiographics.12.5.1529133

133. Resnick D (1995.) Diagnosis of bone and joint disorders, 3rd ed. WB Saunders Co

134. Robertson PL, Schweitzer ME, Bartolozzi AR, Ugoni A (1994) Anterior cruciate ligament tears: evaluation of multiple signs with MR imaging. Radiology 193(3):829–834. doi:10.1148/radiology.193.3.7972833

135. Robinson JR, Bull AM, Amis AA (2005) Structural properties of the medial collateral ligament complex of the human knee. J Biomech 38(5):1067–1074. doi:10.1016/j.jbiomech.2004.05.034

136. Roos H, Ornell M, Gardsell P, Lohmander LS, Lindstrand A (1995) Soccer after anterior cruciate ligament injury—an incompatible combination? A national survey of incidence and risk factors and a 7-year follow-up of 310 players. Acta Orthop Scand 66(2):107–112

137. Rosenthal MD, Rainey CE, Tognoni A, Worms R (2012) Evaluation and management of posterior cruciate ligament injuries. Phys Ther Sport 13(4):196–208. doi:10.1016/j.ptsp.2012.03.016

138. Roychowdhury S, Fitzgerald SW, Sonin AH, Peduto AJ, Miller FH, Hoff FL (1997) Using MR imaging to diagnose partial tears of the anterior cruciate ligament: value of axial images. AJR Am J Roentgenol 168(6):1487–1491. doi:10.2214/ajr.168.6.9168712

139. Rubin DA, Kettering JM, Towers JD, Britton CA (1998) MR imaging of knees having isolated and combined ligament injuries. AJR Am J Roentgenol 170(5):1207–1213. doi:10.2214/ajr.170.5.9574586

140. Sanchez AR 2nd, Sugalski MT, LaPrade RF (2006) Anatomy and biomechanics of the lateral side of the knee. Sports Med Arthrosc Rev 14(1):2–11
141. Sekiya JK, West RV, Ong BC, Irrgang JJ, Fu FH, Harner CD (2005) Clinical outcomes after isolated arthroscopic single-bundle posterior cruciate ligament reconstruction. Arthroscopy 21(9):1042–1050. doi:10.1016/j.arthro.2005.05.023
142. Shelbourne KD, Davis TJ, Patel DV (1999) The natural history of acute, isolated, nonoperatively treated posterior cruciate ligament injuries. A prospective study. Am J Sports Med 27(3):276–283
143. Shelbourne KD, Nitz PA (1991) The O'Donoghue triad revisited. Combined knee injuries involving anterior cruciate and medial collateral ligament tears. Am J Sports Med 19(5): 474–477
144. Shelbourne KD, O'Shea JJ (2002) Revision anterior cruciate ligament reconstruction using the contralateral bone-patellar tendon-bone graft. Instr Course Lect 51:343–346
145. Shimokochi Y, Shultz SJ (2008) Mechanisms of noncontact anterior cruciate ligament injury. J Athl Train 43(4):396–408. doi:10.4085/1062-6050-43.4.396
146. Silva L, Pereira H, Monteiro A, Pereira de Castro A, Piedade SR, Ripoll PL, Oliveira JM, Reis RL, Espregueira-Mendes J (2015) Allografts in PCL reconstructions. In: Doral MN, Karlsson J (eds) Sports Injuries: prevention, diagnosis, treatment and rehabilitation. Springer, Berlin
147. Smith HC, Vacek P, Johnson RJ, Slauterbeck JR, Hashemi J, Shultz S, Beynnon BD (2012) Risk factors for anterior cruciate ligament injury: a review of the literature—part 2: hormonal, genetic, cognitive function, previous injury, and extrinsic risk factors. Sports Health 4(2):155–161. doi:10.1177/1941738111428282
148. Smith HC, Vacek P, Johnson RJ, Slauterbeck JR, Hashemi J, Shultz S, Beynnon BD (2012) Risk factors for anterior cruciate ligament injury: a review of the literature—part 1: neuromuscular and anatomic risk. Sports Health 4(1):69–78. doi:10.1177/ 1941738111428281
149. Smith TO, Postle K, Penny F, McNamara I, Mann CJ (2014) Is reconstruction the best management strategy for anterior cruciate ligament rupture? A systematic review and meta-analysis comparing anterior cruciate ligament reconstruction versus non-operative treatment. Knee 21(2):462–470. doi:10.1016/j.knee.2013.10.009
150. Sonnery-Cottet B, Lavoie F, Ogassawara R, Scussiato RG, Kidder JF, Chambat P (2010) Selective anteromedial bundle reconstruction in partial ACL tears: a series of 36 patients with mean 24 months follow-up. Knee Surg Sports Traumatol Arthrosc 18(1):47–51. doi:10.1007/ s00167-009-0855-4
151. Staubli HU, Noesberger B, Jakob RP (1992) Stressradiography of the knee. Cruciate ligament function studied in 138 patients. Acta Orthop Scand Suppl 249:1–27
152. Stiell IG, Wells GA, McDowell I, Greenberg GH, McKnight RD, Cwinn AA, Quinn JV, Yeats A (1995) Use of radiography in acute knee injuries: need for clinical decision rules. Acad Emerg Med 2(11):966–973
153. Sutton KM, Bullock JM (2013) Anterior cruciate ligament rupture: differences between males and females. J Am Acad Orthop Surg 21(1):41–50. doi:10.5435/JAAOS-21-01-41
154. Swain MS, Henschke N, Kamper SJ, Downie AS, Koes BW, Maher CG (2014) Accuracy of clinical tests in the diagnosis of anterior cruciate ligament injury: a systematic review. Chiropr Man Ther 22:25. doi:10.1186/s12998-014-0025-8
155. Tashman S, Kolowich P, Collon D, Anderson K, Anderst W (2007) Dynamic function of the ACL-reconstructed knee during running. Clin Orthop Relat Res 454:66–73. doi:10.1097/ BLO.0b013e31802bab3e
156. Thomas NP, Kankate R, Wandless F, Pandit H (2005) Revision anterior cruciate ligament reconstruction using a 2-stage technique with bone grafting of the tibial tunnel. Am J Sports Med 33(11):1701–1709. doi:10.1177/0363546505276759
157. Tung GA, Davis LM, Wiggins ME, Fadale PD (1993) Tears of the anterior cruciate ligament: primary and secondary signs at MR imaging. Radiology 188(3):661–667. doi:10.1148/ radiology.188.3.8351329

158. Uribe JW, Hechtman KS, Zvijac JE, Tjin ATEW (1996) Revision anterior cruciate ligament surgery: experience from Miami. Clin Orthop Relat Res 325:91–99
159. Van Dyck P, De Smet E, Veryser J, Lambrecht V, Gielen JL, Vanhoenacker FM, Dossche L, Parizel PM (2012) Partial tear of the anterior cruciate ligament of the knee: injury patterns on MR imaging. Knee Surg Sports Traumatol Arthrosc 20(2):256–261. doi:10.1007/s00167-011-1617-7
160. Van Dyck P, Vanhoenacker FM, Gielen JL, Dossche L, Van Gestel J, Wouters K, Parizel PM (2011) Three tesla magnetic resonance imaging of the anterior cruciate ligament of the knee: can we differentiate complete from partial tears? Skeletal Radiol 40(6):701–707. doi:10.1007/s00256-010-1044-8
161. Van Dyck P, Vanhoenacker FM, Lambrecht V, Wouters K, Gielen JL, Dossche L, Parizel PM (2013) Prospective comparison of 1.5 and 3.0-T MRI for evaluating the knee menisci and ACL. J Bone Joint Surg 95(10):916–924. doi:10.2106/JBJS.L.01195
162. von Porat A, Roos EM, Roos H (2004) High prevalence of osteoarthritis 14 years after an anterior cruciate ligament tear in male soccer players: a study of radiographic and patient relevant outcomes. Ann Rheum Dis 63(3):269–273
163. Voos JE, Mauro CS, Wente T, Warren RF, Wickiewicz TL (2012) Posterior cruciate ligament: anatomy, biomechanics, and outcomes. Am J Sports Med 40(1):222–231. doi:10.1177/0363546511416316
164. Vyas S, van Eck CF, Vyas N, Fu FH, Otsuka NY (2011) Increased medial tibial slope in teenage pediatric population with open physes and anterior cruciate ligament injuries. Knee Surg Sports Traumatol Arthrosc 19(3):372–377. doi:10.1007/s00167-010-1216-z
165. Walden M, Hagglund M, Magnusson H, Ekstrand J (2011) Anterior cruciate ligament injury in elite football: a prospective three-cohort study. Knee Surg Sports Traumatol Arthrosc 19(1):11–19. doi:10.1007/s00167-010-1170-9
166. Warren LF, Marshall JL (1979) The supporting structures and layers on the medial side of the knee: an anatomical analysis. J Bone Joint Surg 61(1):56–62
167. Wijdicks CA, Griffith CJ, LaPrade RF, Spiridonov SI, Johansen S, Armitage BM, Engebretsen L (2009) Medial knee injury: Part 2, load sharing between the posterior oblique ligament and superficial medial collateral ligament. Am J Sports Med 37(9):1771–1776. doi:10.1177/0363546509335191
168. Winters K, Tregonning R (2005) Reliability of magnetic resonance imaging of the traumatic knee as determined by arthroscopy. N Z Med J 118(1209):U1301
169. Wirth CJ, Kohn D (1996) Revision anterior cruciate ligament surgery: experience from Germany. Clin Orthop Relat Res 325:110–115
170. Wolf RS, Lemak LJ (2002) Revision anterior cruciate ligament reconstruction surgery. J South Orthop Assoc 11(1):25–32
171. Wordeman SC, Quatman CE, Kaeding CC, Hewett TE (2012) In vivo evidence for tibial plateau slope as a risk factor for anterior cruciate ligament injury: a systematic review and meta-analysis. Am J Sports Med 40(7):1673–1681. doi:10.1177/0363546512442307
172. Wright RW, Gill CS, Chen L, Brophy RH, Matava MJ, Smith MV, Mall NA (2012) Outcome of revision anterior cruciate ligament reconstruction: a systematic review. J Bone Joint Surg 94(6):531–536. doi:10.2106/JBJS.K.00733
173. www.internetlivestats.com/internet-users/ (2015)
174. Yagi M, Wong EK, Kanamori A, Debski RE, Fu FH, Woo SL (2002) Biomechanical analysis of an anatomic anterior cruciate ligament reconstruction. Am J Sports Med 30(5):660–666
175. Yamamoto Y, Hsu WH, Woo SL, Van Scyoc AH, Takakura Y, Debski RE (2004) Knee stability and graft function after anterior cruciate ligament reconstruction: a comparison of a lateral and an anatomical femoral tunnel placement. Am J Sports Med 32(8):1825–1832
176. Yang JH, Yoon JR, Jeong HI, Hwang DH, Woo SJ, Kwon JH, Nha KW (2012) Second-look arthroscopic assessment of arthroscopic single-bundle posterior cruciate ligament reconstruction: comparison of mixed graft versus achilles tendon allograft. Am J Sports Med 40(9):2052–2060. doi:10.1177/0363546512454532

177. Yasuda K, van Eck CF, Hoshino Y, Fu FH, Tashman S (2011) Anatomic single- and double-bundle anterior cruciate ligament reconstruction, part 1: Basic science. Am J Sports Med 39(8):1789–1799. doi:10.1177/0363546511402659
178. Yoshioka T, Kanamori A, Washio T, Aoto K, Uemura K, Sakane M, Ochiai N (2013) The effects of plasma rich in growth factors (PRGF-Endoret) on healing of medial collateral ligament of the knee. Knee Surg Sports Traumatol Arthrosc 21(8):1763–1769. doi:10.1007/s00167-012-2002-x
179. Yunes M, Richmond JC, Engels EA, Pinczewski LA (2001) Patellar versus hamstring tendons in anterior cruciate ligament reconstruction: a meta-analysis. Arthroscopy 17(3):248–257
180. Zaffagnini S, Bignozzi S, Martelli S, Imakiire N, Lopomo N, Marcacci M (2006) New intraoperative protocol for kinematic evaluation of ACL reconstruction: preliminary results. Knee Surg Sports Traumatol Arthrosc 14(9):811–816. doi:10.1007/s00167-006-0057-2
181. Zaffagnini S, Grassi A, Marcheggiani Muccioli GM, Tsapralis K, Ricci M, Bragonzoni L, Della Villa S, Marcacci M (2014) Return to sport after anterior cruciate ligament reconstruction in professional soccer players. Knee 21(3):731–735. doi:10.1016/j.knee.2014.02.005
182. Zionts LE (2002) Fractures around the knee in children. J Am Acad Orthop Surg 10(5):345–355

Chapter 16
Clinical Management of Ligament Injuries of the Knee and Postoperative Rehabilitation

Sebastián Irarrázaval, Z. Yaseen, D. Guenther and Freddie H. Fu

Abstract Ligament injuries of the knee are common and occur via a variety of mechanisms. The medial collateral ligament is most commonly injured but injury to the anterior cruciate ligament is most often highlighted in the media due to its prevalence in sports. However, there are other less common but important ligamentous injuries that can also occur. The long term impact on overall function of the knee depends on the injury pattern and severity. Each ligament is distinct with respect to anatomy. Therefore, mechanism of injury, diagnosis and treatment are based on physical exam as well as imaging. Treatment and overall outcome depends on type of injury and time from the lesion. Surgical management involves restoring specific anatomy and biomechanical properties of these ligaments. Rehabilitation is tailored to the specific injury pattern and should be individualized for each patient with the hope to provide a stable functional knee.

16.1 Anterior Cruciate Ligament

16.1.1 Introduction

The anterior cruciate ligament (ACL) functions to prevent anterior tibial translation and internal rotation of the knee. Rupture of the ACL is a common injury worldwide. Estimates suggest an annual incidence of thirty-five per 100,000 people of all ages, with approximately two to eight-times higher risk in female athletes than in male athletes [1–3].

S. Irarrázaval (✉)
Department of Orthopaedic Surgery, Pontificia Universidad Católica de Chile,
Diagonal Paraguay 362, Santiago, Chile
e-mail: sirarraz@med.puc.cl

Z. Yaseen · D. Guenther · F.H. Fu (✉)
Department of Orthopaedic Surgery, University of Pittsburgh,
3471 Fifth Avenue, Suite 1011, Pittsburgh, PA 15213, USA
e-mail: ffu@upmc.edu

© Springer International Publishing AG 2017

323

J.M. Oliveira and R.L. Reis (eds.), *Regenerative Strategies for the Treatment of Knee Joint Disabilities*, Studies in Mechanobiology, Tissue Engineering and Biomaterials 21, DOI 10.1007/978-3-319-44785-8_16

16.1.2 Anatomy

The ACL consists is an intrarticular ligament that consists of two bundles, the anteromedial (AM) and the posterolateral bundle (PL) (Figs. 16.1, 16.2). The bundles are named according to their insertion at the tibia. The femoral origin of the AM bundle is located posterior and more superior to the PL bundle on the medial wall of the lateral femoral condyle. The two bundles cross each other and the AM bundle inserts anteromedial to the PL bundle on the tibia with a close relationship to the anterior horn of the lateral meniscus. The femoral origin of the PL bundle is located anterior and distally of the AM bundle. The PL bundle inserts posterolateral to the AM bundle on the tibia with a close relationship to the posterior horn of the lateral meniscus. The two bundles are separated by a distinct septum, containing vascular derived stem cells and each bundle is are covered by a thin membrane [4]. With the knee in full extension, the AM and the PL bundle are parallel. During knee flexion the AM bundle tightens and twists around the PL bundle. The AM bundle is the primary restraint against anterior tibial translation at 90 degree of knee flexion. With increased internal or external rotation, the ACL tightens so that it may operate as a major restraint against rotational moments as well [5].

16.1.3 Biomechanics

The biomechanical properties of the native ACL have been analyzed in ex vivo studies using universal force-moment sensors. The in situ force of the intact ACL is largest at 15° of knee flexion and decreases until 90° of knee flexion [6]. During normal walking, ACL loads of 169 N may be expected, while the activation of the

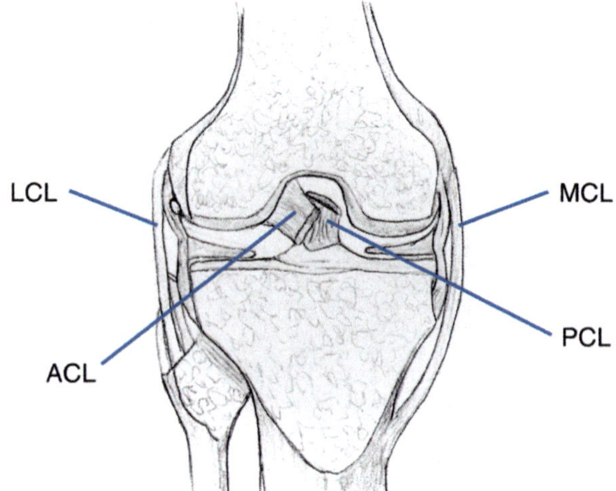

Fig. 16.1 Anterior view of a right knee with a coronal cut that shows both cruciate ligaments and both collateral ligaments. *ACL* anterior cruciate ligament, *PCL* posterior cruciate ligament, *LCL* lateral collateral ligament, *MCL* medial collateral ligament. © Pontificia Universidad Católica de Chile

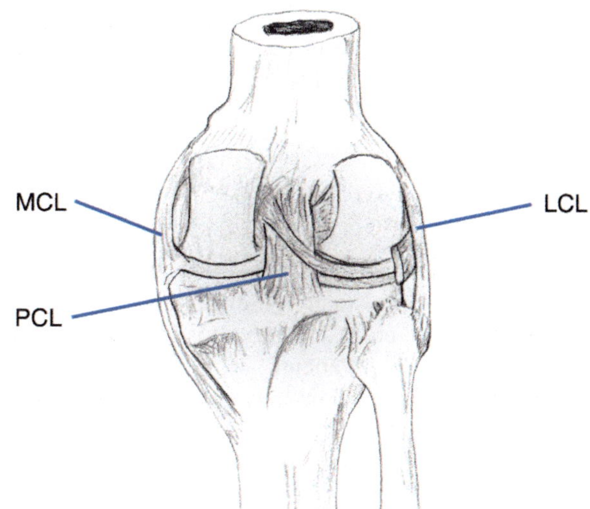

Fig. 16.2 Anterior view of a dissected right knee that shows both cruciate ligaments and both collateral ligaments. *ACL* anterior cruciate ligament (*AM* anteromedial bundle, *PL* posteromedial bundle), *PCL* posterior cruciate ligament, *LCL* lateral collateral ligament, *MCL* medial collateral ligament. © University of Pittsburgh Medical Center

knee extensor apparatus during descending stairs can lead to 445 N in situ force [7]. When applying an anterior tibial load of 110 N with the knee in full extension, the in situ force of the AM bundle significantly differs from the in situ force of the PL bundle [6]. The in situ force of the AM bundle throughout the range of knee motion does not change in a significant manner. The in situ force of the PL bundle is significantly lower at 90 degree of knee flexion when compared to full extension. Considering this, the role of the PL bundle in response to anterior tibial load may be more important near full extension. Most recent evidence suggest that both the AM and PL bundles behave synergistically to control anterior translation and rotational stability, with the AM bundle being more tight in flexion and the PL bundle being more tight in extension [8–12]. The ultimate failure rate of the bone-ligament-bone complex in young cadaveric specimens was reported with 2160 ± 157 N with a mean ACL stiffness of 242 ± 28 N/mm [13, 14]. Ultimate failure load and linear stiffness significantly decreases with age and with the axis of loading.

Disruption of the ACL changes knee kinematics significantly. The transfer of forces can be effective only if the joint is mechanically stable [15]. ACL deficiency leads to an increased anterior tibial translation and an increased internal tibial rotation. Insufficiency of the ACL will cause recruitment of secondary structures like the medial meniscus to resist external forces. This can lead to increased degeneration or secondary failure of those structures [16].

16.1.4   Mechanism of Injury

ACL injuries occur most often from non-contact injuries in sports that involve pivoting and cutting such as basketball, soccer, or lacrosse [17]. This injury occurs most frequently in a flexed knee when combined valgus (angulation of the inferior tibia away from midline) and internal rotational moment is applied to the fixed tibia during landing.

16.1.5 Diagnosis

A complete medical history including mechanism of injury, symptoms such as instability, as well as activity level of the patient have to be obtained in a patient with suspicion of an ACL injury. It is important to distinguish between an acute and a chronic injury. Patients with an acute injury mostly complain of a "giving-way-phenomenon" after feeling a "pop" during the injury mechanism. In patients with chronic injuries the presentation can vary widely due to secondary injuries and recurrent instability associated with sports or daily activity.

The clinical evaluation begins when the patient enters the room. Gait pattern and overall limb alignment is assessed and recorded. The entire lower extremity is evaluated and compared to the other uninjured leg. The physician inspects for any swelling or effusion around the knee. The joint is palpated for tenderness. The range of motion is evaluated because a decreased range of motion can be a sign for cartilage, meniscal or ACL injuries. The three most common tests for ACL injury are the Lachman test, anterior drawer test and pivot shift test.

The uninvolved knee is examined first, followed by examination of the injured knee. Results are documented in comparison to the uninjured side.

The Lachman test is performed with the knee at 30 degrees of flexion. The examiner secures the femur (thigh) with one hand and with the second hand applies a gentle force that pulls the superior tibia anteriorly. Anterior tibial translation relative to the femoral condyles is evaluated. Additionally, the examiner determines if a firm or soft endpoint is present. A soft endpoint with increased anterior tibial translation is indicative for ACL injury. Grading ranges from 1 to 3 with grade 1 being a stable knee and grade 3 injuries indicating significant ACL injury combined with other structural injury. The Lachman test has been shown to be the most sensitive test for injuries to the ACL [18].

The anterior drawer test is performed with the hip at 45 degrees of flexion and the knee at 90 degrees of flexion. The examiner applies a force that pulls the superior tibia anteriorly with the foot fixed on the examination table. Anterior tibial translation relative to the femoral condyles is evaluated. An increased anterior tibial translation without a firm endpoint indicates a potential ACL tear.

The pivot shift test is the most specific test for ACL injuries. However, multiple techniques are being employed and the interpretation of clinical grading varies greatly amongst examiners [19–21]. To improve comparability a standardized pivot shift test has been proposed [21]. The examiner controls the patient's leg with his hand at the heel level keeping the leg internally rotated. The leg is lifted off the table and the hip is slightly rolled to the side. The examiners' other hand is placed on the lateral side of the superior tibia to apply a gentle valgus stress. The knee is taken through range of motion from extension until approximately 40 degree with combined internal rotation and valgus stress. In a positive test the lateral side of the tibia suddenly reduces by gravity and the tension of the iliotibial band around 20–40 degree of knee flexion. This reduction movement is gently supported by the examiners' hand.

Plain radiographs (X-rays) are obtained to look for associated pathology. The radiographs are inspected for overall alignment, fractures, and skeletal maturity (for younger patients). A Segond fracture (lateral capsule avulsion off the tibia) is often related to ACL injury [22]. Patients with knee osteoarthritis should be further evaluated with a long cassette X-ray to determine alignment. Magnet resonance imaging (MRI) is highly sensitive for confirming the diagnosis and evaluating the tear pattern of the ACL. The MRI is also useful to identify any pathology of the cartilage or the menisci.

16.1.6 Treatment

Treatment strategies are dependent upon the age, activity level, presence of advanced arthritis and the patient's desire. These injuries in general are treated operatively but can also be treated non-operatively. For those that choose or attempt a trial of non-operative management, an intense physical therapy protocol is performed to strengthen the knee muscles and improve proprioceptive capability. This may be ideal for the patient who does not complain about "giving way" during day-to-day activities and who is not involved in high frequency pivoting, and cutting sports typically associated with ACL injuries. However, if the patient fails nonoperative management or is active, and still desires to continue in cutting sports, surgery is recommended.

Anatomic ACL reconstruction can best be defined as the ability to restore the ACL to its native dimensions, collagen orientation, and insertion sites [23, 24]. At our institution, a three portal technique is used with a high anterolateral (AL), anteromedial (AM), and accessory anteromedial (AAM) portal. First, a diagnostic arthroscopy is performed to evaluate the rupture pattern and to determine any concomitant injury to the meniscus or cartilage [25]. Based on the native anatomy of the ACL, different surgical approaches have to be considered to achieve good clinical outcome and patients' satisfaction. In the case of a complete rupture of both bundles either a single or double bundle reconstruction can be performed. The final decision is made intraoperative and is based on the overall knee anatomy including ACL insertion site [26]. There are multiple graft options that can be used to reconstruct the ACL. The graft can be from the patient's own tissue (autologous) or cadaver graft (allograft). The choice of the graft depends on donor site morbidity, healing potential, and patient age and activity level. Autologous graft choices include the patellar tendon-bone graft, hamstrings (semitendinosus/gracilis tendon) or the quadriceps tendon graft. Load to failure of each type of autologous graft has been studied at length. Bone-patellar tendon-bone experiences failure at 2977 N, quadrupled hamstring tendon at 4140 N, and quadriceps tendon-bone graft at 2353 N. However, harvest of the graft and graft fixation significantly decreases the ultimate strength and the linear stiffness [27, 28]. Grafts are cyclically preconditioned prior to implantation in order to decrease the viscoelastic elongation behavior post surgery [29].

There are numerous techniques for tunnel placement and fixation of the reconstructed graft. Tunnel placement is a critical step of ACL surgery since failures correlate with tunnel position most frequently. Anatomic positioning of the tunnel is highly important for patients' outcome.

16.1.7   Rehabilitation

In the early post-operative period, goals and milestones of therapy include: control of pain and edema, graft protection, comparable extension of both knees, knee flexion of at least 100°, maintenance of quadriceps strength, and achievement of full weight-bearing and normal gait [30]. Typically this period spans the first 4–6 weeks post-operatively. Crutches with instructions to weight bear as tolerated help to protect the graft and to minimize pain. Excluding cases of concomitant meniscal injury and repair, bracing after ACL reconstruction is not indicated. Many randomized controlled trials have determined that bracing does not provide a benefit in pain reduction, range of motion, graft stability, or rate of re-injury [31, 32]. Similarly, no clear benefit is seen with the use of a continuous passive motion (CPM) machine [30]. Conversely, cryotherapy improves outcomes by reducing pain and swelling [30, 32]. After the early post-operative period, rehabilitation should be geared at strengthening and neuromuscular control before progression of returning to activity and sports. These phases are cautiously given the time frames of 9–16 weeks and 16–22 weeks [32], respectively. Achievement of certain criteria should serve as the benchmarks for progression. During the strengthening and neuromuscular control stage, patients should perform activities of daily living without difficulty, tolerate exercises testing flexion and strength without pain or edema, and jog 2 miles without difficulty [30]. Returning to activity and sport should not be allowed unless patients achieve quadriceps index of at least 85 % and can handle sprinting, cutting, pivoting, jumping, and hopping at full exertion [30]. At our institution we usually recommend return to play at 9 months or greater. If a patient desires to return to play earlier, functional testing is issued, and an MRI may be administered to evaluate graft healing although studies are inconclusive to date. Ultimately, the decision to return to play is an interplay between subjective, and objective components by the physician and patient.

16.2   Posterior Cruciate Ligament

16.2.1   Introduction

The posterior cruciate ligament (PCL) functions primarily to prevent 95 % of posterior translation of the tibia relative to the femur in flexion [33]. It also functions as a secondary restraint to external rotation particularly between 90° and 120° of flexion. Injuries to the ligament can be due to trauma (e.g. motor vehicle accidents or falls) or sports participation. It is estimated that PCL injuries account for between 1 and 40 % of all knee injuries and 3 and 20 % of ligament injuries [34, 35]. It is challenging to truly estimate the incidence of the injury due to lack of recognition of the injury or under reporting. The ligament can be injured in isolation but most commonly it is injured in conjunction with other ligament or cartilage injuries (meniscus or articular surfaces).

16.2.2   Anatomy

The PCL is an intraarticular ligament that originates on the lateral border of the medial femoral condyle and inserts on the posterior aspect of the tibia and extends approximately 1 cm below the joint line (Figs. 16.1, 16.2, 16.3). It also blends into the posterior horn of the lateral meniscus. The PCL is 50 % larger than the ACL at the femoral insertion and approximately 20 % larger at the tibial insertion [35]. Biomechanical studies demonstrated the magnitude of the in situ forces of the PCL ranged from 6.1 to 112.3 N with the knee at 0–90 degrees of knee flexion [36]. The maximum tensile strength of the PCL has been reported to be 739–1627 N which is due in part to the large cross sectional area as well as extensive bony attachments [37, 38].

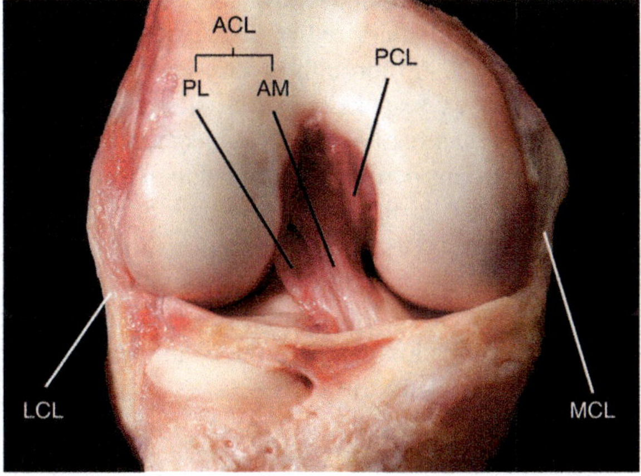

Fig. 16.3 Posterior view of a right knee that shows posterior cruciate ligaments and both collateral ligaments. *PCL* posterior cruciate ligament, *LCL* lateral collateral ligament, *MCL* medial collateral ligament. © Pontificia Universidad Católica de Chile

The ligament is composed of two distinct bundles, an anterolateral (AL) and posteromedial (PM) bundle which are named for their tibial insertion. The AL bundle has greater linear stiffness and ultimate load than the PL bundle [39]. In recent studies, examination of the PCL has determined that each bundle work in conjunction to resist posterior translation of the knee from full extension (0°) through flexion of 120° [40]. However it is generally agreed that in full extension, the PM bundle is tight and resists hypertension while the AL bundle is lax. The PM fibers then relax as the knee begins to flex and the AL bundle becomes taught in flexion [34]. There are other ligaments that also contribute to functionality but their discussion is beyond the scope of this text.

16.2.3 Mechanism of Injury

Injury to the PCL is usually caused by one of three mechanisms: hyperflexion, hyperextension or a posterior directed force to the proximal tibia [41].

A posterior directed force occurs for instance when the knee strikes a dashboard or a football players knee strikes another player pushing the tibia posterior. Hyperextension can cause injury to the PCL as well as the posterior capsule. A classic mechanism for an isolated PCL injury, in sports, is falling onto a flexed knee with the foot plantarflexed (toes pointing downward). If there is any rotation to the tibia or direction of force, then other structures are generally injured as well.

16.2.4 Diagnosis

The diagnosis of a PCL injury is determined carefully by a complete history, physical exam and imaging. It is important for the examiner to interview the patient to determine mechanism of injury and current activity level. Patients generally present with pain on the front of the knee and around the knee cap and complaints of difficulty with stairs rather than instability and even loss of 10–20 degrees of flexion [34, 42]. A thorough history is obtained from the patient including specific activities that cause pain including stair ascending and running.

Physical examination involves inspection of the knee looking for abrasions or bruising patterns particularly in front of the knee. If the patient is able to bear weight, the gait is then analyzed to look for abnormalities such as favoring the hyperextension of the knee or a varus thrust. Varus thrust gait refers to when the foot strikes the ground, the affected knee thrusts outside of the central axis. The patient has to readjust their body weight to reduce the knee back into position. The gait examination can help the examiner determine if there is an injury to other structures aside from the PCL.

The knee is assessed to see if there is swelling or effusion. The knee is then examined for atrophy of the quadriceps muscle, range of motion, and ligamentous stability. The most sensitive test for PCL injury is the posterior drawer [43]. This is performed with the knee in 90 degrees of flexion and thumbs placed gently along the anterior joint line. The examiner measures the amount of step off of the joint with the knee reduced and gently applies a posterior directed force on the tibia. The amount of step off correlates with the grade of injury with the scale being from 1 to 3 in increasing severity. The knee must also be examined for meniscal injuries or other ligament injuries including ACL or collateral ligament injuries. Lastly, it is pertinent to confirm sensation and vascular supply to the leg as in high-energy injuries such as motor vehicle accidents; there could be damage to the nerves and vessels. The examination is also completed on the other uninjured extremity for comparison purposes.

Imaging consists of radiographic (X-ray) images of the affected knee, MRI and at times long leg alignment radiographs of both limbs. The X-ray images are useful to note fractures such as a bony avulsion of the tibial insertion of the PCL. Stress X-ray images can also obtained to determine objectively the amount of posterior translation the tibia experiences relative to the femur. These films are taken and compared to the other, uninjured knee. A displacement of 8 mm or more indicates complete PCL rupture (grade III injury) [44]. If the posterior displacement is more than 12 mm, then likely other ligaments are injured. MRI imaging is utilized to confirm the injury and identify injury to the articular surface, menisci and other ligaments.

16.2.5 Treatment

The healing potential of the PCL is quite high with the ligament being more vascularized than the ACL. Treatment strategies depend on the severity of the injury and whether there are concomitant injuries. PCL injuries with a bony avulsion from the tibia can be treated without surgery and instead a period of immobilization to allow the bone to heal to its fracture site. In general, grade I and II isolated injuries are treated non-operatively with therapy. Grade III PCL injuries, if treated non-operatively, are immobilized or brace for a period to allow for healing of the ligament in the reduced position and then therapy. In those patients who have PCL injury associated with meniscal or other ligamentous injury, there is no consensus on treatment. Some surgeons attempt a trial of non-operative management to give the PCL a chance to heal and then attempt reconstruction of the other injuries. If the PCL remains insufficient, reconstruction can be done at that time. Others feel that if there are associated injuries including meniscal injury or injury to the posterolateral corner structures, surgical management is indicated more acutely [34, 45].

Patients with isolated injuries often rehabilitate to a functional knee and often are not aware of any biomechanical differences in their knee motion or functionality. Kinematic studies of knees with PCL deficiency have demonstrated increased forces in the medial and patellofemoral compartments which is suspected to lead to arthritis in the respective compartments [42]. However, long-term studies of PCL insufficient knees demonstrate variable outcomes. Some patients develop significant arthritic changes in the knee as early as 7–10 years from injury particularly in the patellofemoral and medial compartment of the knee [45, 46]. Non-operative management of PCL with concomitant ligament injuries in general yields worse clinical results particularly if the lateral structures are damaged.

As previously stated, surgical indications include concomitant injuries particularly injury to lateral structures, the meniscus or the articular surface. Patients who have failed a trial of non-operative management also can elect to pursue operative management. There are a variety of surgical techniques that can be utilized such as single bundle or double bundle reconstruction or inlay versus transtibial drilling. There are also multiple graft options including autologous graft or cadaver graft. No one technique has been deemed superior to other techniques. The goal however is to recreate the anatomy. After surgery, the knee is immobilized for a period and protected in extension. The patients begin therapy to work on quadriceps strengthening [34, 45, 47].

16.2.6 Rehabilitation

Regardless of surgical or nonsurgical management, physical therapy involves progressive weight bearing and restoration of normal motion while minimizing strain of the graft. Rehabilitation after surgical reconstruction is generally longer and slower than that for ACL reconstruction. During the first few weeks, knee ROM exercises are initiated in the prone position with motion restricted to a maximum of 90° of flexion. The safest arc of motion is from 40° to 90° of flexion. Patients with grade III injuries or surgically reconstructed PCLs undergo bracing or immobilization for several weeks to protect the PCL/graft when ambulating. Therapy focuses on strengthening the quadriceps. Care should be taken to avoid active contraction of the hamstrings which could translate the tibia posteriorly thereby causing the PCL to not heal properly or lead to an elongated PCL [47, 48]. Hamstring exercises can begin generally 6 weeks post operatively. Patients can expect to resume low impact activities around 12 weeks. Running and proprioceptive exercises can begin around 6 months after surgery. The patient is released to sport and all activity between 9 and 12 months, after completing balance, strength and endurance testing [47]. In non-operatively managed patients, return to sport is avoided until full quadriceps strength is achieved which can range from 6 to 8 weeks in elite athletes up to several months.

16.3 Medial Collateral Ligament and Posteromedial Corner

16.3.1 Introduction

The medial collateral ligament (MCL) functions as a resistant to valgus (angulation of the inferior tibia away from midline) stress and external rotation. It is the most commonly injured ligament of the knee [49]. The MCL can be injured in isolation or in addition to other structures around the knee. The true incidence of MCL injuries may be underestimated because of under-reporting of the lower grades of injury. The most common concomitant ligamentous injury of the knee is the ACL injury (95 %). Concomitant ligamentous injuries occur in 20 % of grade I, 52 % of grade II and 78 % of grade III injuries [49]. Studies have demonstrated that there is an 80 % incidence of combined ligament injury with grade III MCL tears [50].

16.3.2 Anatomy

The medial capsular-ligamentous complex comprises a three-layered sleeve of static and dynamic stabilizers extending from the midline anteriorly to the midline posteriorly [51]. The static stabilizers are the superficial MCL (sMCL), the posterior oblique ligament (POL) and the deep MCL (dMCL) (also called the deep medial ligament or middle capsular ligament). The POL has been identified as a thickening of the posterior medial capsule (PMC), and its importance in medial stability has become increasingly recognized [52]. The POL attaches superiorly and posteriorly to the attachment site of the sMCL on the femur. The dMCL has meniscofemoral and meniscotibial attachments but no attachment to the overlying sMCL.

The dynamic stabilizers provide stability under dynamic conditions. These are the semimembranous complex (composed of 5 insertional attachments), the pes anserinus muscle group (sartorius, gracilis and semitendinous muscles), the vastus medialis and the medial retinaculum.

16.3.3 Biomechanics

The main function of the medial capsular-ligamentous complex, as a whole, is to resist valgus and external rotations loads. Biomechanical studies have demonstrated that the sMCL is the primary restraint to valgus loads and external rotation. The sMCL can be thought of as two divisions, superior which inserts on the femur and inferior which inserts on the tibia [53]. The superior division is the main stabilizer to valgus stress and the inferior division functions to stabilize external rotation at 30° of flexion [54]. The anterior aspect of the sMCL remained taut throughout motion, while the PMC consistently loosened in flexion and tightened in full extension and internal rotation [55]. The ligament has a stiffness of 63 N/mm and mean load to failure of the sMCL 557 N [56].

The POL serves as a restraint to internal rotation and valgus both at and approaching full extension. It exhibits a flexion-dependent reciprocal role in resistance to internal rotation with the sMCL. The dMCL contributes to internal rotation stability depending on the flexion angle [54]. Mean load to failure of the POL and dMCL are 256 N and 101 N respectively and stiffness is 38 and 27 N/mm [56]. The POL, dMCL, and the cruciate ligaments are secondary restraints to valgus stress.

16.3.4 Mechanism of Injury

The mechanism of injury of the MCL is generally by a valgus load and external rotation of the knee. MCL sprains results from a noncontact valgus stress and complete MCL disruption usually results from a direct blow to the lateral aspect of the knee such as in sports [57].

The ability to ambulate and/or continue to participate in athletic activities depends on the degree of disruption, the player's position and the presence of any concurrent injuries.

16.3.5 Diagnosis

Similar to evaluation of ACL and PCL injuries, physical examination begins with a thorough history to attempt to determine mechanism of injury as well as symptoms. The gait should be examined. Both the injured and uninjured knee should be examined for comparison purposes. The knee should be inspected for ecchymosis, localized tenderness and an effusion. Valgus stress testing of the knee should be performed with the knee at 0° and 30° of flexion. Valgus laxity with the knee at or near full extension implies concurrent injury to the posteromedial capsule and/or cruciate ligaments. The superficial MCL is isolated with a valgus stress at 30° of flexion. Pathologic laxity is indicated by the amount of increased medial joint space separation compared with the opposite, normal knee: Grade I (mild): 1–4 mm, grade II (moderate): 5–9 mm, grade III (severe): ≥10 mm [57]. The knee should also be examined for other ligamentous injury as the MCL is often injured in conjunction with other ligaments.

Plain X-rays are typically normal but should be inspected for fractures and Pellegrini-Stieda lesions (calcification at the femoral origin of the MCL, indicative of prior MCL injury). MRI has become the imaging modality of choice to evaluate the injured MCL. The advantage is that it can identify the location and extent of injury as well as rule out associated meniscal, chondral and cruciate ligament injuries [58].

16.3.6 Treatment

Most MCL injuries can be treated without surgery with an expectation of good functional results at completion of rehabilitation [59]. Indications for nonsurgical treatment include isolated grade I and II injuries. Grade III injuries can be treated without surgery only if they are stable in extension and without associated cruciate ligament injury. The patients are placed in a knee brace to protect against valgus stress for 4 or more weeks but encouraged to walk and remain active. If the injury is severe the patient is also provided crutches. Several studies have shown that protection knee braces decrease strain on the MCL [60]. Surrogate modeling in vitro testing demonstrates that bracing decreases forces across the MCL by 20–30 %, and that custom bracing provides improved protection over off-the-shelf versions [61].

Patients treated nonoperatively with grade I or II MCL injury can expect good return of function, normal to near normal stability, and no increased risk of osteoarthritis at 10-year follow-up [62]. Holden et al treated 51 football players with grade I and II MCL injuries nonoperatively, and 80 % of the players returned to sport in an average of 21 days. The 20 % of players who failed to complete rehabilitation were found to have previously unrecognized injury associated with the ACL and/or the medial meniscus [63]. Femoral sided MCL injuries can be associated with difficulty obtaining full motion whereas tibial sided injuries can be associated with residual laxity. This laxity is typically asymptomatic however and is typically seen following nonsurgical treatment of grade II and III injuries. Pellegrini-Stieda lesions may develop overtime from overuse and may become tender from direct pressure.

Contraindications for nonsurgical treatment are laxity to valgus stress at 0° and 30° of flexion. This implies grade III injuries with the ligament displaced into the joint or concurrent ligamentous (ACL or PCL) and/or capsular damage.

Surgical treatment indications is reserved for isolated grade III injuries with persistent instability despite supervised rehabilitation and bracing, grade III injuries with valgus laxity in full extension, ligament entrapment in the joint, chronic valgus instability with associated cruciate ligament deficiency or meniscal/articular cartilage injury, or MCL combined with 1 or more other ligament injuries.

The surgical management varies depending upon the acuity of the injury. In the acute setting, primary repair can be attempted. Diagnostic arthroscopy is recommended to rule out associated damage. Ligament avulsions should be reattached directly and secured with the knee at 30° of flexion. Once the MCL is repaired, the posterior oblique ligament can also be redirected anterosuperiorly on the femur to the adductor tubercle and inferiorly to the tibial metaphysis.

In the chronic setting, surgery includes reconstruction with either autologous hamstring graft or cadaver tissue (hamstring, tibialis anterior or Achilles tendon). The goal is to reconstruct the superficial MCL with isometric fixation to the medial epicondyle (20°–30° of flexion). Knee motion should be checked after fixation of the graft. Limitation of motion or disruption of fixation indicates nonisometric graft placement.

As the anatomy of the posteromedial corner has become better defined, attention has been turned to reconstructing its functional components, namely, the POL. Because of this, some discussion is ongoing as to the best site of attachment for the posterior limb of the reconstruction. Support exists for routing the posterior limb beneath the direct head of the semimembranosus, through a posterior tibial tunnel, or directly onto the proximal sMCL tibial attachment [64]. All techniques have yielded similar clinical results. The most common complication of surgical treatment is loss of motion. For MCL injuries that are associated ligament injuries, the timing of ACL reconstruction with a concurrent MCL injury should be delayed proportional to the degree of the MCL injury to allow for ligamentous healing. That's approximately 3–4 weeks for grade I, 4–6 weeks for grade II and 6–8 weeks for grade III.

The patient is started on a rehabilitation program that involves early motion and progressive weight bearing in a long, hinged knee brace. The brace is continued for a total of 3 months to give added stability to the collateral repair.

16.3.7 Rehabilitation

Nonoperatively managed MCL injuries are initially treated bracing and crutches to allow for increased weight bearing. Physical therapy is initiated to focus on quadriceps strengthening and straight leg raises. When the knee is less uncomfortable, the patient can be transitioned to cycling and progressive resistance exercises. Therapy begins to focus on thigh adduction exercises. The patient then transitions to jogging, agility training and sport-specific drills to prepare the patient for a return to sports. At this point, a functional low profile brace is often provided for use in sports. A general guideline of return to sports is related to the degree of injury. Grade I: 5–7 days, grade II: 4–6 weeks, grade III: 6–8 weeks. Post-operative rehabilitation is managed similarly but at a slower pace to ensure healing of the graft. The therapy protocol is also modified depending on repair or reconstruction of concomitant injuries.

16.4 Lateral Collateral Ligament and Posterolateral Corner

16.4.1 Introduction

Injuries to the lateral or fibular collateral ligament (LCL) and posterolateral corner (PLC) are reported less commonly than injuries to the medial side of the knee, in part due to lack of recognition.

It has been reported that the lateral ligamentous complex is the site of 7–16 % of all knee ligament injuries [65].

16.4.2   Anatomy

The lateral compartment of the knee is supported by dynamic and static stabilizers. The dynamic stabilizers are the biceps femoris, the iliotibial band, the popliteus muscle and the lateral head of the gastrocnemius muscle. The static ligamentous structures consider the arcuate complex: LCL, popliteus tendon and arcuate ligament. The LCL is the primary stabilizer to varus stress of the knee [66].

The lateral capsular complex of the lateral aspect of the knee is divided into thirds. The anterior third attached to the lateral meniscus anterior to the LCL. The middle third attaches to the femoral epicondyle and travels inferiorly to insert on the superior tibia. The posterior third sits posterior to the LCL.

16.4.3   Biomechanics

The structures of the lateral knee and PLC provide the restraint to varus (the inferior tibia deviates inward towards the body) stress of the knee [66]. The structures of the lateral knee and PLC also function to limit external tibial rotation [67].

The LCL is the primary restraint to varus stress across the knee, providing 55 % of restraint at 5° and 69 % at 25° [66]. The mean tensile strength of the LCL is 295 N and the popliteus tendon is 700 N [68]. The popliteus, popliteofibular ligament (PFL), iliotibial band, lateral gastrocnemius tendon, and short and long heads of the biceps tendon, as well as the cruciate ligaments, are secondary restraints to varus force [65].

The LCL in conjunction with the popliteus complex become the primary restraint to tibial external rotation, providing more restriction at 30°–40° of flexion [69]. Just as the LCL and popliteus complex act as a secondary restraint to the posterior cruciate ligament (PCL) in resisting posterior tibial translation, the PCL acts as a secondary restraint to prevent external rotation of the tibia on the femur [65]. The contribution of the PLC increases in the PCL-deficient knee, in which the popliteus tendon appears to contribute the most to secondary stability of the structures of the PLC [70]. Because of this relationship, combined PCL and PLC injuries are particularly unstable to external rotation forces.

The LCL and PLC structures are also secondary restraints to internal rotation. Their contribution becomes better appreciated in the ACL deficient knee. In ACL-deficient knees, the structures of the PLC provide an important secondary anterior translation stabilization role [67]. This is particularly true during the first 40 degrees of knee flexion.

16.4.4 Mechanism of Injury

Injuries to the LCL most frequently result from motor vehicle accidents and athletic injuries. The mechanism of injury is the result of a direct blow or force to the weight-bearing knee, resulting in excessive varus stress, external tibial rotation and/or hyperextension. The most common mechanism is a posterolaterally directed force to the medial tibia with the knee in extension [71]. Isolated injury to the LCL resulting in straight instability is rare. Usually, injuries to the PLC are associated with ACL or PCL tears because only approximately 25 % of PLC injuries are isolated knee ligament tears [22]. Instability in the active patient usually is noted with the knee near full extension. Patients may experience difficulty ascending and descending stairs and during cutting or pivoting.

16.4.5 Diagnosis

Similar to evaluation of ACL and PCL injuries, physical examination begins with a thorough history to determine mechanism of injury as well as symptoms. The gait should be examined specifically for a varus thrust gait or hyperextension of the affected knee. Both the injured and uninjured knee should be examined for comparison purposes. The knee should be inspected for ecchymosis, localized tenderness and an effusion. Varus stress testing is performed at 0° and 30° of knee flexion. Isolated laxity at 30° is consistent with injury to the LCL. Laxity at both 0° and 30° is seen with additional injury to the ACL, PCL or PLC [68]. The most common classification used for the LCL injuries is based on the quantification of varus stress compared to the contralateral knee. Grade I: 0–5 mm of lateral joint opening, grade II: 6–10, grade III >10 without an end point.

The posterolateral drawer test at 30° and 90° of flexion is specific for rotatory injury to the PLC. A positive test has a soft end point with more than 3 mm of side-to-side difference compared with the uninjured knee. A positive test result at 30° is most consistent with posterolateral injury, and at 90° implies an associated PCL injury [72].

The dial test is used to determine the amount of external rotation is present compared to the uninjured side. The test is performed at 30° and 90° of flexion with the patient lying flat. The examiner then gently externally rotates the tibia while stabilizing the femur. The test is considered positive when the involved foot and ankle exhibit more than 10° of external rotation compared with the normal side. A positive test result at 30° indicates a PLC injury and at 90° a combined PCL and PLC injury [70].

The external recurvatum test is performed by the examiner lifting the great toes of both feet with the knees in full extension. The result is positive when the knee shows hyperextension and external tibial rotation.

In the reverse pivot shift test the knee is moved from flexion to extension with the knee held in valgus and the foot in external rotation. A positive test result is indicated by the reduction of the tibia with a shift or jump from its posteriorly subluxated position at 20°–30° of flexion [73].

The evaluation of neurovascular structures is imperative because up to one third of patients with acute PLC injuries have peroneal nerve deficits. If a knee dislocation is suspected, vascular studies should be performed.

As with other knee injuries, X-rays should be obtained for all patients with suspected injury to the PLC. The objective is to rule out associated injuries like osteochondral fractures, fibular head avulsion, avulsion of Gerdy's tubercle, or fracture of the tibial plateau.

Varus and posterior stress radiographs of both the injured and uninjured knee are also obtained in patients with a suspected LCL or PLC injury. LaPrade and associates have demonstrated that sectioning of the LCL results in 2.7 mm of increased lateral gapping with varus stress and that sectioning of the entire PLC allows 4.0 mm of increased lateral gapping [74]. Combined injuries to the posterolateral corner and posterior cruciate ligament should be suspected when posterior stress X-rays demonstrate more than 12 mm of posterior translation of the tibia. Full leg length X-rays should be obtained also to evaluate overall alignment, particularly in chronic cases as this can change surgical management.

MRI is the imaging exam of choice to evaluate the LCL, popliteus tendon and cruciate ligaments. It provides information about the severity and location of injury with high sensitivity and specificity [75].

16.4.6 Treatment

Nonsurgical treatment is limited to grade I or II isolated LCL injuries without involvement of the remaining lateral complex (rare). These patients have little functional instability especially if they have an overall valgus knee alignment. Treatment consists of limited immobilization with protected weight bearing for the first 2 weeks and then physical therapy. Contraindications to nonsurgical treatment include grade III injuries or avulsions of the LCL and combined rotatory instabilities involving the LCL and posterolateral compartment structures. The most common complication of nonsurgical treatment is progressive varus/hyperextension laxity due to unrecognized associated injuries to the PLC [76].

The indications for surgical treatment are complete injuries or avulsions of the LCL and rotatory instability due to concomitant ligament injury like the ACL or PCL. Primary repair of the LCL is reserved to avulsion fractures or acute injuries with substantial good quality tissue for repair. The poor quality specimens or chronic injuries are treated with reconstruction. A prospective review by Stannard et al. [77] compared repair with reconstruction. In their review, repair was inferior to reconstruction in the management of acute injuries. The reconstruction group fared better and had fewer failures (9 vs 37 %) than the repair group.

Lateral reconstruction can be performed utilizing a variety of techniques with autologous (hamstring) or cadaver grafts or even a combination of both grafts. These include nonanatomic and anatomic techniques with no differences in outcomes of either technique. However, ligamentous reconstruction of the LCL should involve placement of the graft tissue directly to the fibular head rather than to the lateral tibia to optimize graft isometry [78].

For those patients with chronic instability, the previously mentioned full length leg X-rays are evaluated for deformities of alignment. In such cases, a tibial osteotomy (cutting and realigning the tibia) is recommended before ligamentous reconstruction.

Surgery performed acutely (less than 2 weeks) has a more favorable outcome than surgery performed for chronic laxity. The main complication of surgical treatment is insufficiency of the reconstruction as demonstrated by persistent varus or hyperextension. Another complication is injury to the peroneal nerve, which can occur during surgical exposure of the fibular neck or during drilling or graft passage through the transfibular tunnel. Loss of knee motion can also occur with LCL reconstruction especially in the setting of reconstruction of multiple ligaments, especially the ACL [45].

16.4.7 Rehabilitation

Nonoperatively treated isolated LCL injuries are treated with physical therapy focusing on progressive range of motion, quadriceps strengthening and functional rehabilitation as tolerated. Return to sports can be expected in 6–8 weeks for these injuries.

Postoperatively, a rigid bracing with an extended foot piece is recommended for 4–6 weeks following surgery to prevent external tibial rotation that may occur with the use of a simple hinged knee brace. Weight bearing is limited for 6 weeks. Quadriceps sets and straight leg raises are initiated immediately postoperatively in the knee immobilizer only [79].

After a few weeks, therapy focusing on range of motion exercises is initiated. Closed-chain strengthening exercises are not initiated until 6 weeks. This focuses on quadriceps strength. Hamstring strengthening is limited so as not to stress the repair or reconstruction until at least 4 months postoperatively. An exercise bike is added when enough knee flexion is present to allow for rotation of the pedals. Sport-specific training is initiated at 4 months, with a return to sports or activity allowed when normal knee range of motion, and normal strength and stability comparable to the contralateral side, have been achieved (frequently at 6–12 months) [80].

Finally, the athlete should have completed sport-specific therapy prior to returning to competitive athletics.

16.5 Multi-ligament Knee Injuries

16.5.1 Introduction

Multi-ligament knee injuries usually are caused by high-energy trauma and are considered knee dislocations [81]. Less frequently, low-energy trauma or ultra-low-velocity trauma can result in this injury pattern in obese patients [82].

A bicruciate (2) ligament injury or a multi-ligament knee injury involving three or more ligaments should be considered a spontaneously reduced knee dislocation [83]. Knee dislocations are thought to account for less than 0.5 % of all joint dislocations [84] but should be considered a limb threatening injury. Therefore, careful monitoring of vascular status after multi-ligament knee injuries is imperative.

High-energy knee trauma can result in multi-ligament knee injury as well as additional injury to the quadriceps or patellar tendon and even result in patellar dislocation. Associated fractures are often present in injured knee and definitive fixation of unstable fractures should be performed before or concomitantly with the ligament surgery depending on the fracture pattern.

16.5.2 Anatomy

The anatomy of the ligaments of the knee has been discussed previously in this chapter. Multi-ligament knee injuries usually involve a partial or complete rupture of both cruciate ligaments and additional injury to either the medial or lateral side of the knee.

The popliteal artery courses in the posterior aspect of the knee and is a branch from the femoral artery. It is the primary blood supply to the knee and remainder of the leg. Popliteal artery injury is present in approximately one third of the cases and peroneal nerve injury can occur in 20–40 % of the cases [85].

16.5.3 Diagnosis

In the emergency room setting, acutely dislocated knees require emergent vascular exam and reduction. The limb must be reassessed after dislocation to ensure that there is blood flow to the extremity. The ankle-brachial index should be obtained in addition to the standard vascular exam. An index less than 0.9 should be considered abnormal. If any concern exists about an abnormal vascular examination, angiography should be performed. Timing is critical, and the presence of vascular compromise following reduction mandates emergent exploration by a vascular surgeon to restore arterial flow [83]. If pulses are still abnormal or absent after the knee reduction, immediate vascular surgical exploration should be indicated. If blood flow is not restored or delayed, the limb could lose perfusion resulting in loss of the limb.

In the clinical setting, an evaluation similar to the previously mentioned ligament injuries should be performed. The physical examination begins with a thorough history to determine mechanism of injury as well as symptoms. Examination of the gait is generally deferred due to pain and instability. Both the injured and uninjured knee should be examined for comparison purposes. The knee should be inspected for ecchymosis, localized tenderness and an effusion. A complete ligamentous examination should be performed to identify injured structures. If the patient is too uncomfortable, examination can be deferred until an examination under anesthesia can be done.

Neurologic examination is also critical because peroneal nerve injury can occur with multi-ligament injuries, particularly in concomitant lateral/posterolateral corner injuries.

Plain radiographs are essential in the initial evaluation of multi-ligament knee injuries. Associated fractures, fractures to the tibia plateau, fibular head or PCL avulsions may affect the timing of surgery and early open reduction and internal fixation of these fractures may improve healing. In chronic multi-ligament-injured knees, weight bearing alignment radiographs should be obtained to evaluate lower limb alignment.

MRI is useful for determining the amount of and extent of ligament injuries and thereby assist in surgical planning. This is particularly useful in severe injuries, when physical examination is often difficult because of pain and guarding.

16.5.4 Treatment

As stated previously, if a patient presents to the emergency department with an acutely dislocated knee, the knee must be reduced and vascular exams performed. If no acute vascular injury is apparent and the patient does not require emergent surgery, treatment can either be immobilization of the knee in a brace or operative spanning external fixation [86]. If the patient requires vascular surgery, the knee is then stabilized with a spanning external fixator device. External fixation allows anatomic reduction of tibiofemoral joint, but arthrofibrosis may result in substantial range of motion problems, and pin tracts may obstruct future planned ligament surgery. X-ray confirming reduction must be obtained. Patients are then discharged from the hospital and follow up in outpatient clinics for evaluation and surgical planning.

Nonsurgical management of a multi-ligament injury is rare and is usually reserved for elderly low-demand patients and patients with comorbidities that would increase surgical risks [87]. These patients can develop persistent knee instability, knee stiffness, loss of motion and gait abnormalities such as fixed varus deformity or dynamic varus thrust.

Surgical treatment, in the non-emergent setting, is indicated for injuries of two or more ligaments that result in an unstable knee [88]. Surgical stabilization is performed often times in a staged fashion after healing of associated fractures or vascular repair or bypass. There are many approaches to the surgical management of multi-ligament-injured knees. No high level of evidence is currently available on which to base definitive recommendations for surgical management [89].

Current controversies about the optimal surgical management of multi-ligament injuries include timing, type of graft, surgical approach, and the selection of ligaments for reconstruction or repair and initial stabilization. Acute reconstruction can restore knee stability early, enabling early protected range of motion, but delayed reconstruction can regain motion and allow swelling and inflammation to subside. The literature suggests that earlier reconstruction may have better outcomes compared with outcomes for chronically reconstructed knees [89]. Autologous grafts are associated with improved healing and ligamentization, but allograft can decrease morbidity considering the many structures requiring reconstruction. Open surgical approach can improve visualization and avoid risk of arthroscopic fluid from spreading throughout the remained of the leg, but arthroscopic approach can decrease morbidity. Reconstructing all ligaments can restore knee stability early, enabling early protected range of motion, but reconstructing certain ligaments and perform staged reconstructions can give gradual restoration of knee stability while limiting morbidity from each procedure and allowing restoration of motion.

Repair versus reconstruction of torn ligaments should be decided depending on the type of ligament injury and timing of surgery. Repair can only be indicated to some medial and lateral side-acute injuries within 2–3 weeks. After 3 weeks, the lateral/posterolateral structures are often scarred and retracting, which often necessitates reconstruction [90]. Missed PLC injuries have been associated with failed ACL and PCL surgery. Any bicruciate or PCL ligament injury should be examined with a high index of suspicion for such injuries.

Complications of surgical treatment are arthrofibrosis with loss of motion, recurrent instability, infection or neurovascular injury, including popliteal artery injury or peroneal nerve injury.

16.5.5 Rehabilitation

Rehabilitation after surgical stabilization is lengthy and slow. It depends largely upon how many ligaments were reconstructed. Initially most multi-ligament reconstructed knees are immobilized in extension for a period of time. Then initiation of range of motion begins with care to not disturb the grafts. Since arthrofibrosis occurs commonly following surgical reconstruction, range of motion should be initiated as soon as the surgeon will allow. Therapy will also focus on strengthening the knee and eventually retraining the patient to walk. Close supervision of the rehabilitation is advised.

References

1. Gianotti SM, Marshall SW, Hume PA, Bunt L (2009) Incidence of anterior cruciate ligament injury and other knee ligament injuries: a national population-based study. J Sci Med Sports 12(6):622–627. doi:10.1016/j.jsams.2008.07.005
2. Hootman JM, Dick R, Agel J (2007) Epidemiology of collegiate injuries for 15 sports: summary and recommendations for injury prevention initiatives. J Athl Train 42(2):311–319
3. Agel J, Arendt EA, Bershadsky B (2005) Anterior cruciate ligament injury in national collegiate athletic association basketball and soccer: a 13-year review. Am J Sports Med 33 (4):524–530. doi:10.1177/0363546504269937
4. Matsumoto T, Ingham SM, Mifune Y, Osawa A, Logar A, Usas A, Kuroda R, Kurosaka M, Fu FH, Huard J (2012) Isolation and characterization of human anterior cruciate ligament-derived vascular stem cells. Stem Cells Dev 21(6):859–872. doi:10.1089/scd. 2010.0528
5. Muller W (1983) Form, function, and ligament reconstruction. In: The knee. Springer, New York, pp 8–75
6. Sakane M, Fox RJ, Woo SL, Livesay GA, Li G, Fu FH (1997) In situ forces in the anterior cruciate ligament and its bundles in response to anterior tibial loads. J Orthop Res 15(2):285–293. doi:10.1002/jor.1100150219
7. Morrison JB (1969) Function of the knee joint in various activities. Biomed Eng 4(12):573–580
8. Yagi M, Wong EK, Kanamori A, Debski RE, Fu FH, Woo SL (2002) Biomechanical analysis of an anatomic anterior cruciate ligament reconstruction. Am J Sports Med 30(5):660–666
9. Kopf S, Musahl V, Bignozzi S, Irrgang JJ, Zaffagnini S, Fu FH (2014) In vivo kinematic evaluation of anatomic double-bundle anterior cruciate ligament reconstruction. Am J Sports Med. doi:10.1177/0363546514538958
10. Zantop T, Herbort M, Raschke MJ, Fu FH, Petersen W (2007) The role of the anteromedial and posterolateral bundles of the anterior cruciate ligament in anterior tibial translation and internal rotation. Am J Sports Med 35(2):223–227. doi:10.1177/0363546506294571
11. Gardner EJ, Noyes FR, Jetter AW, Grood ES, Harms SP, Levy MS (2015) Effect of anteromedial and posterolateral anterior cruciate ligament bundles on resisting medial and lateral tibiofemoral compartment subluxations. Arthroscopy. doi:10.1016/j.arthro.2014.12.009
12. Amis AA (2012) The functions of the fibre bundles of the anterior cruciate ligament in anterior drawer, rotational laxity and the pivot shift. Knee Surg Sports Traumatol Arthrosc 20 (4):613–620. doi:10.1007/s00167-011-1864-7
13. Woo SL, Debski RE, Withrow JD, Janaushek MA (1999) Biomechanics of knee ligaments. Am J Sports Med 27(4):533–543
14. Woo SL, Hollis JM, Adams DJ, Lyon RM, Takai S (1991) Tensile properties of the human femur-anterior cruciate ligament-tibia complex. The effects of specimen age and orientation. Am J Sports Med 19(3):217–225
15. Dargel J, Gotter M, Mader K, Pennig D, Koebke J, Schmidt-Wiethoff R (2007) Biomechanics of the anterior cruciate ligament and implications for surgical reconstruction. Strategies Trauma Limb Reconstr 2(1):1–12. doi:10.1007/s11751-007-0016-6
16. Noyes FR, Mooar PA, Matthews DS, Butler DL (1983) The symptomatic anterior cruciate-deficient knee Part I: the long-term functional disability in athletically active individuals. J Bone Joint Surg Am 65(2):154–162
17. Prodromos CC, Han Y, Rogowski J, Joyce B, Shi K (2007) A meta-analysis of the incidence of anterior cruciate ligament tears as a function of gender, sport, and a knee injury-reduction regimen. Arthroscopy 23(12):1320–1325 e1326. doi:10.1016/j.arthro.2007.07.003
18. Prins M (2006) The Lachman test is the most sensitive and the pivot shift the most specific test for the diagnosis of ACL rupture. Aust J Physiother 52(1):66

19. Noyes FR, Grood ES, Cummings JF, Wroble RR (1991) An analysis of the pivot shift phenomenon. The knee motions and subluxations induced by different examiners. Am J Sports Med 19(2):148–155
20. Kuroda R, Hoshino Y, Kubo S, Araki D, Oka S, Nagamune K, Kurosaka M (2012) Similarities and differences of diagnostic manual tests for anterior cruciate ligament insufficiency: a global survey and kinematics assessment. Am J Sports Med 40(1):91–99. doi:10.1177/0363546511423634
21. Musahl V, Hoshino Y, Ahlden M, Araujo P, Irrgang JJ, Zaffagnini S, Karlsson J, Fu FH (2012) The pivot shift: a global user guide. Knee Surg Sports Traumatol Arthrosc 20(4):724–731. doi:10.1007/s00167-011-1859-4
22. LaPrade RF, Gilbert TJ, Bollom TS, Wentorf F, Chaljub G (2000) The magnetic resonance imaging appearance of individual structures of the posterolateral knee. A prospective study of normal knees and knees with surgically verified grade III injuries. Am J Sports Med 28:191–199
23. van Eck CF, Lesniak BP, Schreiber VM, Fu FH (2010) Anatomic single- and double-bundle anterior cruciate ligament reconstruction flowchart. Arthroscopy 26(2):258–268. doi:10.1016/j.arthro.2009.07.027
24. Fu FH, van Eck CF, Tashman S, Irrgang JJ, Moreland MS (2014) Anatomic anterior cruciate ligament reconstruction: a changing paradigm. Knee Surg Sports Traumatol Arthrosc. doi:10.1007/s00167-014-3209-9
25. Mifune Y, Ota S, Takayama K, Hoshino Y, Matsumoto T, Kuroda R, Kurosaka M, Fu FH, Huard J (2013) Therapeutic advantage in selective ligament augmentation for partial tears of the anterior cruciate ligament: results in an animal model. Am J Sports Med 41(2):365–373. doi:10.1177/0363546512471614
26. Kopf S, Pombo MW, Szczodry M, Irrgang JJ, Fu FH (2011) Size variability of the human anterior cruciate ligament insertion sites. Am J Sports Med 39(1):108–113. doi:10.1177/0363546510377399
27. Hamner DL, Brown CH Jr, Steiner ME, Hecker AT, Hayes WC (1999) Hamstring tendon grafts for reconstruction of the anterior cruciate ligament: biomechanical evaluation of the use of multiple strands and tensioning techniques. J Bone Joint Surg Am 81(4):549–557
28. Staubli HU, Schatzmann L, Brunner P, Rincon L, Nolte LP (1996) Quadriceps tendon and patellar ligament: cryosectional anatomy and structural properties in young adults. Knee Surg Sports Traumatol Arthrosc 4(2):100–110
29. Schatzmann L, Brunner P, Staubli HU (1998) Effect of cyclic preconditioning on the tensile properties of human quadriceps tendons and patellar ligaments. Knee Surg Sports Traumatol Arthrosc 6(Suppl 1):S56–61
30. Yabroudi MA, Irrgang JJ (2013) Rehabilitation and return to play after anatomic anterior cruciate ligament reconstruction. Clin Sports Med 32(1):165–175. doi:10.1016/j.csm.2012.08.016
31. Wright RW, Fetzer GB (2007) Bracing after ACL reconstruction: a systematic review. Clin Orthop Relat Res 455:162–168. doi:10.1097/BLO.0b013e31802c9360
32. van Grinsven S, van Cingel RE, Holla CJ, van Loon CJ (2010) Evidence-based rehabilitation following anterior cruciate ligament reconstruction. Knee Surg Sports Traumatol Arthrosc 18 (8):1128–1144. doi:10.1007/s00167-009-1027-2
33. Kannus P, Bergfeld J, Järvinen M, Johnson RJ, Pope M, Renström P, Yasuda K (1991) Injuries to the posterior cruciate ligament of the knee. Sports Med 12:110–131
34. Wind WM (2004) Evaluation and treatment of posterior cruciate ligament injuries: revisited. Am J Sports Med 32:1765–1775
35. Fanelli GC, Giannotti BF, Edson CJ (1994) The posterior cruciate ligament arthroscopic evaluation and treatment. YJARS 10:673–688
36. Fox RJ, Harner CD, Sakane M, Carlin GJ, Woo SL (1998) Determination of the in situ forces in the human posterior cruciate ligament using robotic technology. A cadaveric study. Am J Sports Med 26:395–401
37. Kennedy JC, Hawkins RJ, Willis RB, Danylchuck KD (1976) Tension studies of human knee ligaments. Yield point, ultimate failure, and disruption of the cruciate and tibial collateral ligaments. J Bone Joint Surg 58:350–355

38. Amis AA, Gupte CM, Bull AMJ, Edwards A (2006) Anatomy of the posterior cruciate ligament and the meniscofemoral ligaments. Knee Surg Sports Traumatol Arthrosc 14:257–263
39. Harner CD, Xerogeanes JW, Livesay GA, Carlin GJ, Smith BA, Kusayama T, Kashiwaguchi S, Woo SL (1995) The human posterior cruciate ligament complex: an interdisciplinary study. Ligament morphology and biomechanical evaluation. Am J Sports Med 23:736–745
40. Ahmad CS, Cohen ZA, Levine WN, Gardner TR, Ateshian GA, Mow VC (2003) Codominance of the individual posterior cruciate ligament bundles. An analysis of bundle lengths and orientation. Am J Sports Med 31:221–225
41. Cosgarea AJ, Jay PR (2001) Posterior cruciate ligament injuries: evaluation and management. J Am Acad Orthop Surg 9:297–307
42. Montgomery SR, Johnson JS, McAllister DR, Petrigliano FA (2013) Surgical management of PCL injuries: indications, techniques, and outcomes. Curr Rev Musculoskelet Med 6:115–123
43. Covey CD, Sapega AA (1993) Injuries of the posterior cruciate ligament. J Bone Joint Surg 75:1376–1386
44. Hewett TE, Noyes FR, Lee MD (1997) Diagnosis of complete and partial posterior cruciate ligament ruptures. Stress radiography compared with KT-1000 arthrometer and posterior drawer testing. Am J Sports Med 25:648–655
45. LaPrade CM, Civitarese DM, Rasmussen MT, LaPrade RF (2015) Emerging updates on the posterior cruciate ligament: a review of the current literature. Am J Sports Med 43(12):3077–3092
46. Boynton MD, Tietjens BR (1996) Long-term followup of the untreated isolated posterior cruciate ligament-deficient knee. Am J Sports Med 24:306–310
47. Pierce CM, O'Brien L, Griffin LW, LaPrade RF (2013) Posterior cruciate ligament tears: functional and postoperative rehabilitation. Knee Surg Sports Traumatol Arthrosc 21:1071–1084
48. Colvin AC, Meislin RJ (2009) Posterior cruciate ligament injuries in the athlete: diagnosis and treatment. Bull NYU Hosp Jt Dis 67:45–51
49. Schein A, Matcuk G, Patel D, Gottsegen CJ, Hartshorn T, Forrester D, White E (2012) Structure and function, injury, pathology, and treatment of the medial collateral ligament of the knee. Emerg Radiol 19:489–498
50. Fetto JF, Marshall JL (1978) Medial collateral ligament injuries of the knee: a rationale for treatment. Clin Orthop Relat Res (132):206–218
51. Warren LF, Marshall JL (1979) The supporting structures and layers on the medial side of the knee: an anatomical analysis. J Bone Joint Surg 61:56–62
52. Hughston JC, Eilers AF (1973) The role of the posterior oblique ligament in repairs of acute medial (collateral) ligament tears of the knee. J Bone Joint Surg 55:923–940
53. Wymenga AB, Kats JJ, Kooloos J, Hillen B (2005) Surgical anatomy of the medial collateral ligament and the posteromedial capsule of the knee. Knee Surg Sports Traumatol Arthrosc 14:229–234
54. LaPrade RF, Engebretsen AH, Ly TV, Johansen S, Wentorf FA, Engebretsen L (2007) The anatomy of the medial part of the knee. J Bone Joint Surg 89:2000–2010
55. Robinson JR, Sanchez-Ballester J, Bull AMJ, Thomas RdWM, Amis AA (2004) The posteromedial corner revisited. An anatomical description of the passive restraining structures of the medial aspect of the human knee. J Bone Joint Surg Br 86:674–681
56. Wijdicks CA, Griffith CJ, Johansen S, Engebretsen L, LaPrade RF (2010) Injuries to the medial collateral ligament and associated medial structures of the knee. J Bone Joint Surg Am 92:1266–1280
57. Phisitkul P, James SL, Wolf BR, Amendola A (2006) MCL injuries of the knee: current concepts review. Iowa Orthop J 26:77–90
58. Rubin DA, Kettering JM, Towers JD, Britton CA (1998) MR imaging of knees having isolated and combined ligament injuries. Am J Roentgenol 170:1207–1213
59. Ellsasser JC, Reynolds FC, Omohundro JR (1974) The non-operative treatment of collateral ligament injuries of the knee in professional football players. An analysis of seventy-four injuries treated non-operatively and twenty-four injuries treated surgically. J Bone Joint Surg 56:1185–1190

60. France EP, Paulos LE, Jayaraman G, Rosenberg TD (1987) The biomechanics of lateral knee bracing. Part II: Impact response of the braced knee. Am J Sports Med 15:430–438
61. Brown TD, Van Hoeck JE, Brand RA (1990) Laboratory evaluation of prophylactic knee brace performance under dynamic valgus loading using a surrogate leg model. Clin Sports Med 9:751–762
62. Lundberg M, Messner K (1997) Ten-year prognosis of isolated and combined medial collateral ligament ruptures. A matched comparison in 40 patients using clinical and radiographic evaluations. Am J Sports Med 25:2–6
63. Holden DL, Eggert AW, Butler JE (1983) The nonoperative treatment of grade I and II medial collateral ligament injuries to the knee. Am J Sports Med 11:340–344
64. Yoshiya S, Kuroda R, Mizuno K, Yamamoto T, Kurosaka M (2005) Medial collateral ligament reconstruction using autogenous hamstring tendons: technique and results in initial cases. Am J Sports Med 33:1380–1385
65. Levy BA, Stuart MJ, Whelan DB (2010) Posterolateral instability of the knee: evaluation, treatment, results. Sports Med Arthrosc 18:254–262
66. Gollehon DL, Torzilli PA, Warren RF (1987) The role of the posterolateral and cruciate ligaments in the stability of the human knee. A biomechanical study. J Bone Joint Surg 69:233–242
67. LaPrade RF, Tso A, Wentorf FA (2004) Force measurements on the fibular collateral ligament, popliteofibular ligament, and popliteus tendon to applied loads. Am J Sports Med 32:1695–1701
68. LaPrade RF, Johansen S, Wentorf FA, Engebretsen L, Esterberg JL, Tso A (2004) An analysis of an anatomical posterolateral knee reconstruction: an in vitro biomechanical study and development of a surgical technique. Am J Sports Med 32:1405–1414
69. Ranawat A, Baker CL, Henry S, Harner CD (2008) Posterolateral corner injury of the knee: evaluation and management. J Am Acad Orthop Surg 16:506–518
70. Harner CD, Höher J, Vogrin TM, Carlin GJ, Woo SL (1998) The effects of a popliteus muscle load on in situ forces in the posterior cruciate ligament and on knee kinematics. A human cadaveric study. Am J Sports Med 26:669–673
71. LaPrade RF, Terry GC (1997) Injuries to the posterolateral aspect of the knee. Association of anatomic injury patterns with clinical instability. Am J Sports Med 25:433–438
72. LaPrade RF, Bollom TS, Wentorf FA, Wills NJ, Meister K (2005) Mechanical properties of the posterolateral structures of the knee. Am J Sports Med 33:1386–1391
73. Cooper DE (1991) Tests for posterolateral instability of the knee in normal subjects. Results of examination under anesthesia. J Bone Joint Surg 73:30–36
74. LaPrade RF, Heikes C, Bakker AJ, Jakobsen RB (2008) The reproducibility and repeatability of varus stress radiographs in the assessment of isolated fibular collateral ligament and grade-III posterolateral knee injuries. An in vitro biomechanical study. J Bone Joint Surg Am 90:2069–2076
75. LaPrade RF (1997) Arthroscopic evaluation of the lateral compartment of knees with grade 3 posterolateral knee complex injuries. Am J Sports Med 25:596–602
76. Djian P (2015) Posterolateral knee reconstruction. Orthop Traumatol Surg Res 101:S159–602
77. Stannard JP, Brown SL, Farris RC, McGwin G, Volgas DA (2005) The posterolateral corner of the knee: repair versus reconstruction. Am J Sports Med 33:881–888
78. Jakobsen BW, Lund B, Christiansen SE, Lind MC (2010) Anatomic reconstruction of the posterolateral corner of the knee: a case series with isolated reconstructions in 27 patients. Arthroscopy 26:918–925
79. Geeslin AG, LaPrade RF (2011) Outcomes of treatment of acute grade-III isolated and combined posterolateral knee injuries. Prospect Case Ser Surg Tech 93:1672–1683
80. LaPrade RF, Griffith CJ, Coobs BR, Geeslin AG, Johansen S, Engebretsen L (2014) Improving outcomes for posterolateral knee injuries. J Orthop Res 32:485–491
81. Kendall RW, Taylor DC, Salvian AJ, O'Brien PJ (1993) The role of arteriography in assessing vascular injuries associated with dislocations of the knee. J Trauma 35:875–888

82. Shelbourne KD, Porter DA, Clingman JA, McCarroll JR, Rettig AC (1991) Low-velocity knee dislocation. Orthop Rev 20:995–1004
83. Montgomery JB (1987) Dislocation of the knee. Orthop Clin NA 18:149–156
84. Richter M, Lobenhoffer P, Tscherne H (1999) [Knee dislocation. Long-term results after operative treatment]. Chirurg 70:1294–1301
85. Wascher DC, Dvirnak PC, DeCoster TA (1997) Knee dislocation: initial assessment and implications for treatment. J Orthop Trauma 11:525–529
86. Harner CD, Waltrip RL, Bennett CH, Francis KA, Cole B, Irrgang JJ (2004) Surgical management of knee dislocations. J Bone Joint Surg 86-A:262–273
87. Taylor AR, Arden GP, Rainey HA (1972) Traumatic dislocation of the knee. A report of forty-three cases with special reference to conservative treatment. J Bone Joint Surg Br 54:96–102
88. Richter M, Bosch U, Wippermann B, Hofmann A, Krettek C (2002) Comparison of surgical repair or reconstruction of the cruciate ligaments versus nonsurgical treatment in patients with traumatic knee dislocations. Am J Sports Med 30:718–727
89. Sisto DJ, Warren RF (1985) Complete knee dislocation. A follow-up study of operative treatment. Clin Orthop Relat Res (198):94–101
90. Shapiro MS, Freedman EL (1995) Allograft reconstruction of the anterior and posterior cruciate ligaments after traumatic knee dislocation. Am J Sports Med 23:580–587

Chapter 17
Biomaterials as Tendon and Ligament Substitutes: Current Developments

Mariana L. Santos, Márcia T. Rodrigues, Rui M.A. Domingues, Rui Luís Reis and Manuela E. Gomes

Abstract Tendon and ligament have specialized dynamic microenvironment characterized by a complex hierarchical extracellular matrix essential for tissue functionality, and responsible to be an instructive niche for resident cells. Among musculoskeletal diseases, tendon/ligament injuries often result in pain, substantial tissue morbidity, and disability, affecting athletes, active working people and elder population. This represents not only a major healthcare problem but it implies considerable social and economic hurdles. Current treatments are based on the replacement and/or augmentation of the damaged tissue with severe associated limitations. Thus, it is evident the clinical challenge and emergent need to recreate native tissue features and regenerate damaged tissues. In this context, the design and development of anisotropic bioengineered systems with potential to recapitulate the hierarchical architecture and organization of tendons and ligaments from nano to macro scale will be discussed in this chapter. Special attention will be given to the state-of-the-art fabrication techniques, namely spinning and electrochemical alignment techniques to address the demanding requirements for tendon substitutes, particularly concerning the importance of biomechanical and structural cues of these tissues. Moreover, the poor innate regeneration ability related to the low cellularity and vascularization of tendons and ligaments also anticipates the importance of cell based strategies, particularly on the stem cells role for the success of tissue engineered therapies. In summary, this chapter provides a general overview on tendon and ligaments physiology and current conventional treatments for injuries caused by trauma and/or disease. Moreover, this chapter presents tissue engineering approaches as an alternative to overcome the limitations of current therapies, focusing on the discussion about scaffolds design for tissue substitutes to meet the regenerative medicine challenges towards the functional restoration of damaged or degenerated tendon and ligament tissues.

M.L. Santos · M.T. Rodrigues · R.M.A. Domingues · R.L. Reis · M.E. Gomes (✉)
3B's Research Group, Biomaterials, Biodegradables and Biomimetics, European Institute of Excellence on Tissue Engineering and Regenerative Medicine, Avepark—Parque de Ciência E Tecnologia, Zona Industrial Da Gandra, 4805-017 Barco GMR, Portugal
e-mail: megomes@dep.uminho.pt

M.L. Santos · M.T. Rodrigues · R.M.A. Domingues · R.L. Reis · M.E. Gomes
ICVS/3B's—PT Government Associate Laboratory, Braga, Guimarães, Portugal

© Springer International Publishing AG 2017   349
J.M. Oliveira and R.L. Reis (eds.), *Regenerative Strategies for the Treatment of Knee Joint Disabilities*, Studies in Mechanobiology, Tissue Engineering and Biomaterials 21, DOI 10.1007/978-3-319-44785-8_17

17.1   Tendon/Ligament Physiology and Properties

Ligaments and tendons are connective tissues formed by dense bands of collagenous fibers. The well-ordered arrangement of these fibers within the tissue confers highly anisotropic mechanical properties and parallel fiber arrangements from the nano to the microscale. Tendons connect muscle to bone, allowing movement by transmitting forces generated by the muscle to the bone, while ligaments connect bone to bone stabilizing movement when forces are applied. Therefore these tissues of the human body play important roles in musculoskeletal biomechanics, especially at the joint level [1–3].

Tendons and ligaments vary in size, shape, orientation and anatomical location, although, in general, they are characterized by the presence of an abundant extracellular (ECM) with a few cells, the tenocytes, dispersed in rows in between the collagen fibers. The major compound of tendon ECM is type I collagen representing 60–85 % of the tendon dry weight. Type I collagen confers stiffness and strength to the tissue but other types of collagen exist in minor amounts, namely type III, V, X, XI, XII and XIV collagens. Type V collagen has been associated to type I collagen in the regulation of the collagen fibril diameter while type III collagen is functionalized in tendon repair. Type XII collagen is present in the surface of fibrils and bonds the fibrils with other matrix components such as decorin and fibromodulin [4, 5]. Due to the similarities between tendon and ligament composition as well as the complementary functions they perform in joint movement, biomechanics and stability, they are often described as the same type of tissue in tissue engineering strategies.

Tendon architecture presents a unique hierarchical organization where collagen molecules assemble and form subunits of increasing diameter and complexity (Fig. 17.1). The highly aligned collagen fibers arranged in a longitudinal way and parallel to the mechanical axis confer high tensile strength to the structure of these tissues.

The mechanical role of tendons and ligaments is based on their visco-elasticity properties allowing these structures to regain the original shape after deformation, when the deformation load resultant from the application of an external force is removed. This phenomenon occurs because of the high degree of resilience of these tissues, characterized by the capacity to absorb and store energy within the elastic range (understretch) and release that energy (when load is removed) so the tissue matrix recoils back and restores its original shape. These tissues are also temperature sensitive, which affects the rate of creep (slow elongation). For an effective elongation, tendons and ligaments should be heated and subjected to a significant load over a long time period to produce creep. If the load is sudden or excessive, the elastic limits of tendon and ligament may be exceeded and the tissue enters the plastic domain (Fig. 17.2). In this domain, the original shape cannot be restored and the tissue is permanently deformed. If loading forces continue over the plastic domain, the tissue reaches the failure point, which results in fiber rupture (Fig. 17.2) [1–3].

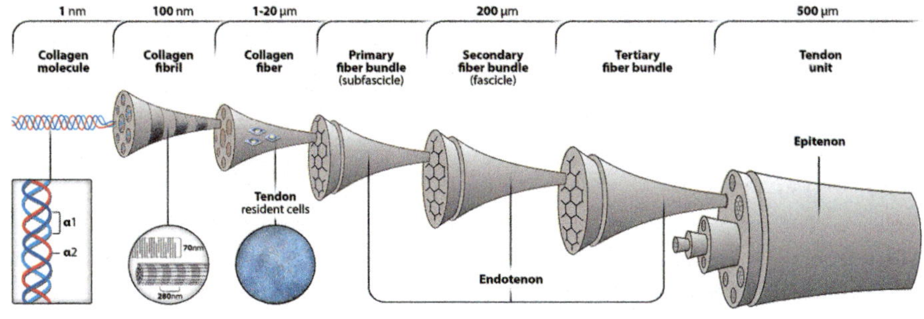

Fig. 17.1 Schematic representation of tendon/ligament architecture hierarchically organized into structural units. The tendon unit is composed of multiple fascicles of collagen fibers that result from the assembly of collagen fibers. These fibers are formed by fibrils that are formed by the aggregation of collagen molecules. This well-organized hierarchical structure from the nano to the microscale confers the anisotropic nature of tendons and ligaments and is responsible for the biomechanical properties of these tissues. Adapted from Ref. [6]. Copyright 2015, with permission of Springer

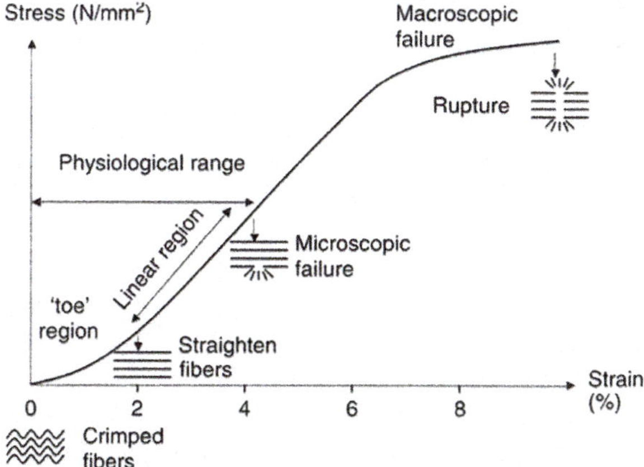

Fig. 17.2 Tendon stress-strain curve. The mechanism of tendon deformation expresses a nonlinear behavior consisting of a toe, linear, and yield regions in a stress-strain curve. Tissue fibers present a linear region until approximately 4 % strain. After that value fibers have microscopic failures and rupture by 10 % strain, showing macroscopic failure. Reprinted from Ref. [7]. Copyright 2006, with permission from Elsevier

Table 17.1 Mechanical properties of tendon and ligaments of the knee

Tissue	Elastic modulus (MPa)	Ultimate tensile strength (MPa)	Ultimate strain (%)	References
MCL	332.3 ± 58.3	38.6 ± 4.8	17 ± 2	[8, 9]
ACL	65–447	13–46	15–44	[8, 9]
PSCL	150–447	30–36	11–19	[8, 9]
LCL	345 ± 22.4	36.4 ± 2.5	16 ± 0.8	[10]

ACL anterior cruciate ligament, *MCL* medial collateral ligament, *PSCL* posterior cruciate ligament, *LCL* lateral collateral ligament

Furthermore, the strength response of tendons and ligaments under loading is determined by two main factors: size and shape of the tissue and speed of loading. The higher the number of oriented fibers in the direction of the loading the stronger with be the tissue, especially if they are wider and thicker [2]. The biomechanical properties of tendons and ligaments within the same joint vary significantly accordingly to their specific function. Knee joint is stabilized by four main ligaments: anterior cruciate ligament (ACL) which restrains anterior translation of the tibia relative to the femur, posterior cruciate ligament (PSCL) restraining posterior tibial displacement, medial collateral ligament (MCL) which restrains *valgus* angulation and lateral collateral ligament (LCL) with the function of restraining *varus* angulation. ACL presents the highest value of elastic modulus, ultimate tensile strength and ultimate strain (Table 17.1) among the described tissues evidencing more resistance to deformity when forces are applied to [8, 9].

17.2 Tendon and Ligament Response to Injury and Regeneration

Tendon and ligament injuries and associated diseases are a common problem and a leading cause of joint disability affecting athletes, active working people and the elder population worldwide. This problematic influences the quality of life of the affected population, once locomotion and local structure integrity are severely compromised.

Tendon and ligament lesions can occur through acute or chronic changes or a combination of both. Intrinsic factors as genetics, age, nutrition or misalignments are more often associated to chronic injuries while extrinsic factors namely pharmacological drug treatment or excessive or absence of mechanical loading have been related to acute injuries. Acute injuries are more frequently related to sports injuries, being the ACL the most affected to rupture in the knee joint [9, 11].

Repetitive trauma, traumatic or chemically induced injury and disuse have been described as major causes for pathogenic conditions. The healing or fail to heal capacity of tendon and ligament has a dependent correlation with intrinsic and extrinsic factors. A failed mechanism will induce molecular and histological changes affecting cellularity, tissue extracellular matrix and vascularity resulting in mechanical weakness, pain and eventual tear or complete rupture [12].

Tissue injury, characterized by three main stages (inflammation, necrosis and pain), can progress toward repair or regeneration. In tissue repair scar tissue will be formed with inflammatory cell infiltration, fibroplasia, disorderly collagen disposition and consequently, impaired mechanical properties. On the other hand, if regeneration occurs the tissue to heal will present few inflammatory cells, absence of fibroplasia, orderly collagen deposition and consequently, restoration of mechanical properties and absence of fibrotic tissue. Therefore regeneration is the desired evolution of injury for a complete regain of tissue functionality.

17.3 Current Conventional Treatments

Current available management of tendon and ligament injuries rely on conservative treatments and or surgically interventions (Table 17.2), which depend on the physio-anatomy of tissue, symptoms and clinical findings on the type and severity of the damage. Despite treatments, the mid to long term outcomes are not completely successful and tendon and ligament injuries will likely progress to nearby tissues and ultimately evolve into mild to severe forms of osteoarthritis (OA).

Because of the limitations and frequent failure of nonsurgical approaches, surgery remains the treatment of choice, especially for athletes suffering from ligament injuries, who want to remain competitively active.

Despite clinical advances and knowledge on surgical management, the replacement of damaged tissue with tissue grafting is still a gold standard despite the morbidity and functional disability of donor tissue that may have severe consequences in the long term that include pain, instability, loss of mechanical competence and degeneration of both tissue and joint.

The most commonly used grafts in anterior cruciate ligament (ACL) reconstruction are the hamstring tendon and patellar tendon. Hamstring and bone-patellar tendon-bone autografts are described to allow approximately 50 % of patients to return to their pre-injury sporting activity level. Hamstring grafts lead to better preservation of extension, higher patient-reported outcome scores, and less radiographic evidence of OA [17]. A recent study also reported a prevalence of patellofemoral OA in 26 % of the patients 12 years after ACL reconstruction function [18]. The prevalence of patellofemoral OA for the contralateral knee was

Table 17.2 Conventional treatments and associated limitations in the management of tendon and ligament injuries

Type of treatment	Aim of the treatment	Approach	Limitation	Ref.
Conservative management	– Control of pain – Reduction of inflammation and swelling	– Rest – Cryotherapy – Injection therapy (corticosteroids, sclerotherapy and hemoderivatives) – Orthontics – Continuous passive motion – Restrictive bracing – Ultrasounds – Laser treatment – Electrotherapy – Exercises at strengthening and balance	– Initial phase of damage – Limited success – Fail to regenerate tissue – Risk of disease/injury progression	[13–15]
Surgery intervention	– Reduce the symptoms – Stabilize and improve articular function	– Removal of damaged tissue (areas of failed healing, fibrosis and pathological nerve ingrowth) – Application of augmentation devices or patches or recurring to auto and allo-grafts to replace damaged tissue – Long rehabilitation	– Instability – Increased risk of failure and recurrence – Formation of scar tissue and or adhesions – Fail to regenerate tissue – Loss of tissue mechano-competence associated to functionality – Mechanical mismatch and tissue laxity – Risk of nerve damage and infection – Expiration date – No protection against long term degenerative changes *In autografts*: – Morbidity and functional disability at the harvesting site – Poor tissue integration/non-anatomic placement – Graft impingement or tension *In allografts*: – Need for immunosuppressive drugs to avoid tissue rejection – Poor tissue integration/non-anatomic placement – Risk of pathogen transmission – Graft impingement or tension	[13–16]

6 %, but only 2.5 % for uninjured contralateral knee. Significant associations were also found between patellofemoral OA and increased age, tibiofemoral OA and impaired function [18].

Other approaches involving artificial augmentation devices are also available for tissue reconstruction [19]. These include commercially available devices as LARS™, Leeds-Keio, Kennedy ligament augmentation device, Dacron, Gore-Tex and Trevira [20] being LARS one of the mostly used for ACL and PSCL. They offer advantages over tissue grafts by avoiding donor tissue morbidity and providing improved knee stability [19] and full weight bearing. Nevertheless they also have limitations and several complications have been associated to the long term follow up with artificial devices, namely, mechanical failure or mechanical mismatch with native tissues, synovitis, chronic effusions, recurrent instability and early knee OA.

Although these systems have been used for decades now, their outcomes are still controversial. Some studies with a 10 year follow up refer that the LARS™ system should not be suggested as a potential graft for primary reconstruction of the ACL [21] while others validate their application. Moreover, special indications have been described in literature for the effective and safe use of some of these devices [21].

Despite the potential and interest generated using biological augmentation for tendon and ligament reconstruction, surgeons do not seem convinced of their bio-mimicry benefits for the knee joint and preferentially choose artificial over biological devices. Biological augmentation is often mediated by decellularized mammalian-derived tissues, mainly from human (GraftJacket®), porcine (Restore™), equine (OrthADAPT®) or bovine (TissueMend®) origin [22]. The risk of immune-rejection and of zoonose transmission together with the lack of publications in recent years, limits the knowledge and clinical outcomes of patients treated with these matrices.

Tendon morphology and functionality are intrinsically associated and function depends on the highly organized hierarchy of parallel, crosslinked fibrils of collagen assembled from nano to macro structures. Partial or total loss of tendon and ligament functionality is mainly caused by a poor alignment of collagen fibrils in scar tissue, resulting in significant mechanical limitation of repaired tendons that never match the properties of healthy non-injured tissue. Thus, the creation of artificial 3D highly sophisticated and complex systems that recapitulate this hierarchical and anisotropic architecture to support a complete regeneration of damaged tissues while remaining mechanically competent is a challenge to overcome by tissue engineering technologies.

17.4   Tissue Engineering Strategies for Tendon and Ligament Regeneration

17.4.1   Tissue Engineering and Regeneration

Regeneration represents one of the most important biological processes, assisting the renewal and remodeling of tissues and organs which have suffered physical damage or injury. With regeneration the normal structure and function of the tissue/organ are completely restored into a functional level and homeostasis is reestablished. On the other hand, this does not occur in tissue repair, a failed attempt to regenerate, resulting in the synthesis of disorganized fibrotic tissue with inferior properties to the original healthy tissue [23]. Thus, tissue engineering and regenerative medicine proposes alternative approaches combining living agents, the cells, with 3D structures to mimic the biophysical and chemical cues of native extracellular matrix, and/or bioactive molecules to biochemically stimulate cells and the tissue milieu, to meet the demanding requirements of tissue regeneration.

17.4.2 Cell-Based Strategies for Tendon and Ligament Tissues

The hypocellular and hypovascular nature of tendons and ligaments comparatively to other tissues has been associated to a very limited natural healing capacity with significant drawbacks for a successful regeneration, especially when severe injuries occur. Moreover, failure to regenerate increases the risk for progression of associated diseases into nearby tissues, inflicting more pain and degeneration to the already injured joint [5, 24]. Thus, it is not surprising that some potential regenerative approaches for tendon and ligament focus on cell based strategies to overcome these limitations and accelerate a tissue regenerative response.

Tendon and ligament resident cells are an obvious choice [25] since these cells are harvested from the target tissue and an eventual level of epigenetic memory could match the desired cell response to meet regeneration in damaged tendons or ligaments. In 2007 Bi and co-workers discovered a tendon stem/progenitor cell population with functionally attractive features including universal stem cell characteristics such as clonogenicity, multipotency and self-renewal capacity, and with the capability to generate a tendon-like tissue after in vitro expansion and in vivo transplantation [26]. Although isolating autologous cells from tendons and ligaments is a feasible process, the harvesting of resident cells, even in limited number, is not the most adequate option, as it may interfere with donor tissue homeostasis, causing severe tissue morbidity. Cells harvested from a different donor are a valid but not so desirable alternative, since tissue supply is limited and there is an associated risk of rejection or disease transmission.

Pluripotent embryonic stem cells (ESCs) are an alternative source to tendon cells, whose potential for the treatment of tendon injuries has been demonstrated in a patellar defect of a rat model [27], resulting in improved mechanical and structural properties without teratoma formation. Nevertheless, the ethical issues associated to the manipulation of human embryos, and the risk of tumor formation post implantation limits advances in human ESC knowledge and prevents new insights for regenerative medicine.

Induced pluripotent stem cells (iPSCs) technology also presents value for tendon and ligament regeneration, as iPSCs can be reprogrammed into a wide range of cell types providing an inexhaustible source of autologous cells without the ethical considerations of ESCs but still resembling and sharing some ESC characteristics. A recent study reported that a treatment with iPSCs derived from neural crest stem cells significantly enhanced tendon healing in a window defect of a rat patellar tendon with improvement in matrix synthesis and mechanical properties [28]. Despite interesting outcomes and a promising future for clinical applications, it is necessary to improve the production efficiency of human iPSCs and assess human safety application for cell therapy.

Adult tissues may also be interesting alternatives as stem cell sources. It is widely described that practically all tissues in the human body have a stem cell population that participates in the endogenous regeneration of the tissue. The role of these stem cells is mediated by the release of trophic factors that influence tissue milieu. Adult stem cells are not pluripotent as ESCs but can commit and differentiate into several tissue lineages and have a high self-renewal capacity.

Bone marrow stem cells are the most studied stem cells of adult origin and were shown to have tenogenic differentiation potential [29, 30]. Furthermore, human bone marrow mesenchymal stem cells (MSCs) supported tendon healing when implanted into artificially induced tendinitis in rat Achilles tendon, promoting neovascularization and produced larger deposits of type I collagen and type III collagen and better organization of the extracellular matrix [31]. No tumor formation or excessive inflammatory reaction was locally detected at the rat tendon [31].

Adipose tissue [32] and amniotic fluid [32] have been also reported to be promising for tendon repair, having the ability to commit into a tenogenic phenotype as measured by increased genetic and protein expression of tendon related markers, namely type I and III collagen, decorin, tenascin C and scleraxis under supplemented culture medium. In comparison to other adult stem cell sources, adipose tissue offers a more abundant source and less invasive procedures for harvesting stem cells with immunomodulatory properties and long-term genetic stability.

Adipose tissue-derived MSCs were applied to the treatment of induced tendinitis of the superficial digital flexor tendon in the horse [33]. The lesions that received treatment with these cells presented a more organized and uniform tissue repair as compared with the control limb, including less inflammatory infiltrate, greater parallel arrangement of the fibers, larger extracellular matrix deposits, and greater type I collagen expression [33].

Despite the growing knowledge on regeneration mediated by stem cells and the fact that bone marrow stem cells have already found a clinical niche in cell therapies for the treatment of several (non-tendon/ligament) diseases, stem cell therapies for tendon and ligament require further research in order to understand the mechanisms of regeneration, recapitulate them in vitro and translate the appropriate stimuli in a spatial-temporal manner toward successful cell-based therapeutic tools. The still limited knowledge about the tenogenic process and associated markers together with the lack of standardization of biochemical cocktails to induce in vitro tenogenesis are holding back the understanding and recapitulation of tissue regeneration. However, it is expected that these drawbacks will be overcome in the next few years and will bring significant insights in therapeutics toward tendon and ligament tissues.

17.4.3 Design and Fabrication of 3D Sophisticated Scaffolds

As mentioned above, the complex hierarchical ECM is essential for tendon and ligament functionality and responsible to be an instructive microenvironment for resident tendon cells.

A key challenge in tendon and ligament TE is exactly the recreation of 3D scaffolding biomaterials that can mimic this unique architecture and support tissue regeneration while remaining mechanically competent [6, 7].

Being mechano-sensitive and mechano-responsive tissues, the cell response may be assisted and guided by 3D structures that would recreate tendon microenvironment with specific topographical and biophysical cues such as the substrate geometry and topography of fiber based scaffolds. The incorporation of growth factors (GFs) and other bioactive molecules within a 3D scaffold can also improve the biofunctionality of the system, once GFs were shown to play a crucial part in tendon and ligament repair [34–41].

However, the development of tendon/ligament scaffolds is a nontrivial issue as they should match the mechanical properties of the targeted tissue in order to allow appropriate functionality, but progressively degrade over time at a rate matching tissue regeneration while preserving the overall construct tensile properties and reduce the risk of premature rupture. Thus, a suitable scaffold should be tailored considering the properties of the biomaterials they are produced from, the scaffold design and architecture as well as the processing technique [42].

Scaffold biomaterials can be synthetic, natural based or a combination of both. Synthetic polymers are known for their higher processability being more versatile in fitting a wide range of properties and structural features, while natural polymers, such as alginate, chitosan, hyaluronic acid are obtained from renewable and abundant sources and may be biodegraded by enzymes naturally present in the body. Moreover, the fact that the biological and chemical properties of the latter share similarities to living tissues, in particular to the extracellular matrix, can be an advantageous parameter for cellular recognition in TE strategies.

Several potential biomaterials for the development of tendon or ligament scaffolds and associated advantages are summarized in Table 17.3.

Aligned fibrous materials have been among the preferred options as potential scaffolds in tendons and ligaments [64–66] mainly due to the linear and fibrillar organization of collagen molecules into fibrils, fiber bundles, fascicles and tendon units. These materials can be obtained through different fiber fabrication technologies, but in the past few years electrospinning [64, 65] and electrochemical alignment technique [46, 67] combined with textile techniques have been in the

Table 17.3 Examples of scaffold biomaterials that have been studied for tendon and ligament TE strategies

Origin	Biomaterial	Processing method	Advantages	References
Natural	Collagen	Wet-spinning Electrochemical alignment	Reasonable mechanical properties Relatively slow rate of degradation Main component of T/L	[43–46]
	Silk	Electrospinning Knitting	Good mechanical properties Slow rate of biodegradation	[39, 47–51]
	Alginate/CHT	Wet-spinning	Provide a proper substrate for fibroblast growth with dense type I collagen production	[52]
Synthetic	PLGA	Electrospinning	Easier to process through different techniques Large scale production with lower cost Higher mechanical properties comparing with natural polymers	[39, 53]
	PLLA	Melt-spinning		[54–57]
	PGA			[55]
	PCL	Freeze drying Electrospinning		[58, 59]
Synthetic/naturals	PLCL/Collagen	Electrospinning	Combination of the best properties of natural and synthetic polymers	[60, 61]
	PCL/CHT			[62]
	PCL/CHT/CNC			[63]

T/L tendon/ligament, *PLGA* poly(lactic-co-glycolic acid), *PLLA* poly-L-lactic acid, *PGA* poly (glycolic acid), *PCL* polycaprolactone, *PLCL* poly(L-lactide-co-ε-caprolactone), *CHT* chitosan, *CNC* cellulose nanocrystals

development forefront of hierarchical scaffolds for the proposed TE applications. These strategies have been showing promising results in this field and will be discussed in more detail in the following sections.

17.4.3.1   Electrospinning

Electrospinning produces long continuous fibers with controlled diameter from nanometers to microns. The advantage of electrospinning comparing with other conventional techniques is that the produced fibrilar systems better mimic the nanoscale morphological structure of tendon and ligament ECM, in order to provide the topographical cues and promote cells contact guidance, increasing the potential for regeneration.

Following this strategy it is possible to artificially reproduce the characteristic anisotropic alignment of collagen fiber bundles in these tissues [68, 69].

This technique also enables the production of fibers from different polymers including those from natural origin, such as collagen, chitosan, hyaluronic acid and silk fibroin; or synthetic, for example poly(ε-caprolactone) (PCL), poly(glycolide) (PLGA), poly(L-lactide) (PLA). Using natural and synthetic polymeric blends it is possible to combine in a single system the adequate mechanical properties from synthetics and the favorable biological response of cells/tissue from natural based biomaterials aiming to mimicking the natural ECM of tendon and ligament [53, 70, 71]. These include e.g., poly(L-lactide-co-ε-caprolactone)/collagen [60, 61] or poly-ε-caprolactone/chitosan (PCL/CHT) [62, 72] aligned nanofibrous scaffolds.

Several systems have been tested to produce aligned nanofiber mats, but high speed rotating collectors forcing to an aligned nanofiber deposition is usually the preferred strategy to produce T/L nanofiber scaffolds [65].

The resulting nano/microtopography of the scaffolds fabricated through this technique has proven advantageous. Anisotropically aligned nanofibrous scaffolds provide tendon biomimetic cues that induced cell alignment through contact guidance mechanisms, resulting in proved impact on cell alignment along the nanofiber aligned axis, but also over stem cells differentiation, phenotype maintenance as well as matrix deposition, while recreating the anisotropic mechanical behavior of tendon tissues [24, 53, 62, 71, 73, 74]. Moreover, this type of well-aligned fiber scaffolds has recently demonstrated to enable the multistep tenogenic differentiation of hiPSCs in vitro and the resulting tissue engineered constructs promoted tendon repair in vivo in a rat Achilles tendon model [66]. The tenogenic commitment of these cells was assigned to the activation of the mechanic-signal pathway resulting from the cytoskeletal rearrangement induced by the scaffold's topography [66].

Recreating the characteristic non-linear tri-phasic deformation behavior of tendons and ligaments is also a very important feature to consider. The toe region under low deformation (typically between 0–2 % strain), which results from the uncramping of collagen fibrils, is important in tendon/ligament biomechanics as shock-absorbing feature to prevent tissue damage. Recent studies developed crimped electrospun scaffolds [75–77]. The crimped pattern on electrospun fiber mats provide a closer mimic of the nonlinear biomechanical behavior of collagen fibrils, which facilitate nonlinear stiffening of the tissue under increasing tensile strains [75]. Furthermore, tendon cells cultured on crimped nanofibers showed a higher level of tolerance toward externally applied strain than those cultured on the straight nanofibers, suggesting that the crimping feature in nanofiber-based scaffolds has a high potential for tendon and soft connective tissue repair.

Considering the anatomic load-bearing function of tendon and ligaments, the tensile behavior of the proposed scaffolds are also critical parameters. Scaffold's design for tendon/ligament applications should match the mechanical properties of native tissues, not only to minimize stress-shielding effects that may lead to disorganized tissue growth, but also to provide a temporary replacement for immediate function with reduced probability of premature rupture after repair, while allowing tissue regeneration over time [78].

The tensile properties reported for native tendons and ligaments are in the range of 5–100 MPa of ultimate tensile strength and 20–1200 MPa of Young's modulus, while strain at failure varies much less and typically fall in the range of 10–15 % [79]. A recurrent concern about the relevance of electrospun scaffolds in tendon/ligament TE strategies is related with their reported limited mechanical performance. Electrospun scaffolds have been preferentially fabricated based on semi-crystalline biodegradable polyesters and their tensile mechanical properties generally range from 1 to 25 MPa in ultimate tensile strength and 1–350 MPa in elastic modulus [14]. While it may be sufficient when applied in comparatively low tensile demanding tendon/ligament tissues such as those from human shoulder, they are not satisfactory when targeting high tensile demanding tissues such as patellar, Achilles tendons or ACL. Our research group has recently addressed this issue by reinforcing tendon/ligament fibrous scaffolds made of a natural/synthetic polymer blend of PCL/CHT with the incorporation of cellulose nanocrystals (CNC) which are bioderived nanofillers [63]. The incorporation of low CNC contents (up to 3 wt %) into PCL/CHT polymer blend to form nanofiber bundles had a significant toughening effect (increased 132 % the Young's modulus and 83 % ultimate tensile strength without significantly affecting strain at failure). Moreover, this reinforcement was achieved while maintaining the structural cues for the superior biological performance (Fig. 17.3). This may thus be a suitable strategy to explore in order to fabricate tendon/ligament mimetic nanofibrous structures balanced with appropriate mechanical performance and expand their potential of applications in this field.

Nonetheless, electrospinning typically produces 2D fiber mats, restricting their further processing into higher hierarchical 3D structures. Thus, scaffolds are limited in terms of dimensions, handling and load-bearing capacity. Different strategies have been devised in recent years [80], including rolling [81] or twisting [82] pre-cut sections from aligned nanofiber to produce nanofiber bundles, or their assembly into higher 3D hierarchical structures through standard textile such as e.g. weaving or braiding [24, 74]. These strategies allow producing aligned nanofiber scaffolds of relevant dimensions, while enabling to tune their general mechanical properties (stiffness, strength, strain and maximum load) to mimic the typical tri-phasic biomechanical behavior of tendons and ligaments by artificially recreating the characteristic toe region in a load-displacement curve.

However, these fabrication strategies are less practical in terms of clinical translation, as well as on their scale up production and standardization because they are based on the typical rotating collection drums or wheels which produce a 2D nanofiber sheet or bundle of limited dimensions. Therefore, the production of continuous electrospun nanofiber yarns as mimicry of tendon fascicles is suitable for further assembling into higher hierarchical 3D structures through standard textile techniques and devices would be a significant breakthrough in T/L TE. Several strategies for their production have been developed in the past few years [65] and recently Mouthuy et al. devised an automated method that enables the manufacture of continuous electrospun filaments with the potential to be integrated

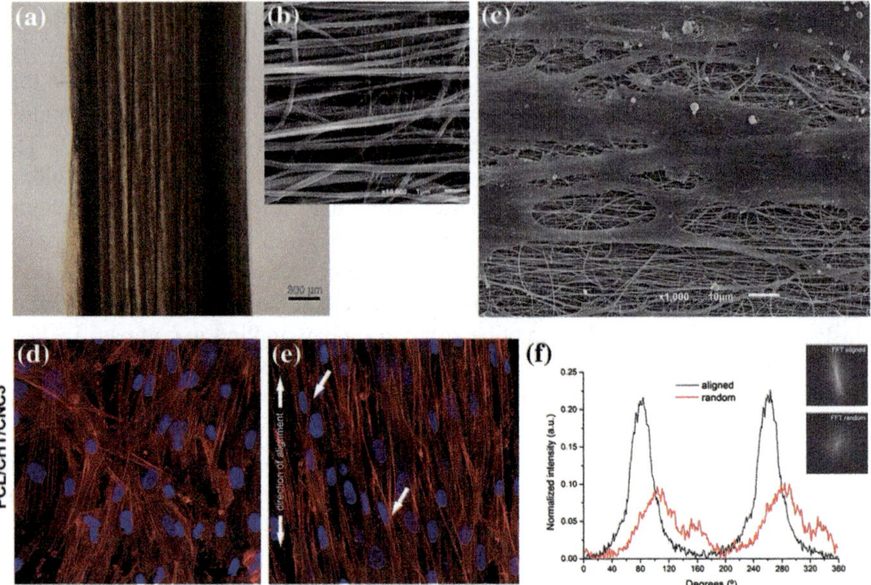

Fig. 17.3 Optical (**a**) and high magnification SEM (**b**) micrographs of aligned nanofiber bundles of PCL/CHT/CNC3. **c** SEM micrographs of hTDCs seeded on PCL/CHT/CNC3 nanofibrous scaffolds with aligned topography after 10 days of culture. Confocal microscopy micrographs of the hTDCs seeded on PCL/CHT/CNC3 nanofibrous scaffolds with random (**d**) and aligned (**e**) topography (*blue* nuclei stained with DAPI; *red* actin filaments stained with rhodamine-conjugated phalloidin). **f** Respective 2D FFT frequency plots (*insets*) and normalized radial intensity plotted against the angle of rotation for hTDCs cultured on random and aligned nanofibers. *Scale bar* 300 µm (**a**), 1 µm (**b**), 10 µm (**c**). Reprinted from Ref. [63]. Copyright 2016 WILEY VCH Verlag GmbH & Co. KGaA, Weinheim. With permission from John Wiley and Sons

into existing textile production lines [65, 83]. In practice the proposed technology is a multistep process (Fig. 17.4) relying on the use of a wire guide to collect a nanofiber mesh which is then detached as a long and continuous thread, the thread is drawn to align the nanofibers and then is twisted to create multifilament yarns.

This concept, the first of his kind, is still in the early stages of development and lack optimization. However, further developments are expected in coming years on biotextile scaffolds based in continuous nanofiber yarns for T/L TE.

In a different strategy to produce 3D hierarchical scaffolds for T/L TE, Yang et al. recently proposed a multilayered fiber-hydrogel composite approach [84]. The concept consisted in simultaneously co-electrospin PCL and methacrylated gelatin using the typical rotating collection drum. The 2D fiber mat sheets are wet with photoinitiator solution and then photocrosslinked to produce the fiber-hydrogel composite scaffolds. Stacking multiple sheets prior to photocrosslinking allows the production of multilayered scaffolds as well as the encapsulation of cells, if desired, within layers. Although the results support that a combination of nanofibrous

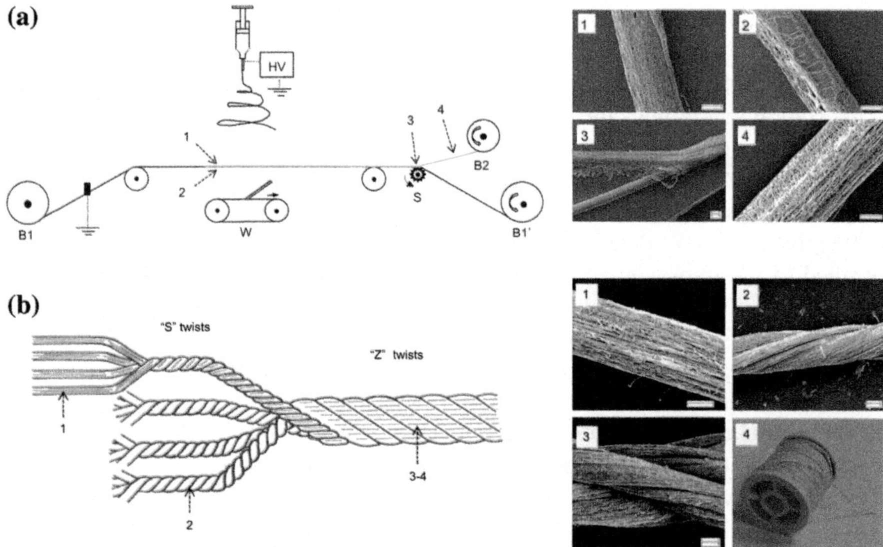

Fig. 17.4 Method used for the fabrication of continuous electrospun filaments and multi filament nanofiber yarns. **a** Sketch of the manufacturing process. The method consisted in spinning the polydioxanone (PDO) fibres on a stainless steel wire progressing at a speed of 0.6 mm s^{-1} underneath the electrospinning nozzle (*B1* wire supply, *B1′* wire collection, *B2* electrospun filament collection, *S* cutter wheel, *W* wiper). The electrospun material was then separated from the wire in the form of a continuous filament. (*1–4*) Scanning electron microscope images taken at different positions in the process. Fibers are mostly collected on the side of the wire exposed to the electrospinning jet (*1*) compared to the hidden side, (*2*) the mesh can be separated from the wire, (*3*) as one continuous thread of randomly oriented submicrofibres (*4*). **b** The stretched filaments (*1*) were assembled into 4-plied yarns by manually twisting four filaments together in a *right-hand direction* ('S' turn) at 400 twists/m. (*2*) Four of these were then twisted together in a *left-hand direction* ('Z' turn) at 200 twists/m to fabricate a cord yarn (*3, 4*). Adapted from Ref. [83] with permission

structures and photocrosslinked hydrogels may closer mimic the T/L structure, providing the aligned topographical cues and contributing for the tenogenic differentiation of hASCs [84], the final mechanical properties are far from ideal (ultimate tensile strength of 1.45 ± 0.19 MPa), which restricts their potential application in this field.

17.4.3.2 Electrochemical Alignment Technique

The so called electrochemical alignment technique is an interesting strategy that allows producing anisotropically aligned collagen bundles through a process based on the pH gradient created between two parallel electrodes [46, 67, 85]. This strategy was firstly proposed for TE of connective tissues by Akkus group [86], that have been developing this technique, culminating in a recent proposed system for the production of continuous electrochemically aligned collagen (ELAC) threads [67].

Their pioneer study has shown that the ELAC threads exhibit a collagen fibrillar organization that mimics the packing density, alignment and strength of T/L tissues [86]. Following studies evaluated the effect of several processing parameters in the mechanical and structural properties of ELAC threads, such as buffer concentration and incubation time [87], crosslinking degree (with genipin) [88] and addition of proteoglycan (decorin) to the collagen matrix [89]. Ultimately, the optimization resulted in the improvement of the of ELACs wet ultimate tensile strength to the range of 80–110 MPa, elastic modulus of 600–900 MPa and strain of 10–15 %, which approximates the values for several native tendons.

Gurkan et al. compared the ability of tendon-derived fibroblasts (TDFs) and bone marrow stromal cells (MSCs) to migrate and populate single and braided ELAC threads crosslinked with genipin [46]. The results support the non-cytotoxicity of crosslinked ELACs and that in vitro both cell types colonize and migrate over ELAC threads more successfully than in singles threads [46]. It was also demonstrated that different crosslinking degrees and threads coagulation treatments have an impact over hMSCs adhesion and proliferation, probably resulting from the different threads stiffness [88]. Moreover, the anisotropically aligned topography of the genipin crosslinked ELAC threads proved to promote tenogenic differentiation of hMSCs [90] and ELAC braided scaffolds showed in vivo biocompatibility and biodegradability after 8 months in a rabbit patellar tendon model [91].

Recently Younesi et al. developed a custom made rotating electrode electro-chemical alignment device able to produce ELAC threads in a continuous mode [67]. Applying biotextile techniques, woven 3D-biotextile scaffold were fabricated with these threads (Fig. 17.5A). The 3D woven structure not only mimics the hierarchical structure and non-linear tensile behavior of native tendons, as MSCs seeded on the scaffold also express increased tendon specific markers when compared to randomly oriented collagen gels. This 3D-biotextile scaffold woven purely from collagen has a remarkable high porosity (80 %) which promotes cell seeding across the bioscaffold (Fig. 17.5B), while the anisotropically aligned substrate texture topographically stimulates tenogenesis [67].

These woven 3D-biotextile scaffolds were tested to span a gap defect between the *infraspinatus* muscle and humerus in rabbit [92]. Although the main objective of the study was to develop the suturing scheme for grafting the scaffold, it was shown that the graft repair was able to withstand similar load as the direct tendon reattachment, thus demonstrating the potential of this system as a full load-bearing construct for segmental defects.

Overall, the previously described outcomes from the in vitro and in vivo studies on ELAC threads scaffolds are encouraging and promising for T/L TE applications. However, it is also important to refer that, considering the basic principles of this fabrication technology, there might be a particular limitation in terms of polymer matrix option. In fact, with the exception of one study [89], scaffolds have been exclusively based on type I collagen. Also, the performance of ELAC based

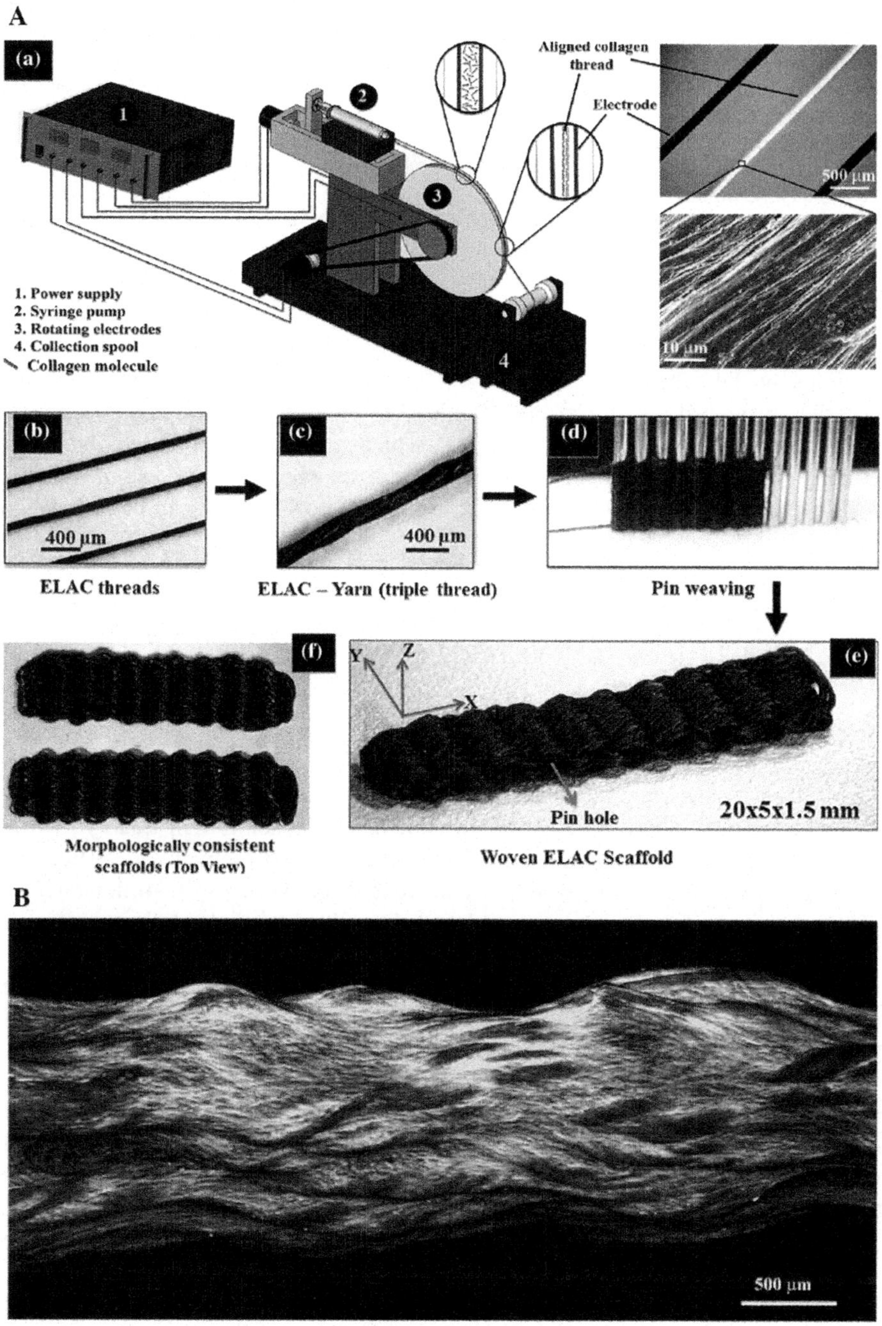

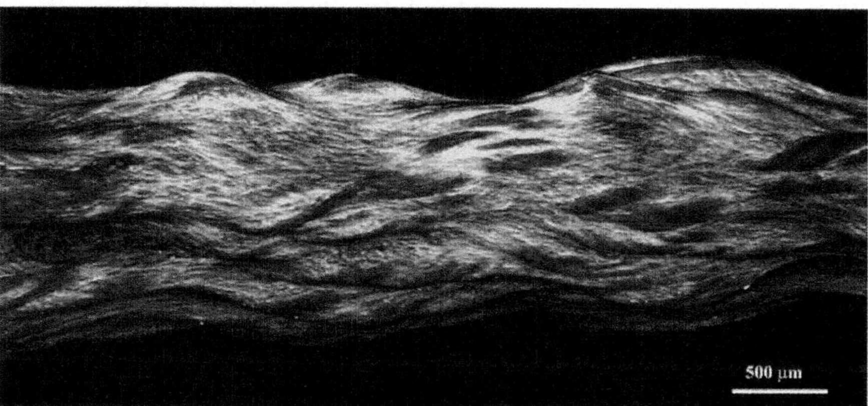

A
(a)
1
2
3
4
Aligned collagen thread
Electrode
500 µm
10 µm
1. Power supply
2. Syringe pump
3. Rotating electrodes
4. Collection spool
Collagen molecule
(b)
400 µm
ELAC threads
(c)
400 µm
ELAC – Yarn (triple thread)
(d)
Pin weaving
(f)
Morphologically consistent scaffolds (Top View)
(e)
Y Z
X
Pin hole
20x5x1.5 mm
Woven ELAC Scaffold
B
500 µm

◀ **Fig. 17.5** **A** Schematic of the rotating electrode electrochemical alignment device (**a**). The main parts of the device are: power supply for providing voltage for the electrochemical cell, the syringe pump, rotating electrodes wheel and collection spool. Compensated polarized image in the *top left* inset demonstrates the collagen molecules to be aligned parallel to the longer axis of the thread as manifested by the *blue color*. Closely packed and aligned topography of the fiber surface is evident from the electron microscopy image. **b** Collagen fiber made by a rotating electrode electrochemical aligning device (REEAD), **c** yarn made by twisting three collagen threads, **d** pin-setup for weaving the collagen scaffold, **e** the resulting woven collagen scaffold, **f** and two scaffolds to demonstrate the consistency of fabrication. **B** A macrograph of cell-seeded scaffold where the cellular F-actin cytoskeleton is labeled. Cells have profusely covered the entire scaffold with elongated morphology. Reprinted from Ref. [67]. Copyright 2014 WILEY-VCH Verlag GmbH & Co. KGaA, Weinheim. With permission from John Wiley and Sons

scaffolds in T/L TE is still limited and has not yet been compared with other materials produced e.g. by electrospinning, as discussed on previous section, or decellularized tendon-derived matrix [93], which share the same basic composition of ELAC scaffolds. This lack of versatility may reveal critical in case of failure at any developmental stage because it restricts the approaches that can be followed to solve possible drawbacks.

17.5 Conclusions

In this chapter, it was presented the important role of tendon and ligament tissues in articular biomechanics and their unique hierarchical organization. Injuries associated to these tissues are a common problem and a leading cause of joint disability affecting the population in general, consequently influencing their quality of life.

The current conventional techniques to manage this problematic, in particular in knee ligaments and tendons, were presented and discussed. In spite of the reduction of the symptoms, i.e. the pain control and the reduction of the inflammation and swelling, the mid and long term outcomes may progress to knee OA and/or T/L weakening. Moreover, due to the limitations of nonsurgical approaches, the resource to surgery remains the most recurrent option to reduce the symptoms and provide some short to mid term quality of life to patients.

In this sense, TE may become the most promising alternative therapy to achieve a complete regeneration of the damaged tissue, recovering its native biomechanical functionality. Nevertheless, despite the remarkable progresses made in the last few years, there are still significant challenges to overcome in this field. It spans from the establishment of cell sourcing and differentiation protocols, to the refinement of 3D scaffold designs that can simultaneously mimic the hierarchical and anisotropic architecture of T/L and match their biomechanical behavior.

The continuous developments and promising outcomes are expected to bring T/L TE strategies a step closer to clinic practice in a near future.

Acknowledgments The authors wish to acknowledge the financial support of the Portuguese Foundation for Science and Technology for the post-doctoral grant (SFRH/BPD/111729/2015) and for the projects Recognize (UTAP-ICDT/CTM-BIO/0023/2014) and POCI-01-0145-FEDER-007038.

References

1. Frank CB (2004) Ligament structure, physiology and function. J Musculoskelet Neuronal Interact 4(2):199–201
2. Weintraub W (2003) The nature of tendons and ligaments. In Tendon and ligament healing: a new approach to sports and overuse injury. P. Publications, pp 5–24
3. Thorpe CT et al (2015) Chapter 1—Tendon physiology and mechanical behavior: structure–function relationships. In: Gomes ME, Reis RL, Rodrigues MT (eds) Tendon regeneration. Academic Press, Boston, pp 3–39
4. Woo SL-Y et al (2007) Chapter 9—Functional tissue engineering of ligament and tendon injuries. In: Mao JJ et al (eds) Transitional approaches in tissue engineering and regenerative medicine. Artech House Publishers, Norwood
5. Hsu S-L, Liang R, Woo SL (2010) Functional tissue engineering of ligament healing. BMC Sports Science, Medicine and Rehabilitation, pp 2–12
6. Costa-Almeida R et al (2015) Tendon stem cell niche. In: Turksen K (ed) Tissue engineering and stem cell niche. Springer, Berlin, pp 221–244
7. Wang JHC (2006) Mechanobiology of tendon. J Biomech 39(9):1563–1582
8. Andrades JA et al (2011) Chapter 5—Skeletal regeneration by mesenchymal stem cells: what else? In: Eberli D (ed) Regenerative medicine and tissue engineering—cells and biomaterials. InTech, Morn Hill
9. Ghosh KM, Deehan DJ (2010) Soft tissue knee injuries. Surg Oxf Int Ed 28(10):494–501
10. Cowin SC, Doty SB (2007) The constituents of tendons and ligaments. In: Cowin SC, Doty SB (eds) Tissue mechanics. Springer, New York, p 562
11. Sharma P, Maffulli N (2006) Biology of tendon injury: healing, modeling and remodeling. J Musculoskelet Neuronal Interact 6(2):181–190
12. Riggin CN, Morris TR, Soslowsky LJ (2015) Chapter 5—Tendinopathy II: etiology, pathology, and healing of tendon injury and disease. In: Gomes ME, Reis RL, Rodrigues MT (eds) Tendon regeneration. Academic Press, Boston, pp 149–183
13. Ackermann PW (2015) Chapter 4—Tendinopathy I: understanding epidemiology, pathology, healing, and treatment. In: Gomes ME, Reis RL, Rodrigues MT (eds) Tendon regeneration. Academic Press, Boston, pp 113–147
14. Rodrigues MT, Reis RL, Gomes ME (2013) Engineering tendon and ligament tissues: present developments towards successful clinical products. J Tissue Eng Regen Med 7(9):673–686
15. Siegel L, Vandenakker-Albanese C, Siegel D (2012) Anterior cruciate ligament injuries: anatomy, physiology, biomechanics, and management. Clin J Sport Med 22(4):349–355
16. Kiapour AM, Murray MM (2014) Basic science of anterior cruciate ligament injury and repair. Bone Joint Res 3(2):20–31
17. Mascarenhas R et al (2012) Bone-patellar tendon-bone autograft versus hamstring autograft anterior cruciate ligament reconstruction in the young athlete: a retrospective matched analysis with 2–10 year follow-up. Knee Surg Sports Traumatol Arthrosc 20:1520–1527
18. Øiestad BE, Holm I, Engebretsen L, Aune AK, Gunderson R, Risberg MA (2013) The prevalence of patellofemoral osteoarthritis 12 years after anterior cruciate ligament reconstruction. Knee Surg Sports Traumatol Arthrosc 21(4):942–949
19. Liu Z-T et al (2010) Four-strand hamstring tendon autograft versus LARS artificial ligament for anterior cruciate ligament reconstruction. Int Orthop 34(1):45–49
20. Batty LM et al (2015) Synthetic devices for reconstructive surgery of the cruciate ligaments: a systematic review. Arthroscopy 31(5):957–968

21. Tiefenboeck TM et al (2015) Clinical and functional outcome after anterior cruciate ligament reconstruction using the LARS™ system at a minimum follow-up of 10 years. Knee 22 (6):565–568
22. Chen J et al (2009) Scaffolds for tendon and ligament repair: review of the efficacy of commercial products. Expert Rev Med Devices 6(1):61–73
23. Yannas IV (2001) Tissue and organ regeneration in adults. Springer, New York
24. Czaplewski SK et al (2014) Tenogenic differentiation of human induced pluripotent stem cell-derived mesenchymal stem cells dictated by properties of braided submicron fibrous scaffolds. Biomaterials 35(25):6907–6917
25. Steinert AF et al (2011) Mesenchymal stem cell characteristics of human anterior cruciate ligament outgrowth cells. Tissue Eng Part A 17(9–10):1375–1388
26. Bi Y et al (2007) Identification of tendon stem/progenitor cells and the role of the extracellular matrix in their niche. Nat Med 13(10):1219–1227
27. Chen X et al (2009) Stepwise differentiation of human embryonic stem cells promotes tendon regeneration by secreting fetal tendon matrix and differentiation factors. Stem Cells 27 (6):1276–1287
28. Xu W et al (2013) Human iPSC-derived neural crest stem cells promote tendon repair in a rat patellar tendon window defect model. Tissue Eng Part A 19(21–22):2439–2451
29. Otabe K et al (2015) The transcription factor Mohawk controls tenogenic differentiation of bone marrow mesenchymal stem cells in vitro and in vivo. J Orthop Res 33(1):1–8
30. Tan S-L et al (2015) Identification of pathways mediating growth differentiation factor5-induced tenogenic differentiation in human bone marrow stromal cells. PLoS ONE 10(11):e0140869
31. Urdzikova LM, Sedlacek R, Suchy T, Amemori T, Ruzicka J, Lesny P, Havlas V, Jendelova P (2014) Human multipotent mesenchymal stem cells improve healing after collagenase tendon injury in the rat. Biomed Eng Online 13(42). doi:10.1186/1475-925X-13-42, http://biomedical-engineering-online.biomedcentral.com/articles/10.1186/1475-925X-13-42
32. Gonçalves AI et al (2014) Understanding the role of growth factors in modulating stem cell tenogenesis. PLoS ONE 8(12):e83734
33. de Mattos Carvalho A et al (2011) Use of adipose tissue-derived mesenchymal stem cells for experimental tendinitis therapy in equines. J Equine Vet Sci 31(1):26–34
34. Molloy T, Wang Y, Murrell GAC (2003) The roles of growth factors in tendon and ligament healing. Sports Med 33(5):381–394
35. Temenoff JS, Mikos AG (2000) Review: tissue engineering for regeneration of articular cartilage. Biomaterials 21(5):431–440
36. Rizzello G et al (2012) Growth factors and stem cells for the management of anterior cruciate ligament tears. Open Orthop J 6:525–530
37. Klein MB et al (2002) Flexor tendon healing in vitro: effects of TGF-β on tendon cell collagen production. J Hand Surg 27(4):615–620
38. Chan BP et al (2000) Effects of basic fibroblast growth factor (bFGF) on early stages of tendon healing: a rat patellar tendon model. Acta Orthop Scand 71(5):513–518
39. Sahoo S, Toh SL, Goh JCH (2010) A bFGF-releasing silk/PLGA-based biohybrid scaffold for ligament/tendon tissue engineering using mesenchymal progenitor cells. Biomaterials 31 (11):2990–2998
40. Murray MM et al (2007) Enhanced histologic repair in a central wound in the anterior cruciate ligament with a collagen-platelet-rich plasma scaffold. J Orthop Res 25(8):1007–1017
41. Murray M et al (2006) Use of a collagen-platelet rich plasma scaffold to stimulate healing of a central defect in the canine ACL. J Orthop Res 24(4):820–830
42. Ricchetti ET et al (2012) Scaffold devices for rotator cuff repair. J Shoulder Elbow Surg 21 (2):251–265
43. Dunn MG et al (1995) Development of fibroblast-seeded ligament analogs for ACL reconstruction. J Biomed Mater Res 29(11):1363–1371
44. Bellincampi LD et al (1998) Viability of fibroblast-seeded ligament analogs after autogenous implantation. J Orthop Res 16(4):414–420

45. Weadock K et al (1995) Physical crosslinking of collagen fibers: comparison of ultraviolet irradiation and dehydrothermal treatment. J Biomed Mater Res 29(11):1373–1379
46. Gurkan UA et al (2010) Comparison of morphology, orientation, and migration of tendon derived fibroblasts and bone marrow stromal cells on electrochemically aligned collagen constructs. J Biomed Mater Res, Part A 94A(4):1070–1079
47. Altman GH et al (2003) Silk-based biomaterials. Biomaterials 24(3):401–416
48. Vepari C, Kaplan DL (2007) Silk as a biomaterial. Prog Polym Sci 32(8–9):991–1007
49. Chen JL et al (2010) Efficacy of hESC-MSCs in knitted silk-collagen scaffold for tendon tissue engineering and their roles. Biomaterials 31(36):9438–9451
50. Chen X et al (2014) Scleraxis-overexpressed human embryonic stem cell-derived mesenchymal stem cells for tendon tissue engineering with knitted silk-collagen scaffold. Tissue Eng Part A 20(11–12):1583–1592
51. Chen J et al (2003) Human bone marrow stromal cell and ligament fibroblast responses on RGD-modified silk fibers. J Biomed Mater Res, Part A 67(2):559–570
52. Majima T et al (2007) Chitosan-based hyaluronan hybrid polymer fibre scaffold for ligament and tendon tissue engineering. Proc Inst Mech Eng Part H 221(5):537–546
53. Moffat KL et al (2009) Novel nanofiber-based scaffold for rotator cuff repair and augmentation. Tissue Eng Part A 15(1):115–126
54. Fan H et al (2009) Anterior cruciate ligament regeneration using mesenchymal stem cells and silk scaffold in large animal model. Biomaterials 30(28):4967–4977
55. Laitinen O et al (1992) Mechanical properties of biodegradable ligament augmentation device of poly(L-lactide) in vitro and in vivo. Biomaterials 13(14):1012–1016
56. Cooper JA et al (2007) Biomimetic tissue-engineered anterior cruciate ligament replacement. Proc Natl Acad Sci USA 104(9):3049–3054
57. Freeman JW, Woods MD, Laurencin CT (2007) Tissue engineering of the anterior cruciate ligament using a braid-twist scaffold design. J Biomech 40(9):2029–2036
58. Petrigliano FA et al (2007) The effects of local bFGF release and uniaxial strain on cellular adaptation and gene expression in a 3D environment: implications for ligament tissue engineering. Tissue Eng 13(11):2721–2731
59. Leong NL et al (2015) Evaluation of polycaprolactone Scaffold with basic fibroblast growth factor and fibroblasts in an athymic rat model for anterior cruciate ligament reconstruction. Tissue Eng Part A 21(11–12):1859–1868
60. Xu Y et al (2013) Fabrication of electrospun poly(L-lactide-co-ε-caprolactone)/collagen nanoyarn network as a novel, three-dimensional, macroporous, aligned scaffold for tendon tissue engineering. Tissue Eng Part C Methods 19(12):925–936
61. Xu Y et al (2014) The effect of mechanical stimulation on the maturation of TDSCs-poly(L-lactide-co-e-caprolactone)/collagen scaffold constructs for tendon tissue engineering. Biomaterials 35(9):2760–2772
62. Leung M et al (2013) Tenogenic differentiation of human bone marrow stem cells via a combinatory effect of aligned chitosan-poly-caprolactone nanofibers and TGF-[small beta]3. J Mater Chem B 1(47):6516–6524
63. Domingues RMA et al (2016) Enhancing the biomechanical performance of anisotropic nanofibrous scaffolds in tendon tissue engineering: reinforcement with cellulose nanocrystals. Adv Healthc Mater. doi:10.1002/adhm.201501048
64. Wang L et al (2015) Nanofiber yarn/hydrogel core-shell scaffolds mimicking native skeletal muscle tissue for guiding 3D myoblast alignment, elongation, and differentiation. ACS Nano 9 (9):9167–9179
65. Shuakat MN, Lin T (2014) Recent developments in electrospinning of nanofiber yarns. J Nanosci Nanotechnol 14(2):1389–1408
66. Zhang C et al (2015) Well-aligned chitosan-based ultrafine fibers committed teno-lineage differentiation of human induced pluripotent stem cells for Achilles tendon regeneration. Biomaterials 53:716–730
67. Younesi M et al (2014) Tenogenic induction of human MSCs by anisotropically aligned collagen biotextiles. Adv Funct Mater 24(36):5762–5770

68. Domingues RMA, Gomes ME, Reis RL (2014) The potential of cellulose nanocrystals in tissue engineering strategies. Biomacromolecules 15(7):2327–2346
69. Nivison-Smith L, Weiss AS (2012) Alignment of human vascular smooth muscle cells on parallel electrospun synthetic elastin fibers. J Biomed Mater Res, Part A 100A(1):155–161
70. Spanoudes K et al (2014) The biophysical, biochemical, and biological toolbox for tenogenic phenotype maintenance in vitro. Trends Biotechnol 32(9):474–482
71. Yin Z et al (2010) The regulation of tendon stem cell differentiation by the alignment of nanofibers. Biomaterials 31(8):2163–2175
72. Zhou C et al (2013) Electrospun bio-nanocomposite scaffolds for bone tissue engineering by cellulose nanocrystals reinforcing maleic anhydride grafted PLA. ACS Appl Mater Interfaces 5(9):3847–3854
73. Abbah SA, Spanoudes K, O'Brien T, Pandit A, Zeugolis DI (2014) Assessment of stem cell carriers for tendon tissue engineering in pre-clinical models. Stem Cell Res Ther 5(38). doi:10.1186/scrt426, http://stemcellres.biomedcentral.com/articles/10.1186/scrt426
74. Barber JG et al (2011) Braided nanofibrous scaffold for tendon and ligament tissue engineering. Tissue Eng Part A 19(11–12):1265–1274
75. Liu W et al (2015) Generation of electrospun nanofibers with controllable degrees of crimping through a simple, plasticizer-based treatment. Adv Mater 27(16):2583–2588
76. Surrao DC et al (2012) A crimp-like microarchitecture improves tissue production in fibrous ligament scaffolds in response to mechanical stimuli. Acta Biomater 8(10):3704–3713
77. Chen F, Hayami JWS, Amsden BG (2014) Electrospun poly(L-lactide-co-acryloyl carbonate) fiber scaffolds with a mechanically stable crimp structure for ligament tissue engineering. Biomacromolecules 15(5):1593–1601
78. Lomas AJ et al (2015) The past, present and future in scaffold-based tendon treatments. Adv Drug Deliv Rev 85:257–277
79. LaCroix AS et al (2013) Relationship between tendon stiffness and failure: a metaanalysis. J Appl Physiol 115(1):43–51
80. Domingues RMA et al (2015) Chapter 10—Fabrication of hierarchical and biomimetic fibrous structures to support the regeneration of tendon tissues. In: Gomes ME, Reis RL, Rodrigues MT (eds) Tendon regeneration. Academic Press, Boston, pp 259–280
81. Pauly HM et al (2016) Mechanical properties and cellular response of novel electrospun nanofibers for ligament tissue engineering: effects of orientation and geometry. J Mech Behav Biomed Mater 61:258–270
82. Bosworth LA et al (2013) Investigation of 2D and 3D electrospun scaffolds intended for tendon repair. J Mater Sci - Mater Med 24(6):1605–1614
83. Mouthuy P-A et al (2015) Fabrication of continuous electrospun filaments with potential for use as medical fibres. Biofabrication 7(2):025006
84. Yang G et al (2016) Multilayered polycaprolactone/gelatin fiber-hydrogel composite for tendon tissue engineering. Acta Biomater 35:68–76
85. Aibibu D et al (2016) Textile cell-free scaffolds for in situ tissue engineering applications. J Mater Sci - Mater Med 27:63
86. Cheng X et al (2008) An electrochemical fabrication process for the assembly of anisotropically oriented collagen bundles. Biomaterials 29(22):3278–3288
87. Uquillas JA, Kishore V, Akkus A (2011) Effects of phosphate-buffered saline concentration and incubation time on the mechanical and structural properties of electrochemically aligned collagen threads. Biomed Mater 6(3):035008
88. Uquillas JA, Kishore V, Akkus O (2012) Genipin crosslinking elevates the strength of electrochemically aligned collagen to the level of tendons. J Mech Behav Biomed Mater 15:176–189
89. Kishore V et al (2011) Incorporation of a decorin biomimetic enhances the mechanical properties of electrochemically aligned collagen threads. Acta Biomater 7(6):2428–2436
90. Kishore V et al (2012) Tenogenic differentiation of human MSCs induced by the topography of electrochemically aligned collagen threads. Biomaterials 33(7):2137–2144

91. Kishore V et al (2012) In vivo response to electrochemically aligned collagen bioscaffolds. J Biomed Mater Res B Appl Biomater 100B(2):400–408
92. Islam A et al (2015) Biomechanical evaluation of a novel suturing scheme for grafting load-bearing collagen scaffolds for rotator cuff repair. Clin Biomech 30(7):669–675
93. Cheng CW, Solorio LD, Alsberg E (2014) Decellularized tissue and cell-derived extracellular matrices as scaffolds for orthopaedic tissue engineering. Biotechnol Adv 32(2):462–484

Chapter 18
Ligament Tissue Engineering

Wasim Khan

Abstract Ligaments are commonly injured in the knee joint, and have a poor capacity for healing due to their relative avascularity. Ligament reconstruction is well established for injuries such as anterior cruciate ligament rupture, however the use of autografts and allografts for ligament reconstruction are associated with complications, and outcomes are variable. Ligament tissue engineering using stem cells, growth factors and scaffolds is a novel technique that has the potential to provide an unlimited source of tissue. In this chapter we discuss the role of tissue engineering in dealing with ligament injuries and provide an overview of in vitro and in vivo studies.

18.1 Introduction

The knee joint with its long lever arms is subject to significant forces, and sporting and other daily activities can put it at a higher risk of injury. Ligament injuries account for a significant proportion of musculoskeletal injuries and result in disability and morbidity in patients worldwide [1]. Ligament injuries may ultimately lead to pain, articular cartilage injury, meniscal injury and early osteoarthritis [87]. It has been estimated that the incidence of injuries involving knee ligaments could be as high as 1,193 per 100,000 person-years, with surgery performed in 3.9 % of ligament injury cases [33]. Anterior cruciate ligament (ACL) rupture is one of the most common injuries of the knee [70]. More than 200,000 ACL reconstructions are performed yearly in the United States and the number being performed is increasing in frequency [75]. The estimated cost of an ACL surgical repair and subsequent rehabilitation is between $17,000–$25,000 per injury [22]. Over 200,000 ACL reconstructions per year are carried out in the US [75]. The total

W. Khan (✉)
Institute of Orthopaedics and Musculoskeletal Sciences, Royal National Orthopaedic Hospital, University College London, Stanmore, Middlesex HA7 4LP, UK
e-mail: wasimkhan@doctors.org.uk

© Springer International Publishing AG 2017
J.M. Oliveira and R.L. Reis (eds.), *Regenerative Strategies for the Treatment of Knee Joint Disabilities*, Studies in Mechanobiology, Tissue Engineering and Biomaterials 21, DOI 10.1007/978-3-319-44785-8_18

expenditure on ACL reconstructions in a year has been estimated as exceeding $5 billion [83].

In this chapter we will discuss limitations of current treatment strategies, ligament structure and ligament healing before taking an in depth view of tissue engineering for ligament repair and reconstruction.

18.2 Limitations of Current Treatment Options

Current treatment strategies for ligament injuries depend on the degree of injury and the patient's activity level and symptoms [57]. Grade I ligament injuries are mild sprains that are not associated with ligament laxity. Grade II injuries demonstrate moderately increased joint laxity. Grade III injuries are severe and associated with complete ligament disruption and significant laxity [88]. Non-operative management consists of pain relief and rehabilitation. Ligaments are poorly vascularized and have a limited capacity for healing. When healing does occur the composition of the healed tissue is different to normal tissue and the biomechanical properties of the healed tissue are usually inferior [72]. Operative management with hamstring or patellar tendon autografts, allografts and synthetic grafts is often undertaken [36] but the reconstructive surgery also may be associated with disadvantages. Autografts are associated with donor site morbidity, weakness, reduced range of movement and anterior knee pain in the case of patellar tendon donor tissue. Laboute et al. [47] found a re-rupture rate of 12.7 % for ACL reconstructions performed with hamstring tendon autografts. Allografts carry the risk of immunological reactions, disease transmission and infection. Synthetic grafts including carbon fibre, polypropylene, Dacron and polyester are associated with a high failure rate due to wear debris, foreign body tissue reactions and synovitis [23, 72].

18.3 Ligament Structure

Ligaments passively stabilize joints by connecting one bone to another and allow smooth motion. They are subject to multidirectional forces depending on activities. They also have a role in joint proprioception. The four main ligaments around the knee are the cruciates and the collaterals. Microscopically the ligament is composed of specialized fibroblasts that account for approximately 20 % of the tissue, and produce the extracellular matrix (ECM) that accounts for approximately 80 % of the tissue. The ECM consists of approximately 70 % water and 30 % organic tissue. The collagen accounts for 75 % of the dry weight with the remaining 25 % consisting of proteoglycans, elastin and other proteins and glycoproteins such as actin, laminin and integrin. Although there are 16 types of collagen, type I collagen accounts for 85 % of the collagen in ligaments. Type I collagen has an enormous tensile strength enabling fibrils to be stretched without being broken. Less than 10 % of the collagen in ligaments is type III. This is more often found in healing tissues before most of it is converted to type I collagen. Very small amounts of collagen types IV, V, XI and XIV are also present. The basic structural unit of collagen is a triple-stranded helical molecule packed together side by side. The collagen bundles are aligned along the long axis into bundles of parallel fibres. The fibres have a periodic change in direction along the length known as the crimp pattern. It is likely that with increased loading, some areas of the ligament 'uncrimp' which allows the ligament to elongate without sustaining damage. The ligaments are covered by an outer layer known as the epiligament. This merges into the periosteum of the bone around the attachment site of the ligament. The epiligament is more vascular and cellular with a greater number of sensory and proprioceptive nerves [28, 29, 53, 72].

18.4 Ligament Healing

As mentioned earlier, regeneration and healing of ligaments after injury is often poor due their relatively avascular nature and low metabolic rate. Healing of ligaments can be divided into four stages. Firstly, there is a haemorrhagic stage in which the ligament ends retract and a blood clot forms and fills the gap. Cytokines are released within the clot and a heavily cellular infiltrate of polymorphonuclear leucocytes and lymphocytes appear within several hours. The macrophages appear by twenty-four to forty-eight hours in the inflammatory stage. By seventy-two hours the wound also contains platelets and multipotential mesenchymal cells. Macrophages phagocytose necrotic tissues as well as secreting growth factors such as basic fibroblast growth factor (FGFβ), transforming growth factor alpha and beta (TGFα and TGFβ) and platelet derived growth factor (PDGF) that stimulate fibroblast proliferation and synthesize types I, III and V collagen and non-collagenous proteins, as well as inducing neovascularization and formation of granulation tissue. During the proliferative stage, fibroblasts produce dense, cellular, collagenous connective tissue binding the torn ligament ends. This 'scar tissue' is initially disorganized and contains more type III and type V collagen, and smaller diameters collagen fibrils. This is followed by remodeling and maturation of the tissue. There is a gradual decrease in the cellularity of the tissue. The matrix becomes denser and longitudinally orientated. The matrix continues to mature for at least a year [28, 72, 88, 90].

The repair tissue never achieves the morphological or mechanical characteristics of normal ligament. There remains an increased vascularity and cellularity, abnormal innervation, decreased collagen fibril diameter, altered relative collagen type proportions with inadequate cross-linking, and altered proteoglycan profiles. The ligaments recover only up to twenty percent of their viscolelastic properties.

The repair tissue also has inferior creep properties (i.e. deformation properties under constant or cyclic loading) that could result in joint laxity. The resultant tissue has half the normal failure load and absorbs less energy before failing [28, 88].

18.5 Tissue Engineering

Tissue engineering involves the use of appropriate cells, growth factors and scaffolds, either in isolation or combination to repair and regenerate tissue, and has a role in musculoskeletal tissue repair and regeneration [41, 43]. Tissue engineering has a potentially useful role in ligament surgery as these structures are often injured and demonstrate limited healing potential [90]. Tissue engineering could be used to repair and regenerate tissue. In vivo injection of appropriate cells into the injured ligament in conjunction with the use of biomimetic scaffolds and bioreactors is a strategy that could potentially accelerate the process of tissue repair [90]. We have previously described a role for stem cells in the tissue engineering of ligaments [18]. Below, we will discuss cell sources, growth factors and scaffolds before considering the role of bioreactors.

18.6 Cell Sources for Ligament Tissue Engineering

Although reparative cells could be recruited from host tissue through the specific attachment of tissue engineered scaffolds, seeding cells could further improve the functionality of tissue engineered constructs [31]. The seeded cells lay down ECM and recruit reparative and/or progenitor cells through chemotaxis through growth factors and cytokines accelerating ligament repair. Additionally, they incorporate and release endogenous growth factors to elicit an immune response [31]. It is important to select the appropriate cell type for the specific application in order for the tissue engineered product to have the best outcome. However, little is known about the optimal cell source for ligament tissue engineering. The cell type selected must show enhanced proliferation and production of an appropriate ECM and must be able to survive in the relevant knee environment, intraarticular in the case of the cruciates [83]. Primary fibroblasts can be derived from ligaments such as the ACL, or from the skin. ACL fibroblasts can be harvested in diagnostic arthroscopic procedures following ACL rupture. The medial collateral ligament (MCL) is extraarticular, and it could be easily harvested partially without impairing its function in the long term [31]. Mesenchymal stem cells (MSCs) have the ability to proliferate and differentiate into a variety of mesenchymal cell phenotypes including osteoblasts, chondroblasts, myoblasts and fibroblasts [76, 90]. Culture conditions can be designed to direct MSC differentiation into the desired mesenchymal phenotype [83].

18.6.1 Fibroblasts

Fibroblasts are a choice of cells that can be harvested from different sources. Bellincampi et al. [10] investigated skin fibroblasts as a potential source for ligament tissue engineering as skin fibroblasts are known to have a greater healing potential and may be easily retrieved in a clinical setting. ACL and skin fibroblasts were harvested, cultured, labeled, seeded on collagen fibre scaffolds in vitro and implanted into the autogenous knee joint in a rabbit model. The cells remained viable for at least four to six weeks after implantation. They concluded that both skin and ACL fibroblasts survived in an intraarticular environment, but the potential of ACL fibroblasts to improve neoligament formation may be limited by a poor intrinsic healing capacity. Cooper et al. [20] investigated the cellular response of primary rabbit connective tissue fibroblasts from four sources (Achilles tendon, patellar tendon, MCL and ACL) to a novel three-dimensional poly-L-lactic acid (PLLA) braided scaffold for ACL tissue engineering. The fibroblasts from all four sources had similar morphological appearances on tissue culture polystyrene. However, the cellular growth differed for cell sources. They concluded that ACL fibroblasts were the most suited for ACL tissue engineering. Tremblay et al. [81] implanted a bioengineered ACL graft seeded with autologous living dermal fibroblasts into goat knee joints for six months. Histological and ultrastructural analysis demonstrated a highly organized ligamentous structure with vascularization, innervation and organized Sharpey's fibres and collagen at the osseous insertion sites of the grafts. Morbidity associated with harvesting of the skin is a potential limitation of using skin fibroblasts as a source for ligament tissue engineering. Additionally, the performance of skin fibroblasts for ligament tissue engineering may be affected as the physiological environment of skin fibroblasts is different to that of ligaments [31].

18.6.2 Mesechymal Stem Cells

Although the use of primary fibroblasts for ligament tissue engineering is a logical approach, the use of MSCs may be more efficient [67]. MSCs are naturally occurring cells that have the ability to both self-replicate as well as differentiate into another cell types [43]. Their capacity to repair is due to the secretion of factors that alter the tissue microenvironment. MSCs may be isolated from a variety of adult tissues including the bone marrow, adipose tissue, cord blood and synovial fluid [24, 45, 54]. MSCs are easily obtainable form bone marrow by a minimally invasive approach and can be expanded in tissue culture and encouraged to differentiate into the desired lineage [77]. Although Cheng et al. [17] reported benefits of stem cells derived from the ACL compared to bone marrow derived MSCs, the ligament has fewer MSCs. Most studies look at MSCs of an earlier passage before they lose their ability to proliferate and differentiate [74]. MSCs are positive for a set of cell surface markers including CD105, CD73, and are negative for the haematopoietic markers CD34, CD45, and CD14 [44]. There is evidence that pericytes may represent MSC in different tissues, and indeed tissues that are vascular have a higher proportion of MSCs [42]. The differentiation into desired lineages is achieved by the use of bioactive signaling molecules, specific growth factors and appropriate environmental conditions [43, 80]. An alternative approach is the use of embryonic stem cells which are derived from the inner cell mass of the blastocyst and are capable of unlimited undifferentiated proliferation and have been shown to differentiate into all types of somatic cells. However, the use of embryonic stem cells is associated with several disadvantages including technical difficulties, immunogenicity, tumour formation in vivo, uncertainty regarding the long-term outcome and ethical considerations [19]. Adult MSCs possess immunomodulatory properties, making them potential candidates for cellular therapy in an allogeneic setting. Transplantation of MSCs into an allogeneic host may not require immunosuppressive therapy. Adult MSCs express intermediate levels of class I major histocompatibility complex proteins but do not express human leucocyte antigen (class II) antigens on the cell surface [14, 48]. MSCs have been shown to have an indirect inhibitory effect on T cells which is mediated by regulatory antigen-presenting cells with T cell suppressive properties [11].

Oe et al. [66] studied ligament regeneration in rats following intra-articular injection of either fresh bone marrow cells (BMCs) or cultured MSCs 1 week after partial ACL transection. At 4 weeks donor cells were detected within the transected ACLs in both the groups and the ACLs exhibited almost normal histology. They concluded that direct intra-articular bone marrow transplantation is an effective treatment for partial ruptures of the ACL. Lim et al. [52] performed ACL reconstructions in adult rabbits using hamstring tendon autografts which were coated with MSCs in a fibrin glue carrier. At 8 weeks good osteointegration was observed and they performed significantly better on biomechanical testing than the controls.

18.6.3 Studies Comparing Fibroblasts and Mesenchymal Stm Cells

There are few studies comparing fibroblasts with MSCs. MSCs may differentiate into ligament fibroblasts after two weeks [90]. It has been shown in a rabbit model, that MSCs have a significantly higher proliferation rate and collagen production than ACL and MCL fibroblasts, and that MSCs could survive for at least six weeks in the knee joint [31]. Van Eijk et al. [82] seeded bone marrow stromal cells, skin fibroblasts and ACL fibroblasts at different seeding densities onto braided poly(L-lactide/glycolide) scaffolds. The cells were cultured for up to 12 days. All cell types readily attached to the scaffold. On day 12, the MSC-seeded scaffolds showed the highest DNA content and collagen production. Scaffolds seeded with ACL fibroblasts showed the lowest DNA content and collagen production. The ideal cell type selected for ligament tissue engineering must be readily available, have excellent proliferative and differentiation ability, be capable of producing an organized ECM, and have a good affinity for the scaffold.

18.7 Growth Factors Through Gene Transfer Technology

Growth factors are polypeptides that support various terminal phenotypes and regulate stem cell differentiation and proliferation. Growth factors such as transforming growth factor-β (TGF-β), bone morphogenic proteins (BMPs), fibroblast growth factors (FGFs), epidermal growth factor (EGF), insulin like growth factor (IGF-I), vascular endothelial growth factor (VEGF), platelet derived growth factor (PDGF), and growth and differentiation factor (GDF) can, in isolation or various combinations, expedite the MSC proliferation and fibroblast differentiation [40, 90].

Gene transfer technology may be used to deliver genetic material and information to cells to alter their synthesis or function [90]. Genes can be introduced into cells using retroviral and adenoviral vectors as carriers, liposomes or with a gene gun. The genes can be placed in the cell ex vivo or in vivo. The target cells can be made to produce or increase expression of growth factors or suppress the synthesis of endogenous proteins or growth factor within the local tissue [88]. When Wei et al. [85] surgically implanted bone marrow derived MSCs transfected with adenovirus vector encoding TGF-β1, VEGF or TGF-β1/VEGF into experimental ACL grafts in rabbits, this significantly promoted angiogenesis compared to non-transfected control cells, and improved the mechanical properties. Hildebrand et al. [35] used a retroviral ex vivo and an adenoviral in vivo technique to introduce and express the LacZ marker gene in the MCL and ACL of rabbits. LacZ gene expression was detected and shown to last between 10 days and 3 weeks in the MCL and ACL with the use of the retrovirus and between 3 and 6 weeks in the MCL and at least 6 weeks in the ACL with the adenoviruses. Menetrey et al. [63] showed the feasibility of gene transfer to a normal ACL using direct, fibroblast mediated and myoblast mediated approaches. Either adenoviral particles were directly injected into the ACL of rabbits, or myoblasts or ACL fibroblasts transduced with recombinant adenoviral particles carrying the LacZ reporter gene were injected. Direct and myoblast mediated gene transfers demonstrated persistence of gene expression up to 6 weeks, but fibroblast mediated gene transfer showed gene expression for only 1 week. A number of other studies have indicated that using gene therapy to improve ligament healing is a promising approach [56, 71, 78, 91].

We conducted a review of growth factors in ACL reconstruction [12]. When TGF-β is added directly to the canine tibial bone tunnel it increases the ultimate load complex and richly generates perpendicular collagen fibres connecting the tendon graft and bone as early as 3 weeks post operatively [89]. As predicted, TGF-β increases proliferation and migration of ACL fibroblasts and stimulates matrix protein deposition thus enhancing wound repair [84] and significantly increasing maximum load and stiffness compared to control groups in rabbits [46]. In humans,

TGF- β increases cell number, increases collagen production and increases expression of alpha-smooth muscle actin in ACL defects [62]. The direct application of a virus vector mediated gene transfer of bFGF (both in vitro and in experimentally injured human ACLs) significantly enhances levels of type I and type II collagen production [58]. This could be the result of enhanced neovascularization and the formation of granulation tissue in injured ACLs in response to bFGF. Weiler et al. [86] found that the local addition of PDGF coated sutures in hamstring tendon ACL reconstruction lead to significantly higher load to failure, crimp length, vascular density and collagen fibril at varying time periods (3–12 weeks) post operatively. PDGF does not however appear to affect inflammatory parameters, MRI appearance of the graft or clinical evaluation scores when injected into the graft and tibial tunnel [64]. Letson and Dahners [50] compared PDGF alone or in combination groups (PDGF + IGF-1, PDGF + bFGF) and found that all three groups improved strength, stiffness and the breaking force of ligaments, suggesting that it is PDGF that is the most important growth factor in ligament healing. Indeed other studies confirm that PDGF has a role in accelerating ligament and tendon healing [6]. VEGF is highly expressed in the early post-operative phase (2–3 weeks) of patellar tendon ACL reconstruction implying VEGF is predominantly involved in the graft remodeling process at this stage [92]. However in trials with rabbits, it seems to work by promoting angiogenesis in the grafts rather than directly affecting the mechanical properties such as anterior-posterior translation, tensile strength, cross-sectional area or strain at failure [38].

18.8   Scaffolds

Biomaterial scaffolds provide a structural and logistic template for new tissue formation and remodeling [83]. Scaffolds are designed to support cell attachment, survival, migration and differentiation as well as control transport of nutrients, growth factors, metabolites and regulatory molecules to and from the cells [18]. A scaffold should be made of a biocompatible, biodegradable material and should be able to bridge any complex three-dimensional anatomical defect. The scaffold should ideally possess adequate strength post-implantation to be effective as a load-bearing construct and degrade at a rate matching the rate of new tissue deposition. The scaffold should have sufficient pore sizes to allow cell infiltration, and sufficient void volume to allow extracellular matrix formation to promote gradual load transfer from the scaffold to the neotissue [79]. Porous scaffolds enhance tissue regeneration by delivering biofactors, however pores that are too large could compromise the mechanical properties of the scaffold [90]. Polymers used in ligament tissue engineering [32] may be naturally derived e.g. gelatin, small intestine submucosal extracellular matrix and silk, or synthetic e.g. polyesters such as polyglycolic acid [59].

18.8.1 Natural Scaffolds

Collagen used in laboratories is usually derived from the bovine submucosa and intestine from rats tails in small quantities. The derived collagen requires processing to remove foreign antigens, improve its mechanical strength and sometimes to slow down the degradation rate by crosslinking. The crosslinking can be performed using chemical agents e.g. glutaraldehyde, formaldehyde, polyepoxy compounds, acylazide, carboiimides and hexmethylenediisocyanate risking potential toxic residues, or physically using drying, heat or exposure to ultraviolet or gamma radiation [32]. The resorption rate and mechanical properties of scaffolds can be altered through cross-linking. Fibroblasts have been shown to attach, proliferate and secrete new collagen when seeded on collagen fibre scaffolds [25]. In vivo, it has been demonstrated that fibroblast seeded collagen scaffolds may remain viable after implantation into the knee joint for prolonged periods [10]. Examples of commercially available biological collagen-based scaffolds include Restore (derived from porcine small intestine), Graftjacket (from human cadaver dermis), Permacol (from porcine dermis) and Bio-Blanket (from bovine dermis) [15]. The scaffolds demonstrate an early decrease in mechanical strength followed by tissue remodeling resulting in a strength gain similar to autografts by 20 weeks [29].

Silk has the advantage of possessing good biocompatibility, slow biodegradability and excellent tensile strength and toughness [91]. Silk fibroin is a protein excreted by silkworms and isolated from sericin [91], and has similar mechanical properties to functional ACL when organized into appropriate wire-rope geometry. Silk scaffolds also support cell attachment and spreading by providing an appropriate three-dimensional culture environment. Silk fibres lose the majority of their tensile strength within one year in vivo and fail to be recognized in two years [32]. Silk-fibre matrices have been shown to support adult stem cell differentiation towards ligament lineages [2]. A composite scaffold fabricated from silk and collagen tested in a rabbit MCL defect model was shown to improve structural and functional ligament repair by regulating ligament matrix gene expression and collagen fibril assembly [16]. Fan et al [26] examined ACL regeneration with MSCs on silk scaffolds in vivo. The lapine model demonstrated that the regenerated ligaments exhibited essential ligament ECM components including collagen I and collagen III in significant amounts, and direct ligament–bone insertion was reconstructed exhibiting the four zones typical of native ACL–bone insertions; bone, mineralized fibrocartilage, fibrocartilage and ligament. The tensile strength of the regenerated ligaments was assessed to be biomechanically adequate.

18.8.2 Synthetic Scaffolds

Synthetic polymers that have been investigated for ligament repair include poly glycolic acid (PGA), polylactic acid (PLA), their copolymers and poly caprolactone

(PCL) [55]. PLA is a commonly used synthetic scaffold that easily degrades within the human body by forming lactic acid. PCL and PGA degrade in a similar way to PLA but exhibit different rates of degradation. Synthetic polymers are not limited by donor source, have no risk of disease transmission and are designed to degrade over time. Their mechanical properties may be controlled by altering the degree of polmer crystallinity, changing the polymer molecular weight or changing the ratio of each polymer in the copolymer [29].

18.8.3 Preclinical and Clinical Studies Using Scaffolds

We performed a systematic review [13] to examine and summarize the preclinical in vivo studies and limited clinical studies on the use of scaffolds in the treatment of ligamentous injuries. We identified eight studies looking at collagen platelet composite (CPC), two studies on collagen in isolation, two on silk and one study each for Poly-L-Lactic acid (PLLA) and small intestinal mucosa. The studies involving CPC were on porcine or canine models of ACL injuries, and all had variable time frames for examination from one to fifteen weeks. The concentration of PRP varied from two to five times the physiological level. All found that CPC had a significant effect on healing. Mastrangelo et al. [61] found in their porcine model that higher PRP concentrations produced greater cellular densities at thirteen weeks. They also found that skeletally immature animals had a greater intrinsic capacity to heal compared to adolescents and adults. Palmer et al. [69] determined that increased temperature of CPC decreased strength and yield load in the porcine model and Magarian et al. [60] determined that decreased yield load occurred when the repair was delayed either two or six weeks, with no difference between the delay groups. Joshi et al. [37] performed ACL repairs in twenty-seven immature pigs, with 14 having a repair augmented with CPC. The CPC augmented group had better functional, load, and stiffness measurements in addition to improved structural properties at 3 months. Nishmoto et al. [65] treated rabbit medial collateral ligament defects using PLLA scaffolds. Fibrocartilage alignment and morphology increased in a time dependent manner, but PLLA fibres were not absorbed after the sixteen-week assessment, raising potential concerns regarding synovitis. Badylak et al. [8] reconstructed goat ACL injuries with either a porcine small intestinal sub mucosa scaffold or a more conventional patella tendon graft and a found no difference in functional testing. The small intestine submucosa group did however show transient weakening early with variable strength over time in comparison to the patella tendon group which increased in strength. We identified three studies on synthetic scaffolds. Cooper et al. [21] presented data on a comparison between seeded and unseeded biomimetic ligament generated by using 3D braiding technology to reconstruct rabbit ACL defects and determined that seeding ligaments with ACL cells resulted in better histological and mechanical evaluation. Liljensten et al. [51] assessed poly urethane urea (PUUR) in thirty-five rabbits and two pigs and found that at six, twelve, and twenty-four months there was no synovial reaction or joint damage in any knees and all had an integrated ligament. Argona et al. [7] demonstrated that carbon fibre polylactic acid polymer ligaments allowed for more stable medial collateral ligament constructs with time related collagen ingrowth in the beagle model at a maximum of twenty-six weeks. There were no biological studies available. Four studies assessed the effects of absorbable copolymer carbon fibre ligaments (ABC) and three found that while tissue ingrowth into the ligaments was found there were unacceptable short and long-term failure rates even after a change in technique in the early 1990's. Petrou et al. [73] prospectively followed up seventy-one patients for a minimum of five years and found that while there was evidence of recurrent synovitis and stiffness there was a one hundred percent survival rate. This discrepancy may have been due to relatively short follow up as it has been previously noted that after a technique and equipment change in the early 1990's the early failure rates of the ABC ligament were replaced by mid to late term failures.

18.9 Bioreactor Systems

The differentiation of MSCs into fibroblasts may be accelerated by the use of a bioreactor that provides a controlled biomimetic optimum environment for cell functions. Bioreactors are a key component of tissue engineering [3, 68]. They use various combinations of chemical, mechanical, electrical or magnetic stimulation to guide differentiation, proliferation and tissue development. In the case of ligament tissue engineering, a bioreactor may be used to accelerate the process of differentiation of MSCs into the fibroblastic lineage [90]. The body may be used as a bioreactor when a cell-scaffold composite is implanted directly into the injured site. Another approach is to culture the cell-scaffold composite in a bioreactor ex vivo for a period of time before transplantation [34, 91].

In order for a bioreactor to function successfully, there are several basic design principles that need to be fulfilled. Firstly, a bioreactor should maintain precise control of the physiological environment of the tissue culture, including control of variables such as temperature, oxygen concentrations, pH, nutrients, media flow rate, metabolite concentrations and specific tissue markers within close limits. Bioreactors should also be able to support the culture of two or more cell types simultaneously particularly when engineering complex tissues. It is also essential that the bioreactor is designed to operate under strict aseptic conditions in order to prevent any contamination of the tissues by influx of microorganisms [68].

Chemical stimulation techniques are employed by using chemicals such as growth factors described in the section above. Mechanical stimulation techniques involve subjecting a scaffold to mechanical stresses resembling the in vivo environment. Intracellular signaling cascades are activated by triggering the cell surface stretch receptors leading to synthesis of the necessary extracellular matrix proteins [90]. The effects of mechanical stimulation are dependent on the magnitude, duration and frequency of mechanical stress [49]. Additionally, mechanical stimulation has been shown to affect extracellular matrix synthesis and remodeling. Enzyme activity and growth factor expression, collagen type I, collagen type III, elastin and tenascin-C expression in MSCs have been shown to be increased with the application of mechanical loads [91]. Electromagnetic stimulation has been shown to have positive results. For example, Fung et al. [30] showed that low energy laser therapy can enhance the mechanical strength of healing MCL in rats and increase collagen fibril size. Co-culture may also be used to induce differentiation of MSCs because of its ability to promote cell communications [90]. Direct co-culture of MSCs with fibroblasts induces MSCs to differentiate into fibroblast-like cells [9]. Cell-to-cell interactions in the microenvironment play a key role in regulating the differentiation of MSCs in the healing process. Additionally, specific regulatory signals released from fibroblasts have been shown to support the selective differentiation of MSCs towards ligament fibroblasts in a two-dimensional transwell insert co-culture system [49]. Fan et al. [27] demonstrated that specific regulatory signals released from fibroblasts in a three-dimensional co-culture also enhanced the differentiation of MSCs for ligament tissue engineering.

Although various commercial bioreactor systems are available, some may not be applicable to ligament tissue engineering as the design lacks the specificity to meet the requirements for engineering of ligament tissue [83]. Altman et al. [4, 5] designed a bioreactor to permit the controlled application of ligament-like multidimensional mechanical strains to undifferentiated cells embedded in a collagen gel. They used mechanical stimulation in vitro to induce the differentiation of mesenchymal progenitor cells from bone marrow into a ligament cell lineage in preference to bone or cartilage cell lineages. Kahn et al. [39] designed a bioreactor for tissue engineering of ligament tissue that imposed mechanical conditions close to the physiological movement of the ACL. The bioreactor consisted of a mechanical part allowing movement to be applied on scaffolds, two culture chambers, a perfusion flow system to renew nutrients in the culture medium, a heating enclosure as well as an electronic component to manage movement and to regulate heating.

18.10 Conclusions

Ligament injuries are common in the knee, and can be challenging to treat with the current nonoperative and operative treatment options available. Considerable progress has been made in generating tissue engineered ligaments. The following requirements were noted by Vunjak-Novakovic et al. [83] as being key to the success of tissue engineered ligaments:

- Autologous source of MSCs that is easily accessed with no associated morbidity to eliminate concerns such as infection, immune rejection or disease;
- Biomaterial scaffold with mechanical properties matching the native ligament, biodegradation to match tissue formation, and porosity to allow for cell infiltration;

- Biochemical and biophysical regulation of MSC differentiation;
- Quantitative methods of measuring success.

Studies on the generation of tissue engineered ligaments have generally been in vitro preliminary studies or trials in animal models. At this stage we appear to be moving closer to achieving the above aims, but human trials need to be conducted in addition to a cost benefit analysis to determine the appropriateness of treatment. Engineering ligaments that have the appropriate mechanical properties is the significant challenge. However, advances in cell biology, understanding of the roles of growth factors, scaffold engineering and mechanical conditioning using bioreactors may be able to provide a viable long-term alternative to current autografts and allografts in the future. The use of tissue engineered ligaments would potentially have significant health care implications. In view of the more active aging population, the number of patients who will benefit from the use of tissue engineered ligaments is likely to increase with time.

References

1. Al-Rashid M, Khan WS (2011) Stem cells and ligament repair. In: Berhardt L (ed) Advances in medicine and biology. Nova Science Publishers Inc, New York, pp 343–347
2. Altman GH, Horan RL, Lu HH, Moreau J, Martin I, Richmond JC, Kaplan DL (2002) Silk matrix for tissue engineered anterior cruciate ligaments. Biomaterials 23(20):4131–4141
3. Altman G, Lu H, Horan R, Calabro T, Ryder D, Kaplan D (2002) Advanced bioreactor with controlled application of multi-dimensional strain for tissue engineering. J Biomech Eng 124:742–749
4. Altman GH, Horan RL, Martin I, Farhadi J, Stark PR, Volloch V, Richmond JC, Vunjak-Novakovic G, Kaplan DL (2002) Cell differentiation by mechanical stress. FASEB J 16(2):270–272
5. Altman G, Lu H, Horan R, Calabro T, Ryder D, Kaplan D (2002) Advanced bioreactor with controlled application of multi-dimensional strain for tissue engineering. J Biomech Eng 124:742–749
6. Anitua E, Andia I, Ardanza B, Nurden P, Nurden AT (2004) Autologous platelets as a source of proteins for healing and tissue regeneration. Thromb Haemost 91(1):4–15
7. Argona J, Parsons JR, Alexander H, Weiss AB (1984) Medial collateral ligament replacement with a partially absorbable tissues scaffold. Am J Sport Med 11(4):228–233
8. Badylak S, Arnoczky S, Plouher P, Haut R, Mendenhall V, Clarke R, Horvath C (2009) Naturally occurring extracellular matrix as a scaffold for musculoskeletal repair. Clin Orthop Relat Res 367S:S333–S343
9. Ball SG, Shuttleworth AC, Kielty CM (2004) Direct cell contact influences bone marrow mesenchymal stem cell fate: Int.J. Biochem Cell Biol 36(4):714–727
10. Bellincampi LD, Closkey RF, Prasad R, Zawadsky JP, Dunn MG (1998) Viability of fibroblast-seeded ligament analogs after autogenous implantation. J Orthop Res 16(4):414–420
11. Beyth S, Borovsky Z, Mevorach D, Liebergall M, Gazit Z, Aslan H, Galun E, Rachmilewitz J (2005) Human mesenchymal stem cells alter antigen-presenting cell maturation and induce T-cell unresponsiveness. Blood 105(5):2214–2219
12. Bissell L, Tibrewal S, Sahni V, Khan WS (2015) Growth factors and platelet rich plasma in anterior cruciate ligament reconstruction. Curr Stem Cell Res Ther 10(1):19–25

13. Caudwell M, Crowley C, Khan WS, Wong JM (2015) Systematic review of preclinical and clinical studies on scaffold use in knee ligament regeneration. Curr Stem Cell Res Ther 10 (1):11–18
14. Chamberlain G, Fox J, Ashton B, Middleton J (2007) Concise review: mesenchymal stem cells: their phenotype, differentiation capacity, immunological features, and potential for homing. Stem Cells 25(11):2739–2749
15. Chen J, Xu J, Wang A, Zheng M (2009) Scaffolds for tendon and ligament repair: review of the efficacy of commercial products. Expert Rev Med Devices 6(1):61–73
16. Chen X, Qi YY, Wang LL, Yin Z, Yin GL, Zou XH, Ouyang HW (2008) Ligament regeneration using a knitted silk scaffold combined with collagen matrix. Biomaterials 29 (27):3683–3692
17. Cheng MT, Liu CL, Chen TH, Lee OK (2010) Comparison of potentials between stem cells isolated from human anterior cruciate ligament and bone marrow for ligament tissue engineering. Tissue Eng Part A 16(7):2237–2253
18. Chimutengwende-Gordon M, Khan WS (2012) Stem cells in ligament tissue engineering. In: Mahato RI, Danquah M (eds) Emerging trends in cell and gene therapy. Springer, Heidelberg. ISBN 978-1-62703-417-3
19. Chimutengwende-Gordon M, Khan WS (2012) Advances in the use of stem cells and tissue engineering applications in bone repair. Curr Stem Cell Res Ther 7(2):122–126
20. Cooper JA Jr, Bailey LO, Carter JN, Castiglioni CE, Kofron MD, Ko FK, Laurencin CT (2006) Evaluation of the anterior cruciate ligament, medial collateral ligament, achilles tendon and patellar tendon as cell sources for tissue-engineered ligament. Biomaterials 27(13):2747–2754
21. Cooper JA, Sahota JA Jr, Gorum WJ 2nd, Carter J, Doty SB, Laurencin CT (2007) Biomimetic tissue-engineered anterior cruciate ligament replacement. Proc Natl Acad Sci USA 104(9):3049–3054
22. de Loes M, Dahlstedt LJ, Thomee R (2000) A 7-year study on risks and costs of knee injuries in male and female youth participants in 12 sports. Scand J Med Sci Sports 10(2):90–97
23. Dheerendra SK, Khan WS, Singhal R, Shivarathre DG, Pydisetty R, Johnstone DJ (2012) Anterior cruciate ligament graft choices: a review of current concepts. Open Orthop J 6 (2):281–286
24. Dhinsa BS, Mahapatra AN, Khan WS (2015) Sources of adult mesenchymal stem cells for ligament and tendon tissue engineering. Curr Stem Cell Res Ther 10(1):26–30
25. Dunn MG, Liesch JB, Tiku ML, Zawadsky JP (1995) Development of fibroblast-seeded ligament analogs for ACL reconstruction. J Biomed Mater Res 29(11):1363–1371
26. Fan H, Liu H, Wong EJW, Toh SL, Goh JCH (2008) In vivo study of anterior cruciate ligament regeneration using mesenchymal stem cells and silk scaffold. Biomaterials 2008 (29):3324–3337
27. Fan H, Liu H, Toh SL, Goh JC (2008) Enhanced differentiation of mesenchymal stem cells co-cultured with ligament fibroblasts on gelatin/silk fibroin hybrid scaffold. Biomaterials 29 (8):1017–1027
28. Frank C (2004) Ligament structure, physiology and function. J Musculoskelet Neuronal Interact 4(2):199–201
29. Freeman J, Kwansa A (2008) Recent advancements in ligament tissue engineering: the use of various techniques and materials for ACL repair. Recent Pat Biomed Eng 1:18–23
30. Fung DT, Ng GY, Leung MC, Tay DK (2003) Effects of a therapeutic laser on the ultrastructural morphology of repairing medial collateral ligament in a rat model. Lasers Surg Med 32(4):286–293
31. Ge Z, Goh JC, Lee EH (2005) Selection of cell source for ligament tissue engineering. Cell Transpl 14(8):573–583
32. Ge Z, Yang F, Goh JC, Ramakrishna S, Lee EH (2006) Biomaterials and scaffolds for ligament tissue engineering. J Biomed Mater Res A 77(3):639–652
33. Gianotti SM, Marshall SW, Hume PA, Bunt L (2009) Incidence of anterior cruciate ligament injury and other knee ligament injuries: a national population-based study. J Sci Med Sports 12:622–627

34. Goh JC, Ouyang HW, Teoh SH, Chan CK, Lee EH (2003) Tissue-engineering approach to the repair and regeneration of tendons and ligaments. Tissue Eng 9(Suppl 1):S31–S44

35. Hildebrand KA, Deie M, Allen CR, Smith DW, Georgescu HI, Evans CH, Robbins PD, Woo SL (1999) Early expression of marker genes in the rabbit medial collateral and anterior cruciate ligaments: the use of different viral vectors and the effects of injury. J Orthop Res 17 (1):37–42

36. Hoffman A, Gross G (2006) Tendon and ligament engineering:from cell biology to in vivo application. Regen Med 1(4):563–574

37. Joshi S, Mastrangelo A, Magarian E, Fleming BC, Murray MM (2009) Collagen-platelet composite enhances biomechanical and histologic healing of the porcine ACL. Am J Sports Med 37(12):2401–2410

38. Ju YJ, Tohyama H, Kondo E, Yoshikawa T, Muneta T, Shinomiya K, et al (2006) Effects of local administration of vascular endothelial growth factor on properties of the in situ frozen-thawed anterior cruciate ligament in rabbits. Am J Sports Med 34(1):84–91. PubMed PMID: 16210580. Epub 2005/10/08. eng

39. Kahn CJ, Vaquette C, Rahouadj R, Wang X (2008) A novel bioreactor for ligament tissue engineering. Biomed Mater Eng 18(4–5):283–287

40. Kanitkar M, Tailor HD, Khan WS (2011) The use of growth factors and mesenchymal stem cells in orthopaedics. Open Orthop J 5:271

41. Khan WS, Malik AA, Hardingham TE (2009) Stem cell applications and tissue engineering approaches in surgical practice. J Perioper Pract 19(4):130–135

42. Khan WS, Adesida AB, Tew SR, Lowe ET, Hardingham TE (2010) Bone marrow derived mesenchymal stem cells express the pericyte marker 3G5 in culture and show enhanced chondrogenesis in hypoxic conditions. J Orthop Res 28(6):834–840

43. Khan WS, Hardingham TE (2012) Mesenchymal stem cells, sources of cells and differentiation potential. J Stem Cells 7(2):75–85

44. Khan WS, Hardingham TE (2012) The characterisation of mesenchymal stem cells: a stem cell is not a stem cell is not a stem cell. J Stem Cells 7(2):87–95

45. Khan WS, Adesida AB, Tew SR, Longo UG, Hardingham TE (2012) Fat pad-derived mesenchymal stem cells as a potential source for cell-based adipose tissue repair strategies. Cell Prolif 45(2):111–120

46. Kondo E, Yasuda K, Yamanaka M, Minami A, Tohyama H (2005) Effects of administration of exogenous growth factors on biomechanical properties of the elongation-type anterior cruciate ligament injury with partial laceration. Am J Sports Med 33(2):188–196

47. Laboute E, Savalli L, Puig P, Trouve P, Sabot G, Monnier G (2010) Analysis of return to competition and repeat rupture for 298 anterior cruciate ligament reconstructions with patellar or hamstring tendon autograft in sportspeople. Ann Phys Rehabil Med 53(10):598–614

48. Le Blanc K, Ringden O (2005) Immunobiology of human mesenchymal stem cells and future use in hematopoietic stem cell transplantation: Biol. Blood Marrow Transpl 11(5):321–334

49. Lee IC, Wang JH, Lee YT, Young TH (2007) The differentiation of mesenchymal stem cells by mechanical stress or/and co-culture system. Biochem Biophys Res Commun 352(1):147–152

50. Letson AK, Dahners LE (1994) The effect of combinations of growth factors on ligament healing. Clin Orthop Relat Res 308:207–212

51. Liljensten E, Gisselfält K, Edberg B, Bertilsson H, Flodin P (2002) Studies of pol-yurethane urea bands for ACL reconstruction. J Mater Sci 13(4):351–359

52. Lim JK, Hui J, Li L, Thambyah A, Goh J, Lee EH (2004) Enhancement of tendon graft osteointegration using mesenchymal stem cells in a rabbit model of anterior cruciate ligament reconstruction. Arthroscopy 20(9):899–910

53. Lodish H, Berk A, Zipursky SL, Matsudaira P, Baltimore D, Darnell J (2000) Section 22.3 Collagen: the fibrous protein of the matrix. In: Molecular cell biology, 4th edn. Freeman WH, New York

54. Giuseppe Longo U, Loppini M, Berton A, La Verde L, Khan WS, Denaro V (2012) Stem cells from umbilical cord and placenta for musculoskeletal tissue engineering. Curr Stem Cell Res Ther 7(4):272–281

55. Giuseppe Longo U, Rizzello G, Berton A, Fumo C, Maltese L, Khan WS, Denaro V (2013) Synthetic grafts for anterior cruciate ligament reconstruction. Curr Stem Cell Res Ther 8 (6):429–437

56. Lou J, Tu Y, Burns M, Silva MJ, Manske P (2001) BMP-12 gene transfer augmentation of lacerated tendon repair. J Orthop Res 19(6):1199–1202

57. Mabvuure NT, Malahias M, Haddad B, Hindocha S, Khan WS (2014) State of the art regarding the management of multiligamentous injuries of the knee. Open Orthop J 8:215

58. Madry H, Kohn D, Cucchiarini M (2013) Direct FGF-2 gene transfer via recombinant adeno-associated virus vectors stimulates cell proliferation, collagen production, and the repair of experimental lesions in the human ACL. Am J Sports Med 41(1):194–202. PubMed PMID: 23172005. English

59. Mafi P, Hindocha S, Mafi R, Khan WS (2012) Evaluation of biological protein-based collagen scaffolds in cartilage and musculoskeletal tissue engineering—a systematic review of the literature. Curr Stem Cell Res Ther 7(4):302–309

60. Magarian EM, Fleming BC, Harrison SL, Mastrangelo AN, Badger GJ, Murray MM (2010) Delay of 2 or 6 weeks adversely affects the functional outcome of augmented primary repair of the porcine anterior cruciate ligament. Am J Sports Med 38(12):2528–2534

61. Mastrangelo AN, Vavken P, Fleming BC, Harrison SL, Murray MM (2011) Reduced platelet concentration does not harm PRP effectiveness for ACL repair in a porcine in vivo model. J Orthop Res 29(7):1002–1007

62. Meaney Murray M, Rice K, Wright RJ, Spector M (2003) The effect of selected growth factors on human anterior cruciate ligament cell interactions with a three-dimensional collagen-GAG scaffold. J Orthop Res 21(2):238–244

63. Menetrey J, Kasemkijwattana C, Day CS, Bosch P, Fu FH, Moreland MS, Huard J (1999) Direct-, fibroblast- and myoblast-mediated gene transfer to the anterior cruciate ligament. Tissue Eng 5(5):435–442

64. Nin JR, Gasque GM, Azcarate AV, Beola JD, Gonzalez MH (2009) Has platelet-rich plasma any role in anterior cruciate ligament allograft healing? Arthroscopy 25(11):1206–1213

65. Nishmoto H, Kokubu T, Inui A, Mifune Y, Nishida K, Fujioka H, Yokota K, Hiwa C, Kurosaka M (2012) Ligament regeneration using an absorbable stent-shaped poly-L-lactic acid scaffold in a rabbit model. Int Orthop (SICOT) 36:2379–2386

66. Oe K, Kushida T, Okamoto N, Umeda M, Nakamura T, Ikehara S, Iida H (2011) New strategies for anterior cruciate ligament partial rupture using bone marrow transplantation in rats. Stem Cells Dev 20(4):671–679

67. Ong E, Chimutengwende-Gordon M, Khan W (2013) Stem cell therapy for knee ligament, articular cartilage and meniscal injuries. Curr Stem Cell Res Ther 8(6):422–428

68. Oragui E, Nannaparaju M, Khan WS (2011) The role of bioreactors in tissue engineering for musculoskeletal applications. Open Orthop J 5(Suppl 2):267–270

69. Palmer MP, Abreu EL, Mastrangelo A, Murray MM (2009) Injection temperature significantly affects in vitro and in vivo performance of collagen-platelet scaffolds. J Orthop Res 27(7):964–971

70. Papoutsidakis A (2011) Predisposing factors for anterior cruciate ligament injury. Br J Sports Med 45:e2

71. Pascher A et al (2004) Enhanced repair of the anterior cruciate ligament by in situ gene transfer: evaluation in an in vitro model. Mol Ther 10(2):327–336

72. Pastides P, Khan W (2011) Tendon and ligament injuries: the evolving role of stem cells and tissue engineering. Br J Med Med Res 1(4):569–580

73. Petrou G, Chardouvelis C, Kouzoupis A, Dermon A, Petrou H, Tilkeridis C, Gavras M (2006) Reconstruction of the anterior cruciate ligament using the polyester ABC ligament scaffold. J Bone Joint Surg [Br] 88-B:893–899

74. Rizzello G, Longo UG, Petrillo S, Lamberti A, Khan WS, Maffulli N, Denaro V (2012) Growth factors and stem cells for the management of anterior cruciate ligament tears. Open Orthop J 6:525–530

75. Shelton WR, Fagan BC (2011) Autografts commonly used in anterior cruciate ligament reconstruction. J Am Acad Orthop Surg 19(5):259–264
76. Siddiqui NA, Wong JML, Khan WS, Hazlerigg A (2010) Stem cells for tendon and ligament tissue engineering and regeneration. J Stem Cells 5(4):187–194
77. Singh J, Onimowo J, Khan WS (2015) Bone marrow derived stem cells in trauma and orthopaedics: a review of the current trend. Curr Stem Cell Res Ther 10(1):37–42
78. Steinert AF, Weber M, Kunz M, Palmer GD, Noth U, Evans CH, Murray MM (2008) In situ IGF-1 gene delivery to cells emerging from the injured anterior cruciate ligament. Biomaterials 29(7):904–916
79. Teh TK, Toh SL, Goh JC (2011) Aligned hybrid silk scaffold for enhanced differentiation of mesenchymal stem cells into ligament fibroblasts. Tissue Eng Part C Methods 17(6):687–703
80. Thanabalasundaram G, Arumalla N, Tailor HD, Khan WS (2012) Regulation of differentiation of mesenchymal stem cells into musculoskeletal cells. Curr Stem Cell Res Ther 7(2):95–102
81. Tremblay P et al (2011) Potential of skin fibroblasts for application to anterior cruciate ligament tissue engineering. Cell Transpl 20(4):535–542
82. Van Eijk F, Saris DB, Riesle J, Willems WJ, Van Blitterswijk CA, Verbout AJ, Dhert WJ (2004) Tissue engineering of ligaments: a comparison of bone marrow stromal cells, anterior cruciate ligament, and skin fibroblasts as cell source. Tissue Eng 10(5–6):893–903
83. Vunjak-Novakovic G, Altman G, Horan R, Kaplan DL (2004) Tissue engineering of ligaments. Annu Rev Biomed Eng 6:131–156
84. Wang Y, Tang Z, Xue R, Singh GK, Lv Y, Shi K et al (2011) TGF-beta1 promoted MMP-2 mediated wound healing of anterior cruciate ligament fibroblasts through NF-B. Connect Tissue Res 52(3):218–225
85. Wei X, Mao Z, Hou Y, Lin L, Xue T, Chen L, Wang H, Yu C (2011) Local administration of TGFbeta-1/VEGF165 gene-transduced bone mesenchymal stem cells for Achilles allograft replacement of the anterior cruciate ligament in rabbits. Biochem Biophys Res Commun 406 (2):204–210
86. Weiler A, Forster C, Hunt P, Falk R, Jung T, Unterhauser FN et al (2004) The influence of locally applied platelet-derived growth factor-BB on free tendon graft remodeling after anterior cruciate ligament reconstruction. Am J Sports Med 32(4):881–891
87. Wong JM, Khan T, Jayadev CS, Khan WS, Johnstone DJ (2012) Anterior cruciate ligament rupture and osteoarthritis progression. Open Orthop J 6(2):295–300
88. Woo SL, Hildebrand K, Watanabe N, Fenwick JA, Papageorgiou CD, Wang JH (1999) Tissue engineering of ligament and tendon healing. Clin Orthop Relat Res 367(Suppl):S312–S323
89. Yamazaki S, Yasuda K, Tomita F, Tohyama H, Minami A (2005) The effect of transforming growth factor-beta1 on intraosseous healing of flexor tendon autograft replacement of anterior cruciate ligament in dogs. Arthroscopy 21(9):1034–1041
90. Yates EW, Rupani A, Foley GT, Khan WS, Cartmell S, Anand SJ (2012) Ligament tissue engineering and its potential role in anterior cruciate ligament reconstruction. Stem Cells Int, p 438125
91. Yilgor C, Yilgor HP, Huri G (2012) Tissue engineering strategies in ligament regeneration. Stem Cells Int, p 374676
92. Yoshikawa T, Tohyama H, Enomoto H, Matsumoto H, Toyama Y, Yasuda K (2006) Expression of vascular endothelial growth factor and angiogenesis in patellar tendon grafts in the early phase after anterior cruciate ligament reconstruction. Knee Surg Sports Traumatol Arthrosc 14(9):804–810

Chapter 19
Regenerative Engineering of the Anterior Cruciate Ligament

Paulos Y. Mengsteab, Mark McKenna, Junqiu Cheng, Zhibo Sun and Cato T. Laurencin

Abstract Anterior cruciate ligament (ACL) injuries, both acute and chronic, are common in sport injuries. The presence of the synovial fluid in the knee joint inhibits the spontaneous healing of the ACL, thus requiring surgical intervention. Although current methods to reconstruct the ACL can stabilize the knee joint, the progression of osteoarthritis is not halted. This chapter describes the current clinical methods to reconstruct an injured ACL and new methods to enhance the healing process. Three therapeutic strategies will be discussed in this chapter on the repair of ACL: (1) single bundle versus double bundle surgical techniques, (2) biodegradable matrices for ACL repair, and (3) biological adjuvants to enhance ACL repair. These strategies are promising clinically translatable methods to allow patients to return to normal activity levels and to alleviate pain and discomfort caused by osteoarthritis.

P.Y. Mengsteab · M. McKenna · C.T. Laurencin (✉)
Department of Biomedical Engineering, University of Connecticut, Storrs, CT 06269, USA
e-mail: pmengsteab@uchc.edu

P.Y. Mengsteab · J. Cheng · Z. Sun · C.T. Laurencin
The Raymond and Beverly Sackler Center for Biomedical, Biological, Engineering, and Physical Sciences, University of Connecticut Health, Farmington, CT 06030, USA

P.Y. Mengsteab · Z. Sun · C.T. Laurencin
The Institute for Regenerative Engineering, University of Connecticut Health, Farmington, CT 06030, USA

P.Y. Mengsteab · Z. Sun · C.T. Laurencin
Department of Orthopedic Surgery, University of Connecticut Health, Farmington, CT 06030, USA

C.T. Laurencin
Department of Materials Science and Engineering, University of Connecticut, Storrs, CT 06269, USA

C.T. Laurencin
Department of Chemical and Biomolecular Engineering, University of Connecticut, Storrs, CT 06269, USA

© Springer International Publishing AG 2017   391
J.M. Oliveira and R.L. Reis (eds.), *Regenerative Strategies for the Treatment of Knee Joint Disabilities*, Studies in Mechanobiology, Tissue Engineering and Biomaterials 21, DOI 10.1007/978-3-319-44785-8_19

19.1 Introduction

The human skeletal system is a joint assembly, and the linchpins of it are ligaments. Comprised of nine hundred ligaments, six hundred in the arms and legs, two hundred and thirty in the torso, and seventy above the shoulder, ligaments allow for the integration of two hundred and six bones to form the internal framework of the body. By forming linkage points, ligaments limit the degrees of freedom of the skeletal system and stabilize joints, preventing damage of soft tissue through inhibition of unnecessary movement. Disruption of this internal framework, due to ligament lesions, may lead to osteoarthritis. Thus, therapeutic strategies to heal damaged ligaments are necessary to allow patients to return to normal activity levels and to alleviate pain and discomfort caused by osteoarthritis.

A common ligament that is injured is the anterior cruciate ligament (ACL). An ACL injury is a momentous event in the career of athletes and overall health of non-athletes. Typically, an ACL injury is associated with sports injuries where the ACL is overloaded in tension or the knee twists causing a high torsional load, which predominately results in a ruptures of the intra-articular region. As the major intra-articular ligament of the knee, the ACL stabilizes the knee by controlling the anterior to posterior translation of the femur and tibia. The loss of ACL function causes joint instability, which leads to damage of meniscus and cartilage due to mechanical distortion.

Annually, approximately 400,000 ACL injuries occur, necessitating surgical intervention [1]. Unlike other ligaments in the body, a torn ACL does not have the capacity to heal due to the presence of the synovial fluid in the knee joint. In the case of a medial cruciate ligament tear, a blood clot forms and serves as a scaffold to allow the healing of the lesion without the need of surgery. However, in the knee joint the synovium environment inhibits the formation of a blood clot, leaving patients with an unstable knee. Therefore, ACL reconstruction is performed in order to regain the proper kinematic function of the knee with the overall goal to recapitulate the native ACL biomechanical properties. Although surgical reconstruction of the ACL is routinely performed and does allow for the stabilization of the knee, recovery of an ACL injury is a long process (approximately 8 months), and patients are at high risk for osteoarthritis. This is due to anterior subluxation of the tibia, leading to compression of the posterior lateral tibial plateau against the anterior lateral femoral condyle [2]. For these reasons, new methods to enhance the healing process of an ACL and to prevent osteoarthritis are of high interest. Three strategies will be discussed in detail: (1) single bundle versus double bundle surgical techniques, (2) biodegradable matrices for ACL repair, and (3) biological adjuvants to enhance ACL repair.

19.1.1 Structure of the Anterior Cruciate Ligament

On average, the human ACL is approximately 27–32 mm in length and has a cross-sectional area of 44.4–57.5 mm^2 [2, 3]. Macroscopically, the gross structure of the ACL appears as a band-like structure, which connects the femur and the tibia. From the femur, the ACL travels anteriorly, medially, and distally to its attachment at the tibia, and is characterized by a 180° twist between its bony attachment ends and its flexible collagenous intra-articular region (Fig. 19.1) [4]. The structure of the ACL is irregular in that the cross-sectional area is not a simple geometric shape and experiences deformation when the knee undergoes flexion [5]. Furthermore, the ACL is defined by two bundles, the anteromedial (AM) and posterolateral (PL) bundle, which act as the functional components of the ligament [6, 7]. The AM and PL bundle are characterized by the location of their insertion into the tibial tunnel. The AM bundle originates in the most proximal part of the femoral origin and inserts at the anteromedial tibial insertion site, whereas, the PL bundle originates distal to the femoral origin of the AM bundle and inserts into the posterolateral part of the tibial insertion site. These two bundles have contrasting orientations, which are dependent on the extension or flexion of the knee. In the case of knee extension, the PL bundle is seen to be in tension while the AM bundle is moderately relaxed (Fig. 19.1a). During flexion of the knee, a 110° bend, the AM

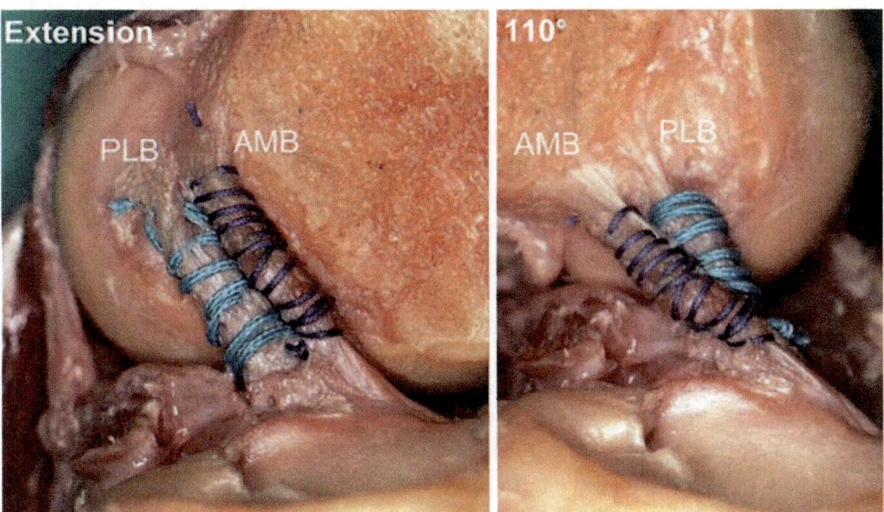

Fig. 19.1 *Diagram of the anteromedial and posterolateral bundle of the ACL.* In extension, the PL bundle is seen to be in tension while the AM bundle is moderately lax. The opposite effect is seen when the knee is in flexion at 110°. (Reprinted from [4] with permission form Elsevier)

bundle is in tension and the PL bundle becomes relaxed (Fig. 19.1b). During passive knee flexion the two bundles experience different patterns of length change and are not isometric in either flexion or extension. Furthermore, the two bundles are distinguished by their individual structures. In comparison to the AM bundle, the PL bundle is comprised by a larger number of fascicles [8]. The structure and anatomical placement of these two bundles help to stabilize the knee joint in differing physiological movements and loads.

19.1.2 Constituent Components of Ligaments

Ligaments are dense, complex, tissues composed of collagens (type I, III, and V), elastin, proteoglycans, water, and cells [4, 9]. Ligaments display a hierarchical structure with collagen molecules, fibrils, fibril bundles, and fascicles that are arranged parallel to the longitudinal axis of the ligament [10]. Microscopically, the ACL has been categorized into three sections: proximal, middle, and distal. Each of these sections are comprised of differing cellular and extracellular matrix components, and are instrumental in the healing of the surgically reconstructed ACL, as well as, the biomechanics of the ACL [4, 5, 7, 8]. The proximal section is characterized by greater cellularity in comparison to the other sections, and is thus less solid. The main components of the proximal part are fibroblasts, type II collagen, and glycoproteins. The middle part contains spindle shaped fibroblast, has a high density of collagen fibers, and a special zone of cartilage and fibrocartilage, which is located at the ligament to bone interface [4]. Furthermore, the importance of the middle zone is that the fusiform and spindle-shaped fibroblasts are longitudinally oriented. This longitudinal organization of the cells contribute to the organization of the deposited collagen fibrils, which is important for the non-linear stress-strain response of the ACL. Finally, the distal part of the ligament is the most solid of the three and is rich in chondroblasts and has a lower density of collagen bundles. In the anterior portion of the ACL, a layer of dense fibrous tissue engulfs the ligament and corresponds to the zone where the ligament is compressed by the anterior rim of the femoral intercondylar fossa. The sections of the ACL correspond to the complex anatomical structure of the tissue, which give rise to the variety of properties necessary for the ACL to comply with the kinematics of the knee.

The bone to ligament interface of the ACL is essential for the motion of the knee. This interface has a unique transitional zone that is defined as the chondral apophyseal enthesis, that guides to transition the ligamentous component of the ACL to rigid bone [8]. This transitional zone allows the ACL and bone tissue to function properly together. During ACL reconstruction the proper healing of this bone to ligament interface is essential for the knee to withstand physiological loads and joint motion. The chondral apophyseal enthesis consists of four layers: the ligament proper, non-mineralized fibroblasts, mineralized fibroblasts, and the subchondral bone plate. The first layer is composed of collagen fibrils that is followed by a second layer of non-mineralized fibroblast cells that are aligned within the collagen bundles. The third layer is composed of mineralized cartilage and facilitates the transition to the subchondral bone plate. The transition zone from ligament to bone serves to distribute the stresses at the insertion site, and therefore decreases the rise of stress concentrations [8, 11, 12]. As such, the transition zone allows for a graduated change in stiffness at the attachment site, and is critical for the stress distribution of the ligament under loads.

The constituent components of the ACL microstructure are similar to that of other soft connective tissues [5]. Of the ACL constituent components, collagen is the major ECM protein that comprises its structure. Collagen fascicles are bundled together to form the band like structure of the ACL. These fascicles range from 250 μm to several millimeters and are connected by paratenon connective tissue. The lateral growth of collagen fascicles are regulated by two extra cellular matrix proteins, decorin and fibromodulin. Within the fascicles are subfascicular groups on the order of 100–250 μm and are enclosed by epitenon tissue. In all, fascicles are composed of approximately 3–20 subfascicular units. The subfascicular groups are undulated, and therefore provide the organization fundamental to the biomechanical response of the ACL. Furthermore, the subfascicular groups are a family of fibers that are composed of collagen fibrils. These collagen fibrils are approximately 25–250 nm in diameter and are the primary component of the ACL structure.

Collagen fibrils have been categorized into two types: fibrils with varying diameter, and uniform diameter fibrils [4, 5]. The inhomogeneous fibrils have varying diameters that peak at 35, 50, and 75 nm and account for 50.3 % of the entire ACL. Biomechanically, the inhomogeneous fibrils have been stated to specialize in resisting high tensile stresses. On the other hand, homogenous fibrils have uniform diameters with a peak diameter of 45 nm and account for 47.3 % of the ACL. The three-dimensional organization of the ligament is provided by these homogenous fibrils, which also serve a critical role in modulating the biomechanical response of the ACL.

The remainder of the ACL is composed of cells and matrix components. The matrix components of the ACL are formed by collagen, glycosaminoglicans (GAGs), glyco-conjugates, and elastic components [4, 5]. There are four types of collagen found in the ACL, type I, III, IV, and VI, of which type I and III collagen are the primary components that affect the biomechanical response of the ACL. At the molecular level, collagen protein are composed of two collagen alpha 1 chains and one collagen alpha 2 chain. These three chains interact together to create a triple helix structure. Collagen type I fibrils are oriented along the longitudinal axis of the ACL, provide the tensile strength of the ligament, and are the aforementioned homogenous fibrils of the ligament. On the other hand, type III collagen is the connective tissue that connects the type I collagen bundles, serves as the main ground matrix of the ACL, are fundamental in fibril assembly, and are inhomogeneous throughout the ligament. Morphologically, type III collagen is either a single or multi-strand, with a diameter of 2 and 9 μm, respectively [1]. Maximal concentrations of type III collagen are seen near the bony attachment end of the ACL, and biomechanically type III collagen is important for the pliability of the ACL [1].

In addition to collagen, GAGs play an essential role in the biomechanics of the ACL. GAGs are important for the viscoelastic properties of the ACL, and since GAGs are highly negatively charged and contain a large proportion of hydroxyl groups, they attract water through hydrogen binding. Thus, GAGs recruit water into the ACL, which comprises 60–80 % of the total wet-weight of the ACL. The GAGs retain water in the ACL, which is released when tensile loads are placed on the ligament. In comparison with tendon, ligaments have a higher proportion of GAGs, approximately two to fourfold greater. Importantly, these GAGs act as a shock-absorber in the ligament. These constituent components work together to allow for proper knee kinematics, yet when the ACL is ruptured and reconstruction of the ACL is needed other factors need to be considered to understand the biomechanics of the healing process. Further insight into the genetic make-up of the ACL is needed to further delineate the components that regulate the post reconstruction biomechanics of the ACL.

Gene analysis of the ACL has been conducted to elucidate the differences between the ACL and tendon. To date, the discrepancy between ligament and tendon is not well understood. Therefore, in engineering approaches to study ligament many of the cellular properties of the tissue are neglected and simplified to describe the ligament as tendon. Ligament and tendon share a common progenitor marker, scleraxis, a transcription factor that promotes the production of collagen extra cellular matrix [13]. To describe the differences between ligament and tendon, Pearse II et al. conducted a microarray analysis of porcine ACL, posterior cruciate ligament (PCL), medial cruciate ligament (MCL), patellar tendon (PT), and Achilles tendon (AT). In the ACL and PCL, it was found that the genes tenascin-C and aggrecan core protein were upregulated in comparison to the PT and AT [14]. Tenascin-C functions as an adhesion-modulatory extracellular matrix protein and plays a fundamental role in regulating fibroblast extra cellular matrix deposition and the ability of fibroblasts to contract their matrix. Relative to ACL reconstruction, tenascin-c is known to be greatly upregulated in extra cellular matrix remodeling during wound repair and neovascularization, and therefore plays an impactful role in the restoration of ACL biomechanics post reconstruction [15].

Aggrecan core protein is a critical component of cartilage structures and has been noted to affect the stiffness of cartilage. In the context of the ACL, aggrecan is found in the bone insertion site. At the bone-ligament interface, aggrecan is most prominent in the mineralized fibrocartilage region. Given that the transitional zone plays a key role in distributing stresses, aggrecan would be important in the biomechanical response. Interestingly, Majima et al. conducted tensile and compressive tests on the MCL and found that the mRNA levels of aggrecan were elevated due to cyclic hydrostatic compression and cyclic tension [16]. Through the comparison of gene expression novel insights on the constituent components of the ACL led to a greater understanding of the cellular components and their role in the biomechanics of the ligament.

19.1.3 ACL Reconstruction: Graft Placement and Their Role in the Kinematics of the Knee

The surgical reconstruction techniques and their outcome are affected by the placement of graft fixation into the femur and tibia. Research has been conducted to determine the correct placement of the ACL. The anatomy of the femoral insertion site is characterized by its length and width, which has been found to be approximately 18 mm in length, 10 mm in width, and 4 mm from the articular cartilage (Fig. 19.2a). In addition, the insertion site is characterized in the sagittal plane, where it is rotated relative to the axis of the femur and reflects the insertion sites congruity to the posterior border of the femoral condyle. Generally, the rotation is 25°–35° relative to the femur (Fig. 19.2a). The AM and PL bands are characterized for their insertion site based on a 90° flexion of the knee. In this reference, the proximal and distal margins of the ACL are approximated at 11 and 10 o'clock, respectively (Fig. 19.2b). In the tibial insertion site, the ACL inserts into the intercondylar eminence of the tibia [17]. Studies have reported measurements for the precise insertion site of the ACL into the tibia with an oval geometric shape. The approximate length and width of the oval site is 18 and 10 mm, respectively. The midline of the oval attachment site can be described from the posterior tibial plateau, and its distance from the plateau is approximately 6 mm (Fig. 19.3) [17]. In ACL reconstruction, it is desirable to mimic the characteristics of the native insertion site to preserve the kinematics of the knee.

A debate on whether a single bundle technique adequately maintains the kinematic integrity of the knee began in the early 2000s. As previously stated, the native ACL exhibits two bands, the AM and PL. Therefore, the use of a double bundle

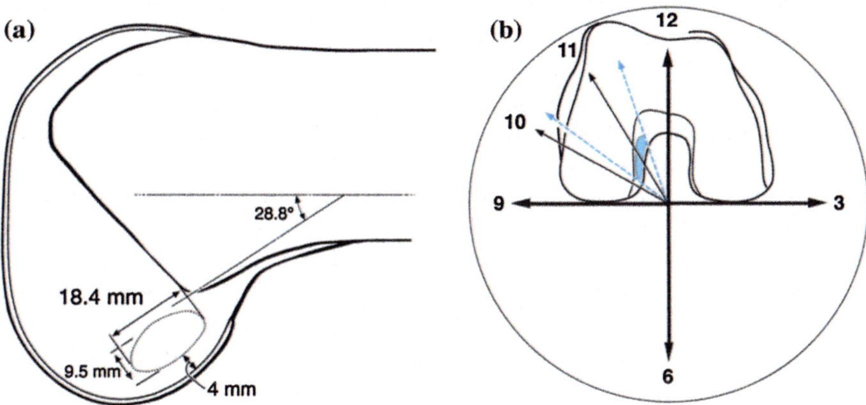

Fig. 19.2 **a** Sagital view of femoral insertion site. **b** Anterior view of the knee at 90° flexion. (Adapted from [17] with permission from SAGE Publications Inc.)

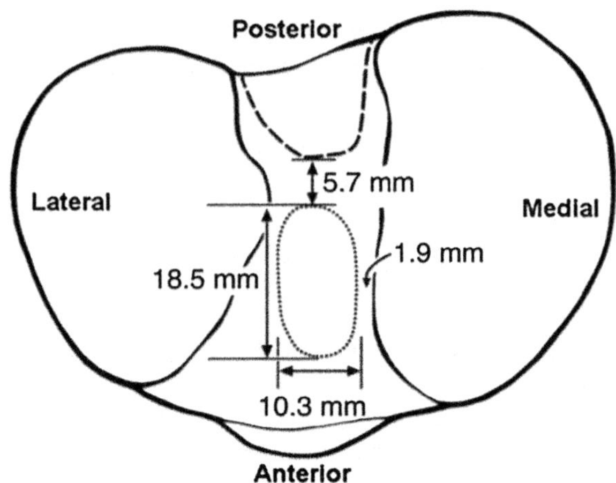

Fig. 19.3 Cross-sectional view of the tibial insertion site for ACL reconstruction. (Reprinted from [17] with permission from SAGE Publications Inc.)

graft for ACL reconstruction has been investigated and used clinically, so as to recapitulate the gross anatomy of the ACL. In the double bundle technique the femoral and tibial insertion sites are reamed. The femoral insertion site allows for the reaming of two tunnels with ease, however due to the geometry of the tibial insertion (Fig. 19.3), considerable variability occurs due to the obliquity of the line separating the two bundles [17, 18]. Although variability will occur in the tibial insertion site, studies have shown the efficacy of a double bundle reconstruction technique. Cadaveric studies on the kinematics of single and double reconstructions demonstrated that the double bundle technique provides better stabilization of the knee in the anterior translation direction when exposed to valgus-internal rotation [19]. Furthermore, Koga et al. recently conducted a clinical trial to compare the results of single bundle versus double bundle ACL reconstruction. Seventy-eight patients were included in the study, in which the ruptured ACL were replaced with an autologous semitendinosus tendon. Of the seventy-eight patients, fifty-three were evaluated for 3 years. The results demonstrated that the double bundle technique lead to greater results in the Lachman, pivot-shift test, and KT-1000 arthrometer measurements [20]. The Lachman test is a physical examination, performed by a clinician, which gains insight in the anterior translation of the tibia in comparison to the tibia. The pivot-shift test is also a physical examination in which a clinician tests the instability of the knee. Finally, the KT-1000 measurement provides an objective means to measure the anterior tibial motion relative to the femur, and this test validates the clinician's assessment. The research in the field of graft placement and reconstruction technique, single versus double bundle grafts, provide insight in choosing a surgical technique to replace a ruptured ACL for a more desirable clinical outcome in regards to the knees function.

19.2 Regenerative Engineering Approaches to Enhance Ligament Regeneration

19.2.1 Criteria for Ligament Scaffolds

Ligament scaffolds should seek to mimic the architecture and behavior of native ligament structures, and as such these scaffolds must be biocompatible, biodegradable, and mechanically competent. With regard to ligament regeneration, biocompatibility implies that fibroblasts can adhere, proliferate, and secrete natural extracellular matrix (ECM) components. The production of native ECM components (such as collagen type I, elastin, and proteoglycans) is especially important for biodegradable scaffolds, which act as a temporary structure to provide mechanical strength and appropriate cellular interactions as fibroblasts create natural ligamentous tissue. The development of a suitable scaffold for ligament regeneration also requires careful observation of the mechanical properties of the natural ligament tissue in its native environment. The mechanical properties of the ACL will be highlighted within because of the extensive research in this area. ACL tissue must be able to provide differential load support [21], resist plastic deformation, and have high tensile strength. It has been discovered that high tensile strength is not sufficient to recapitulate the mechanical behavior of the natural ACL because the ligaments of the knee encounter a variety of different forces and torsions during normal knee movement [8]. Notably, the ACL has been subdivided into an anteromedial (AM) band and a posterolateral (PL) band [22]. These distinct bands are named according to their tibial insertion points, and are generally accepted in the literature in spite of their anatomical simplification.

Natural ACL tissue has a unique microstructural feature, typically referred to as the crimp pattern, which endows the ligament with a non-linear stress-strain behavior [23]. Figure 19.4 displays the three regions of this mechanical behavior. The toe region represents the application of force to collagen fibrils within the

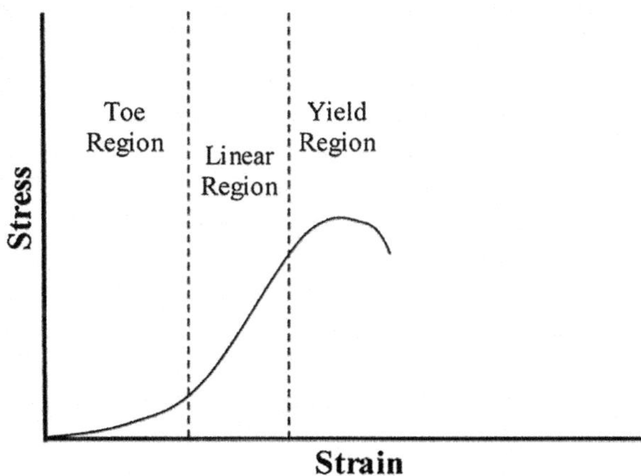

Fig. 19.4 Stress-strain behavior of natural ACL tissue. (Reprinted from [9] with permission from Elsevier)

ligament, causing a straightening of the crimp pattern. The linear region starts once the crimp is fully elongated, and is representative of collagen molecular strain [9]. Finally, the yield region signifies the failure of the ligament as the collagen fibers defibrillate. The understanding of natural ligament tissue properties has enabled researchers to fabricate natural and synthetic ligament replacement materials.

Autografts and allografts were explored due to the inherent mechanical strength and biocompatibility that was offered from natural human ligamentous tissue. Each of these systems has its own drawbacks. Autografts are taken from the body of the patient, and as such require at least two surgeries. This leads to the issue of donor site morbidity, the possibility that the first surgery may impair the viability of the surrounding donor site tissue. Allograft tissue is not harvested from the patient, but there is a limited supply of healthy human ligament tissue. In addition, the ligament from another human may cause an immunogenic response in the patient receiving the allograft, ultimately leading to graft rejection. The shortcomings of autografts and allografts have led researchers to study natural and synthetic scaffolding materials that mimic the properties of the native ligament.

19.2.2 Natural Scaffolds

Ligament scaffolds comprised of naturally occurring materials are attractive due to the biocompatible microenvironment provided by such materials. As the predominant component of native ligament tissue, collagen type I has been extensively researched in the literature [24, 25]. Dunn et al. showed that fibroblasts harvested from rabbit ACL tissue were able to orient along the long axis of collagen fibers and synthesize collagen to a greater degree than fibroblasts adhering to tissue culture plastic [24]. Subsequent in vivo studies confirmed the viability of the fibroblast-seeded collagen constructs, but complete resorption by week 8 limited to impact of the collagen construct [26]. In addition, collagen scaffolds cannot match the mechanical properties of native ligaments, leading groups to search for ways to enhance said properties.

Common techniques reiterated throughout this chapter are the braiding, twisting, and weaving of fibers to enhance the mechanical properties of the respective materials. Walters et al. braided and crosslinked Sprague-Dawley-harvested collagen type I fibers to match the mechanical behavior of native ligaments [25]. The collagen fibers were crosslinked using 1-Ethyl-3-(3-dimethylaminopropyl)carbodiimide (EDC) chemistry, a standard amide-bond forming reagent. Walters et al. reports a Young's modulus of 148 ± 17 MPa and an ultimate tensile strength (UTS) of 19.3 ± 3.1 MPa for the braided and crosslinked scaffolds. Though these values are close to results obtained from native human ACL by Noyes and Grood (Young's modulus: 111 ± 26 MPa, UTS: 37.8 ± 9.3 MPa) [27], there are always concerns about the immunogenicity of harvested materials and the use of harmful crosslinking agents in implanted scaffolds.

Alternative natural materials that exhibit sufficient mechanical integrity without the need for crosslinking have also appeared extensively in literature. Silk fiber is a well-known clinical suturing material with a high UTS and favorable biocompatibility [28]. Altman et al. developed a ligament scaffold using twisted silk fibers arranged into a 6-cord matrix. This silk scaffold is reported to have an UTS of 2337 ± 72 N, a stiffness of 354 ± 26 N/mm, a yield point of 1262 ± 36 N, and an elongation percentage of 38.6 ± 2.4 %. These mechanical properties are similar to those reported for a human ACL. Woo et al. reported that native human ACL tissue has an UTS of 2160 ± 157 N, a stiffness of 242 ± 28 N/mm, a yield point of approximately 1200 N, and an elongation percentage of approximately 33 % [29]. The silk scaffold may not be able to induce infiltration of human bone marrow stromal cells (BMSCs), but the comparable mechanical properties suggest that silk could be a viable material for ligament scaffolds. Chen et al. modified the same silk scaffold with arginylglycylaspartic acid (RGD) peptides to promote cell attachment and collagen production, suggesting that future modification technologies could enhance certain properties of materials that are initially unfavorable [30]. Collagenous matrix production via fibroblast seeded on silk scaffolds should be designed to match the rate of degradation of silk, which is reported to degrade within 1–2 years [28]. The efficacy of proteinaceous fibers for ligament regeneration has led to the development of various naturally derived non-protein fibers.

Chitin is a natural polysaccharide obtained from crustacean shells. Chitosan is formed when chitin is sufficiently deacetylated (>50 % deacetylation), and as such a copolymer containing N-acetylglucosamine and N-glucosamine remains. Polysaccharide fibers have not received the same attention as protein fibers in ligament regeneration applications because proteinaceous polymers are more mechanically competent. However, Irie et al. have produced chitosan-hyaluronan (chitosan-HA) hybrid polymer fibers that match the failure load of rabbit MCLs. The chitosan-HA fibers were produced using a wet-spinning technique [31]. The resulting fibers were then braided into a scaffold using a 30° angle between the braided fibers and the longitudinal line. The chitosan-HA scaffolds had the most favorable mechanical properties when seeded with Achilles tendon fibroblasts (isolated from the same rabbit to be used for subsequent surgeries). Surgical insertion of the cell-seeded scaffolds was followed by surgical removal at 3, 6, and 12 weeks. After 12 weeks, the cell-seeded constructs had a failure load of 125.2 ± 28.4 N and stiffness of 31.5 ± 8.7 N/mm (compared to a failure load of 106.1 ± 27.5 N and stiffness of 92.8 ± 26.5 N/mm for natural rabbit MCL tissue). Chitosan-HA ligament scaffolds cannot match the stiffness of natural ligament tissues, which will eventually lead to unfavorable deformation in vivo. These cell-seeded scaffolds were able to enhance type I collagen production (when compared to similar non-cell-seeded scaffolds) and were not shown to elicit any inflammatory response 12 weeks after surgery. In addition, this group claims that chitosan-HA scaffolds support cell proliferation and extracellular matrix production due to significant swelling of the scaffold cross sections after surgery.

Tamura et al. experimented with chitosan-coated alginate filaments using the aforementioned wet-spinning technique [32]. Increasing chitosan content led to increased tensile strength, but alginate braided constructs were not compared to other leading materials for ligament engineering. As such, further work needs to be done to identify whether alginate has any promising properties to contribute to the regeneration of ligament tissues.

19.2.3 Synthetic Scaffolds

Synthetic scaffolds are an attractive opportunity for ligament tissue engineering approaches because synthetic materials can be tailored to suit the properties of the desired scaffold. Many early ligament replacements utilized non-degradable materials, such as poly(tetrafluoroethylene) (GoreTex), polypropylene (Kennedy Ligament Augmentation Device), and polyethylene terephthalate (Leeds-Keio ligament) [33–35]. Although initially mechanically competent, permanent plastic deformation of these devices is prevalent following surgical implantation. Additional failure mechanisms related to these devices include creep, fatigue, fragmentation, and stress shielding [36].

Next generation synthetic materials for ligament engineering include poly (L-lactic acid) (PLLA), poly(glycolic acid) (PGA), poly(lactic-co-glycolic acid) (PLGA), and polycaprolactone (PCL). These materials improve upon past scaffold designs by better mimicking the architecture and mechanical properties of natural ligaments. Ligament scaffolds composed of these materials also tend to incorporate chemical modifications, such as fibronectin or RGD sequences, to improve cell proliferation and ECM production.

PLLA is a synthetic polymer composed of lactic acid monomers. This polymer is reported to take approximately 2 years to degrade within the body, which is an ideal length of time to permit tissue ingrowth of implanted constructs composed of this material while also maintaining mechanical integrity. Laurencin et al. have developed a braided scaffold with controllable porosity and mechanical properties similar to that of a natural ACL [3]. The braided scaffold has two bony attachment ends to resist bone tunnel-associated wear, and an intra-articular region that is sandwiched by the bony attachment ends. The intra-articular region has pore diameters on the range of 200–250 μm to allow for soft tissue ingrowth. PLLA ligament scaffolds are reported to have favorable cell growth and collagen type I production with the addition of fibronectin [3, 37].

Scaffolds utilizing materials derived from glycolic acid have previously been reported in the literature [38]. Lin et al. have developed a scaffold composed of Dexon II, a material made up of PGA homopolymer coated with polycaprolate. The scaffold was capable of supporting early cell growth of human ligament fibroblasts, especially when these cells were seeded with growth factors. Recent literature has been increasingly critical of growth factor-assisted cell proliferation because of the inherent instability of proteins and immunogenic reactions of such growth factors [39]. The United States Food and Drug Association has strict standards for the approval of growth factors for clinical use, and therefore treatments involving growth factors are not yet clinically viable. Additional concerns associated with PGA-based ligament scaffolds include rapid degradation (complete degradation within 1 month) and loss of mechanical strength.

PLGA has been extensively studied in the literature because of the tunable properties associated with this material. The degradation and mechanical properties can be modified by changing the ratio of lactic acid to glycolic acid monomers when synthesizing the polymer. As such, the degradation of glycolic acid based polymers can be delayed by incorporating more lactic acid into the polymer product. Slow degradation of ligament scaffolds is preferable, but certain scaffolds may benefit from regions of faster degradation. Lu et al. report that PLGA scaffolds have decreased strength and lower rates of ACL fibroblast proliferation compared to PLLA scaffolds, which appears to be a direct result of the incorporation of glycolic acid monomers [37].

The importance of ligament scaffold degradation cannot be overstated. Despite having the slowest degradation of all of the previously discussed synthetic materials (approximately 4 years), PCL has received increased attention in tissue engineering applications because of the superior rheological and viscoelastic properties of the material [40]. Leong et al. report using a PCL scaffold with 31.3 % of the stiffness and 28.2 % of the peak load of native ACL tissue [41]. The stiffness and peak load values were reported to improve when basic fibroblast growth factor (bFGF) was incorporated into the PCL grafts. PCL is commonly characterized as a bioinert material, which implies that the material lacks the ability to induce cell proliferation. Many groups have surface modified PCL to improve the hydrophilicity of the polymer and therefore increase the proliferation of cells seeded onto this scaffolding material [42–44]. Surface modification approaches have been shown to enhance cell proliferation with varying degrees of success, but the implication of such modifications has yet to be translated into a PCL-based ligament scaffold with favorable properties that match native ligaments.

Any tissue engineered ligament scaffold must meet an extensive list of criteria to optimally rejuvenate the injured ligament site. Overall, synthetic materials have a promising future for ligament scaffold applications because of the favorable mechanical properties possessed by these materials. More extensive in vivo studies will be required to verify the efficacy of synthetic scaffolding materials and confirm the clinical potential of the aforementioned tissue engineering constructs.

19.3 Biological Adjuvants for Enhanced Ligament Regeneration

ACL reconstruction contains two main biological processes, ligamentization of the intra-articular region and graft-to-bone healing in the femoral and tibial tunnels. Efforts have been made to enhance the healing of the gold standard ACL reconstruction (bone-patellar tendon-bone graft), through the use of stem cell populations and platelet rich plasma. Efficacy of biological adjuvants to enhance ligamentization and graft-to-bone healing has been heavily researched. Two methods that can be utilized in the operating room is the use of platelet rich plasma and bone marrow aspirate from the iliac crest, without the need to go through further FDA approval. Additionally, there has been further interest in the use of adipose-derived stem cells to aid in the healing of musculoskeletal injuries. Herein, the use of these therapeutic strategies to aid in ACL healing are discussed.

19.3.1 Platelet Rich Plasma to Enhance ACL Repair

Platelet Rich Plasma (PRP) is an autologous plasma suspension enriched with platelets. Normally it is prepared from peripheral blood through a two- phase centrifugation process called plasmapheresis, in which liquid and solid components of anti-coagulated blood are separated, leading to a concentrate with 3–5 times as many platelets as normal blood [45].

PRP contains platelets and a high concentration of the fundamental growth factors proved to be actively secreted by platelets to initiate wound healing, including platelet-derived growth factor (PDGF), vascular endothelial growth factor (VEGF), epithelial growth factor, basic fibroblast growth factor, and Transforming Growth Factor-β1 and β2 (TGF-β) [46]. Additionally PRP also contains three proteins found in blood, fibrin, fibronectin, and vitronectin, which are known to act as cell adhesion molecules for osteoconduction and as a matrix for bone, connective tissue, and epithelial migration [45].

The success of ACL reconstruction depends heavily on biological processes that could improve the outcomes and ensure optimal clinical results [47]. Most of the factors released by PRP are involved in the repair of tendon and ligament injuries, and high concentrations of these growth factors are considered to accelerate tendon/ligament healing [47–49]. During the past decades, the application of PRP has been used as a strategy to enhance the healing of injured tendons and ligaments.

The first study regarding PRP in ACL surgery was published by Ventura et al. in 2005 [50]. In the study, 20 patients were randomly assigned to receive ACL hamstring reconstruction with or without PRP. Three milliliters of PRP was placed in both tunnels directly with autologous thrombin, though the concentration of PRP used was unclear. Clinical outcomes showed that the transformation from autologous quadrupled hamstring tendon graft to new ACL was faster in the PRP treated group than in controls. This suggests that growth factors contained in PRP could accelerate the integration of the new ACL in the femoral and tibial tunnels.

Although many studies have been carried out with PRP in ACL repair, the effectiveness of PRP is still up for debate. Recent works from Murray and associates have analyzed the efficacy of PRP treatment in combination with ACL suture repair [51, 52]. The use of PRP in combination with an extracellular matrix protein scaffold containing collagen, referred to as the bio-enhanced ACL repair, has been evaluated in a porcine model and compared with ACL reconstruction using an allograft tendon. Both models reported no significant difference in mechanical properties at 3 months and 1 year after the surgery. However, the reason for the lack of statistical significance is due to the inconsistent methodologies of the two groups. Though suture repair of the ACL is not improved with the use of PRP alone, the ACL can be effectively repaired with the use of whole blood containing a physiological concentration of platelets in an extracellular matrix-based scaffold. The results of this bio-enhanced repair technique are similar to ACL reconstruction in terms of the mechanical properties of the healing tissue and graft, but the bio-enhanced repairs resulted in less post-traumatic osteoarthritis in large animals. The data from ACL reconstruction study using bio-enhanced ACL in goat and porcine models with whole blood and 5X platelets, respectively, provide encouragement regarding the efficacy of the platelet-enhanced ACL reconstruction approach in immature animals.

Generally speaking, the various systems used to obtain PRP lead to disparities in platelet collection efficiency and repeatability, final leukocyte count, platelet activation and ease of use [53]. Any one of these disparities could lead to controversial results. There is need for standardization of PRP preparation methods, but more evidence is needed to support the routine use of PRP for treating ACL injuries.

19.3.2 Adipose-Derived Stem Cells

Mesenchymal stem cells (MSCs) are adult stem cells from various sources, being multipotent and having the capacity of self-renewal. MSCs can differentiate into mesoderm-associated cell types such as chondrocytes, adipocytes or osteoblasts [53]. Due to ease of harvest and abundance, adipose-derived mesenchymal stem cells (ADSCs) are an attractive, readily available adult stem cell source that has become increasingly popular for use in orthopedic applications [54].

Eagan et al. looked at the in vitro utility of ADSCs for ligament engineering. They treated ADSCs for 4 weeks with TGF-β1 or IGF1 not showing any significant and consistent upregulation in the expression of collagen types I and III, tenascin C, and scleraxis. While treatment with EGF or bFGF resulted in increased tenascin C expression, increased expression of collagens I and III were never observed. Therefore, simple in vitro treatment of human ADSC populations with growth factors may not stimulate ligament differentiation [55]. Little et al. prepared novel ligament derived matrix by mixing phosphate-buffered saline or 0.1 % peracetic acid with a collagen gel. Over 28 days, the matrices were found to promote ADSC differentiation into a ligament fibroblast phenotype [56]. Proffen et al. co-cultured stem cells from both the retropatellar fat pad and peripheral blood, and the results showed stimulated ACL fibroblast proliferation and collagen production in vitro [57]. Further investigation was carried out by adding MSCs obtained from the adipose tissue or peripheral blood to see the in vivo biomechanical properties of bioenhanced ACL repair. After 15 weeks of healing, there were no significant improvements in the biomechanical or histological properties with the addition of

ADSCs. The only significant change with the addition of peripheral blood MSCs was an increase in knee anteroposterior laxity when measured at 30° of flexion, suggesting that the addition of adipose-derived or peripheral blood MSCs to whole blood—before saturation of an extracellular matrix carrier with the blood—did not improve the functional results of bioenhanced ACL repair after 15 weeks of healing in a porcine model [58]. These MSC studies suggest the potential of ADSCs in tendon and ligament repair, but more evidence is needed to fully substantiate these claims.

19.3.3 Alternative Methods for ACL Reconstruction

Murray and associates explored a new paradigm in primary ACL repair. Previous studies for primary repair of the ACL after traumatic rupture have reported unacceptable rates of failure after primary surgical repair, and the poor rate of primary healing is believed to be due to the intra-articular environment and synovial fluid that surrounds the ACL [59]. Through a canine, central ACL wound model, Murray and associates demonstrated the differences in intra-articular (i.e., ACL) versus extra-articular (i.e., MCL) healing. Ligaments which exist outside of joints (extra-articular) heal with an orderly progression of events. The first basic process is bleeding and then formation of a fibrin–platelet clot within the wound site, which fills in the gap between the torn ends of the tissue and forms a provisional scaffold for the surrounding cells to move into and remodel into a functional scar. However, in the intra-articular environment, after an injury, there is an upregulated production of urokinase plasminogen activator by synoviocytes, which converts the inactive plasminogen molecule present in synovial fluid into its active form, plasmin. Plasmin quickly degrades fibrin. Therefore, if a tissue is exposed to synovial fluid after injury, the ends may bleed, but the fibrin is unable to form a stable clot as it is degraded too quickly. The early loss of this provisional scaffold has been thought to be a major reason why tissues within joints, such as the ACL or meniscus, fail to heal after the injury [59–61]. The lack of a scaffold in the intra-articular ligament wounds was also associated with decreased inflammatory cytokines needed for the healing response, including fibrinogen, PDGF, TGF-b, and FGF. However, replacement of the central intra-articular ligament void with a collagen-platelet-rich plasma scaffold resulted in increased filling of the wound with repair tissue that had similar profiles of protein expression to matched, extra-articular ligament wounds. Biomechanical studies of suture ACL repair augmented with a collagen-platelet–rich scaffold in a porcine model have shown significant improvement in load to failure and linear stiffness at 4 weeks compared with control repairs, which lack the collagen-platelet-rich scaffold [58].

19.4 Future Trends

The gold standard to treat ACL ruptures is the use of bone-patellar tendon-bone grafts, and it is understood that the current treatment results in donor site morbidity due to the harvest of the graft. In addition, the current gold standard does not inhibit the progression of osteoarthritis. Given these knowns, researchers have aimed to develop engineered ACL scaffolds that are mechanically competent, biodegradable, and inhibit the progression of osteoarthritis. Current generation engineered scaffolds have shown the capacity to allow for the regeneration of a natural ACL in the intraarticular region after 1.5 years. The next steps are to evaluate the ability of these engineered matrices to inhibit the progression of osteoarthritis and to enhance the ligamentization such that patients can return to their prior levels of activity.

Enhancing the properties of ACL scaffolds can be realized by utilizing biological adjuvants. Current clinical use of PRP and bone marrow aspirate allows for a quicker route to translating new surgical treatment strategies. Further research in the addition of PRP on mechanically competent scaffolds is needed. Additionally, research in prolonging the effects of PRP treatment through the use of carriers may also aid to enhance and accelerate ACL healing.

The use of MSCs from various sources for ACL reconstruction is in its infancy. Bone marrow aspirate and adipose tissue are two sources that are abundant and have stem cells that may aid in the enhancement of ACL healing post reconstruction. Cytokines from these stem cells can help to inhibit inflammation, and these molecules can also provide signals to promote ligamentization and graft-to-bone healing. That being said, the synovial environment presents challenges in delivering these stem cells. Future studies on carriers for stem cells which maintain their stemness and prolong the release of pro-healing cytokines is needed to increase the efficacy of this therapeutic strategy.

References

1. Nau T, Teuschl A (2015) Regeneration of the anterior cruciate ligament: current strategies in tissue engineering. World J Orthop 6:127–36. doi:10.5312/wjo.v6.i1.127
2. Simon D et al (2015) The relationship between anterior cruciate ligament injury and osteoarthritis of the knee. Adv Orthop 2015:1–11
3. Laurencin CT, Freeman JW (2005) Ligament tissue engineering: An evolutionary materials science approach. Biomaterials 26:7530–7536
4. Duthon VB et al (2006) Anatomy of the anterior cruciate ligament. Knee Surg Sports Traumatol Arthrosc 14:204–213
5. Dienst M, Burks RT, Greis PE (2002) Anatomy and biomechanics of the anterior cruciate ligament. Orthop Clin North Am 33:605–620
6. Woo SL-Y, Wu C, Dede O, Vercillo F, Noorani S (2006) Biomechanics and anterior cruciate ligament reconstruction. J Orthop Surg Res 1:2
7. Dargel J et al (2007) Biomechanics of the anterior cruciate ligament and implications for surgical reconstruction. Strateg Trauma Limb Reconstr 2:1–12

8. Zantop T, Petersen W, Sekiya JK, Musahl V, Fu FH (2006) Anterior cruciate ligament anatomy and function relating to anatomical reconstruction. Knee Surg Sports Traumatol Arthrosc 14:982–992

9. Freeman JW, Woods MD, Laurencin CT (2007) Tissue engineering of the anterior cruciate ligament using a braid-twist scaffold design. J Biomech 40:2029–2036

10. Cooper JA, Lu HH, Ko FK, Freeman JW, Laurencin CT (2005) Fiber-based tissue-engineered scaffold for ligament replacement: design considerations and in vitro evaluation. Biomaterials 26:1523–1532

11. Shaw HM, Benjamin M (2007) Structure-function relationships of entheses in relation to mechanical load and exercise. Scand J Med Sci Sports 17:303–315

12. Benjamin M et al (2006) Where tendons and ligaments meet bone: attachment sites ('entheses') in relation to exercise and/or mechanical load. J Anat 208:471–490

13. Huang AH, Lu HH, Schweitzer R (2015) Molecular regulation of tendon cell fate during development. J Orthop Res 33:800–812

14. Pearse RV, Esshaki D, Tabin CJ, Murray MM (2009) Genome-wide expression analysis of intra- and extraarticular connective tissue. J Orthop Res 27:427–434

15. Wenk MB, Midwood KS, Schwarzbauer JE (2000) Tenascin-C suppresses rho activation. J Cell Biol 150:913–920

16. Majima T et al (2000) Compressive compared with tensile loading of medial collateral ligament scar in vitro uniquely influences mRNA levels for aggrecan, collagen type II, and collagenase. J Orthop Res 18:524–531

17. Steiner ME, Murray MM, Rodeo SA (2008) Strategies to improve anterior cruciate ligament healing and graft placement. Am J Sports Med 36:176–189

18. Takahashi M, Doi M, Abe M, Suzuki D, Nagano A (2006) Anatomical study of the femoral and tibial insertions of the anteromedial and posterolateral bundles of human anterior cruciate ligament. Am J Sports Med 34:787–792

19. Yagi M et al (2002) Biomechanical analysis of an anatomic anterior cruciate ligament reconstruction. Am J Sports Med 30:660–666

20. Koga H et al (2015) Mid- to long-term results of single-bundle versus double-bundle anterior cruciate ligament reconstruction: randomized controlled trial. Arthroscopy 31:69–76

21. Leong NL, Petrigliano FA, McAllister DR (2014) Current tissue engineering strategies in anterior cruciate ligament reconstruction. J Biomed Mater Res A 102:1614–1624

22. Girgis FG, Marshall JL, Monajem A. The cruciate ligaments of the knee joint. Anatomical, functional and experimental analysis. Clin Orthop Relat Res 216–231. http://www.ncbi.nlm. nih.gov/pubmed/1126079

23. Surrao DC, Waldman SD, Amsden BG (2012) Biomimetic poly(lactide) based fibrous scaffolds for ligament tissue engineering. Acta Biomater 8:3997–4006

24. Dunn MG, Liesch JB, Tiku ML, Zawadsky JP (1995) Development of fibroblast-seeded ligament analogs for ACL reconstruction. J Biomed Mater Res 29:1363–1371

25. Walters VI, Kwansa AL, Freeman JW (2012) Design and analysis of braid-twist collagen scaffolds. Connect Tissue Res 53:255–266

26. Bellincampi LD, Closkey RF, Prasad R, Zawadsky JP, Dunn MG (1998) Viability of fibroblast-seeded ligament analogs after autogenous implantation. J Orthop Res 16:414–420

27. Noyes FR, Grood ES (1976) The strength of the anterior cruciate ligament in humans and Rhesus monkeys. J Bone Joint Surg Am 58:1074–1082

28. Altman GH et al (2002) Silk matrix for tissue engineered anterior cruciate ligaments. Biomaterials 23:4131–4141

29. Woo SL-Y, Hollis JM, Adams DJ, Lyon RM, Takai S (1991) Tensile properties of the human femur-anterior cruciate ligament-tibia complex: The effects of specimen age and orientation. Am J Sports Med 19:217–225

30. Chen J et al (2003) Human bone marrow stromal cell and ligament fibroblast responses on RGD-modified silk fibers. J Biomed Mater Res A 67:559–570

31. Irie T et al (2011) Biomechanical and histologic evaluation of tissue engineered ligaments using chitosan and hyaluronan hybrid polymer fibers: a rabbit medial collateral ligament reconstruction model. J Biomed Mater Res A 97:111–117
32. Tamura H, Tsuruta Y, Tokura S (2002) Preparation of chitosan-coated alginate filament. Mater Sci Eng, C 20:143–147
33. Bolton CW, Bruchman WC (1985) The GORE-TEX expanded polytetrafluoroethylene prosthetic ligament. An in vitro and in vivo evaluation. Clin Orthop Relat Res 202–213. http://europepmc.org/abstract/med/3888468
34. Schroven ITJ, Geens S, Beckers L, Lagrange W, Fabry G (1994) Experience with the Leeds-Keio artificial ligament for anterior cruciate ligament reconstruction. Knee Surg Sport Traumatol Arthrosc 2:214–218
35. Kdolsky R, Reihsner R, Schabus R, Beer RJ (1994) Measurement of stress-strain relationship and stress relaxation in various synthetic ligaments. Knee Surg Sport Traumatol Arthrosc 2:47–49
36. Laurencin CT, Ambrosio AM, Borden MD, Cooper JA (1999) Tissue engineering: orthopedic applications. Annu Rev Biomed Eng 1:19–46
37. Lu HH et al (2005) Anterior cruciate ligament regeneration using braided biodegradable scaffolds: in vitro optimization studies. Biomaterials 26:4805–4816
38. Lin VS, Lee MC, O'Neal S, McKean J, Sung KL (1999) Ligament tissue engineering using synthetic biodegradable fiber scaffolds. Tissue Eng 5:443–452
39. Lo KW-H, Jiang T, Gagnon KA, Nelson C, Laurencin CT (2014) Small-molecule based musculoskeletal regenerative engineering. Trends Biotechnol 32:74–81
40. Woodruff MA, Hutmacher DW (2010) The return of a forgotten polymer—polycaprolactone in the 21st century. Prog Polym Sci 35:1217–1256
41. Leong NL et al (2015) In vitro and in vivo evaluation of heparin mediated growth factor release from tissue-engineered constructs for anterior cruciate ligament reconstruction. J Orthop Res 33:229–236
42. Zhang M et al (2012) Small-diameter tissue engineered vascular graft made of electrospun PCL/lecithin blend. J Mater Sci Mater Med 23:2639–2648
43. Zhang H, Hollister S (2009) Comparison of bone marrow stromal cell behaviors on poly (caprolactone) with or without surface modification: studies on cell adhesion, survival and proliferation. J Biomater Sci Polym Ed 20:1975–1993
44. de Luca AC, Terenghi G, Downes S (2014) Chemical surface modification of poly-ε-caprolactone improves Schwann cell proliferation for peripheral nerve repair. J Tissue Eng Regen Med 8:153–163
45. Marx RE et al (2004) Platelet-rich plasma: evidence to support its use. J Oral Maxillofac Surg 62:489–496
46. Marx RE et al (1998) Platelet-rich plasma. Oral Surg Oral Med Oral Pathol Oral Radiol Endodontol 85:638–646
47. Sánchez M, Anitua E, Orive G, Mujika I, Andia I (2009) Platelet-rich therapies in the treatment of orthopaedic sport injuries. Sport Med 39:345–354
48. Zhang J, Wang JH-C (2010) Platelet-rich plasma releasate promotes differentiation of tendon stem cells into active tenocytes. Am J Sports Med 38:2477–2486
49. Chen L et al (2012) Autologous platelet-rich clot releasate stimulates proliferation and inhibits differentiation of adult rat tendon stem cells towards nontenocyte lineages. J Int Med Res 40:1399–1409
50. Ventura A et al (2005) Use of growth factors in ACL surgery: preliminary study. J Orthop Traumatol 6:76–79
51. Fleming BC et al (2015) Increased platelet concentration does not improve functional graft healing in bio-enhanced ACL reconstruction. Knee Surg Sport Traumatol Arthrosc 23:1161–1170
52. Hutchinson ID, Rodeo SA, Perrone GS, Murray MM (2015) Can platelet-rich plasma enhance anterior cruciate ligament and meniscal repair? J Knee Surg 28:19–28

53. McLellan J, Plevin S (2011) Does it matter which platelet-rich plasma we use? Equine Vet Educ 23:101–104
54. Tapp H, Hanley EN, Patt JC, Gruber HE (2009) Adipose-derived stem cells: characterization and current application in orthopaedic tissue repair. Exp Biol Med 234:1–9
55. Eagan MJ et al (2012) The suitability of human adipose-derived stem cells for the engineering of ligament tissue. J Tissue Eng Regen Med 6:702–709
56. Little D, Guilak F, Ruch DS (2010) Ligament-derived matrix stimulates a ligamentous phenotype in human adipose-derived stem cells. Tissue Eng Part A. doi:10.1089/ten.tea.2009.0720
57. Proffen BL, Haslauer CM, Harris CE, Murray MM (2012) Mesenchymal stem cells from the retropatellar fat pad and peripheral blood stimulate ACL fibroblast migration, proliferation, and collagen. Gene Expr. doi:10.3109/03008207.2012.715701
58. Proffen BL et al (2015) Addition of autologous mesenchymal stem cells to whole blood for bioenhanced ACL repair has no benefit in the porcine model. Am J Sports Med 43:320–330
59. Spindler KP, Murray MM, Devin C, Nanney LB, Davidson JM (2006) The central ACL defect as a model for failure of intra-articular healing. J Orthop Res 24:401–406
60. Murray MM, Martin SD, Martin TL, Spector M (2000) Histological changes in the human anterior cruciate ligament after rupture*. J Bone Jt Surg Am 82:1387
61. Murray MM et al (2006) Use of a collagen-platelet rich plasma scaffold to stimulate healing of a central defect in the canine ACL. J Orthop Res 24:820–830
62. Freed AD, Doehring TC (2005) Elastic model for crimped collagen fibrils. J Biomech Eng 127:587–593
63. Grytz R, Meschke G (2009) Constitutive modeling of crimped collagen fibrils in soft tissues. J Mech Behav Biomed Mater 2:522–533
64. Figueroa D et al (2014) Anterior cruciate ligament regeneration using mesenchymal stem cells and collagen type I scaffold in a rabbit model. Knee Surg Sports Traumatol Arthrosc 22:1196–1202

Author Index

© Springer International Publishing AG 2017   411
J.M. Oliveira and R.L. Reis (eds.), *Regenerative Strategies for the Treatment
of Knee Joint Disabilities*, Studies in Mechanobiology, Tissue Engineering
and Biomaterials 21, DOI 10.1007/978-3-319-44785-8

Printed by Printforce, the Netherlands